Sciences of Geodesy - II

Guochang Xu
Editor

Sciences of Geodesy - II

Innovations and Future Developments

Springer

Editor
Dr. Guochang Xu
Department 1: Geodesy and
Remote Sensing
GFZ German Research Centre
for Geosciences
Potsdam
Germany

and

Chinese Academy of Space Technology
Beijing
P.R. China

ISBN 978-3-642-27999-7 ISBN 978-3-642-28000-9 (eBook)
DOI 10.1007/978-3-642-28000-9
Springer Heidelberg New York Dordrecht London

Library of Congress Control Number: 2012937728

Printed on acid-free paper

Springer is part of Springer Science+Business Media (www.springer.com)

Preface

This is the second volume of the book series of Sciences of Geodesy. This series of reference books describes different, but complementary fields involving geodesy in seven chapters. Each chapter describes the history, theory, objectives, technology, development and highlights of the research and applications of the individual field. In addition, challenges and future directions are discussed. The subjects covered by this reference book include Computation of Green's Functions for Ocean Tide Loading, General Relativity and Space Geodesy, Global Terrestrial Reference Systems and their Realisations, Photogravimetry, Regional Gravity Field Modelling, Regularisation and Adjustment, and Very Long Baseline Interferometry for Geodesy and Astronomy.

The first volume in this series contains chapters that detail the subjects: Absolute and Relative Gravimetry, Adaptively Robust Kalman Filters with Applications in Navigation, Airborne Gravity Field Determination, Analytic Orbit Theory, Deformation and Tectonics, Earth Rotation, Equivalence of GPS Algorithms and Its Inference, Marine Geodesy, Satellite Laser Ranging, Superconducting Gravimetry and Synthetic Aperture Radar Interferometry.

The above mentioned fields cover the most active areas related to geodesy. These individual subjects are, for the first time, combined in a two-volume series thereby providing a comprehensive overview of the multi-disciplinary nature of geodesy. The series serves as a reference for teaching and learning the basic principles of many subjects related to geodesy. The material is suitable for high-level geodetic researchers, educators as well as engineers and students. Some of the chapters are written to fill voids in the current literature of the related areas. Most chapters are written by international scientists, well known in their specific field of expertise.

The chapters are arranged in alphabetical order of their titles. Summaries of the individual chapters and introductions of their authors and co-authors are as follows:

Chapter 1 "Computation of Green's Functions for Ocean Tide Loading" describes the theory and the methods of the point load problem for a radially symmetric, elastic Earth. A researcher or Ph.D. student who wants to learn more about this classic topic will find in this chapter a good starting point where all assumptions are clearly explained and where enough details are given to implement the equations into a computer program. First, the differential equations for the gravitational elastic deformations are derived from first principles. Then the boundary conditions to solve these equations are presented, and analytical solutions and numerical values of Love numbers for two simple Earth models are discussed. This chapter also contemplates the problem related to periodic loading of a model Earth with a fluid core as the period goes to infinity, the so-called Longman paradox. The degree-1 deformation, the centre-of-mass centre-of-reference problem, receives special attention. Next, several numerical methods to solve the equations are explained. Finally, the formulas for computing Green's functions are listed.

The author and co-authors of Chap. 1 are Dr. Machiel Bos and Dr. Hans-Georg Scherneck.

Machiel Bos studied Aerospace Engineering at Delft University of Technology, The Netherlands. After his graduation in 1996 he performed his Ph.D. research at Proudman Oceanographic Laboratory, Liverpool, United Kingdom. In 2001 he spent 7 months as postdoc at Onsala Space Observatory, Sweden. From 2001 to 2003 he worked as a postdoc at the Faculty of Geodesy of Delft University of Technology. From 2003 to 2008 he held a postdoc position at the Astronomical Observatory of Porto, Portugal and since 2008 he has been working at CIIMAR (Centre of Marine and Environmental Research of the University of Porto). His scientific interests include ocean tide loading, GPS time-series analysis and the geoid.

Hans-Georg Scherneck studied Physics and Geophysics at J. W. Goethe University in Frankfurt/M., Germany. He received a Ph.D. degree in geodesy from Uppsala University in 1986. In 1993 he joined the Department of Earth and Space Science at Chalmers University of Technology, Gothenburg, Sweden. He holds a Docent degree (associate professor) in geodynamic measurement techniques and occupies a position as a Lecturer. His major research interests are the use of gravity and space geodetic techniques in application to solid earth deformation, most prominently Glacial Isostatic Adjustment.

Chapter 2 "General Relativity and Space Geodesy" introduces the general and special relativity theory as it is applied to space geodesy. Section 1 sketches some basic implications of GRT for space geodesy and the need to incorporate GRT in all high accuracy space geodetic applications. Section 2 discusses GRT implications for satellite laser ranging, specifically the Shapiro delay and accelerations as described by the Schwarzschild field, Lense–Thirring precession (frame dragging) and de Sitter (geodesic) precession. A short discussion on using SLR to test the effects of GRT is included and Sect. 2 is concluded with sections on lunar laser ranging and interplanetary laser ranging. Special and general relativity theory considerations for GPS are discussed in some detail in Sect. 3, including reference frame issues, effects on GPS satellites' clocks and how GRT corrections are incorporated. Section 4 consists of a short overview of VLBI estimates of parameterised post-Newtonian parameter Gamma.

The author of Chap. 2 is Dr. Ludwig Combrinck of the Hartebeesthoek Radio Astronomy Observatory (HartRAO) located near Krugersdorp, South Africa, a facility of the National Research Foundation (NRF). Ludwig Combrinck was awarded a Ph.D. by the University of Cape Town in 2000; his thesis focussed on GNSS applications for precise positioning. He is responsible for the Space Geodesy Programme at HartRAO, which includes the NASA satellite laser ranging station, MOBLAS-6. In 2009 he was appointed as Professor-extraordinaire at the University of Pretoria. His main research interests currently include applications of space geodetic techniques, specifically related to tests of general relativity theory, reference frame development for Africa and the development of a new high accuracy satellite and lunar laser ranger for South Africa. His diverse interests in the applications of space geodesy have resulted in the establishment of geodetic stations throughout Africa, Marion Island and Antarctica, in collaboration with international partners.

Chapter 3 is entitled "Global Terrestrial Reference Systems and their Realizations". It is organised in six parts. In Sect. 1 the authors give an introduction and address the key role of geodetic reference systems and frames for measuring the surface structure, the rotation and the gravity field of the Earth along with its variations in time, which is a prerequisite for Earth system studies and for the monitoring of physical processes of global change. The next section provides some basic concepts and fundamentals for the definition and realisation of reference systems. Section 3 deals with the International Terrestrial Reference System (ITRS), its definition and the conventional modelling of station positions and displacements of reference points, which materialise the system. The next two sections focus on its realisation, the International Terrestrial Reference Frame (ITRF), which is the key topic of this chapter. Thereby Sect. 4 provides some general information and gives an overview about the history and the latest developments in the field of global terrestrial reference frame realisations. Section 5 deals with the latest realisation, the ITRF2008, which has been computed from a combination of time series of station positions and Earth orientation parameters from VLBI, SLR, GPS and DORIS observations. In the last section, the present status of the terrestrial reference frame computations is discussed and challenges for future improvements are provided.

The author and co-authors of Chap. 3 are Dr. Detlef Angermann, Dr. Manueal Seitz and Prof. Dr. Hermann Drewes.

Detlef Angermann has been senior research scientist at Deutsches Geodätisches Forschungsinstitut (DGFI) in Munich since 1999. He graduated in geodesy from University Hannover in 1985 and received his Ph.D. from Technical University in Berlin in 1991. He occupied the following positions: Scientific Assistant at Technical University in Berlin (1985–1990); research scientist at DGFI (1990–1992); senior scientist at GeoForschungsZentrum (GFZ) Potsdam (1992–1999); senior scientist at DGFI (since 1999), where he has been head of the research field "Earth system observations" since 2002. Major areas of scientific interests are GNSS and SLR data analysis for geodetic research and geodynamics, the combination of space geodetic techniques and the realisation of geodetic reference systems. He served as chair and as a member in various sub-commissions and working groups of the International Association of Geodesy (IAG) and as a principal investigator of various research projects at DGFI. In 2009, he was nominated as the Secretary of the GGOS Bureau for Standards and Conventions and took over the responsibility of the Director in 2011.

Manuela Seitz studied Geodesy at the Technische Universität Dresden (TUD). After her graduation in 2001 she joined the Deutsches Geodätisches Forschungsinstitut (DGFI) in Munich, where she collaborated on various projects in the field of combination of space geodetic techniques for the realisation of reference systems. She developed strategies for the realisation of the International Terrestrial Reference System (ITRS) on the basis of normal equations for which she obtained her doctorate from TUD in 2008. In addition to her research activities she was strongly involved in the computation of the DGFI solution of the International Terrestrial Reference Frame ITRF2005 and was responsible for the computation of the recent realisation DTRF2008. Her main scientific interests are the global as well as regional realisation of the ITRS as well as the consistent realisation of terrestrial and celestial reference systems. Her focus of attention also comprises the development of combination strategies for the generation of other combined geodetic products, e.g., Earth orientation parameter or tropospheric parameter series.

Hermann Drewes is the Secretary General of the IAG and the past Director of the German Geodetic Research Institute (Deutsches Geodätisches Forschungsinstitut, DGFI), Munich, Germany. He graduated (Dipl.-Ing.) and received his doctor's degree (Dr.-Ing.) from Technische Universität Hannover, Germany, where he worked as assistant professor and chief engineer. From 1977 to 1979 he was a professor at Universidad del Zulia in Maracaibo, Venezuela. His scientific work concentrated at that time on precise gravimetry and geoid determination. In 1979 he moved to DGFI and changed the field of research to geodynamics and geodetic reference systems. In parallel he got a lectureship at Technische Universität München (TUM) and at Universität der Bundeswehr, München. In 1994 he became the Director of DGFI and received an honorary professorship at TUM. From 1995 to 2003 he was at first the Secretary and then the President of the IAG/COSPAR Commission on Space Techniques for Geodesy and Geodynamics (CSTG), and from 2003 to 2007 the President of the IAG Commission on Reference Frames. Since 1994 he has been the IAG representative to the Sistema de Referencia Geocéntrico para las Américas (SIRGAS), and since 2003 the representative of IUGG to the Pan-American Institute for Geography and History (PAIGH). In 2007 he became the IAG Secretary General. In the same year he was awarded the Order of Merit of the Federal Republic of Germany.

Chapter 4 "Photogrammetry" gives an overview about the methods and applications of aerial photogrammetry, focusing on those for geoinformation acquisition. After a short introduction in Sect. 1, three sections follow: Image Acquisition, Image Georeferencing and Image Processing. Large format analogue and digital aerial cameras are described in Sect. 2, as well as the aspects to consider when planning a photo flight. Section 3 deals with several strategies for establishing a georeference for aerial images considering the cases of frame and line scanner images. Spatial resection, ground and GNSS supported triangulation and GPS/IMU supported photo flights are presented in this section. Section 4 describes the most important photogrammetric products and how they are obtained today: line maps, 3-D elevation models, 3-D urban models, orthophotos and realistic virtual models.

The author of Chap. 4 is Prof. Dr. Paula Redweik. She is an Assistant Professor at the Faculty of Sciences of the University of Lisbon, Portugal, in the Department of Geographic Engineering, Geophysics and Energy. She is also a researcher at the Centre of Geology of the same faculty. After obtaining a B.Sc. in Mathematics in 1983 and a degree in Geographic Engineering from the University of Lisbon in 1985, she worked as a researcher in the Institut für Photogrammetrie und Ingenieurvermessungen (IPI) of the University of Hannover, Germany, where during 1993 she obtained a Ph.D. in Photogrammetry. Since 1993 she has been responsible for the education in photogrammetry in several bachelor and master courses at the University of Lisbon; currently, she is the coordinator of the Geographic Engineering bachelor course. She worked in projects for modelling coastal retreat and has supervised M.Sc. theses and co-supervised Ph.D. theses in this subject. She is co-author of one book (two volumes) about topography and several papers on different applications of photogrammetry.

Chapter 5 "Regional Gravity Field Modeling: Theory and Practical Results" gives an overview of high-precision gravity field modelling on a provincial to national and continental scale. In this context, the geoid and quasigeoid are of major interest, e.g., for the transformation between the purely geometric GNSS (Global Navigation Satellite System) ellipsoidal heights and physical heights in geodesy, for the modelling of dynamic ocean topography, as well as for geophysical applications, requiring accuracies at the level of about 1 cm or even below. After the motivation, some fundamentals of physical geodesy are provided, including reference systems, basic gravity field properties, the geoid and height systems, the normal gravity field, as well as some remarks about temporal gravity field variations, tidal systems and atmospheric effects; the intention of this section is to provide the basics for regional gravity field modelling with as few approximations as possible. The next section covers the methodology of gravity field modelling, where the disturbing potential is the primary quantity of interest; in particular, geodetic boundary value problems, the linearisation of the boundary conditions (observation equations), the spherical and constant radius approximations and the associated classical integral formulas of Poisson, Hotine and Stokes, solutions of Molodensky's and Stokes's boundary value problem, the spectral combination approach, least squares collocation, astronomical leveling, as well as the remove-compute-restore technique are described, the latter providing the basis for regional computations. The subsequent section gives some practical results related to the European geoid and quasigeoid calculations carried out at the Institut für Erdmessung (IfE), Leibniz Universität Hannover (LUH), Germany; the data requirements, the collected gravity field data sets and the development and evaluation of the European Gravimetric (Quasi) Geoid model EGG2008 are discussed. Finally, a short summary of the results and an outlook are given.

The author of Chap. 5 is Dr. Heiner Denker, a senior scientist employed at the Leibniz Universität Hannover (LUH), Germany. His major areas of scientific interest are regional and global gravity field modelling (especially geoid and quasigeoid), including the combination of terrestrial and satellite data, vertical reference systems and height determination, as well as geodynamics research. Heiner Denker graduated in 1984 from Universität Hannover (now LUH) and received a Ph.D. in 1988, also from Universität Hannover. In 1989 he was employed as a researcher at The Ohio State University, Columbus, U.S.A., where he investigated the global analysis of satellite altimeter data for dynamic ocean topography estimation. At the end of 1989, Heiner Denker returned to Universität Hannover on a permanent position, where he specialised in gravity field modelling and has given lectures since 1996, covering the areas of physical geodesy, advanced physical geodesy, geometric geodesy and signal analysis. Since 1990, Heiner Denker has been responsible for the computation of the geoid and quasigeoid in Europe, a task supported by the International Association of Geodesy (IAG) in different ways, presently as IAG Sub-Commission 2.4a "Gravity and Geoid in Europe" (Chair: H. Denker). Furthermore, he chaired an IAG Special Study Group, has been a member of several special study groups as well as the advisory boards of some IAG bodies, and since 2008 he has served as Associate Editor for "Geodetic Theory and Applications" of the scientific journal "Marine Geodesy".

Chapter 6 "Regularization and Adjustment" consists of two parts. The first part focuses on regularised solutions for ill-posed problems, while the second provides an overview of the adjustment theory. Following a brief introduction in the first part of the chapter, unstable and ill-posed problems, regularisation algorithms and determination of the regularisation parameters (including suitable examples) are discussed. In the second part, least squares adjustment, sequential application of least squares adjustment via accumulation, sequential least squares adjustment, conditional least squares adjustment, a sequential application of conditional least squares adjustment, block-wise least squares adjustment and a sequential application of block-wise least squares adjustment are described. In addition, an equivalent algorithm to form the eliminated observation equation system and the algorithm to diagonalise the normal equation and equivalent observation equation, a priori constrained adjustment, a priori datum method and a quasi-stable datum method are discussed, before a short summary.

The author and co-author of Chap. 6 are Prof. Dr. Yunzhong Shen and Dr. Guochang Xu.

Yunzhong Shen is a professor in the Department of Surveying and Geo-informatics Engineering of Tongji University where he was the dean from 2003 to 2006. He graduated from Tongji University with a bachelor's degree in Surveying Engineering in 1983, and obtained his master's degree in Geodetic Data Processing in 1986 and a Ph.D. degree in Geophysical Geodesy in 2001 from the Institute of Geodesy and Geophysics. He is an editor of "Acta Geodetica et Cartographica Sinica". His main research interests are theory of geodetic data processing, satellite positioning and satellite gravimetry. He was a visiting member of the staff of Stuttgart University in Germany (1999–2000), visiting scientist of

GeoForschunsZentrum GFZ Potsdam (8.-11.2006), research fellow of Hong Kong Polytechnic University (5.-6.2008) and professorial visiting staff in Queensland University of Technology (5.-6.2009). In the past 5 years, he has published more than 40 refereed journal papers in geophysical geodesy, GNSS theory and application, geodetic data processing theory, of which six papers have appeared in Journal of Geodesy, four in GPS Solutions, and the others in J Surveying Engineering, Chinese Science Bulletin etc.

After graduating in Mathematics and Geodesy from Wuhan University and the Chinese Academy of Sciences (CAS) in 1982 and 1984 respectively, I, Guochang Xu, obtained the Dr.-Ing. degree from the Technical University (TU) Berlin in 1992. Having worked as a research associate at the TU Berlin from 1986 to 1993, as a scientist at the GFZ Potsdam from 1993 to 1998 and as a senior scientist at the National Survey and Cadastre, Denmark, from 1998 to 1999, I returned to GFZ as a senior scientist in 1999. I have authored and co-authored several scientific books and software and acted as supervisor of several Ph.D. and post-doctoral studies. From 2003 to 2008 I was an overseas assessor, adjunct professor, and winner of an overseas outstanding scholar fund of CAS. I have been an overseas communication assessor of Education Ministry China since 2005, adjunct professor of Chang'an University since 2005, National Time Service Center, CAS, Neubrandenburg University of Applied Sciences since 2009, and National Distinguished Expert of Chinese Academy of Space Technology since 2010. In 2011 I was honoured by an appointment as an honorary professor by the South-west Jiaotong University.

Chapter 7 entitled "Very Long Baseline Interferometry for Geodesy and Astrometry" provides an overview of this space geodetic technique which is essential for the determination of the complete set of Earth orientation parameters as well as for the celestial reference frame. After an introduction in Sect. 1 with information about the concept of VLBI and the historical and technological developments, the computation of the delays is discussed in detail in Sect. 2. It covers all models necessary to reach mm-accuracy of the theoretical delays. Section 3 deals with the least squares adjustment which is widely used for the estimation of geodetic parameters in VLBI analysis, such as the Earth orientation parameters, the celestial reference frame expressed by radio source coordinates, or the terrestrial reference frame realized by station coordinates. VLBI observations are coordinated globally by the International VLBI Service for Geodesy and Astrometry (IVS; Sect. 4), and ideas and plans for VLBI2010, the next generation VLBI system, are given in Sect. 5.

The author and co-author of Chap. 7 are Prof. Dr. Harald Schuh and Prof. Dr. Johannes Böhm.

Harald Schuh is a full professor and Director of the Institute of Geodesy and Geophysics, Vienna University of Technology, Austria. Major areas of scientific interest are Very Long Baseline Interferometry (VLBI), Earth rotation, investigations of the troposphere and ionosphere. He graduated in 1979 from Bonn University, Germany and received his Ph.D. in 1986. He occupied the following positions: Scientific assistant and associate professor at Bonn University (1980–1988); program scientist at the German Air and Space Agency (1989–1995), senior scientist and head of the Earth Rotation Division at DGFI, Munich (1995–2000); Chair of the IVS Directing Board since 2007; President of IAU (International Astronomical Union) Commission 19 "Rotation of the Earth" (2009–2012); President of the Austrian Geodetic Commission since 2008 and President of the Austrian National Committee of the IUGG since 2009; Vice-President of the IAG (International Association of Geodesy) since 2011; member of various directing and governing boards; editorial board of the Journal of Geodesy (2003–2007), and served as president, chair, member or consultant of various commissions, sub-commissions and working groups in geodesy (IAG) and astronomy (IAU); coordinator of the German Research Group on Earth Rotation (1999–2003); supervisor, co-supervisor, or examinator of more than 25 dissertations. In 2009 Harald Schuh received the degree of a doctor honoris causa (Dr. h.c.) and in 2011 the Vening-Meinesz Medal of the European Geosciences Union.

Johannes Böhm is associate professor at the Institute of Geodesy and Geophysics, Vienna University of Technology, Austria, where atmospheric effects in space geodesy and very long baseline interferometry (VLBI) are his main fields of interest and research. In 1999 he graduated from the Vienna University of Technology with a thesis about modern geopotential models and received his Ph.D. in 2004 with a dissertation on troposphere delays in VLBI. Troposphere delay modelling for all space geodetic techniques at radio wavelengths with the application of numerical weather models was the topic of his habilitation thesis in 2008. Johannes Böhm is President of IAG Sub-Commission 1.4 "Interaction of Celestial and Terrestrial Reference Frames", and he has been chair or member of various working groups of the IVS and the IAG. He has been on the editorial board of Journal of Geodesy since 2007 and has been leading various research projects related to VLBI at the Vienna University of Technology. Johannes Böhm received the Guy Bomford Prize of the IAG in 2011.

The book has been subjected to an individual review of chapters. I am grateful to reviewers Prof. Trevor Baker of the Proudman Oceanographic Laboratory in the United Kingdom, Dr. Bert Vermeersen of Technical University Delft, Dr. Roberto Peron of the Institute of Physics of Planetary Space (IFSI-INAF) in Rome, Prof. Zhiping Lü and Dr. Xiguang Zhang of Zhengzhou Institute of Surveying and Mapping (ISM), Prof. Shulong Zhu of Zhengzhou ISM, Prof. Rene Forsberg of Danish Space Centre, Dr. Karsten Jacobsen, Dr. Ludger Timmen of the University Hannover, Prof. Wolfgang Torge and Dipl.-Ing. Christian Voigt of the Leibniz Universität Hannover, Prof. Bernhard Heck of Karlsruher Institut für Technologie, Prof. Guigen Ni of the Information Engineering University (IEU) in Zhengzhou, Prof. Yuanxi Yang and Dr. Tianhe Xu of the Institute of Surveying and Mapping (ISM) in Xi'an, Dr. Axel Nothnagel of the University Bonn, Prof. Ludwig Combrinck of the Hartebeesthoek Radio Astronomy Observatory (HartRAO), Dr. Svetozar Petrovic, Dr. Monika Korte, and Dr. Matthias Förster of GFZ. As editor I made a general review of the whole book. A grammatical check of technical English writing has been performed by Springer Heidelberg.

I wish to thank sincerely the key authors of the individual chapters: Dr. Machiel Bos of University Porto, Dr. Hans-Georg Scherneck of Chalmers University of Technology in Sweden, Dr. Ludwig Combrinck of Hartebeesthoek Radio Astronomy Observatory, Dr. Detlef Angermann, Dr. Manueal Seitz and Prof. Hermann Drewes of DGFI in Munich, Prof. Paula Redweik of University of Lisbon, Dr. Heiner Denker of Leibniz University Hannover, Prof. Yunzhong Shen of Tonji University in Shanghai, Prof. Harald Schuh and Prof. Johannes Böhm of Technical University Vienna. Without their consistent efforts such a book would never have become available. I also wish to sincerely thank those scientists who made great efforts for enriching this book. They are Dr. Ludger Timmen of Leibniz University Hannover, Dr. Luisa Bastos of University Porto, Dr. Dietrich Ewert of University Berlin, Prof. Cheinway Hwang and Prof. Tianyuan Shih of Central University of Taiwan.

I wish to express my gratitude towards the former directors Prof. Dr. Ch. Reigber and Prof. Dr. Markus Rothacher of GFZ for their support and trust during my research activities at the GFZ and for granting me special freedom of research. Acting heads Dr. Christoph Förste, Dr. Frank Flechtner and Dr. Jens Wickert of GFZ are thanked for supporting my editorial activities in this book series. I also wish to thank sincerely Prof. Yuanxi Yang of ISM in Xi'an, Prof. Qin Zhang of Chang'an University in Xi'an, Prof. Heping Sun, Prof. Jikun Ou and Prof. Yunbin Yuan of IGG in Wuhan for their friendly support by organising the International Geodetic Forum Xi'an 2006, which is the origin of the idea to write and edit such a series of scientific books. The Chinese Academy of Sciences is thanked for the Outstanding Overseas Chinese Scholars Fund, which greatly supported the valuable scientific activities.

Special thanks go to Springer, Heidelberg; their support and their evaluation for such a series of books are preconditions for successfully organising this publication. I am also grateful to Dr. Chris Bendall of Springer, Heidelberg for his valuable advice.

November 2011 Guochang Xu

Contents

Chapter 1
Computation of Green's Functions for Ocean Tide Loading

M. S. Bos and H.-G. Scherneck

The devil is in the details

1 Introduction

This chapter will discuss the computation of the deformation of the solid Earth due to external forces. It is a classical problem that was studied more than a century ago by famous people such as Thomson and Tait (1867) and Lamb (1895). They were followed by Love (1911) and Hoskins (1920) in the beginning of the twentieth century. Since then it has been studied extensively by seismologists who are interested in modelling the free oscillations of the Earth that occur after large earthquakes. Important contributions to this area were made by Pekeris and Jarosch (1958) and Alterman et al. (1959) which still forms the basis of what we will describe in this chapter. A thorough description of the the theory of the free oscillations of the Earth can be found in the textbook by Dahlen and Tromp (1998). An older but still good reference is the review article by Takeuchi and Saito (1972).

The reader could therefore accuse us of writing about a topic that has already been described. However, we feel that current literature does not pay much attention to the practical details of how a given profile of the density and elastic properties of the Earth are to be used to compute these deformations and it is our objective to fill this gap. We hope that a researcher or Ph.D. student who wants to learn more about this topic finds in our chapter a good starting point where all assumptions are clearly explained and where enough details are given to implement the equations into a computer program.

M. S. Bos (✉)
Centro Interdiscriplinar de Investigacão Marinha e Ambiental, Universidate do Porto, Rua dos Bragas 289, 4050-123, Porto, Portugal
e-mail: mbos@ciimar.up.pt

H.-G. Scherneck (✉)
Chalmers University of Technology, Earth and Space Sciences, SE-412 96, Gothenburg, Sweden
e-mail: hgs@chalmers.se

G. Xu (ed.), *Sciences of Geodesy - II*, DOI: 10.1007/978-3-642-28000-9_1,

We will only look at deformations caused by the varying weight of the ocean tides, also known as ocean tide loading (OTL). With the current accuracy by which these tidal deformations can be observed at the surface, we can ignore the ellipticity of the Earth and its rotation and assume that the mechanical properties of the Earth are the same for all orientations and only vary along the radius of the Earth. With sufficient accuracy we can also assume that the deformation is elastic or at least deviates only slightly from a pure elasticity.

Next, the weight of the ocean tides is normally decomposed into a sum of point loads. The advantage is that, once you know the deformation of the Earth under a single point load, and assuming that the deformations are small enough so that the principle of superposition holds, you can compute the deformation of all point loads in a similar way and add them up to get the total. The deformation due to a point load, which is a Dirac delta function, is called a Green's function. One of the first attempts to compute such a Green's function was given by Slichter and Caputo (1960) although they used a circular disc load instead of the actual limit of reducing the radius of the disc to zero and they ignored any gravity effects due to the mass distribution inside the Earth.

Longman (1962, 1963) was the first to develop the point load into a sum of Legendre polynomials and computed this sum up to degree 40. Farrell (1972) continued the work of Longman and extended the summation up to a degree of 10,000. Farrell's contribution was also a better understanding of the problem at degree 1 where the deformation is invariant with respect to a simple translation of the whole Earth. He also emphasised the use of the analytical solution of the deformation of a half-space as the asymptotic solution of the deformations of the spherical Earth. These asymptotic solutions can not only be used to check the numerical solutions but are also essential to find the value of the infinite sum of Legendre polynomials.

Longman and Farrell used the elastic properties and density profiles of the Earth that were computed by seismologists. An example is the Preliminary Reference Earth Model (PREM) published by Dziewonski and Anderson (1981). The earthquakes that are being studied by seismologists have periods of several seconds and, since tides have a period of several hours, one can wonder whether the same elastic properties should be used. So far, no observations that challenge this assumption have ever been presented.

In this chapter we explain how the these elastic properties and density profiles can be transformed in so-called Love numbers. These numbers can be used to compute the necessary Green's functions. The summation of Love numbers has already been described in detail by, among others, Farrell (1972), Francis and Mazzega (1990), Jentzsch (1997), Guo et al. (2004) and recently by Agnew (2007), and therefore will only be discussed briefly. It is the computation of the Love numbers that we will focus on. We will start at the very beginning, which means we need to start by deriving the set of differential equations that govern the deformation of the solid Earth.

2 Equations of Motions and Rheology

This section will derive the linearized equations of motion in the same way as Dahlen (1974) although we give more attention to the interpretation of all the terms.

We restrict our discussion to models of the Earth that are symmetric, non-rotating and elastic isotropic (SNREI) and everywhere in hydrostatic equilibrium. The positions of the mass particles of the Earth are denoted by $\mathbf{x}$. At the same time we will use these initial locations to label the particles. Let $\mathbf{r}(\mathbf{x}, t)$ be the position of particle with label and initial position $\mathbf{x}$ after the deformation at time t. Now we can write the Lagrangian displacement $\mathbf{s}_L(\mathbf{x}, t)$ as (Dahlen 1974)

$$\mathbf{r}(\mathbf{x}, t) = \mathbf{x} + \mathbf{s}_L(\mathbf{x}, t) \tag{1.1}$$

Instead of following the deformation of a particle with label $\mathbf{x}$ that was initially at position $\mathbf{x}$, one may describe the deformation over time one finds at the fixed position $\mathbf{r}$ inside the Earth. This is the Eulerian description of the deformation.

It will be convenient to write the changes in density ρ and potential ϕ as small perturbations from a reference state. We have (Dahlen 1974)

$$\rho_L(\mathbf{x}, t) = \rho_0(\mathbf{x}) + \rho_1^L(\mathbf{x}, t) \tag{1.2}$$

$$\rho_E(\mathbf{r}, t) = \rho_0(\mathbf{r}) + \rho_1^E(\mathbf{r}, t) \tag{1.3}$$

$$\phi_L(\mathbf{x}, t) = \phi_0(\mathbf{x}) + \phi_1^L(\mathbf{x}, t) \tag{1.4}$$

$$\phi_E(\mathbf{r}, t) = \phi_0(\mathbf{r}) + \phi_1^E(\mathbf{r}, t) \tag{1.5}$$

The subscript or superscript L and E indicate whether we are dealing with a Lagrangian or a Eulerian function. Generally, the coefficients of the functions ρ_L and ρ_E are not equal, neither those of ϕ_L and ϕ_E, because they depend on a different set of variables, the Lagrangian or Eulerian positions. Nevertheless, they describe the same changes in density and potential in the Earth. The subscript 0 represents the reference state. The subscript 1 indicates that it is a perturbed quantity.

It is good to be aware of the difference between the Lagrangian and Eulerian description, especially at the boundaries. However, we will derive here a linearized set of equations that describe small perturbations from the reference state. As a result, we will encounter many situations where this difference of description is of no importance. An example is the case where the reference density is multiplied by a small value ϵ. In these case we have $\rho_0(\mathbf{x})\epsilon \approx \rho_0(\mathbf{r})\epsilon$, where $\mathbf{x}$ and $\mathbf{r}$ are related through (1.1). In addition, for the perturbed density we have $\rho_1^E(\mathbf{x}, t) \approx \rho_1^E(\mathbf{r}, t)$. Similar relations hold for the reference potential ϕ_0 and the perturbed potential ϕ_1.

We will assume that no mass is created or destroyed which leads to the following equation of continuity:

$$\begin{aligned} \rho_1^E(\mathbf{r},t) + \rho_0(\mathbf{r})\nabla\cdot\mathbf{s}_L(\mathbf{x},t) + \mathbf{s}_L(\mathbf{x},t)\nabla\cdot\rho_0(\mathbf{r}) &= 0 \\ \rho_1^E(\mathbf{r},t) &= -\nabla\cdot[\rho_0(\mathbf{x})\mathbf{s}_L(\mathbf{x},t)] \end{aligned} \tag{1.6}$$

Note the change of $\rho_0(\mathbf{r})$ to $\rho_0(\mathbf{x})$ in the second line of this equation which is allowed as long as $\mathbf{s}$ is small.

In words, the first line of (1.6) states that the sum of the perturbed density in a small element plus the density change caused by the deformation of the element plus moving the element to another position where the reference density is different is constant.

Note that we have written the changes in density as the sum of the reference state plus a small perturbation. The small element can thus be considered to have a density ρ_1 and to be floating through a reference density field of ρ_0.

The gradient in density can be smooth or abrupt. At a layer interface the gradient is abrupt. A vertical displacement of the interface implies a density perturbation in the Eulerian system, and this density perturbation appears in Poisson's equation as the source of the perturbed potential to be discussed next.

Poisson's equation relates the gravitational potential to the density inside the Earth. Before we present this equation, the sign convention of the potential must be discussed. Normally, a potential ϕ_0 of a particle represents the amount of energy it contains. Thus, if we consider a particle above the Earth's surface, then the higher it is, the more gravitational potential energy it will have.

To get the reference gravitational force per unit mass, $\mathbf{g_0}$, at a fixed point inside the Earth, one must take the negative gradient of the potential ϕ_0 :

$$\mathbf{g}_0(\mathbf{r}) = -\nabla\phi_0(\mathbf{r}) \tag{1.7}$$

The perturbed gravity force per unit volume:

$$\begin{aligned} \rho_0(\mathbf{r})\mathbf{g}_1^E(\mathbf{r},t) &= -\rho_0(\mathbf{r})\nabla\phi_1^E(\mathbf{r},t) - \rho_1^E(\mathbf{r},t)\nabla\phi_0(\mathbf{r}) \\ &= -\rho_0(\mathbf{x})\nabla\phi_1^E(\mathbf{x},t) - \nabla\cdot[\rho_0(\mathbf{x})\mathbf{s}_L(\mathbf{x},t)]\mathbf{g}_0(\mathbf{x}) \end{aligned} \tag{1.8}$$

Here we have made use of (1.6) to substitute ρ_1 and again replaced $\mathbf{r}$ vectors for $\mathbf{x}$ vectors.

In geodesy, one sometimes inverses the sign of ϕ to make the force equal to the gradient of the potential, without adding a minus sign (Jekeli 2007). Depending on the sign convention of ϕ, Poisson's equation is

$$\nabla^2\phi_0(\mathbf{r}) = \pm 4\pi\rho_0(\mathbf{r})\mathrm{G} \tag{1.9}$$

$$\nabla^2\phi_1^E(\mathbf{r},\mathrm{t}) = \pm 4\pi\rho_1^E(\mathbf{r},\mathrm{t})\mathrm{G} \tag{1.10}$$

where G is the gravitational constant. Farrell (1972), Dahlen (1974), Wu and Peltier (1982) and Dahlen and Tromp (1998) all use the plus sign while Pekeris and Jarosch (1958) and Alterman et al. (1959) used the minus sign in (1.9). Since the work of Alterman et al. was very influential, their convention has been followed by many people such as Kaula (1963), Okubo (1988), Sun and Sjöberg (1999) and Guo et al. (2004). In this chapter we will follow the definition of Dahlen (1974) which means that we keep the potential energy interpretation of ϕ and use the plus sign in (1.9) and (1.10).

Next, since we assume that the Earth is in hydrostatic equilibrium, there is a uniform pressure p_0 at each depth layer in the reference state. This pressure p_0 inside the Earth increases with depth because the weight of the layers of rock above increases. A particle that is displaced to a deeper layer will therefore experience an upward buoyancy force $\mathbf{b}_L$. Remembering that we have to take the negative gradient to compute the force of our potential, the buoyancy force per unit volume to first order is

$$\begin{aligned} \mathbf{b}_L(\mathbf{x},t) &= \nabla[\mathbf{s}_L(\mathbf{x},t)\cdot\rho_0(\mathbf{x})\mathbf{g}_0(\mathbf{x})] \\ &= -\nabla[\mathbf{s}_L(\mathbf{x},t)\cdot\rho_0(\mathbf{x})\nabla\phi_0(\mathbf{x})] \end{aligned} \tag{1.11}$$

In addition, a force is required in a solid body to change the relative distances between the particles. In fact, it is the gradient of the change in distances between the particles, the strain, that relates linearly with the elastic force. This is called Hooke's law, and it is a linear law for small displacements. In three dimensions this linear relation for an isotropic material is given by the Cauchy stress tensor $\mathbf{T}_L$. It requires a constant for the change in volume, the bulk modulus κ, and another constant for the amount of shearing called μ. For our purpose we will assume that we can use the adiabatic bulk modulus. The relation of the Cauchy stress tensor $\mathbf{T}_L$ with the deformations $\mathbf{s}_L(\mathbf{x},t)$, also known as the constitutive law, is given by

$$\mathbf{T}_L(\mathbf{x},t) = \left(\kappa - \frac{2\mu}{3}\right)(\nabla\cdot\mathbf{s}_L(\mathbf{x},t))\mathbf{I} + \mu\left[\nabla\mathbf{s}_L(\mathbf{x},t) + (\nabla\mathbf{s}_L(\mathbf{x},t))^T\right] \tag{1.12}$$

where $\mathbf{I}$ is the identity tensor. We again add a subscript L to $\mathbf{T}$ to indicate it is Lagrangian: The elastic forces act on the deforming body. We implicitly assume that these deformations are so small that there is no significant change in the surface of the body. Otherwise the amount of pressure that is acting on the body would be different before and after the deformation. It is convenient to introduce another variable λ which is defined as $\lambda = \kappa - 2\mu/3$. The pair λ and μ are called the Lamé parameters. The elastic parameters are the entry point where—more generally speaking—the rheology of the Earth can enter. Rheology is the umbrella concept under which elasticity may be generalised to comprise a range of properties of solids describing how they deform, either instantaneously, by creep, or, in the extreme limit, by fluid-like flow or brittle failure. We will remain in the realm of linear laws (ignore stress-dependence of the moduli), avoid the brittle regime, and also ignore heat flow, convective instabilities and phase changes.

From (1.12) it is clear that when there are no displacements, there is no elastic force. However, the Earth is already in a strained situation even without external forcing because of the weight of the layers inside the Earth that are pressing on the layers beneath them (Love 1911). This weight causes the hydrostatic pressure p_0 discussed before for the buoyancy force. Therefore, (1.12) must be interpreted as the deviatoric stress tensor, which is the stress difference with respect to the reference stress state $\mathbf{T}_0$. Any additional stresses introduced into the Earth due to, for example, earthquakes, plate tectonics or mantle convection, which would create a $\frac{1}{3}tr(\mathbf{T}_0) \neq p_0$, are neglected.

The last equation we need is Newton's second law of motion, linearised, that states that the acceleration of a small element is determined by the sum of the gravity force of (1.8), the buoyancy force $\mathbf{b}_L$ of (1.11), the divergence of the stress tensor $\mathbf{T}_L$ and a body force $\mathbf{f}$. It is also known as the momentum equation

$$\begin{aligned} \rho_0(\mathbf{x})D_t^2\mathbf{s}_L(\mathbf{x},t) = & -\rho_0(\mathbf{x})\nabla\phi_1^E(\mathbf{x},t)- \\ & \nabla\cdot[\rho_0(\mathbf{x})\mathbf{s}_L(\mathbf{x},t)]\mathbf{g}_0(\mathbf{x})- \\ & \nabla[\mathbf{s}_L(\mathbf{x},t)\cdot\rho_0(\mathbf{x})\mathbf{g}_0(\mathbf{x})]+\nabla\cdot\mathbf{T}_L(\mathbf{x},t)+\mathbf{f}(\mathbf{x},t) \end{aligned} \tag{1.13}$$

The term D_t^2 on the left is the second order material (or Lagrangian) derivative with respect to time t. The $\mathbf{f}(\mathbf{x},t)$ is body force per volume and assumed to be small enough so that $\mathbf{f}(\mathbf{x},t) = \mathbf{f}(\mathbf{r},t)$.

Equations 1.10 and 1.13 are the same as those presented by Farrell (1972). Note that ϕ_1^E is the only Eulerian variable which will require some attention at the boundaries.

3 Spheroidal and Toroidal Motions

The tensor equations derived in Sect. 1.2 are concise and clear but they are not very convenient for numerical computations. To solve the tensor equations of motions we will chose a reference frame with the origin at the centre of mass of the undeformed Earth and use spherical coordinates (r, θ, λ) containing the radius, co-latitude and longitude, and unit direction vectors $\mathbf{e}_r$, $\mathbf{e}_\theta$ and $\mathbf{e}_\lambda$. This will produce expressions for the gradient, divergence and Laplacian that are more complicated than for a Cartesian coordinate system but it will facilitate the definition of the boundary conditions that will be discussed in Sect. 1.6

Since the east, north and up direction are always orthogonal to each other, one can avoid the theory of general curvilinear tensor components and use the more straightforward method described by Malvern (1969, App. II), Arfken (1985, Chap. 2) and Dahlen and Tromp (1998, App. A) to derive the desired expressions. Malvern and Dahlen and Tromp also list the expression for the Cauchy stress tensor in spherical coordinates. Hoskins (1910, 1920) and Pekeris and Jarosch (1958) present a complete set of all equations of motion expressed in spherical coordinates.

We will now repeat their derivation of these equations, but to do so we first need to put some limits on the shape of our deformation. According to Helmholtz's theorem, any differentiable vector field, thus also our deformations **s**, can be represented as the sum of an irrotational vector field which is the gradient of a scalar potential f plus a solenoidal (equivoluminal) vector field which is the curl of a vector potential **A**; see Arfken (1985, Chap. 1) and Malvern (1969, Chap. 8):

$$\mathbf{s} = \nabla f + \nabla \cdot \mathbf{A} \tag{1.14}$$

with $\nabla \cdot \mathbf{A} = 0$. In the presence of a body force **b** the equation of motion in terms of the potentials is

$$(\lambda + 2\mu)\nabla\, \nabla^2 f + \mu \nabla \times \nabla^2 \mathbf{A} + \rho \mathbf{b} = \rho \nabla \frac{\partial^2 f}{\partial t^2} + \rho \nabla \times \nabla \frac{\partial^2 \mathbf{A}}{\partial t^2} \tag{1.15}$$

This equation is separable into a solenoidal part, independent of f, and a spheroidal part, independent of **A**, if we know how to partition the body force **b** into a curl-free and a divergence-less component (Lamb 1895). If the body force is zero, then (1.15) decouples into the two seismic wave equations, compressional waves with speed $v_\alpha = \sqrt{(\lambda + 2\mu)/\rho}$ and shear waves with speed $v_\beta = \sqrt{\mu/\rho}$.

The division of **s** into a spheroidal part which is both compressible and curl-free, and a complementary solenoidal part affords us a road fork in our story. Before we start to walk down the spheroidal road, let us remind ourselves of the decomposition of the vector potential **A** into a poloidal and a toroidal part according to Backus (1986):

$$\nabla \times \mathbf{A} = \nabla^2(g\mathbf{r}) + \nabla \times (h\mathbf{r}) = \mathbf{S} + \mathbf{T} \tag{1.16}$$

where

$$\mathbf{S} = \nabla \left[\frac{\partial}{\partial r}(r\,g) \right] - \mathbf{r}\nabla^2 g \tag{1.17}$$

$$\mathbf{T} = -\mathbf{r} \times (\nabla h) \tag{1.18}$$

It shows that the divergence-free displacements can themselves be related to scalar potentials g and h. The poloidal part, **S**, will take part in the deformation due to a gravitating surface load with traction along the surface normal; the toroidal part, **T**, is insensitive to potential forces but susceptible to surface shear tractions.

In a radially symmetric planet the body force is due to the gravity potential of the load, and thus the curl of this force is zero. However, this part can be regarded as a particular solution of a non-homogeneous problem. The general problem with zero boundary conditions contains both the spheroidal and the toroidal part, and its solution the full array of free oscillations. We will restrict ourselves to the spheroidal part:

$$\mathbf{s} = \nabla f + \nabla^2(g\mathbf{r}) \tag{1.19}$$

For deformations due to traction, see Merriam (1985, 1986). Expanding (1.19) into its spherical coordinates gives us

$$\mathbf{s} = \begin{pmatrix} u \\ v \\ w \end{pmatrix} = \begin{pmatrix} \frac{df}{dr} - \frac{1}{r}\frac{d^2g}{d\theta^2} - \frac{1}{r\tan\theta}\frac{dg}{d\theta} - \frac{1}{r\sin^2\theta}\frac{d^2g}{d\lambda^2} \\ \frac{1}{r}\frac{df}{d\theta} + \frac{1}{r}\frac{dg}{d\theta} + \frac{d^2g}{drd\theta} \\ \frac{1}{r\sin\theta}\frac{df}{d\lambda} + \frac{1}{\sin\theta}\left(\frac{1}{r}\frac{dg}{d\lambda} + \frac{d^2g}{drd\lambda}\right) \end{pmatrix} \tag{1.20}$$

Owing to radial symmetry, the spheroidal deformation can be decomposed with spherical harmonics as angular base functions and radial factor functions for the depth-dependence:

$$u = \sum_{n=0}^{\infty}\sum_{m=-n}^{n} U_n^m(r) Y_n^m(\theta, \lambda) \tag{1.21}$$

$$v = \sum_{n=0}^{\infty}\sum_{m=-n}^{n} V_n^m(r) \frac{dY_n^m(\theta, \lambda)}{d\theta} \tag{1.22}$$

$$w = \sum_{n=0}^{\infty}\sum_{m=-n}^{n} V_n^m(r) \frac{dY_n^m(\theta, \lambda)}{\sin\theta\, d\lambda} \tag{1.23}$$

We can see that $U(r)$ is associated with the radial deformation and $V(r)$ with the horizontal deformation. We may regard

$$\Psi = \sum_{n=0}^{\infty}\sum_{m=-n}^{n} V_n^m(r) Y_n^m(\theta, \lambda) \tag{1.24}$$

as a potential of horizontal displacement, delivering the vectorial components when we let the horizontal gradient operator $[\hat{\theta} d_\theta,\ \hat{\lambda}(\sin\theta)^{-1} d_\lambda]$ act on it.

The perturbed potential ϕ_1 that appeared in (1.13) can also be written as the sum of spherical harmonics and, following tradition, the part containing the radial function will be represented by $P(r)$. Note that for the horizontal displacement we need to differentiate the spherical harmonics by θ or λ.

As we will argue below, we can restrict our treatment of the Spherical Harmonics of order $m = 0$, i.e. Legendre Polynomials of the first kind. At the same time we can avoid discussing normalisation and in particular the different variants that you may encounter in the literature.

The restriction to $m = 0$ comes without any sacrifice as to physics, since the physically relevant properties relate only to the spherical harmonic degree, while the spherical harmonic order carries information about such arbitrary things like pole location and azimuthal orientation; after all our model planet is radially symmetric (Phinney and Burridge 1973). Thus, for (1.21) we can equally well write

$$u = \sum_{n=0}^{\infty} U_n^m(r) \sum_{m=-n}^{n} C_{nm} Y_n^m(\theta, \lambda)$$

and mutatis mutandis for v and w, where the dimensionless coefficients C_{nm} come from the expansion of the forcing field (so the same set applies to u, v and w). We are only interested in the radial functions, so contemplating the simplest case for m, $m = 0$ suffices.

If we now fill in (1.20) for given degree n and order 0 into the equations of motion, (1.10) and (1.13), in spherical coordinates and drop the subscript n and superscript 0 from the coefficients U_n^0, V_n^0 and P_n^0, we get (Alterman et al. 1959; Wu and Peltier 1982)

$$\omega^2 \rho_0 U - \rho_0 \frac{dP}{dr} + g_0 \rho_0 X - \rho_0 \frac{d}{dr}(g_0 U) + \frac{d}{dr}(\lambda X + 2\mu \frac{dU}{dr}) + \frac{\mu}{r^2}\left[4\frac{dU}{dr} r - 4U + n(n+1)(-U - r\frac{dV}{dr} + 3V)\right] = 0 \tag{1.25}$$

$$\rho_0 \omega^2 V r - \rho_0 P - g_0 \rho_0 U + \lambda X + r\frac{d}{dr}\left[\mu\left(\frac{dV}{dr} - \frac{V}{r} + \frac{U}{r}\right)\right] + \frac{\mu}{r}\left[5U + 3r\frac{dV}{dr} - V - 2n(n+1)V\right] = 0 \tag{1.26}$$

$$\frac{d^2P}{dr^2} + \frac{2}{r}\frac{dP}{dr} - \frac{n(n+1)}{r^2} P = 4\pi G(\frac{d\rho_0}{dr} U + \rho_0 X) \tag{1.27}$$

with

$$X = \frac{dU}{dr} + \frac{2}{r} U - \frac{n(n+1)}{r} V \tag{1.28}$$

Equation 1.28 represents the dilatation of the material. Due to the sign difference in Poisson's equation, Alterman et al. use $-P$ in (1.25)–(1.27). In addition, we have assumed that the deformation is periodic with an angular velocity of ω. The second time derivative of the deformation **s** can in this case be written as $-\omega^2 \mathbf{s}$.

Next, (1.25) and (1.27) have been divided by Y_n^0 and (1.26) has been divided by $dY_n^0/d\theta$. This is important to remember for the case $n = 0$ which results in $dY_n^0/d\theta = 0$. For $n = 0$ one should simply set $V = 0$ and discard (1.26).

To derive (1.25)–(1.27) from (1.10) and (1.13) we not only needed the expressions of the gradient and divergence in spherical coordinates but also made use of the following relation:

$$\frac{d^2Y_n^0}{d\theta^2} + \cot\theta\frac{dY_n^0}{d\theta} = -n(n+1)Y_n^0 \tag{1.29}$$

The result is that we have reduced the set of coupled differential equations from three to one dimensions, although one has to compute them repetitively for all values of degree n. In addition, since we are using spherical coordinates, we can more easily define the boundary conditions.

While some numerical methods, such as the spectral method discussed in Sect. 1.9, may integrate the second-order differential equations (1.25)–(1.28) with sufficient accuracy, we also give the six equations of first order in ∂_r, using the auxiliary variables

$$\begin{gathered} \alpha = \kappa + \frac{4}{3}\mu \qquad \beta = \kappa - \frac{2}{3}\mu \qquad \eta = 3\kappa + 2\mu \\ R = \tau_{rr} \qquad S = \tau_{r\theta} \end{gathered} \tag{1.30}$$

where α and β relate to the seismic longitudinal (compressional) and shear velocities

$$v_\alpha = \sqrt{\alpha/\rho} \qquad v_\beta = \sqrt{\beta/\rho} \tag{1.31}$$

respectively, parameters that are normally tabulated by seismologists for various depths of the Earth. As before, κ is the bulk modulus, which is the inverse of the compressibility, and μ is the shear modulus or rigidity. R and S are two components from our Cauchy stress tensor $\mathbf{T}_L$ and represent the radial and shear stress. Rewriting their definition provides us with two of the six first order differential equations:

$$\frac{dU}{dr} = \frac{1}{\alpha}\left(-\frac{2\beta}{r}U + \frac{n(n+1)\beta}{r}V + R\right) \tag{1.32}$$

$$\frac{dV}{dr} = -\frac{1}{r}U + \frac{1}{r}V + \frac{1}{\mu}S \tag{1.33}$$

Note that to here we deviate from (Dahlen and Tromp 1998, p. 271) who define our scalar V as $n(n+1)V$. The third equation is provided by rewriting the definition of the auxiliary variable Q which denotes the perturbed gravity plus a term $(n+1)P/r$:

$$\frac{dP}{dr} = -4\pi G\rho U - \frac{n+1}{r}P + Q \tag{1.34}$$

In Sect. 1.6 we will see that this auxiliary variable will facilitate defining the boundary condition at the surface. Filling in the definitions of R, S and Q into (1.25, 1.26, 1.27) gives us (Dahlen and Tromp 1998):

$$\frac{dR}{dr} = \left(-\omega^2\rho + \frac{12\kappa\mu}{\alpha r^2} - \frac{4g\rho}{r}\right)U + n(n+1)\left(-\frac{6\kappa\mu}{\alpha r^2} + \frac{g\rho}{r}\right)V \\ -\frac{4\mu}{\alpha r}R + \frac{n(n+1)}{r}S + \rho Q \tag{1.35}$$

$$\frac{dS}{dr} = \left(-\frac{6\kappa\mu}{\alpha r^2} + \frac{g\rho}{r}\right)U - \left(\omega^2\rho + \frac{\mu(n^2+n-2)}{r^2}\right)V \\ +\frac{\rho}{r}P - \frac{\beta}{\alpha r}R - \frac{3}{r}S \tag{1.36}$$

$$\frac{dQ}{dr} = -\frac{4\pi G\rho}{r}[(n+1)U - n(n+1)V] + \frac{n-1}{r}Q \tag{1.37}$$

Gravity acceleration $g = g(r)$ can be computed from the density model. To derive these equations we also made use of the relation:

$$\frac{dg_0}{dr} = -\frac{2g_0}{r} + 4\pi G\rho_0 \tag{1.38}$$

Outside the Earth, only the first term on the right side of (1.38) would be necessary. However, inside the Earth to the second term is also necessary. With the usual notation $\mathbf{y} = [U,\ V,\ P,\ R,\ S,\ Q]^T$:

$$\frac{d\mathbf{y}}{dr} = \mathbf{A}\mathbf{y} \tag{1.39}$$

Another convention followed, for example, by Alterman et al. (1959), Longman (1962, 1963) and Farrell (1972) is to label vector $\mathbf{y}$ as $[y_1, \ldots, y_6]$. However, note that the definition of y_6 by Alterman et al. (1959) is different from our Q because it lacks the $(n+1)P/r$ part and represents the true perturbed gravity value. We prefer our semi-perturbed gravity parameter Q because it simplifies the formulation of the boundary condition at the surface.

At large n the radial functions run over many orders of magnitude, so that the equation system needs stabilisation. One method is to replace r and $\mathbf{Y}$ as follows

$$\tilde{r} = \frac{r}{a}, \quad q = (n+1)\log\tilde{r} \quad \text{and} \quad \mathbf{Y} = \mathbf{LZ} \tag{1.40}$$

respectively, where a is the mean radius of the Earth and

$$\mathbf{L} = \exp\left\{\mathbf{diag}\left[a\sqrt{\tilde{r}},\ na\sqrt{\tilde{r}},\ ag(a)\sqrt{\tilde{r}},\ \frac{\kappa(0)\sqrt{\tilde{r}^3}}{(n+1)},\ \kappa(0)\sqrt{\tilde{r}^3},\ g(a)\sqrt{\tilde{r}^3}\right]\right\} \tag{1.41}$$

where κ_0 is the maximum incompressibility in the Earth, and to transform (1.39) according to Lyapunov (Gantmacher 1950) into

$$\frac{d\mathbf{Z}}{dq} = \mathbf{BZ} \tag{1.42}$$

where

$$\mathbf{B} = \mathbf{L}^{-1}\left(\frac{a}{n+1}\exp\left[\frac{q}{n+1}\right]\mathbf{AL} - \frac{d\mathbf{L}}{dq}\right) \tag{1.43}$$

in which

$$\mathbf{L}^{-1}\frac{d\mathbf{L}}{dq} = \frac{3}{2(n+1)}\mathbf{diag}[1,\ 1,\ 1,\ 3,\ 3,\ 3] \tag{1.44}$$

Matrix **B** has been given in full in Appendix 1. This scaling is particularly useful when one uses a numerical integration method such as Runga–Kutta to solve the differential equations; see Sect. 1.9

4 Fluid Core

So far we have assumed that the Earth is a solid body but seismologists tell us that the Earth has a fluid core. A fluid differs from a solid by having zero rigidity. Thus, by setting the shear modulus μ to zero in the Cauchy stress tensor, the equations presented in Sect. 1.3 continue to be applicable and we are treating the fluid as a very weak solid.

However, problems arise when the forcing period is taken to infinity to simulate static forcing. This phenomenon has received a relatively large amount of attention in the literature. We will now try to point out some main conclusions that have been derived.

It was Longman (1963) who showed that, for the case of $\omega = 0$, the (1.25) and (1.26) are no longer independent in the fluid core. This can be seen by writing these two equations in the following form:

$$\frac{d}{dr}[\rho_0(P - g_0U) + \lambda X] + \rho_0 g_0 X - (P - g_0U)\frac{d\rho_0}{dr} = 0 \tag{1.45}$$

$$\rho_0(P - g_0U) + \lambda X = 0 \tag{1.46}$$

From (1.46) one can deduce that the term within the square brackets of (1.45) must be zero. If (1.46) is then used to rewrite (1.45) we have:

$$\left(\frac{g_0\rho_0}{\lambda} + \frac{1}{\rho_0}\frac{d\rho_0}{dr}\right)X = 0 \tag{1.47}$$

$$-\frac{N^2}{g_0}X = 0 \tag{1.48}$$

Here we made use of the definition of the Brunt–Väisälä frequency $N(r)$ that is related to the stratification of the fluid:

$$N^2(r) = -\frac{g_0^2\rho_0}{\kappa} - \frac{g_0}{\rho_0}\frac{d\rho_0}{dr} \tag{1.49}$$

In a fluid $\lambda = \kappa$. As was explained before, X is the dilatation of the material; see (1.28). For a real Earth the dilatation is not always zero which leads us to the conclusion that $N = 0$ in (1.47) and this puts a new condition on the properties of the fluid that was not needed before. This situation corresponds to the so-called Adams–Williamson or neutral buoyancy condition. It means that the compressibility of the fluid is such that, when a small parcel of liquid is pushed to a deeper and denser layer, it will compress exactly to a volume with the same density as the surrounding fluid. If, however, the parcel afterwards rises up again, then the stratification of the fluid is stable, $N > 0$. If the parcel continues to sink the stratification is unstable, $N < 0$.

The fact that the fluid core can only be in neutral buoyancy seems strange and is called the Longman paradox (Dahlen 1974; Wunsch 1974; Chinnery 1975). One part of the solution of this paradox is that one should be careful when taking the limit of $\omega \to 0$. The result of this limit also depends on the real stratification of the fluid.

If the stratification is unstable, a boundary layer develops that gets thinner for increasing forcing period. In the extreme case of $\omega = 0$, it represents an infinitely thin layer but it still has a finite influence on the dynamics. The radial stress experiences a jump in the boundary layer and is zero in the fluid. Because in a fluid the radial stress is proportional to the dilatation, this means that X is zero in fluid after all and that the Adams–Williams condition, or neutral buoyancy, is no longer necessary to satisfy (1.47). In Sect. 1.7 we will discuss a homogeneous fluid which means $d\rho_0/dr = 0$ and $N < 0$. Thus, the stratification is unstable and, near the boundary of the fluid core with the mantle, such a boundary layer develops. Pekeris and Accad (1972) also discuss the results for a fluid with $N = 0$. In this case no boundary layer develops. For a stable stratified fluid, $N > 0$, core oscillations develop which get shorter and shorter wave-lengths for $\omega \to 0$.

Although Pekeris and Accad (1972) provide analytically correct solutions for the static deformation of the Earth with a fluid core, the fact that for an unstable stratification the horizontal displacement goes to infinity in the boundary layer and the fact that for a stable stratification an infinite amount of core oscillations are produced, indicates that there are still some problems.

Dahlen and Fels (1978) opposed the notion of trying to solve a Fourier-transformed problem in a fluid at the limit $\omega = 0$ from extrapolating solutions for small $|\omega| > 0$. Before we revisit the arguments of Dahlen and Fels (1978) we give our conclusion and recommendation. The static response cannot be obtained from sinusoidal load responses as a limit $\omega \to 0$; we endorse the use of a non-zero frequency when solving the load problem.

Stripping the problem down to the essentials, Dahlen and Fels (1978) showed that the same problem occurs in a stratified fluid in a box with hard side walls and a deformable lid exposed to a laterally homogeneous gravity field. The normal modes of this system pile up around zero frequency. The inverse Fourier transform employs the Cauchy principal value theorem for cases like this; however, as the open interval (0, Ω) contains infinitely many poles, albeit countably many, the principal value does not converge. In fact, if you expect a finite displacement to result, the Fourier integral of such a signal does not exist, since it is not square-integrable. Thus, if you expect a finite response at zero frequency (a doubtful concept per se), or, alternately, a finite response at infinite time, in Fourier the time is indistinguishable whether it is $+\infty$ or $-\infty$. Thus, you need to involve causality. Thus, Laplace transform and a Heaviside load history is the concept that is applicable, not Fourier transform.

Our task is perhaps not to estimate the time it takes for the system to reach the finite state within a given margin, but rather to determine the finite state. For that purpose, Dahlen and Fels (1978) suggest that an ad hoc viscosity be used for the core fluid. This will displace the poles of the inviscid system from the real frequency axis, giving them a slight imaginary part. The system can now be solved using the residual value theorem. The bottom line is that you would continue to exploit the 6×6 differential equations, changing the role of the shear modulus into a viscosity and Laplace-transform the equations such that the constitutive relation is expressed by

$$\sigma = 2\mu\dot{\epsilon} \quad \circ - \bullet \; \tilde{\sigma} = 2s\mu\tilde{\epsilon} \tag{1.50}$$

and the $-\omega^2$ factors are replaced by s^2, s being the Laplace transform parameter.

Farrell (1972) circumvented these difficulties by setting ω equal to the tidal period of harmonic M_2 (12.42 h). Since our main interest is to compute Green's functions for ocean tide loading, this approach is sufficient for us. Thus, it seems more instructive to represent the problem for non-vanishing ω, and again we follow Dahlen and Tromp (1998, Chap. 8).

The vanishing shear stress in a fluid region has the consequence that horizontal displacement becomes directly related to vertical displacement, potential perturbation, and vertical stress:

$$V = \frac{\rho_0 g(r) U + \rho_0 P - R}{\omega^2 \rho_0 r} \tag{1.51}$$

This equation has been derived from (1.26) by setting $\mu = 0$ and using the fact that the radial stress R is in this case equal to λX. One can use (1.51) to substitute V in (1.25) and (1.27) after which we are left with two second order differential equations.

When we use the six first order differential equations, then in the fluid we lose two rows from the differential equations, which reduce to

$$\frac{dU}{dr} = \left(\frac{\omega^2 g_0 n(n+1)}{r^2} - \frac{2}{r}\right)U + \left(\frac{1}{\kappa} - \frac{n(n+1)}{\omega^2 \rho_0 r^2}\right)R + \frac{n(n+1)}{\omega^2 r^2}P \tag{1.52}$$

$$\frac{dR}{dr} = \left(-\omega^2 \rho_0 - \frac{4\rho_0 g_0}{r} + \frac{n(n+1)\rho_0 g_0^2}{\omega^2 r^2}\right)U - \frac{n(n+1)g_0}{\omega^2 r^2}R + \left(\frac{n(n+1)\rho_0 g_0}{\omega^2 r^2} - \frac{\rho_0(n+1)}{r}\right)P + \rho_0 Q \tag{1.53}$$

$$\frac{dP}{dr} = -4\pi G\rho_0 U - \frac{n+1}{r}P + Q \tag{1.54}$$

$$\frac{dQ}{dr} = 4\pi G\rho_0\left(\frac{n(n+1)g_0}{\omega^2 r^2} - \frac{n+1}{r}\right)U - 4\pi G\frac{n(n+1)}{\omega^2 r^2}R + 4\pi G\rho_0\frac{n(n+1)}{\omega^2 r^2}P + \frac{n-1}{r}Q \tag{1.55}$$

At $n = 0$, the matrix elements on the right-hand side simplify considerably. The outcome being fairly obvious, we do not write it out. Since $V = 0$ for $n = 0$ the Earth just inflates or deflates a bit but remains spherically symmetric (Dahlen and Tromp 1998). As a result, the perturbed gravity is zero. If this is so, then we have the following relation for our semi-perturbed gravity parameter Q :

$$Q = -\frac{1}{r}P \tag{1.56}$$

which also provides us with the relation that states that no potential perturbation is possible except for the Bouguer effect due to vertical displacement:

$$\frac{dP}{dr} = -4\pi G\rho_0 U$$

If we do not suppose a solid inner core, the differential equations for the fluid interior can for $n = 0$ be shortened to a 2×2 system in U and R (Longman 1963):

$$\frac{dU}{dr} = -\frac{2}{r}U + \frac{1}{\kappa}R \tag{1.57}$$

$$\frac{dR}{dr} = -\left(\omega^2\rho_0 + \frac{4g_0\rho_0}{r}\right)U \tag{1.58}$$

The general solution in a homogeneous sphere (constant κ) is

$$U(r) = r \exp\left[\frac{-\mathrm{i}\,\omega}{k}\,r\right]\left\{C_2\,L\left(-2-\frac{2\,\mathrm{i}\,g}{k\,\omega},3,\frac{2\,\mathrm{i}\,\omega}{k}\,r\right)\right.$$
$$\left.+C_1\,U\left(2+\frac{2\,\mathrm{i}\,g}{k\,\omega},4,\frac{2\,\mathrm{i}\omega}{k}\,r\right)\right\} \tag{1.59}$$

$$R(r) = \kappa \exp\left[\frac{-\mathrm{i}\,\omega}{k}\,r\right]\left\{2\,C_2\,\kappa\,r\,L\left(-2-\frac{2\,\mathrm{i}\,g}{k\,\omega},3,\frac{2\,\mathrm{i}\,\omega}{k}\,r\right)\right.$$
$$\left.+C_1\times\left(2-\mathrm{i}\frac{\omega}{k}\,r\right)U\left(2+\frac{2\,\mathrm{i}\,g}{k\,\omega},4,\frac{2\,\mathrm{i}\,\omega}{k}\,r\right)\right\} \tag{1.60}$$

where $k = \sqrt{\kappa/\rho}$ the compressional wave speed in the fluid, $L_n^\alpha(z) = L(n,\alpha,z)$ is the generalised Laguerre polynomial and $U(a,b,z)$ the Confluent Hypergeometric function of the second kind. The latter is singular at $r = 0$ so we only need the Ls.

5 Resonance Effects

We will tacitly assume that the Earth–Moon system has reached a stationary situation. If you assume for the moment that there is no Moon and it suddenly appears, you will have some start up effects, among others starting seismic free oscillations which, owing to internal friction, slowly die out, resulting in the periodic tidal deformations that we experience today. So, when we say that we solve the tidal loading problem, we assume that the load acts on the surface with a temporal periodicity sufficiently different from the resonance frequencies that mode excitation can be neglected. In a purely elastic Earth, resonance occurs at sharply defined frequencies; however, in a visco-elastic mantle the resonance loses quality and the susceptible frequencies widen to finite intervals. As much as we are aware of this complication, we will avoid it by restricting the claims of our simplified approach to load frequencies well below one cycle per hour.

However, there is one resonance that needs attention, and it comes from the shape and fluidity of the core in a rotating planet. The core and the mantle rotate around slightly different axes, and the relative motion is known as Free Core Nutation or Nearly-Diurnal Free Wobble. Both astronomical tides of degree two and order one with a nearly-diurnal frequency and the associated ocean tides are able to excite the resonance although none of the forcing frequency exactly matches the 1 + 1/435 cycles per sidereal day frequency of the resonance. Wahr and Sasao (1981) have solved this problem by separating out the resonance in the load Love numbers and adding the effect to the normal Love numbers h_2, k_2 and l_2 (see Sect. 1.7 for their definition). This is possible since the resonance effects are primarily in the degree $n = 2$, order $m = 1$ spherical harmonics, and the excitation is due to the corresponding pro-grade ocean tide harmonic coefficients C_{21}^+ for amplitude and ϵ_{21}^+ for phase; see Lambeck and Balmino (1974) for the notations. It adds a complex-valued contribution that can conveniently be computed for different ocean models with the parametrisation in Scherneck (1991):

$$\left.\begin{array}{c}\Delta h_2(\omega)\\ \Delta l_2(\omega)\\ \Delta k_2(\omega)\end{array}\right\} = -\mathrm{i}\frac{4\pi G\rho_w a\Omega}{5(\omega-\Omega_R)}\frac{C_{21}^{+}}{\Phi_n}\exp(\mathrm{i}\epsilon_{21}^{+})\left\{\begin{array}{c}S_h'\\ S_l'\\ S_k'\end{array}\right. \tag{1.61}$$

where Ω_R is the angular rate of the resonance, Φ_n the potential coefficient of the luni-solar tide that generated the ocean tide whose pro-grade order-1 surface height is represented by $(C_{21}^{+}, \epsilon_{21}^{+})$ and the S' coefficients signify the resonance strength in the respective load Love numbers (Wahr and Sasao specified $S_h' = -2.88 \times 10^{-4}$, $S_l' = 9.16 \times 10^{-6}$ and $S_k' = -1.45 \times 10^{-4}$). Further modification is needed unless an observed tide at the exact frequency ω has been used to compute $(C_{21}^{+}, \epsilon_{21}^{+})$. If we are forced to resort to frequency-domain interpolation, a factor is needed to take the effect of resonance in the body and load tide Love numbers into account at the instance of ocean tide generation, and possibly we have additional knowledge of the variation in ocean dynamics across the resonance band. These are the factors $R(\omega, \omega_0)$ and $D(\theta, \lambda, \omega, \omega_0)$ in Wahr and Sasao (1981, Eqs. 4.5 and 4.6).

6 Boundary Conditions

Now that the differential equations are in place, we will address the boundary conditions that they have to fulfil. Since our set of equations are only valid in material that shows smooth variations in density and elastic properties (their radial derivative must exist), we need to divide our Earth into spherical layers in order to cope with the jumps in density and elastic properties. As a result, we must prescribe boundary conditions at the Earth's centre, at the boundaries between the layers and at the Earth's surface. We will start with the boundary conditions at the centre of the Earth where the solutions are regular. This means that, for $n \neq 1$, the displacements and perturbed potential are zero. Mathematically this statement can be presented as

$$U(0) = 0, \quad V(0) = 0, \quad P(0) = 0 \tag{1.62}$$

In the case $n = 1$ we have a situation where displacements and potential perturbation require an additional constraint owing to the fact that a rigid translation can be added to the displacements. The only effect of this translation is a gravity term $\delta P = -g/u_c$. While this will be dealt with in detail in Sect. 1.8, we note for the conditions in the centre that the particular displacement field that causes no perturbation of gravity potential at both $r = 0$ and $r = a$ does imply a shift of the figure and thus of its centre. The relation with the normal-stress function S is as follows:

$$U(0) = -\frac{48\pi G\rho + 3\omega^2}{(8\pi G\rho)^2\rho - 80\pi G\rho^2\omega^2 - 3\rho\omega^4} S(0) \tag{1.63}$$

at $n = 1$ if $\omega \neq 0$ and

$$U(0) = \frac{3}{4G\pi\rho^2} S(0) \tag{1.64}$$

if $\omega = 0$. That $U(0) = V(0) = 0$ may be deduced from the fact that we have a symmetric loading for $n \neq 1$ which cannot affect the position of the origin. The reason $P(0)$=0 can be seen from Poisson's equation

$$\nabla^2 \left[P(r)Y_n^0(\cos\theta)\right] = -4\pi G\nabla \cdot \left(\rho\left[U(r)Y_n^0(\cos\theta)\hat{r} + V(r)\partial_\theta Y_n^0(\cos\theta)\hat{\theta}\right]\right) \tag{1.65}$$

Lifting the divergence from this equation and working out the components of the gradient, the θ-component of the equation tells us that

$$\frac{1}{r}P(r)\partial_\theta Y_n^0(\cos\theta) \rightarrow V(0)\partial_\theta Y_n^0(\cos\theta) \quad \text{for } r \rightarrow 0 \tag{1.66}$$

If $V(0)$ would some how settle at a non-zero value, the left-hand side would grow to infinity, which is a contradiction. And obviously, horizontal displacement cannot grow as $\mathcal{O}(1/r)$ when $r \rightarrow 0$. Thus, both $V(0)$ and $P(0)$ are zero.

Next, at the interface of two solid layers we have continuity in radial and horizontal displacements, in radial and horizontal stresses and in semi-perturbed gravity and potential. Mathematically this is represented as

$$\begin{aligned} R(r^+) = R(r^-), \quad S(r^+) = S(r^-), \quad Q(r^+) = Q(r^-) \\ U(r^+) = U(r^-), \quad V(r^+) = V(r^-), \quad P(r^+) = P(r^-) \end{aligned} \tag{1.67}$$

where r denotes the radius of the interface, r^+ just above it and r^- just below it. Of course the true perturbed gravity is also continous over the boundary, $y_6(r^+) = y_6(r^-)$. At the boundary of a solid and fluid layer the situation is a little different. If we indicate the radius of this mantle core boundary by c and assume the mantle lies above the core, we have $S(c^+) = 0$ while shear stresses in the fluid core are undefined because $\mu = 0$. Furthermore, continuity in the horizontal displacement V is no longer required, so this equation disappears. Another relation we have at the mantle-core boundary for $n = 0$ is

$$\begin{aligned} Q(c^+) &= \left.\frac{dP}{dr}\right|_{r=c^-} + 4\pi G\rho U(c^-) + \frac{1}{r}P(c^-) \\ &= \frac{1}{r}P(c^-) \end{aligned} \tag{1.68}$$

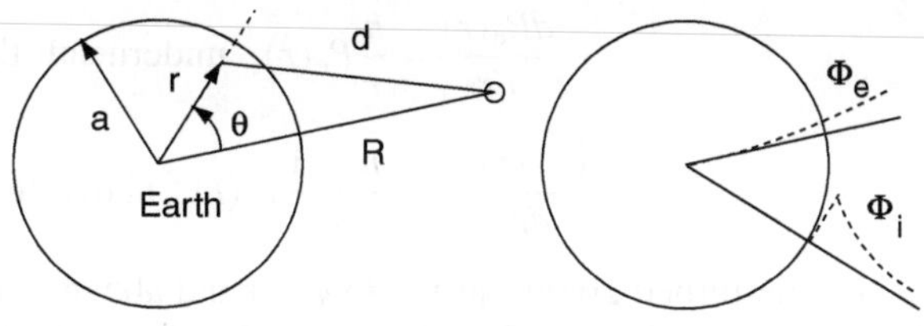

Fig. 1.1 In the *left panel* the definition of angle and distances is given. In the *right panel* we schematically show the behaviour of the external and internal perturbed potential

Finally, we need to describe the boundary conditions at the surface which depend on the type of loading is applied. Since we are interested in ocean tide loading, we assume that we have a parcel of tide-water lying on the Earth's surface. This parcel has a mass that generates a perturbation in the potential field of the solid Earth. Due to its weight, this parcel also presses on the ocean bottom. Therefore, in the ocean loading problem we must prescribe at the surface a perturbation in the potential and a normal stress. For the tidal deformation of the Earth caused by the Moon and Sun, this surface stress is zero.

Now it becomes important to distinguish between the Lagrangian and Eulerian descriptions that were explained in Sect. 1.2 The perturbed potential is a Eulerian function, evaluated at the undeformed boundary layers. Since deformation moves mass, the perturbed potential sees a 'Bouguer' effect. The stresses and displacements are evaluated at the deformed boundaries (Lagrangian) but to second order one can also just evaluate them at the undeformed boundaries.

To define our boundary conditions at the Earth's surface, it is convenient to assume that we have a unit point mass m_u at a distance R away from the Earth's centre; see Fig. 1.1. The external potential ϕ_e of this unit point mass m_u can be written as a sum of spherical harmonics:

$$\phi_e(r,\theta) = -\frac{G}{d} = -\frac{G}{a}\sum_{n=0}^{\infty}\left(\frac{r}{a}\right)^n Y_n^0(\cos\theta) \quad \text{for } r<R \tag{1.69}$$

We have added a minus sign because the potential should increase, become less negative, with increasing distance. Since the Earth is not completely rigid, it deforms due to the presence of this external potential, creating an additional internal potential ϕ_i. Outside the Earth this internal perturbed potential can also be written as a sum of spherical harmonics:

$$\phi_i(r,\theta) = -\frac{G}{a}\sum_{n=0}^{\infty} k_n(a)\left(\frac{a}{r}\right)^{n+1} Y_n^0(\cos\theta) \quad \text{for } r > a \tag{1.70}$$

where $k_n(a)$ are some unknown constants which will be determined later. Inside the Earth (1.70) is not valid. The total perturbed potential is $\phi_1 = \phi_e + \phi_i$. In Sect. 1.3 we have shown that for each degree n the radial part of ϕ_1 can be written as a function $P(r)$. Using the same scaling of Sect. 1.3 and setting $r = a$ we have $P_e = G/a$ and $P_i = k_n'(a)G/a$. At the surface the radial derivatives of these functions are

$$\frac{dP_e(r)}{dr} = \frac{n}{r} P_e(r) \quad \text{underneath the load} \tag{1.71}$$

$$\frac{dP_i(r)}{dr} = -\frac{n+1}{r} P_i(r) \quad \text{above the surface} \tag{1.72}$$

The perturbed gravity just below ($^-$) and above ($^+$) the surface should be equal. Remembering that the Earth's surface has been displaced due to the deformation our equation of continuity of perturbed gravity is

$$\nabla\phi_1^E(\mathbf{x},t)^- + \mathbf{s}_L(\mathbf{x},t)^- \cdot \nabla^2\phi_0(\mathbf{x},t)^- = \nabla\phi_1^E(\mathbf{x},t)^+ + \mathbf{s}_L(\mathbf{x},t)^+ \cdot \nabla^2\phi_0(\mathbf{x},t)^+ \tag{1.73}$$

Using Poisson's relation, one can replace the $\nabla^2\phi_0^-$ on the left side of the equation with $4\pi G\rho_0$ while the same term on the right is zero because we neglect the density of the atmosphere and put $\rho_0 = 0$ outside the Earth.

To first order, we will only need to consider the radial derivative and can replace the ∇ operator by d/dr. If we again decompose (1.73) into spherical harmonics, then for each degree n we have

$$\frac{dP^-}{dr} + 4\pi G\rho_0 U = \frac{dP^+}{dr} \tag{1.74}$$

If for the moment we assume that there is no external potential P_e and use (1.72) to substitute the term on the right:

$$\frac{dP_i}{dr} + \frac{n+1}{r} P_i + 4\pi G\rho_0 U = 0 \tag{1.75}$$

If we add the both the internal and external potential in (1.74), we get at the surface

$$\frac{dP}{dr} + \frac{n+1}{r} P + 4\pi G\rho_0 U \equiv Q = -\frac{2n+1}{a}\left(\frac{G}{a}\right) \tag{1.76}$$

Equation 1.76 provides the boundary condition for the semi-perturbed gravity Q at the surface. The beauty of (1.76) is that it does not contain the unknown internal potential ϕ_i explicitly.

Now we will derive the expression for a unit point load σ. According to Longman (1962), the Legendre expansion of the Dirac δ-function on a sphere with radius a is

$$\begin{aligned} \sigma &= \sum_{n=0}^{\infty} \frac{2n+1}{4\pi a^2} Y_n^0(\cos\theta) \\ &= \left(\frac{G}{a}\right) \sum_{n=0}^{\infty} \frac{2n+1}{4\pi G a} Y_n^0(\cos\theta) \end{aligned} \tag{1.77}$$

Our unit mass exerts a point load of magnitude $-g$ at the surface which means that the boundary condition for the normal stress R for degree n is

$$R = -\frac{2n+1}{a}\frac{g}{4\pi G}\left(\frac{G}{a}\right) \tag{1.78}$$

Together with the boundary condition that the horizontal stress is zero, $S = 0$, (1.76) and (1.78) provide three boundary conditions at the surface.

7 Simple Earth Models and Love Numbers

At this point it is instructive to discuss the deformation of an elastic solid Earth with constant density and constant elastic properties. For this particular situation there exist three analytical solutions for each parameter which, combined, describe the radial and horizontal deformation and perturbed potential throughout the Earth. These analytical solutions are provided by Dahlen and Tromp (1998) and are reproduced in Appendix 2. For example, the radial displacement, for degree n, is

$$\begin{aligned} U(r) &= U^{+}(r) + U^{-}(r) + U^{\oplus}(r) \\ &= y_{11}\, j_n(\gamma^{+}r) + y_{12}\, j_n(\gamma^{-}r) + y_{13}\, r^{n} \end{aligned} \tag{1.79}$$

Note that these solutions automatically produce zero displacements and disturbed gravity at the Earth's centre.

In the second line of (1.79), we have factored out the terms containing the spherical Bessel functions j_n and r^n and formed new coefficients y_{11}, y_{12} and y_{13} (Okubo 1988). This is not necessary but has been done to emphasise the fact that each term depends on a different function. The solutions for the horizontal displacement V and perturbed potential P can be written in the same format. All these coefficients can be grouped in a matrix:

$$\mathbf{D} = \begin{Bmatrix} y_{11} & y_{12} & y_{13} \\ y_{31} & y_{32} & y_{33} \\ y_{51} & y_{52} & y_{53} \end{Bmatrix} \tag{1.80}$$

The displacement vector $\mathbf{s}$ can now be computed as $\mathbf{DJ}\boldsymbol{\theta}$ where $\mathbf{J}$ is a 3×3 matrix with our $j_n(\gamma^{+}r), j_n(\gamma^{-}r)$ and r^n terms on the diagonal. Vector $\boldsymbol{\theta}$ contains scale factors because each separate solution can be multiplied with an arbitrary constant.

From the solutions for U, V and P, we can derive the analytical solutions for the radial stress R, tangential stress S and semi-perturbed gravity Q. Again we can factor out the $j_n(\gamma^{+}r)$, $j_n(\gamma^{-}r)$ and r^n terms and form a new matrix $\mathbf{E}$ in such a way that the vector $(R, S, Q)^T$ is $\mathbf{EJ}\boldsymbol{\theta}$. Matrix $\mathbf{E}$ is defined in a similar ways as matrix $\mathbf{D}$:

$$\mathbf{E} = \begin{Bmatrix} y_{21} & y_{22} & y_{23} \\ y_{41} & y_{42} & y_{43} \\ y_{61} & y_{62} & y_{63} \end{Bmatrix} \tag{1.81}$$

Each row of matrix $\mathbf{E}$ is associated with the radial stress, tangential stress and semi-perturbed gravity. For example, the radial stress is written as

$$R(r) = y_{21}\, j_n(\gamma^+ r) + y_{22}\, j_n(\gamma^- r) + y_{23}\, r^n \tag{1.82}$$

Our next task is to estimate the scale factors $\boldsymbol{\theta}$ in order to fulfil the boundary conditions at the surface described in Sect. 1.6 Following Okubo (1988) we will compute the scaling factors for the three solutions for the body tide and load tide simultaneously. These boundary conditions, for degree n, at the surface are stored in the columns of the following matrix $\mathbf{x}$:

$$\mathbf{x} = \frac{2n+1}{a} \begin{pmatrix} 0 & \frac{g}{4\pi G} \\ 0 & 0 \\ 1 & 1 \end{pmatrix} \tag{1.83}$$

The first column of $\mathbf{x}$ shows that for the body tide, only the potential is non-zero at the surface. In the second column, one can see that for the load tide there is an additional radial stress. Note that the factor $-G/a$ has disappeared. Instead of a unit-mass, we are computing the deformation due to a unit-potential.

The scale factors $\boldsymbol{\theta}$ are determined with $(\mathbf{EJ}(\mathbf{a}))^{-1}\mathbf{x}$. Since the matrix $\mathbf{EJ}$ can be ill-conditioned, it makes sense to scale each row of $\mathbf{EJ}$ in such a way that the largest entry is 1. This will not change the value of $\boldsymbol{\theta}$ if vector $\mathbf{x}$ is scaled by the same factors, but will improve its numerical accuracy.

Now that these scale factors are known, we can compute the deformations U and V and the perturbed potential P at any radius r using $\mathbf{DJ}(r)\boldsymbol{\theta}$. Remember that we have computed the radial deformations $U(r)$ for a unit potential load on the Earth's surface. It was Love who represented these deformations as the product of a function $h_n(r)$ divided by g. For any other external potential ϕ_e, that again can be developed into spherical harmonics with a radial function at the surface $P_e(a)$, the radial deformations are, for degree n

$$U(r) = -h_n(r)\frac{P_e(a)}{g} \tag{1.84}$$

For the tangential displacements a similar function $l_n(r)$ is defined:

$$V(r) = -l_n(r)\frac{P_e(a)}{g} \tag{1.85}$$

The same can be done for the perturbed potential although it is customary to introduce a function k_n that is only associated to the internal perturbed potential ϕ_i :

$$P(r) = (1 + k_n(r))P_e(a) \tag{1.86}$$

Table 1.1 General constants

Constant	Unit	Value
G	$m^3kg^{-1}s^{-2}$	6.673×10^{-11}
a	m	6.371×10^{6}
ω	rad/s	1.40526×10^{-4}

Table 1.2 Properties of a homogeneous Earth (model β) and an Earth with a homogeneous mantle and a fluid core with a radius of $0.55a$ (model α)

		Model β	Model α	
Constant	Unit	All	Core	Mantle
Mean density ρ	(kg/m^3)	5517	11020	4460
Shear modulus μ	(GPa)	146	0	174
Lamé parameter λ	(GPa)	347	950	231

The minus sign in (1.84) and (1.85) is the result of our definition of the potential with the opposite sign as Love (1911) and Alterman et al. (1959). The definitions of the functions $h_n(r)$ and $l_n(r)$ already have a long tradition and it would cause too much confusion if we were to define new Love numbers with the opposite sign. Wu and Peltier (1982) follow the same sign convention of the potential as we use here but compute the deformation of the Earth due to a negative unit potential. This causes the minus sign to disappear in the definition of $h_n(r)$ and $l_n(r)$ but then it reappears in (1.86).

Normally, the values of $h_n(r), l_n(r)$ and $k_n(r)$ are only given for the Earth's surface which turns them into numbers instead of functions. The $l_n(a)$ Love number is also called the Shida number. The Love numbers are needed to compute our Green's functions to compute the ocean tide loading as we announced in Sect. 1.1 and which we will explain in more depth in Sect. 1.11

Love (1911) studied the deformation of the Earth due to the tidal force of the Moon and thus had no pressure forces on the surface. To distinguish between load Love numbers and the body tide Love numbers, the former are normally written as h'_n, l'_n and k'_n, a notation that was introduced by Munk and MacDonald (1960). As an example, we give the values of normal Love numbers and load Love numbers for a homogeneous Earth, called model β. The values for the Gravitational constant G, the mean radius of the Earth a and the angular velocity of the forcing ω (corresponding to the main tidal period of 12.42 h) are given in Table 1.1. The properties of the homogeneous Earth are listed in Table 1.2 and were taken from Alterman et al. (1959). The results are listed in Table 1.3 where we have multiplied the l, l', k and k' numbers by degree n, just to get a convenient size. The functions $h'_n(r)$ and $k'_n(r)$ are plotted in Figs. 1.2 and 1.3 for various degrees n. Note that for high values of degree n, the functions $h'_n(r)$ and $k'_n(r)$ are very small throughout the Earth and only increase near the surface. As a result, the

[1] Almost ignore; you need to assume that $g(r)/r = \text{const.}$ in order to retain the structure of the analytical solution (Vermeersen et al. 1996).

Table 1.3 The normal and load Love numbers for Earth model β for several degrees

Degree	h_n	nl_n	nk_n	h'_n	nl'_n	nk'_n
1	−18.22448	−18.22448	−18.22448	−0.18599	0.14700	0.00000
2	0.52221	0.28413	0.60384	−0.58502	−0.02167	−0.44057
10	0.10622	0.01229	0.14818	−0.88125	0.14981	−0.91403
100	0.01167	0.00014	0.01736	−1.00537	0.22378	−1.14968
1000	0.00118	0.00000	0.00177	−1.02022	0.23250	−1.17926

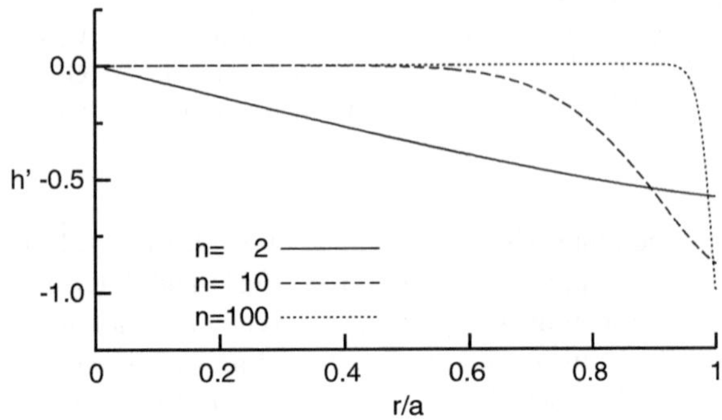

Fig. 1.2 The load love numbers h'_n for $n = 2$, 10 and 100 for a homogeneous Earth as a function of the Earth's radius

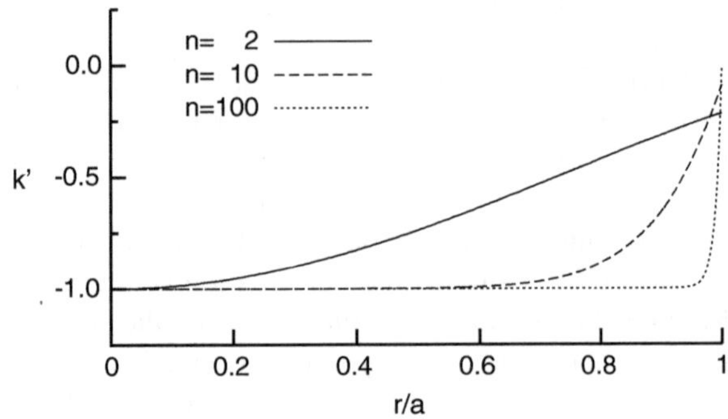

Fig. 1.3 The same as Fig. 1.2 but for k'_n

properties of the Earth just underneath the station increase in importance for increasing degree n.

Another interesting case is the deformation of an Earth with homogeneous mantle and a homogeneous liquid core. Following Alterman et al. (1959), we will call this model α. Its density and elastic properties are given in Table 1.2 . In each layer, analytical solutions for the deformation can be derived; see Appendix 2. However, in contrast to the case of the completely homogeneous Earth, in the mantle we now also need spherical Bessel functions of the second kind and solutions that contain $1/r^n$ terms. Therefore, we must extend our $\mathbf{D}, \mathbf{E}$ and $\mathbf{J}$ matrices discussed before to include these terms; see Martinec (1989).

As we discussed in Sect. 1.4 and Appendix 2, in a fluid we can derive the tangential displacement and stress from the other parameters: U, R, P and Q. In

Table 1.4 Love numbers for our model α Earth for degree $n = 2$ for different periods of forcing ($T = 2\pi/\omega$)

T	h_2	$2l_2$	$2k_2$	h'_2	$2l'_2$	$2k'_2$
6 h	0.69216	0.27579	0.72544	−0.87766	−0.08390	−0.65887
12 h	0.68037	0.27254	0.71343	−0.86444	−0.08128	−0.64730
24 h	0.67741	0.27174	0.71048	−0.86035	−0.08067	−0.64433
∞	0.67633	0.27148	0.70949	−0.85782	−0.08051	−0.64316

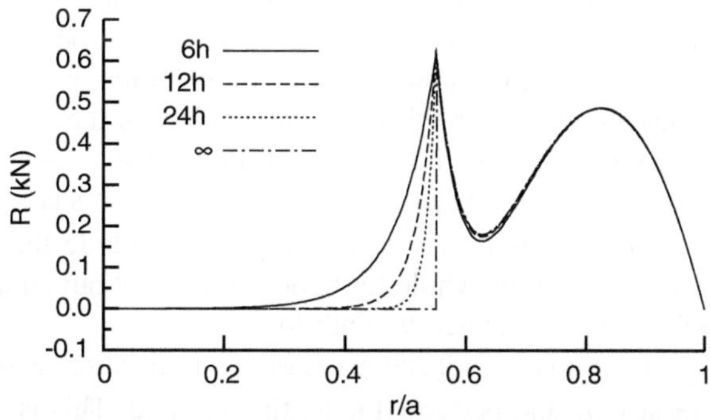

Fig. 1.4 The radial stress R inside the Earth for different values of the forcing period

addition, we only have one solution of the spherical Bessel functions of the first kind. As a result, our **E** becomes a 2×2 matrix.

We will ignore the fact that the gravity can no longer be described as $4\pi G\rho_0 r/3$ throughout the Earth, which was one of the assumptions in deriving these analytical solutions.[1]

Of course we can generalise this procedure and divide the Earth into multiple layers with constant density and constant elastic properties. Describing the problem of the deformation of the Earth as a set of propagating matrices is called the Thomson–Haskell method (Gilbert and Backus 1966) and is popular among post-glacial rebound modellers although they use something more complicated than just constant elastic properties.

Returning to our model α, the normal Love numbers for several values of the forcing period are given in Table 1.4 and the radial stress R is plotted in Fig. 1.4. In this last figure one can see that, for decreasing period, a boundary layer develops underneath the core–mantle boundary. Since in the fluid core the radial stress is related to dilatation through $R = \lambda X$, one can see that in the limit $\omega \to 0, X = 0$ throughout the core and that the Adams–Williams condition is not needed as an extra condition (Pekeris and Accad 1972).

It is interesting to see what these Love numbers would be when the limit of $\omega \to 0$ is taken. Now we should remember that the stratification of our homogeneous fluid is unstable and that a boundary layer develops (Pekeris and Accad 1972). If the jump through the boundary layer is taken into account, then we get the Love numbers listed in the last line of Table 1.4.

8 Degree-1 Response and Translational Invariance

At this point we take the opportunity to look at much discussed problem of separating displacement into a whole-body rigid translation and deformation notably at spherical harmonic degree 1; see, for instance, Blewitt (2003).

For degree 1 the situation is a little different because the load is not symmetric and this causes the Earth to move in space, in addition to deforming it. First, we will discuss the translation of the Earth in space which is equivalent to a constant $\mathbf{s}_L$. As a result, $\nabla \mathbf{s}_L = 0$. Looking at the Cauchy stress tensor, (1.12), we see that a translation of the Earth in space does not introduce any stress.

As a side note, assume for the moment that we have a homogeneous Earth with constant density. In this case, the gradient of the reference density ρ_0 is zero. From the continuity equation, (1.6), it follows that a translation of the solid Earth cannot perturb the density: $\rho_1 = 0$. Applying Poisson's equation we see that the perturbed potential ϕ_1^E is also zero and we can conclude that for a homogeneous Earth, a translation of the whole Earth does not affect our equations although it will have an effect on our boundary conditions.

For a non-homogeneous Earth, a translation will create a non-zero perturbed density ρ_1 and perturbed potential ϕ_1^E field. This is the consequence of defining a reference density ρ_0 and potential ϕ_0 field at the origin of the undeformed Earth, fixed in space, and describing the deformations as perturbations with respect to this reference field. A translation z along the $\theta = 0$ direction causes a perturbation in the potential equal to

$$\phi_1 = (r\omega^2 - g_0)z\cos\theta \tag{1.87}$$

Here we have added the potential produced by the acceleration of the translation. For tidal periods, $r\omega^2$ is much smaller than g_0 and has therefore probably been neglected by Farrell (1972).

So far we have only discussed a translation of the whole Earth. However, there also exists a degree one deformation that will generate a perturbed potential in the same way as we described in the previous sections. The only difference is that, due to the asymmetric loading, we no longer have zero displacements and a zero perturbed gravity value at the centre of the Earth and require three new boundary conditions.

To find these three new boundary conditions at the centre, we must realise that in a small ball with radius δ around this centre the Earth can be considered to be homogeneous. Repeating the results presented in Sect. 1.7 and invoking the associated mathematics from Appendix 2, we note that only the analytical solution that depends on r^n can produce displacements that are non-zero at the centre. This solution has been reproduced here (for $n = 1$) :

Table 1.5 The same as Table 1.3 but for Earth model α.

Degree	h_n	l_n	k_n	h'_n	l'_n	k'_n
1	−12.80564	−14.38607	−13.12949	−0.52853	−0.27453	−0.32385

$$
\begin{aligned}
U^{\oplus} &= c_r \frac{n}{r} r = c_r \\
V^{\oplus} &= c_r \frac{1}{r} r = c_r \\
P^{\oplus} &= c_r(\omega^2 r - \frac{4\pi}{3} G\rho r) = c_r(\omega^2 r - g_0) \\
R^{\oplus} &= c_r \frac{2n(n-1)\mu}{r^2} r = 0 \\
S^{\oplus} &= c_r \frac{2(n-1)\mu}{r^2} r = 0 \\
Q^{\oplus} &= c_r \frac{(2n+1)\omega^2 - 8\pi G\rho n(n-1)/3}{r} r = 3c_r\omega^2
\end{aligned}
\tag{1.88}
$$

where c_r is to be determined from the boundary condition at the surface. From (1.88) we can see that three new possible boundary conditions are: $U(0) = V(0), R(0) = 0$ and $S(0) = 0$. The other analytical solutions for a homogeneous sphere containing the spherical Bessel functions j_1 produce zero displacements and stresses at the Earth's centre. The solutions containing terms with $1/r$ or the spherical Bessel functions y_1 are infinite at the Earth's centre and therefore need to be set to zero.

Now that we know our new boundary conditions at the Earth's centre, let us discuss the Love numbers for our α and β Earth models discussed in Sect. 1.7 For the homogeneous Earth one can see in Table 1.3 that all normal Love numbers are the same. Because the Earth is homogeneous, no differential forces occur and the Earth does not deform but only oscillates back and forth in space. The amplitude of these oscillations is larger the longer the period of forcing. These forces produce the motion of the Earth around the solar system and are not of interest us here where we want to study tidal phenomena and our equations are only valid for small perturbations from the undeformed reference state.

The situation for the load Love numbers is different because, in addition to the gravitational attraction of the unit potential, it exerts a load on the surface in the opposite direction. That this produces a zero internal perturbed potential at the surface is just a peculiarity of a homogeneous Earth. For our Earth with a homogeneous mantle and fluid core the Love numbers for degree one are given in Table 1.5 . One can see that now the normal Love numbers are not all the same because the Earth is no longer homogeneous. Also the k'_1 load Love number is now different from zero.

It is customary to keep the origin of the reference frame fixed to the centre of mass of the deformed solid Earth. For $n \neq 1$ this always coincided with the position of the origin of the undeformed solid Earth which was the origin of our reference frame in the previous sections. However, now we must shift the frame. The centre of mass of the solid Earth has the property that it has a zero value for the perturbed potential at the surface. To achieve this we need to adjust our load Love numbers as follows (Farrell 1972):

$$\begin{aligned} [h_1']_{CE} &= h_1' - k_1' \\ [l_1']_{CE} &= l_1' - k_1' \\ [k_1']_{CE} &= k_1' - k_1' = 0 \end{aligned} \tag{1.89}$$

For other choices for the origin of the reference frame, see Blewitt (2003). We only want to point out that all associated translations of the reference frame and modifications of the load Love numbers can be derived from our original load Love numbers h_1', l_1' and k_1'.

9 Numerical Methods

In Sect. 1.7 we computed the deformation of the Earth using the Thomson–Haskell method that uses the analytical solutions of the deformation inside each layer with constant density and constant elastic properties. We have already briefly mentioned that we ignored the fact that the gravity can no longer be described by $4/3\pi G\rho_0 r$ throughout the Earth. Although there are ways to minimise this last problem, one would still face problems that the deeper layers in most recent Earth models, such as PREM (Dziewonski and Anderson 1981), have density and elastic properties that vary inside each layer. Instead of also trying to minimise this problem, for example by sub-dividing these layers into layers with constant properties, we will now present methods that solve the differential equations numerically. These numerical methods are slightly more elaborate to implement than the Thomson–Haskell method but provide more flexibility. The most popular method of solving the six differential (1.32)–(1.37) is the Runge–Kutta method (Alterman et al. 1959). As with the Tomson-Haskell method, these equations are solved in each layer separately. One starts by integrating the equations from the centre of the Earth upwards to the boundary of the first layer. The computed values for the six parameters U, V, P, R, S and Q at the upper boundary are the starting values for the integration in the next layer. This process is repeated until one reaches the surface.

Starting at the centre of the Earth sounds simple. However, inspection of the differential equations shows that they are singular at $r = 0$. Secondly, we should not forget that the high spherical harmonic degrees for which we seek the load Love numbers imply extremely small deformation in the deep interior. Factoring out a scaling function and mapping the radial coordinate on a logarithmic scale helps to overcome the numerical problems. This is the Lyapunov transformation mentioned in Sect. 1.3. However, the convenience the trick gives with one hand it takes away with the other: we need starting solutions for a tiny homogeneous sphere in order to avoid the singularity problem. This has already been discussed in Sect. 1.7 but we would like to add that because of the small radius, the spherical Bessel functions of the first kind, j_n, may be approximated for radii\epsilon in the range 1–10 km,

$$j_n(\epsilon) = \frac{\sqrt{\pi}}{2\Gamma(n+3/2)} \left(\frac{\epsilon}{2}\right)^n \tag{1.90}$$

where Γ represents the Gamma function if $\epsilon \ll \sqrt{n+3/2}$, which is always fulfilled except at a few small values of n. Below $n = 10$ the spherical Bessel functions are unproblematic.

For the case of an Earth model with a fluid core at its centre we can also compute these analytical solution using a power-series ansatz. First, we replace g_0 in (1.52)–(1.55) with $4\pi G\rho_0 r/3$. With power series

$$\begin{pmatrix} U(r) \\ R(r) \\ P(r) \\ Q(r) \end{pmatrix} = \sum_{j=1}^{\infty} \begin{pmatrix} u_j/r \\ s_j \\ p_j \\ q_j/r \end{pmatrix} r^{n+j-1} \tag{1.91}$$

the differential equations produce a set of coupled recursion relations:

$$-\kappa\rho\left[4\pi G\rho n(n+1) - 3(j+n)\omega^2\right]u_j + 3\left[n(n+1)\kappa(s_j - \rho p_j)\right] = 3\rho\omega^2 s_{j-2}$$

$$\begin{aligned}\rho\left[(4\pi G\rho)^2 n(n+1) - 48\pi G\rho\omega^2 - 9\omega^4\right]u_j \\ - 3\left[4\pi G\rho n(n+1) + 3(j+n-1)\omega^2\right]s_j \\ - 3\rho\left[(n+1)(4\pi G\rho n - \omega^2)p_j + 3\omega^2 q_j\right] = 0\end{aligned}$$

$$4\pi G\rho u_j + (2n+j)p_j - q_j = 0$$

$$\begin{aligned}\frac{4\pi}{3} G\rho(n+1)(4\pi G\rho n - 3\omega^2)u_j \\ - 4\pi G n(n+1)(s_j - \rho p_j) + (1-j)\omega^2 q_j = 0\end{aligned} \tag{1.92}$$

with starting equations

$$\rho\,[(4\pi\,G\,\rho)^2\,n(n+1)\;-\;48\pi\,G\,\rho\,\omega^2-9\,\omega^4]$$
$$-\;3n\,[4\pi\,G\,\rho\,(n+1)+3\,\omega^2]\,s_1$$
$$-\;3\rho\,[(n+1)(4\pi\,G\,\rho\,n\;-\;\omega^2)\,p_1+3\,\omega^2\,q_1]=0$$

$$\rho\,(4\pi\,G\,\rho\,n\;-\;3\,\omega^2)\,u_1-3\,n\,(s_1-\rho\,p_1)=0 \tag{1.93}$$

The resolved equations are shown in Appendix 3. The recursion starts with $u_1=0$, an arbitrary s_1 and a compatible $p_1=s_1/\rho$. From this, q_1 can be computed, and the recursion can step ahead to $j=3,5,\ldots$.

For degree $n\neq 1$, we know that at the Earth's centre $U=V=P=0$. However, we do not know the starting values of the radial and tangential stresses R and S, nor the starting value of the semi-perturbed gravity Q. The solution of this problem is to solve the differential equations three times and each time set another one of these three unknowns to 1 and the other two to zero. These three solutions have to be scaled afterwards to fit the boundary conditions. If we remember that in Sect. 1.3 we had written our six first order linear differential equations as, 1.39:

$$\frac{d\mathbf{y}}{dr}=\mathbf{A}\mathbf{y} \tag{1.94}$$

with $\mathbf{y}=[U,\;V,\;P,\;R,\;S,\;Q]^T=[y_1,\ldots,y_6]$, then for the case of ocean tide loading we can write the three solutions $\mathbf{y}^{(1)},\mathbf{y}^{(2)}$ and $\mathbf{y}^{(3)}$ at the surface as:

$$y_2^{(1)}c_1+y_2^{(2)}c_2+y_2^{(3)}c_3=\frac{2n+1}{a}\frac{g}{4\pi G} \tag{1.95}$$

$$y_4^{(1)}c_1+y_4^{(2)}c_2+y_4^{(3)}c_3=0 \tag{1.96}$$

$$y_6^{(1)}c_1+y_6^{(2)}c_2+y_6^{(3)}c_3=\frac{2n+1}{a} \tag{1.97}$$

Solving (1.95)–(1.97) provides us the scale factors c_1, c_2 and c_3.

We would like to emphasise that we are solving a set of non-homogeneous differential equations. In principle, we can add to these the solutions for the homogeneous differential equations that correspond to the free-oscillation of the Earth. In fact, the procedure described above is exactly how these free-oscillations of the Earth are computed. One computes the solutions of the homogeneous differential equations for various values of the forcing period $T=2\pi/\omega$ until (1.95)–(1.97) become linearly dependent, which indicates that a resonance period has been found.

As before, complications arise due to the existence of a fluid core. If we for the moment we assume that there is no solid inner core, then only need to integrate U, P, R and Q from the centre of the Earth to the core–mantle boundary as was explained in Sect. 1.4. This involves only two unknowns: R and Q. At the bottom of the mantle the tangential stress S is zero, and only the horizontal displacement V is unknown and takes the place of S in the procedure described above. If we have a solid inner core, then the situation is a little more complicated. As before, we need to compute three solutions for unit values of R, S and Q at the Earth's centre. Since we know that the tangential stress is zero at the boundary of the inner and outer core, one of the scale factors can already be written as a function of the other two. Again, we are left with three unknowns and the rest of the procedure remains the same as before.

For degree $n = 1$ we have different conditions at the Earth's centre which causes other problems because now we no longer know the values of $U(0), V(0)$ and $P(0)$. However, now $R(0) = S(0) = 0$ and we know that $U(0) = V(0)$ which is sufficient information to solve the equations.

Another numerical method that is very suitable to solve the differential equations is the spectral method where the solution is approximated by a sum of basis functions (Boyd 2000). In our case, we will use Chebychev polynomials as basis functions and our method is thus better described as the Chebychev collocation method. Its use to compute the deformation of the Earth was pioneered by Guo et al. (2001, 2004).

To explain its principles, assume that the radial displacement function $U(r)$ can be approximated by:

$$U(r) \approx U_N(r) = \sum_{i=0}^{N} a_i \psi_i(r) \tag{1.98}$$

where ψ_i is a Chebychev polynomial of degree i and a_i is a constant. Similar approximations can be made for $V(r)$ and $P(r)$. Since polynomials are easily differentiated, we can insert these approximations into our second order differential equations (1.25)–(1.27). If we evaluate these equations at $N+1$ positions inside the Earth, called nodes, we have created $3(N+1)$ equations with $3(N+1)$ unknowns: the coefficients a_i of U_N, V_N and P_N. This system of linear equations can be solved and, once we know the values of the coefficients a_i, we have an approximation for the solution of the differential equations.

The distribution of these nodes should be done in such a way so as to optimize the accuracy of the approximation. It turns out that there are two good distributions, called Gauss-Radau and Gauss-Lobbato (Boyd 2000). The only significant difference between the two is that the Gauss-Radau distribution includes the end points while Gauss-Lobbato does not. Guo et al. (2001) advocate the use of the latter to avoid the singularities at the Earth's centre. However, it is convenient to replace the differential equations at these start and end nodes with the boundary conditions. At the same time singularity problems are avoided. Thus, we will use the Gauss-Radau distribution of nodes:

$$x_j = \cos\frac{\pi j}{N} \quad , \quad j = 0 \ldots N \tag{1.99}$$

where x lies between -1 and 1. We must thus scale the radius r in each layer to fit this interval. This scaling should not be forgotten when one takes the derivatives of ψ_i.

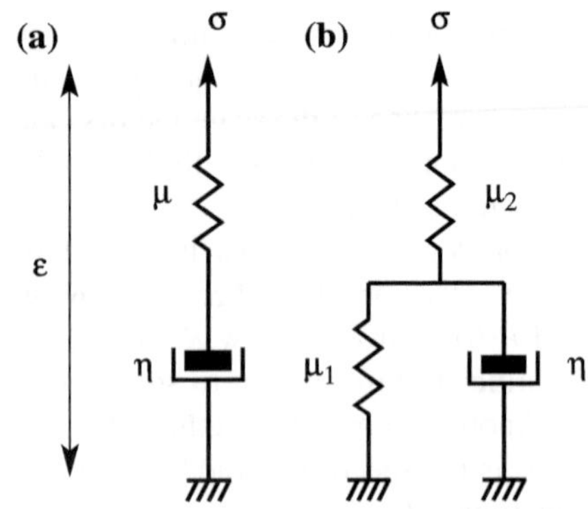

Fig. 1.5 Elementary rheological models, **a** the Maxwell body, **b** the Zener body

A disadvantage of using the three second-order differential equations is that these include the radial derivative of the elastic properties. However, these properties are normally given as low-degree polynomials, which ensures us that these derivatives can be computed easily. Another detail that might require some attention occurs again in the fluid core. Since in a fluid only four boundary conditions are needed, we must evaluate (1.26) on all nodes and not substitute the start and end points with boundary conditions. At the Earth's centre we know that for $n \neq 1, U = V = 0$ so one can make an exception and replace the differential equation on the start node at $r = 0$ with the boundary condition. For degree $n = 1$ we know that $U = V$ which removes any remaining singularities in a fluid at the Earth's centre.

As a final remark, we note that for high values of degree n, the core hardly deforms and one could in principle set the boundary conditions $U = V = P = 0$ at the core–mantle boundary.

10 Rheology: Viscosity and Anelasticity

When the temperature of rock materials is high enough yet still safely below the melting point, elastic stress will relax with time. The material will creep under stress. Whether or not this deformation recovers determines whether the material is termed anelastic or inelastic, respectively (Nowick and Berry 1972).

Basic properties can be illustrated with rheological circuit diagrams. Consider for instance the Maxwell body (Fig. 1.5a) as an example of irreversible deformation. When stress is applied, the viscous element starts creeping, but when the stress is removed it remains in the deformed position. Application of a single Maxwell model to explain delayed recovery from deformation is commonly proposed in the problem of Glacial Iostatic Adjustment (GIA). This is a phenomenon on a time scale of 1,000–100,000 years. At the time scale of tides the viscosity that is inferred from GIA studies produces very small effects of inelasticity, and we may wonder whether other approaches viable in long-period seismology might not be better suited.

We serve ourselves from developments in seismology that were set out to explain Q, the quality factor that describes the attenuation of seismic waves along their paths, or, if you wish, the decay of free oscillations over time (Knopoff 1964). The recipes that follow below will end in what is known in seismology as Generalised Maxwell rheology and the Standard Linear Solid.

Parallel connection of the dash-pot with a spring element is a simple model for recoverable strain, since now elastically stored energy is left to do work on the viscous element. By the same token the body shows stress relaxation when a strain is prescribed as a step. In the model referred to as Zener body or, alternatively, Standard Linear Solid (Fig. 1.5b), shortly after deformation, much force is put on the viscous element. As time goes by, the viscous element relaxes and stress is shared by two elastic elements in series.

We leave the solid state physics of stress relaxation or strain retardation aside and concentrate instead on how rheology enters into our differential equations.

First of all, the temporal aspect adds a phase-shifted relation between stress and strain. Fourier-transforming the shear deformation

$$\frac{\sigma}{2\mu} = \epsilon$$

(elastic), respectively

$$\frac{\dot{\sigma}}{2\mu} + \frac{\sigma}{2\eta} = \dot{\epsilon}$$

(Maxwell) gives

$$\left(\frac{\mathrm{i}\omega}{2\mu} + \frac{1}{2\eta}\right)\sigma = \mathrm{i}\omega\epsilon \tag{1.100}$$

or

$$\mu(\mathrm{i}\omega) = \frac{\mathrm{i}\omega\mu\eta}{\mathrm{i}\omega\eta + \mu} \tag{1.101}$$

which, at $\omega \to \infty$, displays unaltered elasticity μ, but at $\omega \to 0$, the resistance to shear is zero. The quantity η/μ is called relaxation time or Maxwell time. Thus, the only effect is that the equations need to be doubled with an imaginary part. The stress and gravity boundary conditions remain real-valued.

Slightly more complicated, the following exercise considers the Zener body. Here

$$\mu(\tau, \mathrm{i}\omega) = \frac{\mu_2(\mu_1 + \mathrm{i}\omega\eta)}{\mu_1 + \mu_2 + \mathrm{i}\omega\eta} = \frac{\mu_1\mu_2(1 + \mathrm{i}\omega\tau)}{\mu_2 + \mu_1(1 + \mathrm{i}\omega\tau)}$$

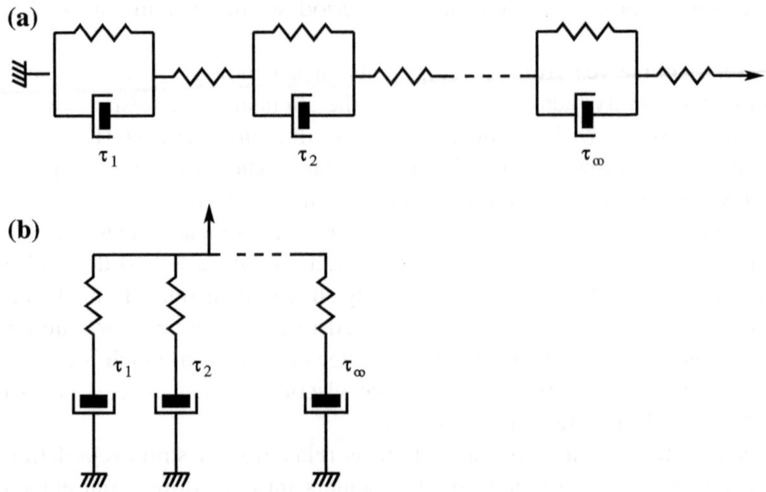

Fig. 1.6 Continuous rheological models, **a** an infinite chain of Zener elements, **b** the Generalised Maxwell Body. If the range of relaxation times is finite, the bodies may serve as models for the seismic absorption band. Extrapolation to non-seismic frequencies might be daring

with relaxation time $\tau = \eta/\mu_1$. Its zero- and infinite-frequency responses are the relaxed compliance $1/\mu_r = 1/\mu_1 + 1/\mu_2$ and the unrelaxed one $\mu_u = 1/\mu_2$, respectively.

In order to widen the discrete circuitry to a continuum, we may imagine a spectrum of relaxation times distributed over an infinite chain of Zener elements. The resulting model body is designated as the Standard Linear Solid (Fig. 1.6). Strain is now described as the integral over this infinite number of elements given the stress

$$\epsilon(i\omega) = \left(\int_{\tau_1}^{\tau_2} A(\tau) \frac{d\tau}{2\mu(\tau, i\omega)} \right) \sigma(i\omega) \tag{1.102}$$

For normalisation of $A(\tau)$, the compliance—the integral in (1.102)—is to evaluate the relaxed and unrelaxed values at zero and infinite frequency, respectively, which amounts to demanding

$$\int_{\tau_1}^{\tau_2} A(\tau)\, d\tau = 1$$

Zschau (1983) gave an expression for the shear compliance of the absorption band model:

$$\frac{1}{\mu(\mathrm{i}\omega)} = \frac{1}{\mu_\infty}\left\{1+\Delta^* - \frac{\mathrm{i}\omega\alpha}{1+\alpha}\,\frac{\Delta^*}{\tau_2^\alpha - \tau_1^\alpha}\left[\tau^{1+\alpha}\mathrm{F}(1+\alpha,1;2+\alpha;-\mathrm{i}\omega\tau)\right]_{\tau_1}^{\tau_2}\right\} \tag{1.103}$$

where F is the hypergeometric function, Δ^* the creep strength of the body

$$\Delta^* = \frac{\epsilon(t\to\infty) - \epsilon(t=0)}{\epsilon(t=0)}$$

(the ratio of after-effect strain to initial strain) and μ_∞ the shear modulus at ultra-seismic frequencies. The elegance of this model lies in the parsimony with three parameters, α, Δ^* and μ_∞.

Let us end this section by mentioning another generalised body, this time based on an infinite parallel coupling of Maxwell bodies with a spectrum of relaxation times, the Generalised Maxwell Body (b in Fig. 1.6). Instead of prescribing stress we now prescribe strain and formulate how the stress relaxes

$$\sigma(\mathrm{i}\omega) = 2\left(\int A(\tau)\mu(\mathrm{i}\omega,\tau)\,d\tau\right)\epsilon(\mathrm{i}\omega)$$

where $\mu(\mathrm{i}\omega,\tau)$ is now taken from (1.101). Closed formulas for the Generalised Maxwell Body are generally not possible; however Müller (1983) has shown such expressions for fractional integer exponents $A \propto \tau^{1/n}$. The banded nature of A as it appears limited between two finite relaxation times might appear as rather artificial. However, this so-called high-temperature background or absorption-band model has been shown to own some realism in laboratory studies of rock samples (Kampfmann and Berckhemer 1985). It can be argued that the absorption band model has an upper limit where Maxwell rheology (irreversible deformation) exceeds the relaxation that the generalised Zener body accomplishes, so that it is the lower band edge that is more critical to determine, but at the same time it is more accessible to observation owing to studies of seismic wave propagation.

11 Green's Functions

Computation of loading effects at a field point due to the tide loads distributed over the ocean surfaces of the planet is conveniently carried out using convolution with a Green's kernel function. The method and its alternative decompose the load into surface spherical harmonics and multiplication of the wave number spectrum with the load Love number spectrum has been presented comprehensively in Agnew (2007), including the extension to infinite harmonic degree in the point load problem and the generation of Green's functions using Kummer transforms. We refer to this work as seminal.

Farrell (1972) also considered disk loads in order to avoid the Gibbs phenomenon when loads are treated as if condensed to singular mass points. This creates a problem particularly when the loads are known at a few locations or on sparse grids only. As it turned out, the assumption of loading mass distributed over a circular disk of certain size attenuates the high-degree terms when summing over the load Love numbers and helps to speed up the convergence of the infinite sums in the Green's functions. The ever increasing spatial resolution of modern ocean tide models render this approach mostly obsolete today. However, the assumption of disk-distributed masses might still be necessary in a few cases.

In gravity, near-by masses exert a notable attraction effect if gravimeter and load mass are located at different heights, more specifically when a big correction ϵ is required in the equation

$$\frac{\cos\phi}{d^2} = \frac{-1}{4a^2 \sin\theta/2} + \epsilon$$

where d is slope distance, ϕ the zenith angle under which the gravimeter "sees" the load, and θ the arc distance of the two points after mapping them on the sphere. An example of this effect is described by Bos et al. (2002). However, cylindrical disk loads will in general be inept to represent the actual mass distribution, and consequently, the specific geometry of the load needs to be accounted for. This amounts to have to sample the land-sea distribution with high resolution while the details of tide height in the wet areas will be uncritical.

In tilt as measured by vertical pendulums or fluid-filled tubes (Ruotsalainen 2001), loads at close distance have both a strong direct attraction effect and sensible influence due to deformation. As a third complication, the infinite sum in the Green's function converges most slowly among those considered by Farrell (1972) and Agnew (2007), even when the Kummer trick is utilised (numerical precision in the then final sums becoming critical).

We exemplify Green's function computation only in the case of tilt. The function is given as

$$\mathcal{G}_{tilt}(\theta) = \frac{GM_l}{g\,a^2} t_0 \sum_{n=1}^{\infty} (t^{-n} - h'_n + k'_n) \frac{dP_n(\cos\theta)}{d\theta}$$

with t the station height ratio $(a+h)/a$ and $t_0 = t$ if $h<0$ else $t_0 = 1$. With the disc factor included, the sum terms receive a factor $D_n = D_n(\delta)$ according to Farrell (1972):

$$D_n = -\frac{1+\cos\delta}{n(n+1)\sin\delta} \frac{dP_n(\cos\delta)}{d\delta}$$

and the sum

Fig. 1.7 Geometry when integrating over a disk load

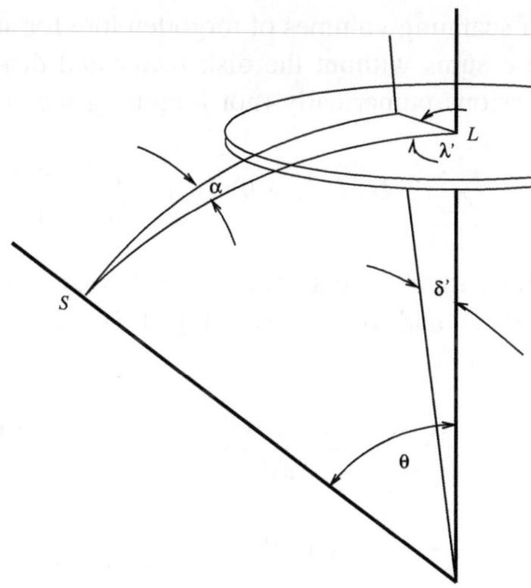

$$\sum_{n=0}^{\infty} D_n P_n(\cos\theta) = \begin{cases} 0, & \theta > \delta \\ 1, & \theta \le \delta \end{cases}$$

Ultimately, the point load problem cannot avoid dealing with the deformation in the asymptotic limit of an infinitely large pressure on an infinitely small surface area. This is the point where the analytical half-space solutions of Farrell (1972) come into play. The load Love numbers obtained are h'_∞ and $(Nl')_\infty$, and

$$(Nk')_\infty \sim \frac{3\rho_0(a)}{2\bar{\rho}}(h'_\infty + (Nl')_\infty) \tag{1.104}$$

appears to be a good approximation ($\bar{\rho}$ is the mean density of the Earth). For exercising the Kummer transformation we utilise that

$$h'_n \to h'_\infty \qquad nk'_n \to (Nk')_\infty$$

as $n \to \infty$, so we need evaluations or expressions of the infinite sums

$$\sum_{n=1}^{\infty} t^{-n} D_n \frac{dP_n(\cos\theta)}{d\theta}$$

$$\sum_{n=1}^{\infty} D_n \frac{dP_n(\cos\theta)}{d\theta}$$

$$\sum_{n=1}^{\infty} \frac{D_n}{n} \frac{dP_n(\cos\theta)}{d\theta}$$

Instead of scanning volumes of forgotten lore for analytical expressions, we can evaluate these sums without the disk factor and disk-integrate the resulting analytical expressions numerically (not forgetting the cosine of the azimuth):

$$\sum_{n=1}^{\infty} t^{-n} D_n\, P_n(\cos\theta) = \int_0^{\delta}\int_0^{2\pi} \frac{\cos\alpha\ \sin\delta'\ d\lambda'\ d\delta'}{\sqrt{1-2t\cos\theta'+t^2}}$$

Since we deal with short distances, the spherical trigonometric relations between the angles θ', α' and δ', λ' (see Fig. 1.7) can be approximated with plane trigonometry.

$$\sum_{n=1}^{\infty} t^{-n} \frac{dP_n(\cos\theta)}{d\theta} = -\frac{t\sin\theta}{(1-2t\cos\theta+t^2)^{3/2}} \tag{1.105}$$

$$\sum_{n=1}^{\infty} \frac{1}{n}\frac{dP_n(\cos\theta)}{d\theta} = -\frac{1}{2}\cot(\theta/2)\,\frac{1+2\sin(\theta/2)}{1+\sin(\theta/2)} \tag{1.106}$$

To be complete, we now give the definitions of the Green's function for the other type of deformations; see also Agnew (2007). In all cases we assume a distribution of mass elements dm' over the oceanic areas $\mathcal{O}$ and a solution to a deformation problem where radial symmetry of the planet's properties causes deformation under the load depending only on the the arc distance $\theta' = \angle\mathbf{r}, \mathbf{r}'$ between load and field point.

Vertical displacement

$$\begin{aligned} u(\mathbf{r}) &= \frac{G}{g\,a}\int_{\mathcal{O}} \mathcal{G}_u(\theta')\ dm' \\ \mathcal{G}_u(\theta) &= \sum_{n=0}^{\infty} h'_n\ P_n(\cos\theta) \end{aligned} \tag{1.107}$$

Horizontal displacement in the directions $\hat{n}$ (north) and $\hat{e}$ (east)

$$\begin{aligned} \left.\begin{matrix} v_n(\mathbf{r}) \\ v_e(\mathbf{r}) \end{matrix}\right\} &= \frac{G}{g\,a}\int_{\mathcal{O}} \mathcal{G}_v(\theta') \left\{\begin{matrix} \cos\alpha' \\ \sin\alpha' \end{matrix}\right\} dm' \\ \mathcal{G}_v(\theta) &= \sum_{n=1}^{\infty} l'_n\ \frac{dP_n(\cos\theta)}{d\theta} \end{aligned} \tag{1.108}$$

Gravity on deforming surface

$$\Delta g(\mathbf{r}) = \frac{G}{a^2} \int_{\mathcal{O}} \mathcal{G}_{\Delta g}(\theta')\, dm'$$
$$\mathcal{G}_{\Delta g}(\theta) = \sum_{n=0}^{\infty} [2h'_n - (n+1)k'_n]\, P_n(\cos\theta) - \frac{z/a + 2\sin^2\theta/2}{[(z/a)^2 + 4(1+z/a)\sin^2\theta/2]^{3/2}} \tag{1.109}$$

where z is the topographic height of the station, (error in Agnew (2007), who has $[\,]^3/2$ instead of $[\,]^{3/2}$ in the denominator; see also Appendix 4).

Geoid height

$$N(\mathbf{r}) = \frac{G}{g\,a} \int_{\mathcal{O}} \mathcal{G}_N(\theta')\, dm'$$
$$\mathcal{G}_N(\theta) = \sum_{n=0}^{\infty} (1 + k'_n)\, P_n(\cos\theta) \tag{1.110}$$

Tide raising potential

$$P(\mathbf{r}) = \frac{G}{a} \int_{\mathcal{O}} \mathcal{G}_P(\theta')\, dm'$$
$$\mathcal{G}_P(\theta) = \sum_{n=0}^{\infty} (1 + k'_n - h'_n)\, P_n(\cos\theta) \tag{1.111}$$

Tilt in the directions $\hat{n}$ (north) and $\hat{e}$ (east)

$$\left.\begin{matrix} t_n(\mathbf{r}) \\ t_e(\mathbf{r}) \end{matrix}\right\} = -\frac{G}{g\,a^2} \int_{\mathcal{O}} \mathcal{G}_t(\theta') \left\{\begin{matrix} \cos\alpha' \\ \sin\alpha' \end{matrix}\right\} dm'$$
$$\mathcal{G}_t(\theta) = \sum_{n=1}^{\infty} (1 + k'_n - h'_n)\, \frac{dP_n(\cos\theta)}{d\theta} \tag{1.112}$$

Astronomical deflection of the vertical in $\hat{n}$ (north) and $\hat{e}$ (east)

$$\left.\begin{matrix} \eta = i_n(\mathbf{r}) \\ \xi = i_e(\mathbf{r}) \end{matrix}\right\} = -\frac{G}{g\,a^2} \int_{\mathcal{O}} \mathcal{G}_t(\theta') \left\{\begin{matrix} \cos\alpha' \\ \sin\alpha' \end{matrix}\right\} dm'$$
$$\mathcal{G}_i(\theta) = \sum_{n=1}^{\infty} (1 + k'_n - l'_n)\, \frac{dP_n(\cos\theta)}{d\theta} \tag{1.113}$$

Strain in $\hat{n}$ (north) and $\hat{e}$ (east)

$$\begin{pmatrix} \epsilon_{nn}(\mathbf{r}) & \epsilon_{ne}(\mathbf{r}) \\ \epsilon_{ne}(\mathbf{r}) & \epsilon_{ee}(\mathbf{r}) \end{pmatrix} = \frac{G}{g\,a^2} \int_{\mathcal{O}} \mathbf{T} \begin{pmatrix} \mathcal{G}_{\theta\theta}(\theta') & 0 \\ 0 & \mathcal{G}_{\alpha\alpha}(\theta') \end{pmatrix} \mathbf{T}^{\top} dm'$$

$$\mathcal{G}_{\theta\theta}(\theta) = \sum_{n=0}^{\infty} [h'_n - n(n+1)l'_n - l'_n \cot\theta \frac{d}{d\theta}]\; P_n(\cos\theta) \tag{1.114}$$

$$\mathcal{G}_{\alpha\alpha}(\theta) = \sum_{n=0}^{\infty} (h'_n + l'_n \cot\theta)\; P_n(\cos\theta)$$

$$\mathbf{T} = \begin{pmatrix} \cos\alpha' & \sin\alpha' \\ -\sin\alpha' & \sin\alpha' \end{pmatrix} \tag{1.115}$$

We give useful formulas for arc distance θ' and azimuth α' :

$$\theta' = 2\arcsin\sqrt{\sin^2\left(\frac{\beta'-\beta}{2}\right) + \sin^2\left(\frac{\lambda'-\lambda}{2}\right)\cos\beta\ \cos\beta'} \tag{1.116}$$

$$\begin{aligned} \alpha' &= \operatorname{atan2}\left(\frac{\sin\beta' - \cos\theta'\sin\beta}{\cos\theta'},\ \cos\beta'\sin(\lambda'-\lambda)\right) \\ &= \operatorname{atan2}(\cos\alpha',\ \sin\alpha') \end{aligned} \tag{1.117}$$

where β is latitude, λ longitude, the dashed coordinates are associated with the load points and the undashed with the field point.

12 Final Remarks

We have presented here the derivation of the differential equations for the elastic-gravitational deformation of the Earth. As we have noted in our introduction, this is a classic topic that has been discussed extensively in the literature. However, we felt that current literature does not provide much information on how these equations can be implemented into a computer program. For that reason, we have tried to put more emphasis on the differences in sign conventions and definitions of variables that one may find in various publications. We also have presented here all boundary conditions explicitly and have paid particular attention to the problems related to the existence of a fluid core. Although we have focussed on ocean tide loading, where the period of the forcing is finite, we have shown which problems occur in the fluid core when the forcing period becomes infinite. We recalled the results of Pekeris and Accad (1972) who showed that for a fluid with an unstable stratification, an infinitely thin boundary layer develops that has a finite influence on the deformation. However, we pointed out that one is interested in computing the static deformation of the Earth, and some kind of dissipation should be taken into account (Wunsch 1974; Dahlen and Fels 1978).

Next, we put more emphasis than usual on the degree 1 deformation. This deformation differs from the other degrees by the fact that it causes a shift of the whole Earth in space. We also explicitly described the new boundary conditions at the centre of the Earth for this degree which is rarely done.

As examples, we presented the deformation of a homogeneous Earth and an Earth with a homogeneous mantle and fluid. For these simple cases analytical solutions exist that involve spherical Bessel functions of the first and second kind. Since we are interested in ocean tide loading where we need to compute load Love numbers up to degree 10,000, we have presented in Appendix 2 an algorithm to achieve this. The results can be used to validate the implementation of other numerical methods such as Runge–Kutta and Chebychev collocation. The case of a homogeneous Earth is also used to solve the singularity problem at the Earth's centre for the Runge–Kutta method.

In our short treatise the ocean tide loading response of the Earth is still assumed to be completely elastic. However, we showed how the elastic properties need to be changed to represent more realistic rheology. Finally, we discussed the formulas that transform the set of Love numbers into Greens functions.

Appendix 1: Lyapunov-Transformed Matrices

The Lyapunov-transformed matrix designated **B** in (1.43) is

$$\begin{pmatrix}
\frac{-2(3\kappa-2\mu)}{(n+1)(3\kappa+4\mu)} & \frac{n^2(3\kappa-2\mu)}{3\kappa+4\mu} & 0 & \frac{3\kappa_0 Z^2}{(1+n)^2(3\kappa+4\mu)} & 0 & 0 \\
\frac{-1}{n(n+1)} & \frac{1}{n+1} & 0 & 0 & \frac{\kappa_0 Z^2}{\mu n(n+1)} & 0 \\
\frac{-4\pi\rho a G Z}{g_0(n+1)} & 0 & -1 & 0 & 0 & \frac{Z^2}{n+1} \\
\frac{a^2}{\kappa_0}\left(\frac{36\kappa\mu}{a^2(3\kappa+4\mu)Z^2}-\omega^2\rho-\frac{4g\rho}{aZ}\right) & \frac{n^2(n+1)}{\kappa_0 Z^2}\left(ag\rho Z-\frac{18\kappa\mu}{3\kappa+4\mu}\right) & \frac{-ag_0(1+n)\rho}{\kappa_0 Z} & \frac{-12\mu}{(n+1)(3\kappa+4\mu)} & n(n+1) & \frac{ag_0\rho Z}{\kappa_0} \\
\frac{1}{(n+1)\kappa_0 Z^2}\left(ag\rho Z-\frac{18\kappa\mu}{3\kappa+4\mu}\right) & \frac{a^2 n}{(n+1)\kappa_0}\left(\frac{6(2n(n+1)-1)\kappa\mu+4(n(n+1)-2)\mu^2}{a^2(3\kappa+4\mu)Z^2}-\omega^2\rho\right) & \frac{ag_0\rho}{(n+1)\kappa_0 Z} & \frac{-3\kappa+2\mu}{(n+1)^2(3\kappa+4\mu)} & \frac{-3}{n+1} & 0 \\
-\frac{4\pi\rho a G}{g_0 Z} & \frac{4\pi\rho a G n^2}{g_0 Z^2} & 0 & 0 & 0 & \frac{n-1}{n+1}
\end{pmatrix} \tag{1.118}$$

where $Z = \exp\frac{q}{n+1}$, and κ and μ have been made dimensionless by scaling with respect to κ_0 at the centre.

The matrix for the fluid case in the variables U, S, P and Q is

$$\begin{pmatrix}
-\frac{5}{2(1+n)}+\frac{gn\omega^2}{aZ} & \frac{\left(\frac{Z^2}{\kappa}-\frac{n(1+n)}{a^2\rho\omega^2}\right)}{(n+1)^2} & \frac{n}{a^2\omega^2} & 0 \\
a^2\rho\left(-\frac{4g}{aZ}+\frac{g^2 n(1+n)}{a^2 Z^2\omega^2}-\omega^2\right) & -\frac{3}{2(1+n)}-\frac{gn}{aZ\omega^2} & (1+n)\rho\left(-1+\frac{gn}{aZ\omega^2}\right) & ag_0 Z\rho \\
-\frac{4a^2 G\pi\rho}{(1+n)} & 0 & -1-\frac{3}{2(1+n)} & \frac{ag_0 Z}{1+n} \\
\frac{4G\pi\rho(gn-aZ\omega^2)}{g_0 Z^2\omega^2} & -\frac{4Gn\pi}{ag_0(1+n)Z\omega^2} & \frac{4Gn\pi\rho}{ag_0 Z\omega^2} & 1-\frac{7}{2(1+n)}
\end{pmatrix} \tag{1.119}$$

Appendix 2: Analytical Solution for a Homogeneous Earth

In Sect. 1.7 we discussed the computation of the Love numbers for a homogeneous Earth. In this case there exist, for each degree n, analytical solutions for the radial and tangential displacements U and V and for the perturbed potential P. These were presented for degree 2 by Love (1911) and Pekeris and Jarosch (1958). For all degrees n these were presented by Okubo (1988) and Dahlen and Tromp (1998) although they both contain small sign errors. Therefore, we will present them again, hopefully without errors, in this appendix.

The analytical solutions contain spherical Bessel functions which are defined as

$$j_n(z) = \sqrt{\frac{\pi}{2z}} J_{n+1/2}(z) \tag{1.120}$$

where n is our usual degree, z a complex number and J the normal Bessel function of the first kind. For $n = 0$ and $n = 1$, these spherical Bessel functions are

$$j_0(z) = \frac{e^{iz} - e^{-iz}}{2iz} = \frac{\sin z}{z} \tag{1.121}$$

$$j_1(z) = \frac{e^{iz} - e^{-iz}}{2iz^2} - \frac{e^{iz} + e^{-iz}}{2z} = \frac{\sin z}{z^2} - \frac{\cos z}{z} \tag{1.122}$$

As it turns out, we shall only be using values of z that are real or purely imaginary. In the first case we are dealing with fractions containing trigonometric functions. In the second we are dealing with fractions containing exponentials. To avoid numerical problems for large values of z, it is convenient to work with the logarithm of $j_n(z)$. Higher orders of $j_n(z)$ can in principle be computed using the following recurrence relation:

$$j_{n+1}(z) = \frac{2n+1}{z} j_n(z) - j_{n-1}(z) \tag{1.123}$$

However, this recursive equation is numerically unstable for increasing values of n. To compute the higher orders, we should use the algorithm of Rothwell (2008) who introduced the ratio R_n :

$$R_n = \frac{j_{n-1}(z)}{j_n(z)} = \frac{1}{(2n-1)/z - R_{n-1}} \tag{1.124}$$

$$R_n(z) = \frac{2n+1}{z} - \frac{1}{R_{n+1}(z)} \tag{1.125}$$

The continued fraction can for each degree n be computed using Lentz's method which is numerically stable (Press et al. 1988). After R_n has been computed using Lentz's method for the largest value of n, the other values of R_n can be computed using (1.125). The logarithm of $j_n(z)$, with $n \geq 1$, can now be computed as

$$\log j_n(z) = \log j_0(z) - \sum_{k=1}^{n} \log R_k \tag{1.126}$$

Note in these equations that the logarithm functions include their analytical continuation in the complex plane because R_n can be complex valued. Since the R_n are either purely real or purely complex in our application we have

$$\begin{aligned} \log i|x| &= \log|x| + i\frac{\pi}{2} \\ \log -|x| &= \log|x| + i\pi \end{aligned} \tag{1.127}$$

so the sum in (1.126) accumulates a factor of $-(\mathrm{i}^n)$. A gain in accuracy can be achieved if the summation in (1.126) is carried out separately on the characteristic and the mantissa.

We will now assume that the solutions are of the form

$$U = A_u r^{-1} j_n(\gamma r) + B_u \gamma j_{n+1}(\gamma r) \tag{1.128}$$

$$V = A_v r^{-1} j_n(\gamma r) + B_v \gamma j_{n+1}(\gamma r) \tag{1.129}$$

$$P = A_p j_n(\gamma r) \tag{1.130}$$

That the solutions contain a combination of j_n and j_{n+1} makes sense because you can write the first and second derivatives of these two functions again as a sum of parts containing j_n and j_{n+1}. Furthermore, one has to use j_n/r and j_{n+1} to ensure that both are of the same order of r. That the perturbed potential solution does not contain a j_{n+1} part seems odd at first glance. However, from Poisson's equation, (1.10), we know that the Laplacian of the perturbed potential depends on the perturbed density. The latter depends on the divergence of the displacement and one can prove that the j_{n+1} terms cancel out in the divergence (1.28).

Our task is to determine the six constants: A_u, B_u, A_v, B_v, A_p and γ. This can be done using a computer algebra program such as Mathematica or Maxima. However, we will directly present the answer in the same format as that of Dahlen and Tromp (1998, Chap. 8) who introduced the following auxiliary variables and solution for γ :

$$k = \sqrt{n(n+1)} \tag{1.131}$$

$$\gamma^2 = \frac{\omega^2}{2v_\beta^2} + \frac{\omega^2 + \frac{16}{3}\pi G\rho}{2v_\alpha^2} \pm \frac{1}{2}\left[\left(\frac{\omega^2}{v_\beta^2} - \frac{\frac{16}{3}\pi G\rho}{v_\alpha^2}\right)^2 + \left(\frac{8\pi Gk\rho}{3v_\alpha v_\beta}\right)^2\right]^{1/2} \tag{1.132}$$

$$\zeta = \frac{3}{4}(\pi G\rho)^{-1} v_\beta^2 (\gamma^2 - \omega^2/v_\beta^2) \tag{1.133}$$

$$\xi = \zeta - (n+1) \tag{1.134}$$

where v_α and v_β are the compressional and shear wave velocities, respectively. Besides solutions that contain spherical Bessel function, there exist solutions containing terms with r^n. To summarise, for each of the radial and tangential displacements U and V and the perturbed potential P we have three independent solutions:

$$U^{\pm} = \frac{n\xi}{r} j_n(\gamma r) - \zeta\gamma j_{n+1}(\gamma r) \ , \quad U^{\oplus} = nr^{n-1} \tag{1.135}$$

$$V^{\pm} = \frac{\xi}{r} j_n(\gamma r) + \gamma j_{n+1}(\gamma r) \ , \quad V^{\oplus} = r^{n-1} \tag{1.136}$$

$$P^{\pm} = -4\pi G\rho\zeta j_n(\gamma r) \ , \quad P^{\oplus} = (\omega^2 - \tfrac{4}{3}\pi G\rho n) r^n \tag{1.137}$$

As explained by Dahlen and Tromp the symbol $^{\pm}$ needs to be understood as two solutions, one for each solution of γ. The $\oplus$ symbol is associated with solutions containing an r^n term. For the radial and tangential stresses, R and S, and perturbed gravity Q we have the following three solutions:

$$R^{\pm} = -\left[(\kappa + \tfrac{4}{3}\mu)\zeta\gamma^2 - 2n(n-1)\mu\xi r^{-2}\right] j_n(\gamma r) \\ + 2\mu(2\zeta + k^2)\gamma r^{-1} j_{n+1}(\gamma r) \tag{1.138}$$

$$R^{\oplus} = 2n(n-1)\mu r^{n-2} \tag{1.139}$$

$$S^{\pm} = \mu\left[\gamma^2 + 2(n-1)\xi r^{-2}\right] j_n(\gamma r) - 2\mu(\zeta+1)\gamma r^{-1} j_{n+1}(\gamma r) \tag{1.140}$$

$$S^{\oplus} = 2(n-1)\mu r^{n-2} \tag{1.141}$$

$$Q^{\pm} = -4\pi G\rho r^{-1}\left[k^2 + (n+1)\zeta\right] j_n(\gamma r) \tag{1.142}$$

$$Q^{\oplus} = \left[(2n+1)\omega^2 - \tfrac{8}{3}\pi G\rho n(n-1)\right] r^{n-1} \tag{1.143}$$

Sometimes it is assumed that the Earth is built up of spherical layers with constant properties. In this case the solutions listed above apply but in addition we need spherical Bessel functions of the second kind, also called spherical Neumann functions y_n, and solutions that depend on $1/r^n$ (represented by the $\ominus$ superscript). The two solutions associated with the spherical Neuman functions can be found by simply replacing the j_n terms with y_n. We indicate this with superscripts $\pm j$ and $\pm y$, respectively in (1.148). These functions may be computed using the relation $y_n(z) = (-1)^{n-1} j_{-n-1}(z)$. Equation 1.123 may be used to compute $j_{-n-1}(z)$ because the index of j is now decreasing. The solutions associated with $1/r$ are

$$U^{\ominus} = -\frac{n+1}{r^{n+2}} \qquad V^{\ominus} = \frac{1}{r^{n+2}} \qquad P^{\ominus} = -\frac{\left(\omega^2 + \frac{4}{3}\pi G\rho(n+1)\right)}{r^{n+1}} \tag{1.144}$$

$$\begin{aligned} R^{\ominus} &= 2\mu\frac{(n+1)(n+2)}{r^{n+3}} \\ S^{\ominus} &= -2\mu\frac{n+2}{r^{n+3}} \\ Q^{\ominus} &= \frac{4\pi G\rho(n+1)}{r^{n+2}} \end{aligned} \tag{1.145}$$

If the Earth is divided up into spherical layers with constant properties, the gravity inside this Earth is mostly different from that of a homogeneous Earth. To keep using the presented above equations, one therefore scales in each layer the term $\frac{4}{3}\pi G\rho$ to the mean value of g/r inside this layer (Vermeersen et al. 1996).

At this point the concept of Haskell propagator matrices (Haskell 1953) can be invoked. We have an analytical expression that relates the values of the radial solution functions between two consecutive interfaces j and $j+1$:

$$\mathbf{y}(r_{j+1}) = \mathbf{P}(r_{j+1}, r_j)\,\mathbf{y}(r_j) \tag{1.146}$$

and

$$\mathbf{y}(r) = \mathbf{A}\,\boldsymbol{\alpha} \tag{1.147}$$

where $\mathbf{A}$ is the 6×6 layer matrix composed of solutions $(U, V, P, R, S, Q)^{\top}$ where

$$\mathbf{A} = \begin{pmatrix} U^{\oplus} & U^{\ominus} & U^{+j} & U^{+y} & U^{-j} & U^{-y} \\ V^{\oplus} & V^{\ominus} & & \cdots & & \\ & & & \cdots & & \\ & & & \cdots & & Q^{-y} \end{pmatrix} \tag{1.148}$$

The propagator matrix is then

$$\mathbf{P}(r_{j+1}, r_j) = \mathbf{A}(r_{j+1})\mathbf{A}^{-1}(r_j) \tag{1.149}$$

so it can be stepped from $j=1$ to the surface. The $\boldsymbol{\alpha}$ vector is finally determined from the boundary conditions. The reader will realise at this point that the account needs substantially more detail, for instance how the layer matrix can be inverted with elegance. We therefore refer him or her to the original source where the method is described at necessary depth, Martinec (1989). We only repeat Martinec's advice on how to treat a fluid layer. In this case, V and S are taken out of the equations, and the characteristic root is single-valued,

$$\gamma^2 = \frac{1}{v_\alpha^2}\left[\omega^2 + \frac{16\pi}{3}G\rho - \left(\frac{4\pi Gk\rho}{3\omega}\right)^2\right] \tag{1.150}$$

so we obtain a 4×4 matrix for $\mathbf{A}$.

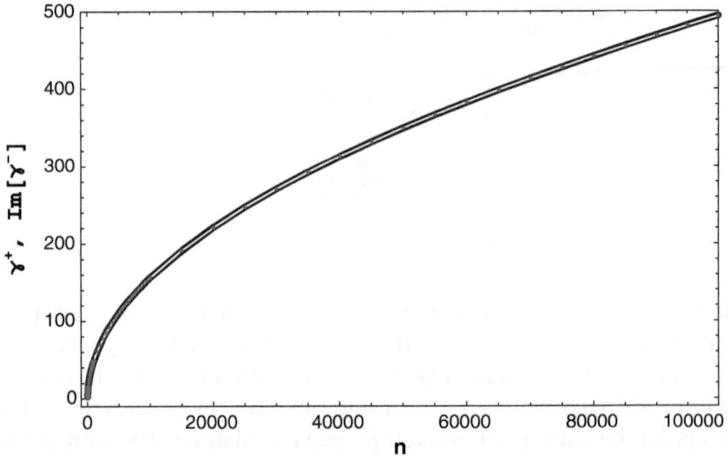

Fig. 1.8 The characteristic roots γ^+ and $\mathrm{Im}(\gamma^-)$, also providing the radial scale for the spherical Bessel functions vs spherical harmonic degree n Both functions are similar and difficult to discern in the diagram

When $\omega \to 0$, these equations no longer hold in the fluid. Pekeris and Accad (1972) showed that a boundary layer develops at the core–mantle interface that becomes infinitely thin but still has a finite effect on the dynamics. Instead of following their derivation, we will compute the jumps in U, R and Q (there is no jump in P) in the boundary layer for $\omega \to 0$ by taking the limit of our equations involving spherical Bessel functions of the first kind. To do so, it will be convenient to define

$$A = \frac{4}{3}\pi G\rho \tag{1.151}$$

Using (1.150), one can derive the limit value of γ :

$$\lim_{\omega\to 0} \gamma = \frac{kAi}{v_\alpha}\frac{1}{\omega} \tag{1.152}$$

Using this result, we can compute the limit of the spherical Bessel function:

$$\lim_{\omega\to 0} j_n(\gamma r) = \pm\frac{v_\alpha \omega}{2kAr} e^{kAr/(v_\alpha \omega)} \tag{1.153}$$

The sign of the limit depends on the degree n. Of course the exponential grows fast to infinity but we are allowed to scale our solutions with any constant so we choose it in such a way to make the $j_n(b\gamma) = 1$ at the boundary, where b is the radius of the boundary (Pekeris and Accad 1972). Below the core–mantle interface, $j_n(r\gamma) = 0$ and it thus nicely represents the jump through the boundary layer. Other results are

$$\zeta = -\frac{\omega^2}{A} \qquad \xi = -(n+1) \tag{1.154}$$

If we now collect all terms which are not zero when $\omega \to 0$, then we have the following solutions for our jump:

$$U^{\diamond}(b) = -\frac{n(n+1)}{b} \tag{1.155}$$

$$R^{\diamond}(b) = -\frac{4}{3}\pi G\rho k\frac{\kappa}{v_{\alpha}^2} \tag{1.156}$$

$$Q^{\diamond}(b) = -\frac{4\pi G\rho k^2}{b} \tag{1.157}$$

These equations should be added to the $\oplus$ solutions to fulfil the boundary conditions at the core–mantle interface. Inside the fluid core, only the + solutions apply.

Finally, we would like to remark that the roots $\gamma^{\pm}$ are nicely bounded. For our homogeneous Earth, model β, their values have been plotted in Fig. 1.8.

Appendix 3: Analytical Solution for a Homogeneous Fluid Inner Sphere

In a fluid the shear modulus $\mu = 0$. Here we distinguish four cases $n = 0, n > 0$, and $\omega = 0, \omega \neq 0$. First $n > 0$. Since the region includes $r = 0$, we can dismiss the irregular solutions involving the Neumann y_n-functions and the $\ominus$ functions.

Dahlen and Tromp (1998) give the solution for $n > 0$. The matrix of the differential equation simplifies to a 4×4 system in the same variable as before except for horizontal displacement and shear stress.

Dahlen and Tromp (1998) give the solution for $n > 0$. The matrix of the differential equation simplifies to a 4×4 system in the same variable as before except for horizontal displacement and shear stress.

$$\frac{d}{dr}\begin{pmatrix} U \\ R \\ P \\ Q \end{pmatrix} =$$

$$\begin{pmatrix} \frac{4\pi\rho G n(n+1)}{3r\omega^2} - \frac{2}{r} & \frac{1}{\kappa} - \frac{n(n+1)}{r^2\rho\omega^2} & \frac{n(n+1)}{r^2\omega^2} & 0 \\ -\frac{\rho}{9}\left(48\pi G\rho - \frac{16\pi^2 G^2\rho^2 n(n+1)}{\omega^2} + 9\omega^2\right) & -\frac{4\pi G_{\rho} n(n+1)}{3r^2\omega^2} & \frac{(n+1)\rho(4n\pi G - 3\omega^2}{3r\omega^2} & \rho \\ -4\pi\rho & 0 & -\frac{n+1}{r} & 1 \\ \frac{-4\pi G\rho(n+1)(4n\pi G\rho - 3\omega^2)}{3r^2\omega^2} & -\frac{4\pi G n(n+1)}{r^2\omega^2} & -\frac{4\pi G\rho n(n+1)}{3r^2\omega^2} & -\frac{n+1}{r} \end{pmatrix}. \tag{1.158}$$

$$\begin{pmatrix} U \\ R \\ P \\ Q \end{pmatrix}$$

with the solutions

$$U^{+} = \frac{n\xi}{r} j_n(\gamma r) - \zeta\gamma j_{n+1}(\gamma r) \tag{1.159}$$

$$R^{+} = -\kappa\zeta\gamma^2 j_n(\gamma r) \tag{1.160}$$

$$P^{+} = -4\pi G\rho\zeta j_n(\gamma r) \tag{1.161}$$

$$Q^{+} = -\frac{4\pi G\rho}{r}(n+1)(n+\zeta) j_n(\gamma r) \tag{1.162}$$

and

$$U^{\oplus} = n r^{n+1}, \quad P^{\oplus} = (\omega^2 - \frac{4\pi}{3} G\rho n) r^n \tag{1.163}$$

$$R^{\oplus} = 0, \quad Q^{\oplus} = [(2n+1)\omega^2 - \frac{8\pi}{3} G\rho n(n+1)] r^{n-1} \tag{1.164}$$

where γ was given in (1.150) and

$$\zeta = -\frac{3\omega^2}{4\pi G\rho} \qquad \xi = \zeta - (n+1) \tag{1.165}$$

These equations follow from what what discussed before, but it becomes interesting when one takes the limit of $\omega \to 0$. Unfortunately, these solutions cannot be used for $\omega \to 0$. Instead, we refer to Longman (1963) who solves a 2×2 system in the gravity variables only. At the core–mantle boundary, vertical and horizontal displacement start with arbitrary values into the mantle, horizontal shear stress is zero, vertical normal stress starts as

$$R = \rho g(r) U - \rho P \tag{1.166}$$

and the gravity variable Q as

$$Q = H - \rho U \tag{1.167}$$

The 2×2 system in the core is

$$\frac{d}{dr}\begin{pmatrix} P \\ H \end{pmatrix} = \begin{pmatrix} 0 & 1 \\ \frac{n(n+1)}{r^2} - \frac{\rho^2}{\kappa} & -\frac{2}{r} \end{pmatrix} \cdot \begin{pmatrix} P \\ H \end{pmatrix} \tag{1.168}$$

The solution of this system is

$$P = C\sqrt{\frac{2\gamma_0}{r}} j_n(\gamma_0 r) \tag{1.169}$$

$$H = C\frac{\sqrt{2\gamma_0}}{r}\left(\frac{n}{\sqrt{r}} j_n(\gamma_0 r) + \gamma_0\sqrt{r} j_{n+1}(\gamma_0 r)\right) \tag{1.170}$$

where

$$\gamma_0 = \rho\sqrt{\frac{4\pi G}{\kappa}} \tag{1.171}$$

and C is an arbitrary constant.

Appendix 4: Tiny Fluid Sphere

The recursion relations for a tiny homogenous fluid sphere in the variables U, S, P and Q (1.91) are

$$u_1 = 0, \qquad p_1 = s_1/\rho = \gamma \tag{1.172}$$

$$q_1 = \frac{(-8\,G\,n\,(1+n)\,\pi\,\rho + (1-2\,n)\,\omega^2)\,p_1}{3\,\omega^2} \tag{1.173}$$

with an arbitrary γ as a start. Then, for $j = 3, 5, \ldots$

$$\begin{aligned} q_j = {} & 4\,G\,(1+n)\,\pi\,\rho\,\big[32\,G^2\,n^2\,(1+n)\,\pi^2\,\rho^2 - 4\,G\,n\,(4-j+j^2 \\ & +2\,n+2\,j\,n)\,\pi\,\rho\,\omega^2 + 3\,(j-1)\,(j+2\,n)\,\omega^4)\,\big]\,s_{j-2} \\ & \frac{1}{3\,(j-1)\,(j+2\,n)\,\kappa\,\omega^2\,[8\,G\,n\,(1+n)\,\pi\,\rho - (j-1)\,(j+2\,n)\,\omega^2]} \end{aligned} \tag{1.174}$$

$$\begin{aligned} u_j = {} & \Big\{32\,G^2\,n^2\,(1+n)^2\,\pi^2\,\rho^2 - 4\,G\,n\,(1+n)\,[4+j^2 \\ & -10\,n + j\,(+2\,n\,-\,7)]\,\pi\,\rho\,\omega^2 \\ & -3\,(j-1)\,[j^2 + 2\,(n-1)\,n + j\,(3\,n\,-\,1)]\,\omega^4\Big\}\,s_{j-2} \\ & \frac{1}{3\,(j-1)\,(j+2\,n)\,\kappa\,\omega^2\,[8\,G\,n\,(1+n)\,\pi\,\rho - (j-1)\,(j+2\,n)\,\omega^2]} \end{aligned} \tag{1.175}$$

$$\begin{aligned} s_j = {} & \rho\,\Big\{128\,G^3\,n^2\,(n+1)^2\,\pi^3\,\rho^3 - 16\,G^2\,n\,(n+1)\,[22+j^2 \\ & -4\,n + j\,(2\,n\,-\,1)]\,\pi^2\,\rho^2\,\omega^2 + 24\,G\,[2\,j^2 + 3\,j\,(n-1) \\ & -4\,n\,(2+n)]\,\pi\,\rho\,\omega^4 + 9\,(j-1)\,(j+2\,n)\,\omega^6\Big\}\,s_{j-2} \\ & \frac{1}{9\,(j-1)\,(j+2\,n)\,\kappa\,\omega^2\,[8\,G\,n\,(n+1)\,\pi\,\rho - (j-1)\,(j+2\,n)\,\omega^2]} \end{aligned} \tag{1.176}$$

$$p_j = -\frac{4\,G\,\pi\,\rho}{(j-1)\,(j+2\,n)\,\kappa}\,s_{j-2} \tag{1.177}$$

Appendix 5: Gravity Green's Function and Kummer Transform

We noted another error in Agnew (2007) where he specified the Kummer transformation of the Green's function for surface gravity. With our notation,

$$\mathcal{G}_{\Delta g}(\theta) = \sum_{n=0}^{\infty} [2h'_n - (n+1)k'_n]\, P_n(\cos\theta) - \frac{z/a + 2\sin^2\theta/2}{[(z/a)^2 + 4(1+z/a)\sin^2\theta/2]^{3/2}} \tag{1.178}$$

Since

$$\lim_{n\to\infty} nk'_n = (Nk')_\infty \neq 0$$

the Kummer transformation of the sum term should read

$$\begin{aligned} &2h'_0 + \sum_{n=1}^{\infty}\left[2(h'_n - h'_\infty) - (n+1)\left(k'_n - \frac{(Nk')_\infty}{n}\right)\right] P_n(\cos\theta) \\ &\quad + \frac{h'_\infty}{\sin\theta/2} - (Nk')_\infty\left[\frac{1}{2\sin\theta/2} - 1 - \log\left(\sin\frac{\theta}{2} + \sin^2\frac{\theta}{2}\right)\right] \end{aligned} \tag{1.179}$$

References

Agnew D (2007) Earth tides. In: Herring T (eds) Treatise on geophysics: geodesy, Elsevier, Amsterdam, pp 163–195

Alterman Z, Jarosch H, Pekeris CL (1959) Oscillations of the Earth. Proc R Soc London, Ser A 252:80–95

Arfken G (1985) Mathematical methods for physicists. 3rd edn. Academic Press Inc, New York

Backus G (1986) Poloidal and toroidal fields in geomagnetic field modeling. Rev Geophys 24:75–109

Blewitt G (2003) Self-consistency in reference frames, geocenter definition, and surface loading of the solid earth. J Geophys Res 108: . doi:10.1029/2002JB002082

Bos MS, Baker TF, Røthing K, Plag HP (2002) Testing ocean tidemodels in the nordic seas with tidal gravity observations. Geophys J Int 150(3):687–694

Boyd JP (2000) Chebyshev and Fourier spectral methods. Dover Publications Inc, New York

Chinnery MA (1975) The static deformation of an Earth with a fluid core: a physical approach. Geophys J R Astron Soc 42:461–475

Dahlen FA (1974) On the static deformation of an Earth model with a fluid core. Geophys J Int 36:461–485. doi:10.1111/j.1365-246X.1974.tb03649.x

Dahlen FA, Fels SB (1978) A physical explanation of the static core paradox. Geophys J R Astr Soc 55:317–331

Dahlen FA, Tromp J (1998) Theoretical global seismology. Princeton University Press, Princeton

Dziewonski AM, Anderson DL (1981) Preliminary reference earth model. Phys Earth Planet Int 25:297–356

Farrell WE (1972) Deformation of the earth by surface loads. Rev Geophys Space Phys 10(3):761–797
Francis O, Mazzega P (1990) Global charts of ocean tide loading effects. J Geophys Res 95(C7):11,411–11,424
Gantmacher F (1950) The theory of matrices. vol 2, Chelsea Publishing Company, New York
Gilbert F, Backus GE (1966) Propagatormatrices in elasticwave and vibration problems. Geophysics 31:326–332
Guo JY, Ning JS, Zhang FP (2001) Chebyshev-collocation method applied to solve ODEs in geophysics singular at the Earth center. Geophys Res Lett 28:3027–3030. doi:10.1029/2001GL012886
Guo JY, Li YB, Huang Y, Deng HT, Xu SQ, Ning JS (2004) Green's function of the deformation of the Earth as a result of atmospheric loading. Geophys J Int 159:53–68. doi:10.1111/j.1365-246X.2004.02410.x
Haskell NA (1953) The dispersion of surface waves in multilayered media. Bull Seism Soc Am 43:17–24
Hoskins LM (1910) The strain of a non-gravitating sphere of variable density. Trans Am Math Soc 11:494–504
Hoskins LM (1920) The strain of a gravitating sphere of variable density and elasticity. Trans Am Math Soc 21:1–43
Jekeli C (2007) Potential theory and static gravity field of the Earth. In: Herring T (ed) Treatise on Geophysics, vol 11. pp 11–42
Jentzsch G (1997) Earth tides and ocean tidal loading. In: Wilhelm H, Zurm W, Gwenzel H (ed) Tidal phenomena. pp 145
Kampfmann W, Berckhemer H (1985) High temperature experiments on the elastic and anelastic behavior of magmatic rocks. Phys Earth Planet Int 40:223–247
Kaula WM (1963) Elastic models of the mantle corresponding to variations in the external gravity field. J Geophys Res 68(17):4967–4978
Knopoff L (1964) Q Rev Geophys 2:625–660
Lamb H (1895) Hydrodynamics. Cambridge University Press, Cambridge
Lambeck Cazenave A K, Balmino G (1974) Solid Earth and ocean tides estimated from satellite orbit analyses. Rev Geophys Space Phys 12:421–434
Longman IM (1962) A Green's function for determining the deformation of the Earth under surface mass loads, 1. theory. J Geophys Res 67(2):845–850
Longman IM (1963) A Green's function for determining the deformation of the Earth under surfacemass loads, 2. computations and numerical results. J Geophys Res 68(2):485–496
Love AEH (1911) Some problems of geodynamics. Dover Publications Inc, New York, 1967
Malvern LE (1969) Introduction to the mechanics of a continuous medium. Prentice Hall Inc, Englewood Cliffs
Martinec Z (1989) Thomson–haskell matrix method for free spheroidal elastic oscillations. Geophys J Int 98:195–199. doi:10.1111/j.1365-246X.1989.tb05524.x
Merriam JB (1985) Toroidal love numbers and transverse stress at the Earth's surface. J Geophys Res 90:7795–7802. doi:10.1029/JB090iB09p07795
Merriam JB (1986) Transverse stress Green's functions. J Geophys Res 91:13903–13913. doi:10.1029/JB091iB14p13903
Müller G (1983) Generalised maxwell bodies and estimates of mantle viscosity. Royal Astron Soc Geophys J 87:1113–1141
Munk WH, MacDonald GJF (1960) The rotation of the Earth; a geophysical discussion. Cambridge [Eng.] University Press, Cambridge
Nowick A, Berry B (1972) relaxation in crystalline solids. Academic, New York
Okubo S (1988) Asymptotic solutions to the static deformation of the Earth-I. Spheroidal mode. Geophys J Int 92:39–51. doi:10.1111/j.1365-246X.1988.tb01119.x
Pekeris CL, Accad Y (1972) Dynamics of the liquid core of the Earth. R Soc London Philos Trans Ser A 273:237–260. doi:10.1098/rsta.1972.0093

Pekeris CL, Jarosch H (1958) The free oscillations of the Earth. In: Benioff H, Ewing M, Howell BF, Press F (eds) Contributions in geophysics in honor of Beno Gutenberg, Pergamon, New York, pp 171–192

Phinney RA, Burridge R (1973) Representation of the elastic—gravitational excitation of a spherical Earth model by generalised spherical harmonics. Geophys J R Astron Soc 34:451–487

Press WH, Flannery BP, Teukolsky SA, Vetterling WT (1988) Numerical recipes in C. Cambridge University Press, Cambridge

Rothwell EJ (2008) Computation of the logarithm of Bessel functions of complex argument and fractional order. Commun Numer Methods Eng 24:237–249. doi:10.1002/cnm.972

Ruotsalainen H (2001) Modernizing the finnish long water-tube tiltmeter. J Geod Soc Jpn 47:28–33

Scherneck H-G (1991) A parametrized solid Earth tide model and ocean tide loading effects for global geodetic baseline measurements. Geophys J Int 106:677–694

Slichter LB, Caputo M (1960) Deformation of an Earth model by surface pressures. J Geophys Res 65:4151. doi:10.1029/JZ065i012p04151

Sun W, Sjöerg LE (1999) Gravitational potential changes of a spherically symmetric Earth model caused by a surface load. Geophys J Int 137:449–468

Takeuchi H, Saito M (1972) Seismic surface waves. In: Bolt BA (eds) Seismology: surface waves and free oscillations, methods computer physics, Academic, San Diego, pp 217–295

Thomson W, Tait PG (1867) A treatise on natural philosophy. Oxford Press, Oxford

Vermeersen LLA, Sabadini R, Spada G (1996) Compressible rotational deformation. Geophys J Int 126:735–761. doi:10.1111/j.1365-246X.1996.tb04700.x

Wahr J, Sasao T (1981) A diurnal resonance in the ocean tide and in the Earth's load response due to the resonant 'free core nutation'. Geophys J R Astr Soc 64:747–765

Wu P, Peltier WR (1982) Viscous gravitational relaxation. Geophys J R Astr Soc 70:435–485

Wunsch C (1974) Simple models of the deformation of an Earth with a fluid core-I. Geophys J Int 39:413–419. doi:10.1111/j.1365-246X.1974.tb05464.x

Zschau J (1983) Rheology of the Earth's mantle at tidal and Chandler wobble periods. In: Kuo J (ed) In: Proceedings of the 9th international symposium on Earth tides, E. Schweizerbart'sche Verlagsbuchhandlung (Nägele und Obermiller), Stuttgart, pp 605–629

Chapter 2
General Relativity and Space Geodesy

Ludwig Combrinck

1 Background

Newton's final version (published in 1726) of *Philosophiae Naturalis Principia Mathematica* was a great scientific achievement of the time and contained sufficient information to allow calculation of the dynamics of terrestrial and celestial bodies; it also expounded on the absolute nature of time and space. As examples, Newton's statements (Newton 1726) that "Absolute, true, and mathematical time, of itself, and from its own nature flows equably without regard to anything external" and "Absolute space, in its own nature, without regard to anything external, remains always similar and immovable" were fundamental to Newtonian calculations. These Newtonian concepts of space and time were challenged and proven to be only approximate by Einstein through his 1905 paper on special relativity, as well as his discovery of general relativity in 1915.

According to Einstein (1920), the geometrical properties of space–time are dependent on the distribution of matter in space–time, so that if the accuracy of our calculations increases, small departures from the theory of Newton become apparent, though they may escape the test of our observations as these deviations are very small. In more modern terms, the geometrical properties of space–time (in particular the space–time metric g) are dependent and determined to some extent, through the field equation, by the distribution of mass-energy and mass-energy currents in space. It is now also true to say that, in many astrophysical systems, the departures from Newtonian predictions are not small, although these effects resort under the strong-field regime of general relativity theory (GRT), whereas space geodesy operates within the weak-field regime. The small

L. Combrinck (✉)
Space Geodesy Programme, Hartebeesthoek Radio Astronomy Observatory,
PO Box 443 Krugersdorp 1740, South Africa
e-mail: ludwig@hartrao.ac.za

G. Xu (ed.), *Sciences of Geodesy - II*, DOI: 10.1007/978-3-642-28000-9_2,

deviations from Newtonian theory are also fundamental to experiments designed to evaluate general relativity in this weak-field regime.

Currently, the four main space geodetic techniques, Very Long Baseline Interferometry (VLBI), Global Navigation Satellite Systems (GNSS), Satellite or Lunar Laser Ranging (SLR/LLR) and Doppler Orbitography and Radiopositioning Integrated by Satellite (DORIS), have improved in accuracy to such an extent that for their optimal use one cannot disregard the effects of GRT. Analyses of the resulting data have to be done within the framework of a post-Newtonian formalism (Klioner 2003). In order to analyse the data correctly, the complete context within which the modelling is performed, i.e. reference and time frames, solar body ephemerides, signal propagation and observables (such as laser pulse travel time and satellite clock frequency) must consider GRT (Müller et al. 2008). These corrections are routinely done in advanced geodetic analysis software.

1.1 Introduction

In this chapter there is no space to describe the various space geodetic techniques in detail; however, the corrections and implications of GRT will be discussed in enough detail so as to allow practical application in software development, with adequate reference material for additional reading. Basically this chapter follows the recommendations of the international earth rotation service (IERS); see IERS Conventions (2003) (McCarthy and Petit 2003), IERS Conventions 2010 (Petit and Luzum 2010).

Solutions of GRT pertaining to space geodesy are weak-field, slow motion approximations. These approximations are valid as the gravitational field the solutions refer to has a potential U of small magnitude and the velocities v involved for any of the satellites are much less than the speed of light c. Therefore one has $U/c^2 \ll 1$ and $v/c \ll 1$ so that Einstein's field equations may be linearised and expressed in a form similar to Maxwellian equations of electromagnetism.

Space geodetic techniques depend to a large extent on the accuracy and stability of clocks; without these clocks the high accuracy measurements obtained currently would not be achieved. The timing aspect is therefore important and the high (atomic) clock accuracies have allowed an increase in measurement accuracy to such an extent that GRT must be applied to exploit the full potential of these techniques. A hydrogen maser clock is a requirement for VLBI (Wei-qun et al. 2001) with typical clock frequency accuracy of $\pm 5 \times 10^{-13}$ and stability of 9.7×10^{-15} in a 24 h period (geodetic VLBI experiments are normally of 24 h duration). Clock stability and accuracy are normally expressed in parts per so many counts; therefore a stability of 9.7×10^{-15} indicates that there are 9.7 excess counts for every 1×10^{15} counts or pulses in a 24 h period.

Apart from certain limitations (set by the atmosphere, spectral purity and stability of electronic equipment such as local oscillators and frequency multipliers), the coherence time of a VLBI system is determined by the frequency standard, which normally acts as the fundamental frequency source for the other equipment.

The impact of clock phase stability can be seen in (2.1) which expresses VLBI signal to noise ratio:

$$\mathrm{SNR} = \frac{SA\sqrt{2B\tau_c}}{kT_s}. \tag{2.1}$$

In (2.1) the source density flux is denoted by S, the geometric mean of the telescope collecting areas is A, B is the bandwidth, τ_c is the coherent integration time, k is Boltzmann's constant and T_s is the geometric mean of the system temperatures of the radio telescopes (Moran 1989).

The coherent integration time τ_c is approximately

$$\omega \times \tau_c \times \sigma_y(T) = 1. \tag{2.2}$$

In (2.2) ω is the local oscillator frequency in radians per second and $\sigma_y(T)$ is the two-sample Allan variance. To provide an example, in order to achieve signal coherence for an observation period of 1,000 s, where the local oscillator frequency has been set to 8.0 GHz, the two frequency standards at each end of the interferometer need to maintain relative stability of $\sim 2 \times 10^{-14}$. As the observables in the VLBI technique are in principle recorded signals measured in the proper time of the station clocks, the influence of clock stability extends throughout the VLBI hardware.

In the IERS GRT model for VLBI time delay as amended 1 June 2004, (McCarthy and Petit 2003), the final result is kept accurate to picosecond level by including all terms of order 10^{-13} s or larger. The component of error in the total delay due to error in gravitational delay ($\sim$2 mm) is therefore a small fraction of the total delay error budget, especially in comparison to the total delay model's errors ascribed to the troposphere ($\sim$20 mm), radio source structure and antenna structure ($\sim$10 mm each) (Sovers et al. 1998). Therefore VLBI implementations of GRT corrections are currently at an appropriate level and their contribution to reduce the total modelling error is clear. The drive towards millimetre precision within the framework of the global geodetic observing system (GGOS) (see Beutler et al. 2005a in terms of historical motivation for GGOS) will require GRT modelling at <1 mm accuracy if its error contribution to the total is to remain proportionally small; GGOS aims for millimetre accuracy, so other error sources will have to be reduced by a factor of at least ten.

A main objective of GGOS is to improve dramatically our understanding of the implications of surface changes and mass transport and how these processes affect the dynamics of our planet. One of the main challenges of GGOS (Drewes 2007) is the combination of geometric and gravimetric methods in a common procedure, which will have to include consistent approaches, constants, conventions, models and parameters. Considering the complexity of the different techniques, processing software and independent research strategies, such a combination will have to be done without constraining independent and original model contributions, as this could lead to scientific stagnation. As time observations are the basic observable in space geodesy, observations should be done using the same time system [e.g. Geocentric Coordinate Time (TCG—temps-coordonnée géocentrique)], and fundamental constants should be referred to the same time system. By implication a consistent GRT approach needs to be implemented in all space geodetic techniques to comply with and support the GGOS initiative.

1.2 Basic Implications of GRT for Space Geodesy Techniques

The basic effects of GRT on space geodetic measurements are related to how GRT affects the different observables and the dynamics of satellite orbits. Table 2.1 contains the main techniques and relativistic implications on their observables. Geometrically speaking, the Earth, Sun, Moon, planets and in fact all mass in the universe cause a curvature of space–time at some level in the immediate vicinity of the mass (in fact, in all the space–time). For instance, the curvature of space–time increases the up and down leg travel time of a laser pulse emitted by an SLR station to a satellite.

The effect on the main observable, time-of-flight (ToF), is an increase of several millimetres in the measured range (derived from the ToF) in the case of a LAGEOS satellite. In the case of near-Earth satellites, the effect of GRT can be modelled in the barycentric celestial reference system (BCRS) or geocentric celestial reference system (GCRS). Here we mainly consider the GCRS. The main relativistic effect on satellites in near-Earth orbit is due to the Schwarzschild field. Using LAGEOS as an example, the largest dynamical effect is the well known perigee advance, which for LAGEOS is $\sim$9 mas/d (milli-arc-second per day).

Another smaller effect is the Lense–Thirring (frame dragging) effect (Lense and Thirring 1918), which causes a precession in the longitude of the ascending node (longitude of Ω) of $\sim$31 mas/year (equal to about 1.8 m/year along-track displacement) and a change in the mean motion of a satellite. This precession is always in the direction of the rotation of the Earth. Frame dragging also advances the argument of perigee by $\sim$31.6 mas/year. De Sitter (or geodesic) precession (de Sitter 1916) is due to the motion of the Earth through the gravitational field of the Sun and leads to a precession of the orbital plane of the satellite.

2 Satellite Laser Ranging

Satellite laser ranging (SLR) is introduced in Sciences of Geodesy I (Chap. 9). Considering SLR there are several GRT effects which need to be taken into account; these will be described in detail and examples of how they influence orbital determination will be given, using the LAGEOS satellites as ranging targets.

Table 2.1 GRT implications on space geodetic techniques

GRT effect	VLBI	GPS	SLR
Schwarzschild field		$3\frac{(GM)^2}{c^2a^3} = 2.8 \times 10^{-10}\ \mathrm{ms}^{-2}$ $\Delta a = -4.4$ mm	(LAGEOS) $\dot{\omega}\,9$ mas/d $3\frac{(GM)^2}{c^2a^3} = 2.9 \times 10^{-9}\ \mathrm{ms}^{-2}$ $\Delta a = -4.4$ mm
Lense–Thirring (frame dragging)		$2\frac{GM}{c^2a^2}nJ = 1.8 \times 10^{-12}\ \mathrm{m.s}^{-2}$ $\Delta a/\cos i = -28\ \mu\mathrm{m}$ $\dot{\Omega} = 8.8\ \mu\mathrm{as/d}$	$\dot{\Omega} \sim 31$ mas/year $\dot{\omega} \sim 31.6$ mas/year $2\frac{GM}{c^2a^2}nJ = 2.7 \times 10^{-11}\ \mathrm{ms}^{-2}$ $\Delta a/\cos i = -42\ \mu\mathrm{m}$
de Sitter precession		$3\frac{GM_S}{c^2R}n_{San} = 2.28 \times 10^{-11}\ \mathrm{m.s}^{-2}$ $\dot{\Omega} = 53\ \mu\mathrm{as/d}$	$3\frac{GM_S}{c^2R}n_{San} = 3.37 \times 10^{-11}\ \mathrm{m.s}^{-2}$ $\dot{\Omega} \sim 17.6$ mas/year
Time-dilation, function of velocity		Onboard clock runs slow relative to a clock on the geoid	
Amplitude of effect on clock at GPS satellite velocity of 3.874 km/s		−7.1 μs/day	
Gravitational red-shift (blue-shift), due to gravitational potential, function of altitude		Onboard clock runs fast relative to a clock on the geoid	
Amplitude of blue-shift at GPS orbital altitude of ∼20 200 km		45.7 μs/day	
Combination of blue-shift and time-dilation		Satellite clock frequency must be reduced by 0.00455 Hz. GPS clock runs faster by 38.6 μs/days	

(continued)

Table 2.1 (continued)

GRT effect	VLBI	GPS	SLR
Additional contribution due to Earth's oblateness, amplitude of periodic effect		38 ps Peak to peak ~76 ps	
Additional contribution due to tidal potentials of Sun and Moon		Amplitude of periodic effect 1 ps	
Sagnac effect, depends on path and direction travelled (rotating frame of reference)		Onboard clock runs fast or slow relative to a clock on the geoid Maximum effect 136 ns	
Maximum effect for a stationary GPS receiver located on the geoid		133 ns	
Residual periodic effect at GPS orbital eccentricity of 0.02		Amplitude of residual periodic effect ~46 ns (Keplerian) peak to peak ~19 mm non-Keplerian <1.2 ns	
Shapiro delay	From 17×10^4 ps for a light ray grazing the Sun's limb to ~17 ps when the direction to the source is opposite to the Sun. Delay due to Earth's gravity field ~21 ps for a baseline of 6000 km	~19 mm	~7 mm
Gravitational light deflection	1.75 arcsec at limb of Sun (8.5 μrad)		

2.1 Shapiro Delay

Data obtained by the SLR system are converted to a normal point (NP) and by utilising the speed of light and incorporating some additional corrections; the Normal Point Range (NPR) can be calculated from the following:

$$\mathrm{NPR}_i = \frac{\left(\frac{\mathrm{NPtof}_i}{1\times 10^{12}} \times c\right)}{2} - \Delta a_i + \Delta CoM_i - \Delta R_{b_i} - \Delta GR_i - \Delta \varepsilon_i. \tag{2.3}$$

In (2.3) the main observable NPtof_i is the normal point time-of-flight (in ps) at a certain epoch and c is (Kaplan 2005) the speed of light (299,792,458.0 m/s). The range given by the first term on the right-hand side of (2.3) needs to be corrected by taking into account the effects of the atmosphere Δa_i; subsequent terms are a satellite dependent centre-of-mass correction ΔCoM_i, SLR station range-bias ΔR_{b_i}, a relativistic correction (Shapiro 1964) (referred to as the Shapiro delay) ΔGR_i and other $(\Delta\varepsilon_i)$ errors. If one uses LAGEOS as an example, the Shapiro delay correction is about 7 mm. According to McCarthy and Petit (2003),

$$t_2 - t_1 = \frac{\left|\vec{x}_2(t_2) - \vec{x}_1(t_1)\right|}{c} + \sum_J \frac{(1+\gamma)GM_J}{c^3} \ln\left(\frac{r_{J1} + r_{J2} + \rho}{r_{J1} + r_{J2} - \rho}\right). \tag{2.4}$$

In (2.4), γ is the parameterised post-newtonian (PPN) (Eddington 1923; Robertson 1962) parameter which should equal unity if GRT is valid, $t_2 - t_1$ denotes the total time delay considering a laser pulse emitted from coordinate x_1 (SLR station) at time t_1 and the return pulse is received at coordinate x_2 (SLR station) at time t_2. The PPN formalism (Will and Nordtvedt 1972) is a framework designed to classify various theories of gravity according to five attributes, which include curvature of space–time and nonlinearity of gravity. This formalism is valuable in tests and evaluations of GRT. In (2.4) the range defined by $\rho = \left|\vec{x_2} - \vec{x_1}\right|$ is the uncorrected (for GRT) range; in addition $r_{J1} = \left|\vec{x_1} - \vec{x_J}\right|$ and $r_{J2} = \left|\vec{x_2} - \vec{x_J}\right|$. This formulation was first derived by Holdridge (1967), which was a more elegant solution than previous solutions which involved angles. The last term in (2.4) describes the (Shapiro) correction for space curvature. Similar to the calculation of the numerator of the first term in (2.4), determination of the relativistically uncorrected range ρ is not simply the subtraction of two vectors, but involves an iterative solution of two light-time equations for the uplink and downlink path. This procedure is described in Montenbruck and Gill (2001) and Combrinck (2010). For the upleg (SLR station to satellite) a fixed-point iteration with

$$\tau_u^{(i+1)} = 1/c \cdot \left|\vec{r}(t - \tau_d) - \vec{R}\left(t - \tau_d - \tau_u^{(i)}\right)\right| \tag{2.5}$$

is executed in a loop until τ_u achieves an accuracy threshold that has been defined in the software algorithm. Four to five iterations are normally adequate. All calculations are done in an inertial (geocentric) reference frame (for example J2000).

Considering the downleg, the initial condition is $\tau_0 = 0$, then consecutive solutions are done using the fixed-point iteration

$$\tau_d^{(i+1)} = 1/c \cdot \left| \vec{r}(t - \tau^i) - \vec{R}(t) \right|. \tag{2.6}$$

The average of upleg range ρ_u and downleg range ρ_d, can then be used to find the range $\rho = \left| \vec{x_2} - \vec{x_1} \right|$ in (2.4) where

$$\rho = 0.5(\rho_u + \rho_d). \tag{2.7}$$

In the formulation given by (2.4), the sum is carried over all bodies J with mass M_J centred at x_J (McCarthy and Petit 2003). In practice, only the Earth needs to be considered as J for near-Earth satellites (including LAGEOS), as analysis is done in the geocentric frame of reference (Ries et al. 1988; Huang et al. 1990).

2.2 GRT Accelerations

A satellite experiences a variety of accelerations when in orbit and accelerations due to GRT can be separated from those that are purely Newtonian (at least in the weak-field and slow motion regime). One can therefore write the perturbing acceleration as

$$\ddot{\vec{r}} = -\frac{GM_{\oplus}}{r^3}\vec{r} + \vec{f}, \tag{2.8}$$

where (extending the notation of Tapley et al. 2004) the total perturbing force $\vec{f}$ is made up of a number of additional forces which perturb the orbit in addition to the first term in (2.8), which can be written as

$$\vec{f} = \vec{f}_{\text{NS}} + \vec{f}_{\text{TC}} + \vec{f}_{3B} + \vec{f}_g + \vec{f}_{\text{Drag}} + \vec{f}_{\text{SRP}} + \vec{f}_{\text{ERP}} + \vec{f}_{\text{other}} + \vec{f}_{\text{Emp}} \tag{2.9}$$

In (2.9) $\vec{f}_{\text{NS}}$ results from the uneven Earth mass-distribution, while the temporal variations of the static gravity field are represented by $\vec{f}_{\text{TC}}$. Perturbations caused by the gravitational forces from the Sun, Moon and planets are denoted by $\vec{f}_{3B}$, GRT is described by $\vec{f}_g$, atmospheric drag is $\vec{f}_{\text{Drag}}$, $\vec{f}_{\text{SRP}}$ is due to solar radiation, $\vec{f}_{\text{ERP}}$ is the Earth radiation pressure and $\vec{f}_{\text{Other}}$ contains other (very small) forces such as thermal, satellite rotation dependent effects. Once per-cycle-per-revolution empirical corrections, usually expressed in a local frame and divided into radial, tangential and normal (RTN) components are given by $\vec{f}_{\text{Emp}}$. A brief discussion of these perturbing forces is made in Book I (Chap. 9) of this series. Additional discussions can be found in the literature (cf. Hoffman-Wellenhof and Moritz 2005).

The IERS 2003 (McCarthy and Petit 2003) recommendations discuss the relativistic correction to the acceleration of a satellite in Earth orbit where

$$
\begin{aligned}
\Delta\ddot{\vec{r}} = \frac{GM_\oplus}{c^2r^3}\left\{\left[2(\beta+\gamma)\frac{GM_\oplus}{r} - \gamma\left(\dot{\vec{r}}\cdot\dot{\vec{r}}\right)\right]\vec{r} + 2(1+\gamma)\left(\vec{r}\cdot\dot{\vec{r}}\right)\dot{\vec{r}}\right\} \\
+ (1+\gamma)\frac{GM_\oplus}{c^2r^3}\left[\frac{3}{r^2}\left(\vec{r}\times\dot{\vec{r}}\right)\left(\vec{r}\cdot\vec{J}\right) + \left(\dot{\vec{r}}\times\vec{J}\right)\right] \\
+ \left\{(1+2\gamma)\left[\dot{\vec{R}}\times\left(\frac{-GM_s\vec{R}}{c^2R^3}\right)\right]\times\dot{\vec{r}}\right\}
\end{aligned}
\tag{2.10}
$$

is the correction which includes as:

- First term, the nonlinear Schwarzschild field of the Earth ($\approx 10^{-9}\,\mathrm{m\,s^{-2}}$)
- Second term, Lense–Thirring precession (frame dragging) ($\approx 10^{-11}\,\mathrm{m\,s^{-2}}$)
- Third term, de Sitter (geodesic) precession ($\approx 10^{-11}\ \mathrm{m\,s^{-2}}$)

Equation (2.10) is due to the formalism of Damour et al. (1994).

In (2.10), the speed of light is denoted by c, and PPN parameters β, γ equal unity if general relativity is valid. The parameter β (Eddington 1923; Robertson 1962) questions how much nonlinearity there is in the superposition law for gravity (refer to Will and Nordtvedt 1972 for a modern description of PPN parameters). The position of the satellite relative to the Earth is $\vec{r}$ and $\vec{R}$ is the position of the Earth relative to the Sun. Earth's angular momentum per unit mass is described (Petit and Luzum 2010) by $\left|\vec{J}\right| \cong 9.8\times10^8\ \mathrm{m^2s^{-1}}$, $\mathrm{GM}_\oplus$ is the gravitational coefficient of Earth (also μ in this chapter or GM) and GM_s the gravitational coefficient of the Sun.

Although these accelerations are small, they must be included for precise orbit determination (POD) purposes as there are some long term periodic and secular effects (Huang and Liu 1992) of the orbit. Precession of perigee results from the Schwarzschild effect, de Sitter precession can lead to long-period variations of some orbital elements (Ω, ω, M) and Lense–Thirring precession leads to secular rates (Ciufolini and Wheeler 1995) in the orbital elements Ω and ω. Refer to Table 2.1 for comparative rates between GNSS and LAGEOS.

Following Hugentobler (2008), it is quite informative to have a closer look at the separate terms of (2.10) in more detail, which will be done in the following sections for the three relativistic components. The Gaussian perturbations of a satellite orbit (Beutler et al. 2005b; Xu 2007) is given by (cf. Vallado 2001, for a discussion on limitations of the Gaussian form of the variation of parameters)

$$\dot{a} = \sqrt{\frac{p}{\mathrm{GM}}}\frac{2a}{1-e^2}\left[e\sin v\cdot R + \frac{p}{r}\cdot S\right], \tag{2.11}$$

$$\dot{e} = \sqrt{\frac{p}{\mathrm{GM}}}[\sin v\cdot R + (\cos v + \cos E)\cdot S], \tag{2.12}$$

$$\dot{i} = \frac{r\cos u}{na^2\sqrt{1-e^2}} \cdot W, \tag{2.13}$$

$$\dot{\Omega} = \frac{r\sin u}{na^2\sqrt{1-e^2}\sin i} \cdot W, \tag{2.14}$$

$$\dot{\omega} = \frac{1}{e}\sqrt{\frac{p}{\mathrm{GM}}}\left[-\cos v \cdot R + \left(1+\frac{r}{p}\right)\sin v \cdot S\right] - \dot{\Omega}\cos i, \tag{2.15}$$

$$\dot{M}_0 = \frac{1-e^2}{nae}\left[\left(\cos v - 2e\frac{r}{p}\right) \cdot R - \left(1+\frac{r}{p}\right)\sin v \cdot S\right] - \frac{3n}{2a}(t-t_0)\dot{a}. \tag{2.16}$$

Here a is the semi-major axis, v is the true anomaly, u is the argument of latitude and the argument of perigee is denoted by ω. The distance from the primary focus to the orbit (semiparameter, also known as "semi-latus rectum") is given by $p = a(1-e^2)$ and the average angular velocity is described by

$$n = \frac{2\pi}{T} = a^{-3/2}\mu^{1/2}, \tag{2.17}$$

where $\mu = \mathrm{GM}$ and the period of the satellite motion T can be written (Xu 2007) as

$$T = \frac{\pi ab}{\frac{1}{2}h} = \frac{2\pi ab}{\sqrt{\mu a(1-e^2)}} = 2\pi a^{0.7ex3/2}\mu^{-0.7ex1/2}. \tag{2.18}$$

Considering secular perturbations and circular orbits ($e = 0$), the radial (R), along-track (S) and cross-track (W) perturbing accelerations are directed parallel to the vectors $\vec{r}, \dot{\vec{r}}$ and $\vec{r} \times \dot{\vec{r}}$ respectively. For non-circular orbits the along-track axis (S) is not parallel (except at perigee and apogee) to the velocity vector. During processing, the position and velocity vectors of a satellite (including gravity gradients etc.) are normally transformed into an inertial reference frame such as ICRF/EME2000. The position, velocity and angular momentum vectors can then be written in a frame with the z axis orthogonal to the orbital plane as

$$\begin{aligned} \vec{r} &= a(\cos v, \sin v, 0) \\ \vec{v} &= an(-\sin v, \cos v, 0) \\ \vec{r} \times \vec{v} &= a^2 n(0,0,1). \end{aligned} \tag{2.19}$$

The unit vectors (direction vectors) of the perturbing accelerations can be written as

$$\begin{aligned} \hat{e}_R &= (\cos v, \sin v, 0) \\ \hat{e}_S &= (-\sin v, \cos v, 0) \\ \hat{e}_W &= (0,0,1). \end{aligned} \tag{2.20}$$

2.2.1 Nonlinear Schwarzschild Field Contribution to Acceleration

The first term of (2.10) is the contribution of the Schwarzschild metric (static part of the field generated by the central mass) to the GRT acceleration. Schwarzschild (1873–1916) was a German astrophysicist who found the first rigorous solution of Einstein's field equations. This solution is suitable (to a good approximation) for application to satellites in orbit around Earth. His original paper (Schwarzschild 1916) utilises as a background theme the excess beyond pure Newtonian motion of the perihelion of Mercury. While located at the Russian front, Karl Schwarzschild solved this problem by taking into account Einstein's requirements for a solution to the motion of the perihelion of Mercury. Consequently he found the solution to "the line element that forms the exact solution of Einstein's problem" and continued to derive the motion of a point in the gravitational field, i.e. the geodesic line corresponding to the line element.

When a topological space is described locally by Euclidean geometry, it is a manifold. In the immediate vicinity of a point in the manifold, there is a neighbourhood of points which is nearly flat. This reminds one that all statements regarding the principle of equivalence are *local* in nature. Einstein's solution of the Mercury perihelion problem was only to first order, whereas Schwarzschild's solution was exact. Mercury's total observed perigee advance is 574 arcsec per century, mostly due to planetary gravitational perturbations. The excess is about 43 arcsec, in agreement with GRT; this was one of the first 'proofs' of the validity of GRT. In space geodesy, when considering satellite orbits in the immediate vicinity of Earth (including high Earth orbiters such as GLONASS and GPS), it is convenient to use the Schwarzschild geometry as Earth's gravity field model even though it excludes the rotational effect of Earth in the gravity field. The metric of a space is basically its distance measure. Considering the metric of the Schwarzschild geometry (static spherically symmetric geometry), it has the form

$$\mathrm{d}s^2 = -\left(1 - \frac{2GM}{rc^2}\right)c^2\mathrm{d}t^2 + \left(1 - \frac{2GM}{rc^2}\right)^{-1}\mathrm{d}r^2 + r^2\mathrm{d}\theta^2 + r^2\sin^2\theta\mathrm{d}\phi^2. \quad (2.21)$$

This form (Schwarzschild coordinates) is only simple when unperturbed and unmodified; furthermore it is not in isotropic coordinates (Misner et al. 1973). In (2.21) the universal gravitational constant is denoted by G, and M is a parameter with dimensions of mass, which for our purposes is the mass of Earth. The quantity s is the proper time in seconds, r is the radial coordinate in metres, t is the time coordinate in seconds, θ is the colatitude and ϕ is the longitude (both in radians). If M was set to zero, the result would be equal to the Minkowskian (gravity-free space–time) four-dimensional metric expressed in spherical polar spatial coordinates. Minkowski space is recovered when $M/r \to \infty$.

The Schwarzschild line element can be modified to isotropic coordinates so that the relationship between r, θ, ϕ and x, y, z can be described in the usual manner, with $r = (x^2 + y^2 + z^2)$ becoming an isotropic coordinate [see (2.85) and (2.86)], which should not be confused with r in the Schwarzschild coordinates (see Misner et al. 1973).

Experimental confirmation during the last 3 centuries using optical instruments and radar observations since 1966 has provided credibility for GRT theory. Some polemic still surrounds the issue, as several other parameters could also affect the exact value of the advance, e.g. the quadrupole, J_2, which is not yet very accurately determined, contribution of the Sun as well as the (small) gravitomagnetic contribution resulting from the angular momentum of the Sun. A detailed account of the drama surrounding the initial proof and resistance to acceptance of GRT can be found in Crelinsten (2006) and more technical details can be obtained from Ciufolini and Wheeler (1995).

Following Hugentobler (2008), if we set $GM = n^2a^3$ the Schwarzschild acceleration for a circular orbit can be written as

$$\begin{aligned}\Delta\ddot{\vec{r}}_S &= \frac{GM}{c^2a^3}\left[4\frac{GM}{a} - v^2\right]\vec{r}\\ &= \frac{GM}{c^2a^3}\left[4\frac{GM}{a} - \frac{GM}{a}\right]\vec{r}\\ &= 3\frac{(GM)^2}{c^2a^4}\cdot\vec{r}.\end{aligned}\tag{2.22}$$

The radial component of the Schwarzschild acceleration is then

$$R = \vec{e}_R \cdot \Delta\ddot{\vec{r}}_S = 3\frac{(GM)^2}{c^2a^3},\tag{2.23}$$

whereas the alongtrack and crosstrack accelerations are both equal to zero:

$$S = \vec{e}_S \cdot \Delta\ddot{\vec{r}}_S = 0 \quad \text{and} \quad W = \vec{e}_W \cdot \Delta\ddot{\vec{r}}_S = 0.\tag{2.24}$$

Considering the sign convention of gravitational acceleration as per (2.8) it is clear that the radial component of the Schwarzschild field imparts an outward acceleration, which reduces the Newtonian component, effectively changing GM so that

$$\begin{aligned}\vec{e}_R \cdot \ddot{\vec{r}}_{\text{tot}} &= -\frac{GM}{a^2} + R\\ &= -\frac{GM}{a^2}\left(1 - \frac{Ra^2}{GM}\right)\\ &= -\frac{GM'}{a^2}.\end{aligned}\tag{2.25}$$

Using Kepler's third law written as (2.17) where $\mu = GM$ and keeping the period T of the satellite motion (2.18) fixed and setting $a' = a + \Delta a$, then

$$a'^3n^2 = GM'\tag{2.26}$$

and to first order

$$\Delta a = -\frac{1}{3}\frac{Ra^3}{GM}. \tag{2.27}$$

Substituting R from (2.23) into (2.27) one can then write (Hugentobler 2008)

$$\Delta a = -\frac{GM}{c^2} = -4.43 \text{ mm}. \tag{2.28}$$

The Schwarzschild acceleration on a satellite in orbit is thus a radial, in-plane effect, with zero magnitude effects in the alongtrack and crosstrack directions.

2.2.2 Lense–Thirring Precession

The second term of (2.10) is due to Lense–Thirring precession; it is clear that the Earth's angular momentum J plays a major role in the generation of the Lense–Thirring effect. In the case of LAGEOS, the effect is very small ($\sim$31 mas/year) and is therefore extremely difficult to measure (Ciufolini 1986). The secular rates of the ascending node and perigee of a body orbiting a central rotating mass is given by (Lense and Thirring 1918)

$$\dot{\Omega}_{\mathrm{LT}} = \frac{2GJ}{c^2a^3(1-e^2)^{3/2}} \tag{2.29}$$

and

$$\dot{\omega}_{\mathrm{LT}} = -\frac{6GJ\cos i}{c^2a^3(1-e^2)^{3/2}}, \tag{2.30}$$

where G is the gravitational constant, c denotes the speed of light, J is the proper angular momentum of Earth, and a, e and i are the semi-major axis, eccentricity and inclination of the orbit respectively. Following Hugentobler (2008) the angular momentum vector can be written in the reference frame as defined by (2.19) as

$$\vec{J} = J(0, \sin i, \cos i). \tag{2.31}$$

In the case of a circular orbit, therefore,

$$\vec{r}\cdot\vec{J} = aJ\sin i \sin v \text{ and } \vec{r}\times\vec{J} = anJ(\cos v\,\cos i, \sin v\,\cos i, -\sin v\,\sin i). \tag{2.32}$$

The radial, alongtrack and crosstrack components are then

$$R = \Delta\vec{r}_{LT}\cdot\vec{e}_R = 2\frac{GM}{c^2a^4}\vec{r}\cdot\left(\dot{\vec{r}}\times\vec{J}\right) = 2\frac{GM}{c^2a^4}\vec{J}\cdot\left(\vec{r}\times\dot{\vec{r}}\right) = 2\frac{GM}{c^2a^2}nJ\cos i, \tag{2.33}$$

$$S = \Delta\vec{r}_{LT}\cdot\vec{e}_S = 0, \tag{2.34}$$

$$
\begin{aligned}
W &= \Delta\vec{r}_{LT} \cdot \vec{e}_W \\
&= 2\frac{GM}{c^2a^3}\left[\frac{3}{a^2}a^2 naJ \sin i \sin v - anJ \sin i \sin v\right] \\
&= 4\frac{GM}{c^2a^2} nJ \sin i \sin v.
\end{aligned}
\tag{2.35}
$$

The semi-major axis undergoes a reduction of

$$
\Delta a = -\frac{2}{3}\frac{anJ}{c^2}\cos i \propto a^{-1/2}, \tag{2.36}
$$

due to the constant radial acceleration. If the orbit is perpendicular to the equatorial plane there is no radial acceleration or change in semi-major axis due to the $\cos i$ factor in (2.36). Changes in the semi-major axis are inversely proportional to the square root of the semi-major axis, so that higher orbit satellites experience a smaller effect.

Considering (2.14) and setting $u = v$ (true anomaly is undefined for circular orbits as they have no periapsis) in the case of a circular orbit,

$$
\begin{aligned}
\dot{\Omega}_{LT} &= \frac{\sin u}{na \sin i} \cdot W \\
&= 4\frac{GM}{c^2a^3 J \sin^2 u} \\
&= 2\frac{GM}{c^2a^3 J(1 - \cos 2u)}.
\end{aligned}
\tag{2.37}
$$

The precession of the node is not dependent on the inclination i of the orbit as it is a frame precession effect; it is inversely related to a^3 so that it decreases very rapidly. Iorio (2007) provides an alternative discussion on the (RTN) radial($\vec{r}$), transverse ($\vec{t}$) and out-of-plane ($\vec{n}$) projections of the perturbing acceleration, commencing with the Gaussian perturbation equations of a satellite orbit.

2.2.3 de Sitter (Geodesic) Precession

Geodesic precession is basically a result of the motion of Earth through the Sun's gravitational field. The consequences are that a satellite's orbital pole precesses about the normal to the ecliptic at a rate of 19.2 milliarcseconds (mas) per year and the ascending node of the satellite's orbit on the ecliptic increases in celestial longitude by 19.2 mas per year (Moyer 2000). In addition, geodesic precession decreases the general precession by 19.2 mas.

Both de Sitter precession and Lense–Thirring precession are manifestations of frame-dragging as clearly shown by Ashby and Shahid-Saless (1990). Within an appropriately chosen coordinate system and without incorporating spatial curvature, geodetic precession of a gyroscope orbiting a spherically symmetric, spinning mass can be remoulded as a Lense–Thirring frame-dragging effect. Geodesic precession and Lense–Thirring precession can therefore be described in terms of two components of a single gravitomagnetic effect. As applied to SLR, the Lense–Thirring component is due to the spin angular momentum of Earth whereas the deSitter precession can be ascribed to the orbital angular momentum of Earth as it revolves around the origin of the appropriately selected quasi-inertial coordinates. This interpretation is still a matter of debate however, (cf. Ciufolini and Wheeler 1995; Ciufolini 2007) where some interpretations regard the de Sitter effect and the Lense–Thirring drag as fundamentally different phenomena.

Following Hugentobler (2008), in curved space–time, a parallel transported (local inertial) reference frame undergoes precession with respect to distant stars. For a satellite the precession rate (19.2 mas/year in the vicinity of Earth) does not depend on the distance from the Earth. The equation describing the resulting acceleration [refer to (2.10)] is

$$\Delta\ddot{\vec{r}}_{dS} = -2\left[-\frac{3}{2}\frac{GM_S}{c^2R^3}\vec{R}\times\dot{\vec{R}}\right]\times\dot{\vec{r}}. \tag{2.38}$$

Equation (2.38) has the form of a Coriolis term. The Coriolis acceleration of a satellite in orbit around the Earth due to geodesic precession is

$$\ddot{\vec{r}} = 2\vec{\Omega}\times\dot{r}, \tag{2.39}$$

where the angular velocity vector due to geodesic precession (Moyer 2000) is

$$\vec{\Omega} = \frac{GM_S\left(\gamma+\frac{1}{2}\right)}{c^2R^3}\left(\vec{R}\times\dot{\vec{R}}\right). \tag{2.40}$$

The formulations at (2.39) and (2.40) [instead of (2.38)] are useful if the PPN parameter γ needs to be included in an estimation process and is equivalent to the third term of (2.10). In (2.38) M_S is the mass of the Sun and R is the distance to it, respectively. The precession vector $\vec{\omega}_{dS}$ of the global frame (the term in brackets) is relative to the precessing (local inertial) frame and it points to the southern ecliptic pole. If the frame defined in (2.19) is redefined by keeping the z-axis the same but rotating the x-axis towards the ascending node with respect to the ecliptic plane (Hugentobler 2008), the precession vector of the global frame can be written as

$$\begin{aligned}\vec{\omega}_{dS} &= -\tfrac{3}{2}\frac{GM_S}{c^2R^3}\vec{R}\times\dot{\vec{R}} \\ &= -\tfrac{3}{2}\frac{GM_S}{c^2R}\sqrt{1-e_S^2}n_S(0,\sin\beta,\cos\beta).\end{aligned} \tag{2.41}$$

In (2.41) the mean motion of Earth around the Sun is denoted by n_S and the inclination of the orbital plane to the ecliptic is represented by β. The inclination is in the range $(i-\varepsilon,\ i+\varepsilon)$ and can be calculated using

$$\cos\beta = \cos\varepsilon\,\cos i + \sin\varepsilon\,\sin i\,\cos\Omega, \tag{2.42}$$

with the obliquity of the ecliptic $\varepsilon = 23°26'$. In the defined frame, for a circular orbit the relation

$$\left(\vec{R} \times \dot{\vec{R}}\right) \times \vec{r} = R^2 n_S \sqrt{1 - e_S^2}\, an\, (-\cos\beta\,\cos v, -\cos\beta\,\sin v, \sin\beta\,\sin v)\,, \tag{2.43}$$

can be written and consequently (Hugentobler 2008) the perturbing accelerations are

$$R = \Delta\ddot{\vec{r}}_{dS} \cdot \vec{e}_R = -3\frac{GM_S}{c^2 R} n_S \sqrt{1 - e_S^2}\, an \cos\beta, \tag{2.44}$$

$$S = \Delta\ddot{\vec{r}}_{dS} \cdot \vec{e}_S = 0, \tag{2.45}$$

$$W = \Delta\ddot{\vec{r}}_{dS} \cdot \vec{e}_W = 3\frac{GM_S}{c^2 R} n_S \sqrt{1 - e_S^2}\, an \sin\beta \sin v. \tag{2.46}$$

Considering the $\cos\beta$ factor in the radial component, orbits which are perpendicular to the ecliptic plane experience no change in semi-major axis and no radial acceleration as the Coriolis acceleration is perpendicular to the orbital plane.

The negative radial acceleration increases the semi-major axis by

$$\Delta a = +\frac{GM_S}{c^2}\frac{a\, n_S}{R\, n}\sqrt{1 - e_S^2}\cos\beta \propto a^{5/2}. \tag{2.47}$$

Using (2.14) and setting $u = v$ for a circular orbit the precession of the ascending node with respect to the ecliptic plane can be calculated as

$$\begin{aligned}\dot{\Omega}_{dS} &= \frac{\sin u}{na \sin\beta} \cdot W \\ &= 3\frac{GM_S}{c^2 R} n_S \sqrt{1 - e_S^2} \sin^2 u \\ &= 3\frac{GM_S}{c^2 R} n_S \sqrt{1 - e_S^2} s(1 - \cos 2u).\end{aligned} \tag{2.48}$$

Figure 2.1 illustrates the acceleration values of the radial components of the three terms of (2.10), as described by (2.23, 2.33) and (2.44). Considering LAGEOS, the acceleration due to the radial component of the de Sitter effect is slightly larger than the component due to the frame-dragging effect. The Lense–Thirring effect dominates the de Sitter effect for semi-major axes that are less than 11,000 km. Table 2.1 summarizes some of the relativistic accelerations and advances of perigee and the node due to the three terms of (2.10).

According to the IERS Conventions 2010 (Petit and Luzum 2010), the magnitude and observable effects of these relativistic components depend to some extent on the satellite orbital characteristics and the analysis setup strategy. For instance, the satellite height changes during its orbital path so that $a \simeq r$ for near circular orbits, so the slightly simplified examples here for circular orbits will provide slightly different answers compared to a rigorous implementation of (2.10),

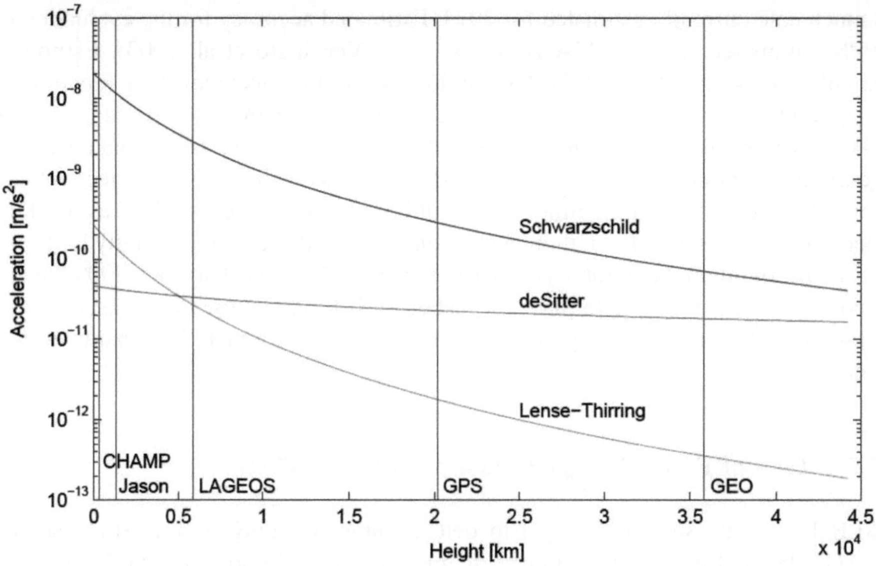

Fig. 2.1 Radial acceleration as a function of satellite height for circular orbits (Hugentobler 2008)

even though LAGEOS and GPS satellites have very nearly circular orbits. If (as suggested by Petit and Luzum 2010) orbital parameters are adjusted and the Schwarzschild term is not taken into account, there would be an apparent 4.4 mm decrease in orbital radius as described using the formulations of Hugentobler (2008). This stresses the necessity for consistent application of the formulation at (2.10), as different strategies (and different values of, for instance, the Earth's angular momentum per unit mass) will lead to different results.

2.3 SLR Tests of General Relativity Theory

In the immediate environment of our solar system, the linearised weak-field and slow-motion approximation is adequate and is the regime where space geodesy can be utilised to test GRT. As mentioned by Einstein (1920), these deviations beyond pure Newtonian dynamics *may escape the test of our observations* as these deviations are very small. This certainly is still the case to some extent, as experiments designed to test GRT are often very difficult and expensive. For example, the next higher accuracy tests are likely to come from GAIA (a space-based astrometric mission). The GAIA mission (Turon et al. 2005) has as its objective the creation of a three-dimensional map of our galaxy, which will improve our knowledge of its composition, formation and evolution. In addition, new tests of general relativity are included in its projected five-year mission, with a launch date currently estimated for 2013. Estimated accuracy for the evaluation of PPN parameter γ with GAIA is $\sim 2 \times 10^{-7}$ (Vecchiato et al. 2003). Estimated angular accuracy [this is a best scenario case as this accuracy is a function of magnitude and ecliptic latitude of the object (Lindegren 2009)] is $\leq 10\,\mu$as (Perryman et al. 2001), which is a factor of ten or more better than what is (currently) routinely achieved with VLBI. Included in its mission objectives are tests for light deflection, time delay, and Doppler frequency shift as well as perihelion precession. Even though missions such as these will constantly be added to the list of attempts to improve the accuracy of GRT tests (Turyshev 2009), space geodesy has its own role to play; for instance, SLR can be used to measure frame dragging, gravitational delay, and can also be used to estimate PPN parameters γ and β.

2.3.1 Tests of Frame Dragging (Lense–Thirring Effect)

SLR has been used in attempts to detect frame dragging. Using SLR and the LAGEOS satellite to detect frame dragging was initially proposed by Cugusi and Proverbio (1977), see also Cugusi and Proverbio (1978). The first reported results were by Ciufolini et al. (1996) who analysed the SLR range observations of satellites LAGEOS and LAGEOS II utilising the software package GEODYN II (Pavlis et al. 2007). They obtained the first direct measurement of the Lense–Thirring effect, or dragging of inertial frames, and the first direct experimental evidence for the gravitomagnetic field. The stated accuracy of their measurement was $\sim 30\%$. This work was very valuable in that it created opportunity for more investigation using improved gravity models and higher accuracy perturbation models; it also started a competitive and critically evaluated research avenue.

This first report was followed by Ciufolini et al. (1998), claiming an improved result of 20%; the parameter μ is introduced, which measures the strength of the Lense–Thirring effect (in GRT, $\mu \equiv 1$). Initially both nodes of LAGEOS I and II were used as well as the argument of perigee of LAGEOS II. Satellite LAGEOS I has a smaller eccentricity (~ 0.004) than LAGEOS II (~ 0.14), which makes detection of the advance in the argument of perigee of LAGEOS I more difficult, and in addition the frame dragging effect is nearly twice the value for LAGEOS II than for LAGEOS I.

Subsequent estimates used only the nodes (Ciufolini and Pavlis 2004, Ciufolini et al. 2006) in a 'butterfly' configuration of the retrograde LAGEOS I ($i = 109.8°$) and the prograde LAGEOS II ($i = 52.6°$) orbits. A very dedicated effort to create a realistic error budget is contained within the Ciufolini et al. (2006) paper, where several gravity models are used. Best estimates were obtained by using the EIGEN-GRACE02S model, where $\mu = 0.99$, with a total error between 5 and 10% of the GRT predicted value of the Lense–Thirring effect. The improved results are directly related to the improved gravity models, as the frame dragging tests can be strongly affected by mismodelling of the even zonal harmonic coefficients. Other sources of error which have a smaller impact result from model imperfections of any factor which causes orbit perturbations including the solid Earth and pole tides as they modulate the static gravity field.

The results of the most recent SLR-based tests in determining frame dragging using the LAGEOS satellites have been questioned by Iorio (2010b). The contributions by Iorio are important in that a careful scrutiny of the techniques employed, evaluation of error sources, and independent tests are necessary to ensure that results are scientifically valid. These tests are not trivial as the small effects which are being investigated are easily obscured by classical phenomena which are explained in Newtonian terms. According to Iorio (2010b), the systematic error $\delta\mu$ in the Lense–Thirring measurements published to date should be increased by a factor of 3–4 times. Following the literature, a healthy debate has developed and continues regarding these tests and there are numerous publications concerning the validity of these tests (Iorio 2006). The effect of Iorio's work has been that much more attention has been given to evaluation of the magnitude and influence of errors on estimates of frame dragging.

2.3.2 Estimation of Perigee Shift in the Schwarzschild Gravitoelectric Field

In the previous section we had a glimpse of the activities of researchers determining the effects of gravitomagnetism. The general relativity shift of the perigee of LAGEOS II resulting from the Schwarzschild gravitoelectric field has been estimated by Lucchesi (2003) during a simulation of a measurement and error budget. Using the gravity model EGM96, Lucchesi demonstrated the potential to estimate LAGEOS II's general relativistic shift with 2% accuracy. Error budget estimation covered an observational period of approximately 7 years. Similar to the determination of frame dragging due to gravitomagnetism, the largest errors are due to the uncertainties in the even zonal harmonics of the Earth's gravity field and to a lesser extent the mismodelling of non-gravitational perturbations. The result obtained may be viewed as a 2% accuracy derivation of the PPN parameters γ and β. Details of error sources and their possible influence are given.

2.3.3 Estimation of Perigee Shift in the Schwarzschild Gravitoelectric and Gravitomagnetic Field

A very interesting approach was taken recently by Lucchesi and Peron (2010), placing moreover new constraints on non-Newtonian gravity. They analysed 13 years of SLR data of the LAGEOS satellites with the GEODYN II (Pavlis et al. 2007) software; the models for general relativity were not included in the orbit determination, thereby obtaining the relativistic signal in the residuals. Utilising LAGEOS II pericentre residuals they were able to obtain a 99.8% agreement with the predictions of Einstein's theory. Basically this approach is a measurement in the field of the Earth of the combination of the γ and β PPN parameters of general relativity. This work is different from other approaches as it measures all the relativistic secular effects simultaneously. It is unfortunately not able to separate the PPN parameters γ and β, as this result, essentially from the "Schwarzschild" signal, places a constraint only on the combination of γ and β. Other, separate constraints would be required to disentangle them. The signals from the Lense–Thirring and de Sitter components could provide information on γ, but their signal-to-noise ratios in the case of LAGEOS I are too small to be really useful in this respect (Peron, 3 Mar 2011, "personal communication"). This unique approach in the PPN framework can be considered as a 0.03% measurement of the combination of γ and β PPN parameters. The results of Lucchesi and Peron (2010) also constrain possible deviations from the gravitational inverse-square law in favour of new weak interactions parameterised by a Yukawa-like potential with strength α and range λ.

2.3.4 Direct Estimates of PPN Parameters

The estimation of PPN parameters γ and β can be done directly within the least-squares solution of precise orbital determination. A first attempt was undertaken by Combrinck (Combrinck 2008), providing an error of $\sim 5 \times 10^{-4}$ on γ. In these tentative initial results, the PPN parameter was evaluated as a solve-for parameter in an analysis of five months of LAGEOS II SLR data. A rejection filter was used to constrain the orbital integration and parameter estimation. However, it was noted that careful analyses of the effects of alternative strategies such as different gravity models and a priori constraints on other solve-for or consider parameters need to be done to evaluate this technique. The consider parameters are parameters which could be estimated, but by setting their a priori constraints very high (i.e. very low error values), they essentially obtain fixed values but are affected by their uncertainties. This evaluation includes the Schwarzschild terms and the effects of rotational frame-dragging (Lense–Thirring precession), de Sitter (geodesic) precession and Shapiro delay. The solved for PPN parameters are fed back into the least-squares process during the analysis.

In this approach, the radial component of the SLR measurements is the strength of the technique; the relativistic acceleration on LAGEOS is mainly a radial component. This preliminary study solved for γ in the least-squares sense utilising SLR data in a strategy where the O–C residuals indicate better observation/ modelling fits, through different levels of O–C residual rejection levels. This strategy assigns greater weight to SLR measurement accuracy than to the modelling parameters. Basically the filter consists of low-pass and high-pass criteria set to an O–C standard deviation based on a selected number of iterations during the least-squares fitting process. This effectively creates a bandpass filter, which rejects observations which fall outside the rejection criteria level.

Additional work done since included estimates of both γ and β using a longer time series of $\sim$4 years (Combrinck 2011). In this work, the suggestion by Iorio (2010a) that possible imprinting of GRT in the gravity field models could adversely affect tests of GRT is taken into account. Therefore certain gravity field spherical harmonic coefficients, $J_2 - J_5$, C_{21} and S_{21}, are estimated. This requires a step by step approach where currently 20 iterations in the least-squares solution are required, with certain parameters being estimated at certain iterations. Coefficients C_{21} and S_{21} together with J_2 are estimated to determine pole tide. In comparison to γ the estimate for β is weaker using this technique, as β is only evaluated in the Schwarzschild term of the GRT acceleration, whereas γ is present in all the terms of (2.10) and in the second term of (2.4). The values obtained in (Combrinck 2011) are somewhat inferior but more rigorous with respect to the initial (Combrinck 2008) tentative results. New results obtained were: values of $\gamma - 1$ are $6.5 \times 10^{-4} \pm 7.4 \times 10^{-4}$ and $9.0 \times 10^{-4} \pm 9.6 \times 10^{-4}$ for LAGEOS 1 and 2 respectively, and values of $\beta - 1$ are $1.2 \times 10^{-3} \pm 1.4 \times 10^{-3}$ and $1.4 \times 10^{-3} \pm 1.5 \times 10^{-3}$ for LAGEOS 1 and 2 respectively. New work is underway, which includes improvement in modelling of the range delay due to the atmosphere, by including an azimuth dependent range delay correction in the SLR analysis software as the atmosphere exhibits nonlinear behaviour (Botai et al. 2010).

2.3.5 Lunar Laser Ranging

Lunar laser ranging (LLR) entails laser ranging to arrays of corner cube reflectors placed on the Moon (see Merkowitz 2010 and references therein for an overview). This high accuracy laser ranging (using equipment similar to SLR, with some system modifications, e.g. using an event timer instead of an interval counter to measure the ToF of the laser pulse) translates to a very accurate orbit determination. The highly accurate orbit can be used for (amongst others) fundamental physics. Placement of the reflectors was done by the Apollo 11, 14 and 15 astronauts, while two French-built reflector arrays were added by the Soviet Luna 17 (the lander carrying robotic rover Lunakhod 1) and Luna 21 missions. Rover Lunakhod 1 was lost in 1971 but relocated using images obtained by the lunar reconnaissance orbiter (LRO). It was consequently ranged to with LLR by Tom Murphy and his team [Apache Point Observatory Lunar Laser-Ranging Operation (APOLLO)], see (Murphy et al. 2007), using the 3.5 m telescope at the Apache Point Observatory in New Mexico. LLR is a sensitive technique to test the equivalence principle (EP).

The EP is the backbone of GRT and involves the equality of gravitational and inertial mass (Newtonian EP). In essence, Einstein's EP (EEP) requires that in local freely falling frames, all physical laws must be independent of the velocity of the frame (i.e. local Lorentz invariance). Furthermore, that two different bodies (such as Earth and the Moon) in a gravitational field (such as that of the Sun) the bodies will experience the same acceleration [weak equivalence principle (WEP)], i.e. EEP requires that the WEP be valid. The strong equivalence principle (SEP) includes the gravitational self-energy of a body in the counting of its total energy content. LLR can also be used to evaluate β, the geodetic precession and $\dot{G}/G$; γ is conveniently used from other high level estimates [such as using the result (Bertotti et al. 2003) of the radiometric tracking data of the Cassini spacecraft on its approach to Saturn, which gave the best results to date of $\gamma - 1$ $(2.1 \pm 2.3 \times 10^{-5})$].

Similar to SLR, LLR has seen a constant improvement in results due to model and technology improvements. The Earth-Moon-Sun system provides the best laboratory for testing the SEP, with LLR being the only available solar system technique at this time, which may be augmented with interplanetary laser ranging (ILR) in the future. The SEP parameter η is related to the PPN parameters:

$$\eta = 4\beta - \gamma - 3, \tag{2.49}$$

where in GRT $\eta = 0$.

Variation of G in time will be reflected in anomalous evolution of the orbital period of the Moon. If G changes, it will affect the monthly lunar orbit as well as the annual Earth-Moon orbit around the Sun. This is quite evident (Merkowitz 2010) considering Kepler's third law

$$P^2 = \frac{4\pi^2 r^3}{Gm}, \tag{2.50}$$

and by taking the time derivative and re-arranging:

$$\frac{\dot{G}}{G} = 3\frac{\dot{r}}{r} - 2\frac{\dot{P}}{P} - \frac{\dot{m}}{m}. \tag{2.51}$$

After considering factors which make up the non-anomalous orbital evolution (solar perturbation, tidal friction, etc.) (see Williams et al. 1996 for more details) and utilising raging data to the Moon, the anomalous orbital evolution can be estimated and an estimate for $\dot{G}$ derived. In the case of a violation of the EP, a displacement of the lunar orbit along the Earth-Sun line will occur, which will be evidenced in a range signature having a 29.53 day synodic period (not the same as the lunar orbit period of 27 days) (Williams et al. 2009). Some of the LLR tests relating to the fundamental nature of gravity are summarized in Murphy (2009), with short descriptions of the phenomenologies related to SEP, time-rate-of-change of the gravitational constant, gravitomagnetism, inverse square law, and preferred frame effects.

Recent results (Williams et al. 2004) for PPN parameter β are based on $\eta = 4\beta - \gamma - 3$, and are very sensitive to β. Utilising the result of the Cassini spacecraft determination (Bertotti et al. 2003) of $\gamma - 1\left(2.1 \pm 2.3 \times 10^{-5}\right)$, in combination with $\eta = 4\beta - \gamma - 3 = (4.4 \pm 4.5) \times 10^{-4}$, results in $\beta - 1 = (1.2 \pm 1.1) \times 10^{-4}$. Their test of temporal variation of the gravitational constant delivered the value

$$\frac{\dot{G}}{G} = (4 \pm 9) \times 10^{-13} \mathrm{yr}^{-1}.$$

The LLR network is currently limited to three operational LLR stations, all in the Northern hemisphere. Two stations are located in the USA, the McDonald laser ranging station (MLRS) near Ft. Davis (0.75 m telescope) and the Texas APOLLO at the Apache Point Observatory in New Mexico (3.5 m telescope). One station (Centre d'Etudes et de Recherche en Géodynamique et Astronomie, CERGA) in France, located at the Observatoire de la Côte d'Azur (OCA) on the Plateau de Calern near Grasse (France), is equipped with a 1.54 m Cassegrain telescope. Even though the LLR network has received a boost through the addition of APOLLO, biases could exist in the ranging solutions as all ranging data are obtained from the Northern hemisphere. In collaboration with OCA and NASA, the Hartebeesthoek Radio Astronomy Observatory (HartRAO) has commenced with a project to develop an LLR system based on an ex-OCA 1 m aperture telescope. This LLR should be operational by 2015 and as it would be dedicated to LLR, should add significantly to the database and strengthen the geometry of the network.

2.3.6 Interplanetary Laser Ranging

Interplanetary laser ranging (ILR) will be able to make a contribution to the dynamics of the solar system and to evaluations of general relativity or alternative theories of gravity by, for instance, ranging to laser transponders placed on suitable planets such as Mars, or ranging to interplanetary probes equipped with laser transponders. In May 2005, timed observations of laser pulses between the Mercury Laser Altimeter (MLA) instrument, which is located onboard the MESSENGER spacecraft, and the Goddard Geophysical Astronomical Observatory (GGAO) (using a 1.2 m telescope) measured the two-way ToF (range) with sub-nanosecond precision (Smith et al. 2006). In addition, a one-way only optical experiment was executed a few months later between GGAO and the Mars Orbiter Laser Altimeter (MOLA) aboard the Mars Global Surveyor (MGS) spacecraft. The distance involved was 81 Gm (0.54 AU). These successful tests demonstrate the possibility of interplanetary communication and precise ranging using modest power (Neumann et al. 2006).

Several possible scenarios utilising ILR have been proposed, including placing an active laser transponder on the Martian moon Phobos with the possibility of millimetre-level ranging resolution (Turyshev et al. 2010). The primary objective of this proposed mission is to measure PPN parameter γ to a level of 2×10^{-7} which would improve today's best result (as determined by radiometric tracking data from the Cassini mission; see Bertotti et al. 2003) by two orders of magnitude. Included in the objectives is a measurement of $\dot{G}$, the time-rate-of-change of the gravitational constant.

It is expected that ILR will make a huge impact on GRT tests, although there will be technical challenges due to the large distances involved. With regard to these future ILR missions, Iorio (2011) has numerically investigated how the ranges between the Earth, the inner planets, as well as Jupiter and Saturn, could be influenced by specified Newtonian and non-Newtonian dynamical effects. This was done by the simultaneous integration of the equations of motion of all the major bodies of the solar system, including Ceres, Pallas, Vesta, Pluto and Eris in the Solar System Barycentric reference frame over a 2 year period, except Mars, Jupiter and Saturn for which a period of 5 years was used.

3 Global Positioning System

The GPS satellites orbit at a height of ∼20,200 km. They have an orbital period of about 718 min and a velocity of about 3,874 m/s. This velocity, relative to the geocentre, Earth's rotation and the difference in gravitational potential between observer and the GPS satellite, make it absolutely essential that GRT needs to be considered in timing (frequency) and orbital parameters. The basic theoretical timescale for geodesy and geophysics is Geocentric Coordinate Time, *TCG*, which is the coordinate time of the GCRS, which has coordinates $(T, \vec{X})$ (Müller et al. 2008; Petit and Luzum 2010).

There are two relativistic effects which affect the clocks of GPS satellites: time dilation and gravitational redshift. The relative motion between the observer (GPS receiver) and the GPS satellite results in special relativistic time dilation, whereas differences in the gravitational potential as experienced by the observer and satellite result in gravitational redshift. As a consequence of these two relativistic effects, so as to align approximately the GPS clocks with terrestrial time (TT), the onboard oscillators require a small frequency (i.e. frequency but not phase aligned, which means a phase delay or advance exists) offset adjustment. Second-order effects resulting from the non-circular orbits have to be corrected in the GPS receiver during processing of observational data by applying a correction of the order of $2(\vec{r} \cdot \dot{\vec{r}})/c^2$, where $\vec{r}$ is the position vector of the satellite (Senior et al. 2008).

3.1 Reference Frame Issues

In the earth-centred inertial (ECI) frame, the special relativistic theory is valid to a high level. The ECI frame is basically a freely falling, local, non-rotating inertial frame with its origin at the centre of the Earth. Although the Earth is accelerating towards the Sun, in this frame, the speed of light can be assumed to be constant. For the purposes of GPS and in general of satellites with clocks on board, it is most convenient (Ashby 2003) to describe their motions in the ECI frame. This approach makes the Sagnac effect irrelevant although the Sagnac effect on Earth-based (moving) receivers must still be taken into account (see Sect. 3.2.3). In the Earth Centred Earth Fixed Frame (ECEF), which is a rotating frame, clock synchronisation is difficult as light travels in a spiral path due to the Sagnac effect. Practically, the ECI is used for the establishment of positions by the GPS; afterwards a rotation to the ECEF is performed.

3.2 Clock and Frequency Effects

Similar to the other space geodetic techniques, the technological basis for GPS is founded on the very stable and accurate atomic clocks used to generate frequencies utilised in the satellite systems, and in the stable quartz oscillators used in GPS receivers. Higher quality receivers are equipped with better (more stable) oscillators, and for certain demanding applications, for instance the international GNSS service (IGS) receiver (HRAO) at HartRAO, the 5 MHz signal of an external Hydrogen MASER is used as clock reference. A comparison of the modified Allan deviation (MDEV) as calculated from IGS (final product) 30 s satellite clocks active over the period 15 June–5 July 2007 was made by Senior et al. (2008). The MDEV is more suitable when estimating short-term stability and allows distinguishing between white and flicker phase noise (see Riley 2007 for definition and formulation). The comparison includes a listing of the atomic frequency standard (AFS) (caesium or rubidium) and PRN numbers (pseudo-random noise) which is very useful. Included in the comparison were five ground clocks, which were among the highest weighted clocks realizing the IGST (IGS Final) timescale for that period. The HartRAO IGS station (HRAO) using an Oscilloquartz EFOS C passive H-maser as external frequency standard was included, as well as BRUS (Quartzlock CH1-75 active H-maser), BREW (Sigma Tau passive H-maser), STJO (passive H-maser) and NRC1 (Kvarz CH-175 active H-maser).

It is clear from the comparison by Senior et al. (2008) that the GPS satellites have instabilities that are five times greater than the ground clocks. In addition, satellite clock behaviour is dependent on AFS type and constellation block. Block IIA (older block) satellites exhibit 12 h variations (there are also shorter period variations but of lower amplitude, of which the 6 h variation is also significant) up to 8 ns, Block IIR 0.1–0.3 ns and Block IIR-M 0.12 ns. For very accurate applications these variations need to be included in modelling. These clock variations are much smaller than the relativistic effects. The stability of the GPS caesium clocks is such that, after initialisation and an interval of 1 day, the clock would still be correct to $\sim$5 parts in 10^{14}, which is about 4 ns (4×10^{-9}s), which is small compared to relativistic effects (Ashby 2003).

3.2.1 Gravitational Redshift

A clock in orbit experiences relativistic shifts which have both constant and time varying components. The constant component can be compensated for by incorporating a fixed offset, which lowers the frequency of the on-board oscillator.

Orbital eccentricity and the quadrupole (including higher order terms) of the Earth's gravity field are primarily responsible for the time varying components (Larson et al. 2007). An arbitrary atomic clock's time τ is associated with TCG in non-rotating GCRS coordinates by (Müller et al. 2008) so that

$$\frac{d\tau}{dTCG} = 1 - \frac{1}{c^2}\left(U + \frac{1}{2}\vec{v}^2\right) + \mathcal{O}(c^{-4}), \tag{2.52}$$

where the speed of light in vacuum is denoted by c, $\vec{v}$ is the GCRS speed of the satellite and the GCRS gravitational potential at the clock is U. Several gravitational components are contained within U. Earth's tidal potential U_E is the main constituent, although also included are the tidal potentials of the Sun, Moon and planets U_{tidal} and U_{iner}, an inertial component. A more detailed description of these components can be found in Müller et al. (2008).

As an example (Müller et al. 2008), if two clocks are located on an equipotential surface, the rates of the clocks will be the same. If however, one clock is moved to a height of 1 km, their rates would differ by about 10^{-13}. The gravitational redshift can then be written (Larson et al. 2007) in terms of the Newtonian gravitational potential $\Phi(\vec{r})$ at the point $\vec{r}$ of the GPS satellite in orbit as

$$\frac{\Delta f}{f} = \frac{\Phi(\vec{r}) - \Phi_0}{c^2}. \tag{2.53}$$

In (2.53) Φ_0 is the gravity potential (including the centrifugal potential resulting from the rotation of Earth) at the reference clock located on Earth's geoid. The International Astronomical Union (IAU) has defined (IAU Resolutions 2000, Resolution B1.9) the relation Φ_0/c^2 by setting the relation between TCG and TT to have a rate of $-6.969\,290\,134 \times 10^{-10}$ (Kaplan 2005; Burša et al. 2007). Considering the reference clock located on the geoid, the potential at Φ_0 is approximately (cf. Cazenave 1995; Kouba 2004; Ashby and Nelson 2009)

$$\Phi_0 = -\frac{GM}{r_\theta}\left(1 - \frac{J_2 \cdot r_{eq}^2}{r_\theta^2} \cdot \frac{1}{2}\left(3\sin^2\theta - 1\right)\right) - \frac{1}{2}\left(\omega r_\theta \cos\theta\right)^2, \tag{2.54}$$

where the first term makes up the static component and the second is the centrifugal component. In (2.54), θ is measured north or south from the equator, r_θ is the radius of Earth at the specified latitude and $r_{eq} = 6.378\,137 \times 10^6\,m$ is the equatorial radius of Earth. The angular velocity of Earth's rotation is $\omega = 7.291\,151\,467 \times 10^{-5}$ rad/s. Earth's quadrupole moment (coefficient) $J_2 = 1.08268 \times 10^{-3}$ accounts for the oblateness of the Earth. Considering a reference clock on the equator (clocks on the equator essentially run at the same rate as clocks that are not on the equator due to higher gravitational redshift at higher latitudes, more time dilation on the equator and a correction due to the quadrupole; these effects compensate to a high level; see Ashby 2006), one can rewrite (2.54) where $\theta = 0$, as

$$\Phi_0 = -\frac{GM}{r_{eq}}\left(1 + \frac{J_2}{2}\right) - \frac{1}{2}\left(\omega r_{eq}\right)^2. \tag{2.55}$$

Earth's quadrupole's effect on the potential at the GPS satellite is approximately one part in 10^{14} (Ashby 2005), so in the case of the potential at the satellite, the contribution of the quadrupole can be ignored in most cases (GPS orbits are high enough to be nearly Keplerian); there is also no centrifugal component so that the gravitational potential at the GPS satellite is to a good approximation

$$\Phi_{GPS} = \frac{-GM}{|\vec{r}|}. \tag{2.56}$$

The total gravitational frequency shift of the clock onboard the GPS satellite can then be calculated using (2.53) as (Ashby 2003, 2006; Larson et al. 2007)

$$\begin{aligned}\frac{\Delta f}{f} &= \frac{\Phi_{GPS} - \Phi_0}{c^2} \\ &= -\frac{GM}{c^2|\vec{r}|} - \left(-\frac{GM}{r_e c^2}\left(1 + \frac{J_2}{2}\right)\right) \\ &\approx 5.288 \times 10^{-10} \approx 45.688\,\mu\text{s.d}^{-1}.\end{aligned} \tag{2.57}$$

Ignoring this increase of the satellite clock frequency will lead to a timing error and consequent navigational (one way range ρ) error (per day) of

$$\begin{aligned}\rho_{error} &= c \times \frac{\Delta f}{f} \\ &= 299792458.0 \times 4.5688 \times 10^{-5} \\ &\approx 13.697\text{ km}.\end{aligned} \tag{2.58}$$

3.2.2 Special Relativity: Second Order Doppler Effect

As a consequence of the high speed ($\sim$3874 m/s) of GPS satellites, the special relativity theory of Einstein needs to be applied. The time dilation effect causes the GPS satellite to appear to run slow by about 7 μs. d^{-1}. Using the mathematical formalism of general relativity, a specified reference system is fixed by the specific form of the metric tensor $g_{\alpha\beta}(t, x^i)$. The metric tensor allows (Soffel et al. 2003) calculation of the 4-distance ds between any two events x^α and $x^\alpha + \mathrm{d}x^\alpha$ following the rule

$$\begin{aligned}\mathrm{d}s^2 &= g_{\alpha\beta}(t, x^i)\mathrm{d}x^\alpha \mathrm{d}x^\beta \\ &\equiv g_{00}c^2\mathrm{d}t^2 + 2g_{0i}c\mathrm{d}t\mathrm{d}x^i + g_{ij}\mathrm{d}x^i\mathrm{d}x^j.\end{aligned} \tag{2.59}$$

In (2.59) Einstein's summation convention is implied. Four coordinates $x^\alpha = (x^0, x^i) = (x^0, x^1, x^2, x^3)$ describe the four-dimensional space–time reference system. Greek indices adopt the values 0, 1, 2 and 3; Latin indices adopt the values 1, 2 and 3. Indices 1, 2 and 3 refer to the three spatial coordinates and index 0 refers to the time variable. For dimensional reasons, $x^0 = ct$ is normally used. Here the speed of light is denoted by c and t is a time variable. Translational and rotational equations of motion of bodies can be derived using the metric, allowing one to describe the propagation of light and to model the process of observation. For instance, one can model the relationship between the observed (proper) time τ of an observer and the coordinate time t. Proper time is the time actually read at the clock. Coordinate time is the time specified by the time coordinate x^0. In terms of a specific model, these components can be combined into a relativistic model. Refer to Soffel et al. (2003) for more details.

Following Nelson and Ely (2006), we derive the relativistic effect (one of the many formulations of it, cf. Larson et al. 2007) on a GPS satellite's clock time in general, including the relativistic time dilation, and conclude with approximate formulations (typically used) of the corrections to be applied to the satellite clocks.

In an ECI frame the components of the metric tensor can be written as (cf. Soffel et al. 2003, discussion of the metric tensor for space–time coordinate systems $(t, \vec{x})$, which are centred at the barycentre of an ensemble of masses)

$$\begin{aligned} -g_{00} &\approx 1 - 2U/c^2 \\ g_{0j} &= 0 \\ g_{ij} &\approx \delta_{ij}, \end{aligned} \tag{2.60}$$

as an approximation in the analysis of clock transport. In (2.60) the Newtonian gravitational potential is denoted by U and δ_{ij} is the Kronecker delta. For a clock on board the GPS satellite, the elapsed coordinate time can be expressed in terms of the proper time by the integral

$$\Delta t = \int_{\tau_0}^{\tau} \left(1 + \frac{1}{c^2} U + \frac{1}{2}\frac{1}{c^2} v^2\right) d\tau. \tag{2.61}$$

In (2.61) under the integral, proper time, gravitational redshift and time dilation are the first, second and third term respectively. The magnitude of the time dilation can be expressed as (Zhang et al. 2006)

$$\Delta f_r \equiv f_s \left(1 - \frac{\vec{\dot{r}}_T}{c}\right)\left(\frac{v^2}{2c^2}\right) = \frac{f_s v^2}{2c^2} - \frac{f_s \vec{\dot{r}}_T v^2}{2c^3} \approx \frac{f_s}{2}\frac{v^2}{c^2}. \tag{2.62}$$

In (2.62) $\vec{\dot{r}}_T$ is the GPS satellite transversal velocity, v is its tangential velocity, f_r is the frequency received at the receiver and f_s is the original frequency of the transmitter. The GPS L1 and L2 frequencies are in the Gigahertz range, so the frequencies are high enough (for L1 $\lambda = 1.57542$ GHz, for L2 $\lambda = 1.22760$ GHz) to make the Doppler effect considerable, as

$$D \equiv f_r - f_s = \frac{\vec{\dot{r}}_T}{c} f_s = \frac{1}{\lambda}\vec{\dot{r}}_T \tag{2.63}$$

Considering that $f_r > 0.1$ Hz, which translates to more than 2 cm/s error in the range rate, the second-order Doppler effect cannot be neglected for geodetic applications. The order of magnitude of the time dilation effect is (Ashby 2006)

$$-\frac{1}{2}\frac{v^2}{c^2} \approx -8.35 \times 10^{-11}. \tag{2.64}$$

As the reference clock (in the receiver) is also moving, although at a lower speed (~465 m/s) relative to the GPS satellite, the fractional frequency difference due to time dilation between a GPS satellite clock and a reference clock on the equator can be obtained by calculating the difference,

$$\frac{\Delta f}{f} = -\frac{1}{2}\frac{v^2}{c^2} - \left(-\frac{1}{2}\frac{(\omega r_e)^2}{c^2}\right) = -8.228 \times 10^{-11}. \tag{2.65}$$

In (2.65) ω is the angular velocity of Earth and r_e is Earth's equatorial radius. This fractional frequency shift, if not considered, would lead to a navigational error of 2.13 km/day (Ashby 2006). The negative value indicates that the satellite clock runs slow relative to the reference clock on the equator.

Continuing with the reasoning of Nelson and Ely (2006), a new coordinate time can be defined (in the ECEF frame) by applying a change of scale:

$$\Delta t' = \left(1 - \frac{1}{c^2}W_0\right)\Delta t = \int_{\tau_0}^{\tau}\left\{1 + \frac{1}{c^2}(U - W_0) + \frac{1}{2}\frac{1}{c^2}v^2\right\}\mathrm{d}\tau. \tag{2.66}$$

In (2.66) W_0 equals Φ_0 as defined in (2.54) and (2.55) and U equals $\Phi(\vec{r})$ as defined in (2.53) and (2.56) where for our purposes $\Phi(\vec{r}) = \Phi_{\mathrm{GPS}}$. Coordinate time $\Delta t'$ represents proper time of the reference clock at rest on the geoid and is therefore the coordinate clock. The elapsed coordinate time for the GPS satellite clock after integration is (Nelson and Ely 2006)

$$\Delta t' = \left(1 + \frac{3}{2}\frac{1}{c^2}\frac{GM}{a} - \frac{1}{c^2}W_0\right)\Delta\tau + \frac{2}{c^2}\sqrt{GMae}\sin E. \tag{2.67}$$

In (2.67) a, e and E are the semi-major axis, eccentricity and eccentric anomaly of the GPS satellite orbit respectively. A constant rate offset is contained within the first term; this offset is between the satellite clock and a reference clock on the geoid whereas the second term results from orbital eccentricity and leads to a small relativistic periodic correction (amplitude of $\sim$30 ns), which has to be corrected by the GPS receiver software. According to Ashby (2003) the clocks in the GLONASS satellite are adjusted before broadcast. It would seem that the GPS system carries some historical baggage, as the decision to have the user make the orbital eccentricity correction was due to the weak computing power available in the early GPS satellite vehicles.

In (2.67) there are two constant rate corrections in the first term. Extending (Ashby 2003) and adding the time dilation contribution as described by (2.65) so that the formulation equals that of Ashby (2006) one has three constant rate terms

$$\frac{3}{2}\frac{1}{c^2}\frac{GM}{a} - \frac{\Phi_0}{c^2} - \frac{1}{2}\frac{(\omega r_e)^2}{c^2} = -4.4647 \times 10^{-10}. \tag{2.68}$$

This constant rate in (2.68) can be explained (Ashby 2006) by reviewing the total contribution of the fractional frequency shift, which is obtained by addition. If one combines (2.57) with (2.65),

$$\left(\frac{\Delta f}{f}\right)_{\text{net}} = -\frac{GM}{c^2|\vec{r}|} - \left(-\frac{GM}{r_e c^2}\left(1+\frac{J_2}{2}\right)\right) + \left(-\frac{1}{2}\frac{v^2}{c^2} - \left(-\frac{1}{2}\frac{(\omega r_e)^2}{c^2}\right)\right), \quad (2.69)$$

considering that the total energy per unit mass of the satellite is

$$\frac{1}{2}v^2 - \frac{GM}{r} = -\frac{GM}{2a} \quad (2.70)$$

In (2.70) $a = 26.562 \times 10^3$ km is the semi-major axis of a GPS satellite orbit. The velocity term in (2.69) can be removed by keeping r and a, which (Ashby 2006) leads to the equation

$$\frac{\Delta f}{f} = -\frac{2GM}{c^2}\left(\frac{1}{r} - \frac{1}{a}\right) - \frac{3GM}{2c^2 a} + \frac{GM}{c^2 r_e}\left(1+\frac{J_2}{2}\right) + \frac{1}{2}\frac{(\omega r_e)^2}{c^2}. \quad (2.71)$$

In (2.71) the first term will disappear when the orbit has zero eccentricity. This leaves us with the constant part as described by (2.68).

The GPS satellite clocks are adjusted for the three constant rate corrections before launching them into orbit, the negative sign of (2.68) implying that the satellite clock has a higher frequency in orbit than on the ground (read geoid, where the clock frequency should be 10.23 MHz) and its proper frequency should therefore be reduced to

$$\left(1 - 4.4647 \times 10^{-10}\right) \times 10.23\,\text{MHz} = 10.229\,999\,995\,43\ \text{MHz}. \quad (2.72)$$

The second term in (2.67) (to be corrected by the user's software) may be written as

$$\Delta t_{\text{rel}} = \frac{2}{c^2}\sqrt{GMae}\,\sin E \quad (2.73)$$

and (2.73) can be written as

$$\Delta t_{\text{rel}} = \frac{2\vec{r}\cdot\dot{\vec{r}}}{c^2} \quad (2.74)$$

The dot product of the position vector $\vec{r}$ and velocity vector $\dot{\vec{r}}$ in (2.74) is a scalar; one can therefore use it in the ECI or in the ECEF coordinate system. Equations (2.73) and (2.74) are formulations (which include only the main monopole contribution of Earth's gravity field) often used in precise geodetic applications and are accurate to a sufficient level for most GPS applications, but are inadequate when evaluating lower orbits such as GRACE (Larson et al. 2007). More exact formulations of (2.73) and (2.74) can be found in Kouba (2004) and Larson et al. (2007).

3.2.3 Sagnac Effect

In addition to the gravitational red (blue)-shift and time-dilation effects, a further effect involves the second postulate of special relativity (the constancy of the speed of light), the fundamental principle on which the Global Positioning System is based. As already mentioned, it is convenient to synchronise clocks in an ECI frame as light does not travel in a straight line in a rotating frame. This excludes an ECEF frame from being used to synchronise clocks, due to the Sagnac effect. A stationary GPS receiver located on the equator will have a velocity of ~465 m/s through the ECI frame as the Earth rotates. The corresponding Sagnac correction can be as large as 133 ns (equal to 86 ms signal propagation). This correction is also applied in the receiver. Allowing for the Sagnac effect in the ECEF is equivalent to correcting for the receiver's motion in the ECI frame (Ashby 2002).

Following Ashby (2006), to determine position using the GPS, three satellites are required for position and four are required to determine position and time. Clocks onboard the satellites are synchronised in the ECI frame. A user $\mathrm{GPS_{user}}$ will receive time signals at a specific time and position, whereas the GPS satellites will transmit signal messages containing the time and position of the transmission events, so that

$$\begin{aligned} \mathrm{GPS_{user}} &= \{t_u, \vec{r}_u\} \\ \mathrm{GPS_{sat}} &= \{t_j, \vec{r}_j\} \qquad j = 1\ldots n, \end{aligned} \tag{2.75}$$

where j is the number of the GPS satellite from which data are being received and n is the total number in view. The constancy of the speed of light is then represented by

$$c(t - t_j) = |\vec{r} - \vec{r}_j|; \quad j = 1\ldots n. \tag{2.76}$$

The nonlinear system (2.76) needs to be solved to provide the user's position; this can be done by linearising the equations and initialising an iterative algorithm with an a priori position. Due to the motion of the GPS receiver, the navigation equations in (2.76) are not valid in the ECEF frame. Most of the time, of course, users would want their positions in the ECEF, not in an ECI frame. In the ECEF, the rotation of Earth will move the GPS receiver while the GPS signal is propagating to Earth, so (2.76) needs to be altered to account for this as

$$t = t_j + \frac{|\vec{r}(t) - \vec{r}_j|}{c} = \frac{|\vec{r}(t_j) + \vec{v} \times (t - t_j) - \vec{r}_j|}{c}. \tag{2.77}$$

In (2.77) the receiver position at time t is denoted by $\vec{r}(t)$ and $\vec{v}$ is the velocity of the receiver at the time of the GPS satellite transmission. The velocity of the receiver is far less than that of c, and therefore the equations can be solved through an iteration algorithm. An iteration algorithm was also required to find the two-range as determined through SLR as discussed in Sect. 2.2.1. The range from GPS satellite to receiver can be defined as

$$\vec{R} = \vec{r}(t_j) - \vec{r}_j \tag{2.78}$$

and excluding the velocity term we get the time of arrival of the signal if it were in the ECI frame:

$$t = t_j + \frac{|\vec{r}(t_j) - \vec{r}_j|}{c} = t_j + \frac{R}{c}. \tag{2.79}$$

If t is substituted back into (2.77) then one can find (Ashby 2006)

$$t = t_j + \frac{\sqrt{R^2 + 2\vec{R} \cdot \vec{v}(t - t_j)}}{c} \approx \frac{R}{c} + \frac{\vec{R} \cdot \vec{v}}{c^2}. \tag{2.80}$$

In the case where the receiver's velocity is only a result of the rotation of Earth, then

$$\vec{v} = \vec{\omega} \times \vec{r}(t_j) \tag{2.81}$$

and one can rewrite the Sagnac correction term as

$$\Delta t_{\text{Sagnac}} = \frac{\vec{R} \cdot \vec{v}}{c^2} = \frac{2\vec{\omega} \cdot \vec{A}}{c^2} \tag{2.82}$$

with the vector area $\vec{A}$ being given by

$$\vec{A} = \frac{1}{2}\vec{r}(t_j) \times \vec{R}. \tag{2.83}$$

In (2.81) and (2.82) $\vec{\omega} = (0, 0, \omega)$. The dot product in (2.82) projects area $\vec{A}$ (the Sagnac correction is proportional to this area) onto a plane that is parallel to the equatorial plane. Area $\vec{A}$ is created by the sweeping vector from the rotation axis to tip of the signal pulse as it propagates from transmitter to receiver (Ashby 2004). Area $\vec{A}$ is therefore swept out by the electromagnetic pulse as it propagates from the GPS satellite transmitter to the receiver.

3.3 General Relativistic Accelerations

The relativistic accelerations in the weak-field and slow motion approximation as described by the standard IERS formulation (2.10) for the Schwarzschild field, frame dragging and de Sitter precession are additional relativistic effects, which should be taken into account during POD. Table 2.1 lists the magnitude of these effects when considering GPS satellites.

3.4 Spatial Curvature Effect on Geodetic Distance

The proper distance between a receiver on the surface of the Earth at radius r_1 and a GPS satellite at radius r_2 is approximately (Ashby 2003)

$$\int_{r_1}^{r_2} \left[1 + \frac{GM}{c^2 r}\right] dr = r_2 - r_1 + \frac{GM}{c^2} \ln\left(\frac{r_2}{r_1}\right). \tag{2.84}$$

Using (2.84) and, in addition, calculating the coordinate distance, the difference between proper and coordinate distance is about 6.3 mm ((2.4) as applied to SLR).

4 Very Long Baseline Interferometry

Observations of compact extragalactic radio sources using the technique of VLBI are very well suited to the study of Earth orientation in space, as these sources serve as an excellent approximation to an inertial frame. The VLBI technique is unique in that it provides Earth orientation measurements of high accuracy in an inertial frame of reference (Sovers et al. 1998). Similar to the other space geodetic techniques, the reduction of VLBI data requires consideration of a large range of effects, which include the effects of the Earth's internal structure on its dynamics, the VLBI site velocity as caused by tectonic plate motion, terrestrial tidal effects, and quantification of tropospheric and ionospheric parameters. In addition, consideration must be given to special relativity in the interpretation of the radio signals travelling from the distant sources, as well as to general relativistic retardation. For a thorough introduction to VLBI the reader should refer to the chapter by Harald Schuh and Johannes Böhm in this volume, in which is included a discussion of the VLBI GRT model for propagation. I will attempt to provide additional information which could be read in the context of and as ancillary material to the Schuh and Böhm chapter without unnecessary repetition.

4.1 Gravitational Delay

According to GRT, an electromagnetic signal will experience retardation in terms of its travel time when propagating in a gravitational potential relative to its propagation in gravity field-free space. This has implications for VLBI, as the value determined for the difference in arrival time at the VLBI stations in question must be corrected for gravitational effects. Furthermore, considering the implications of GRT, one must take into account both a time delay (Shapiro 1964) and a bending delay (deviation from a straight-line path) (Shapiro 1967). The current general relativistic VLBI model for propagation used by international VLBI service (IVS) analysis centres is the IERS (Petit and Luzum 2010) recommended "consensus model" (Eubanks 1991) which provides an accuracy below 1 ps as described inChap.11of the IERS Conventions. According to Petit (2009) there are no changes expected in the short term, although a review is possible in the light of future VLBI accuracy improvements. This is true specifically with regard to VLBI2010, which will require that the consensus model be re-evaluated to ensure that it includes all terms down to the order of 0.3 ps (Heinkelmann and Schuh 2009). Objectives of VLBI2010 include 1 mm position accuracy over a 24 h observing session (on global baselines), 0.1 mm/year station velocity accuracy, continuous observations, and delivery of initial results within 24 h after taking data (MacMillan et al. 2011)

4.2 General Relativistic Tests Using VLBI

VLBI currently achieves very high accuracies, better than 0.1 mas. These high accuracies make VLBI an excellent tool for GRT tests and evaluation and therefore the geodetic VLBI technique has often been used to evaluate the space curvature parameter γ introduced in (2.4). Tests of special and general relativity were quickly launched after Einstein's publications. Acceptance of relativity was not instantaneous and general acceptance was fraught with misunderstanding, political viewpoints, self-serving attitudes and the typical slow acceptance of a new scientific doctrine. A very good review of the early tests and human drama involving astronomers of the early twentieth century is given in Crelinsten (2006). These tests involved gravitational redshift and light bending (primarily light deflection at the Sun's limb during eclipses). Currently one of the most accurate methods to evaluate γ is by utilising VLBI.

The space–time geometry around the Sun can be described by a static and spherically symmetric metric (Schwarzschild 1916). However, Eddington (1923) provided an isotropic formulation of Schwarzschild's original anisotropic version of the metric (2.21) (where, as noted by Eddington, in the original coordinates, the speed of light is not the same for transverse and radial directions), which can be written as

$$ds^2 = -\left(1 - 2\frac{GM}{c^2 r} + 2\left(\frac{GM}{c^2 r}\right)^2\right)(cdt)^2 + \left(1 + 2\frac{GM}{c^2 r}\right)\left[dx^2 + dy^2 + dz^2\right]. \tag{2.85}$$

Here the gravitational constant is G, the speed of light is given by c and M is the mass of the star (Sun). The PPN parameters γ and β are the most physically significant of the ten parameters in the PPN formalism and this is demonstrated by their placement (Margot and Giorgini 2009) in (2.85):

$$ds^2 = -\left(1 - 2\frac{GM}{c^2 r} + 2\beta\left(\frac{GM}{c^2 r}\right)^2\right)(cdt)^2 + \left(1 + 2\gamma\frac{GM}{c^2 r}\right)\left[dx^2 + dy^2 + dz^2\right]. \tag{2.86}$$

In the first term on the right-hand side of (2.86) β describes the degree of nonlinearity in the superposition law for gravity, while the second term (spatial part) contains PPN γ, which describes how much curvature is produced by unit rest mass, and can be tested by deflection of light, bending of radio waves and Shapiro delay experiments. In GRT both parameters γ and β are equal to unity, whereas the other eight PPN parameters are zero (Will and Nordtvedt 1972).

4.2.1 Evaluation of PPN Parameter γ

The classical test by Eddington on the deflection of light by the Sun (Dyson et al. 1920) and the delay of an electromagnetic signal as it propagates near the Sun (Shapiro et al. 1968) essentially measure the propagation of photons in curved space near the Sun; these measurements depend on the PPN parameter γ. The amount of space curvature per unit mass is related to γ through the proportional relationship

$$\delta\theta \propto 1/2(1+\gamma). \tag{2.87}$$

Following Will (2006), an electromagnetic signal (ray of light or radio signal from VLBI source) passing close to the Sun at distance d will be deflected by an angle,

$$\delta\theta = \frac{1}{2}(1+\gamma)(4m_\odot/d)[(1+\cos\Phi)/2], \tag{2.88}$$

where the mass of the Sun is denoted by $m_\odot$ and Φ is the angle formed between the direction of the incoming electromagnetic signal and the line between Earth and the Sun. The relative angular separation may be changed when the line-of-sight of one of the sources moves close to the Sun. This angular separation is given by

$$\delta\theta = \frac{1}{2}(1+\gamma)\left[-\frac{4m_\odot}{d}\cos\chi + \frac{4m_\odot}{d_r}\left(\frac{1+\cos\Phi_r}{2}\right)\right]. \tag{2.89}$$

In (2.89) the points of closest approach to the Sun in terms of distance are given by d and d_r for the source and reference rays, respectively. The angle created by the Sun-source and Sun-reference directions, projected against the plane of the sky, is given by χ and Φ_r denotes the angle between the reference source and the Sun. More details can be obtained from Will (2006). This short introduction describes the basics for the determination of the varying relative angular separation as would be determined using VLBI, when the line-of-sight of a radio source is close to the Sun,

$$d \sim R_{\odot},\ d_r \gg d$$

and χ is varying with time. Of course the Sun could be replaced by a large planet such as Jupiter (see for instance Schuh et al. 1988; Fomalont and Kopeikin 2003) and the radio source could be a transmitter on board an interplanetary probe. The literature abounds with comparisons between the results of different (e.g. radar determined) techniques but here we focus on VLBI (cf. Pitjeva 2005; Turyshev 2009).

Constant instrumental and data analysis upgrades throughout the development of the VLBI technique have delivered a continuous increase in the accuracy of the evaluations of PPN parameter γ. This continuous improvement is illustrated in Fig. 2.2, which is based on the table of γ estimates spanning the period 1972–2009 as provided by Heinkelmann and Schuh (2009). Figure 2.2 contains the standard error of the various γ estimates, reflecting the accuracy of the parameter evaluation. An exponential fit constrained to the first and last estimate provides a value of $\pm\ 2.5 \times 10^{-5}$, when using the fitted function to predict towards 2020. If this predicted accuracy level is achieved by VLBI, perhaps supported by the developments around VLBI2010 in the GGOS framework, it would be comparable to the accuracy (currently the best) of the estimate of γ ($\pm 2.3 \times 10^{-5}$) achieved during the microwave tracking of the Cassini spacecraft on its approach to Saturn (Bertotti et al. 2003). Evaluations represented in Fig. 2.2 are contained in Table 2.2.

The gravitational signal retardation (Shapiro effect) is described by

$$\tau_{\text{grav}} = (1+\gamma) \cdot \frac{GM}{c^3} \cdot \ln \left[\frac{|\vec{X}_1| + \vec{X}_1 \cdot \vec{k}}{|\vec{X}_2| + \vec{X}_2 \cdot \vec{k}} \right], \tag{2.90}$$

where $\vec{X}_i$ is the position vector of the individual VLBI antennas relative to the centre of the gravitating body and the unit vector towards the radio source as viewed from the Earth-bound baseline is denoted by $\vec{k}$. Following Heinkelmann and Schuh (2009), the partial derivative of the delay relative to γ can be written as

$$\frac{\partial \tau}{\partial \gamma} = \frac{GM}{c^3} \cdot \ln \left[\frac{|\vec{X}_1| + \vec{X}_1 \cdot \vec{k}}{|\vec{X}_2| + \vec{X}_2 \cdot \vec{k}} \right], \tag{2.91}$$

which will be required for the estimation of γ utilising the Shapiro delay in a least-squares process. A typical value for the Shapiro delay due to the gravitational field of the Earth (see Table 2.1) for a baseline of 6000 km, is about 21 ps (Klioner 1991). The Shapiro time delay that results from the Sun for the same baseline length can vary from 17×10^4 ps for an electromagnetic ray grazing the Sun's limb to about 17 ps when the rays are incident at about 90° from the Sun. Models for the gravitational delay are continuously improved (cf. Klioner and Kopeikin 1992; Kopeikin and Schäfer 1999). An additional delay, which is caused by the effect due to the finite speed of the propagation of gravity, may have to be included (Kopeikin 2001). This delay could be at the level of several ps; however, recent refinements and the formulation of higher level models (utilising these refinements) have not completely been incorporated into the standard IERS formulation.

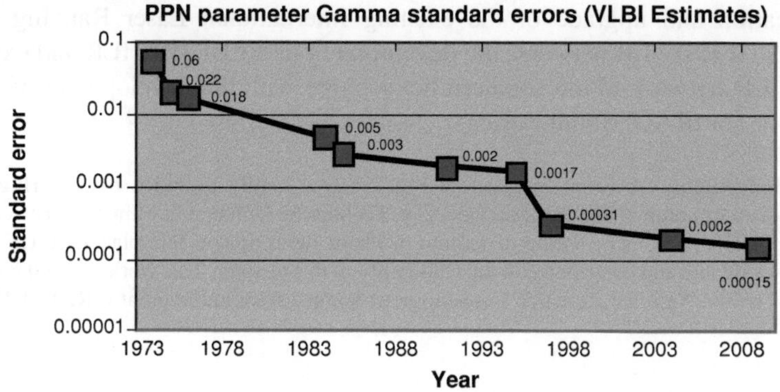

Fig. 2.2 Standard errors associated with geodetic VLBI evaluations of PPN parameter γ. Continuous improvement is due to instrumentation, software and model development as well as longer time series of data

Table 2.2 Standard errors associated with geodetic VLBI evaluations of PPN parameter γ

Authors	Standard error
Counselman et al. (1974)	±0.06
Fomalont and Sramek (1975)	±0.022
Fomalont and Sramek (1976)	±0.018
Robertson and Carter (1984)	±0.005
Carter et al. (1985	±0.003
Robertson et al. (1991)	±0.002
Lebach et al. (1995)	±0.0017
Eubanks et al. (1997)	±0.00031
Shapiro et al. (2004)	±0.00021
Lambert and Le Poncin-Lafitte (2009)	±0.000152

5 Concluding Remarks

It is very rewarding to see, throughout the development of the Space Geodetic techniques, how theory, experiment, human innovation and the constant drive towards better science and higher accuracies have meshed to form global networks of instruments and people. Applications of space geodesy extend from outer space to the core of the Earth; they utilise and are capable of testing and evaluating GRT.

Future improvements, within the framework of GGOS, will provide improved accuracies and a better understanding of the space and world in which we live. General relativity will continue to be tested by scientific experiments in which space geodesy has its own specific role to play, providing certainty on the levels to which GRT can reliably be used; eventually, however, experiments will lead to a post-GRT theory. Exciting scientific projects based on space geodesy are on the horizon which will play a role in the evaluation of GRT. These include densification and upgrade of the existing International Laser Ranging Service Network (ILRS) SLR network, the development of VLBI2010, ILR and expansion of the LLR network to the southern hemisphere with the development of an LLR system at HartRAO, South Africa.

Acknowledgements A script to generate Fig. 2.1 was kindly provided by Urs Hugentobler, Institute of Astronomical and Physical Geodesy, Technische Universität München. The comments and suggestions by Roberto Peron of Istituto di Fisica dello Spazio Interplanetario (IFSI-INAF) were very valuable and contributed to the readability of this chapter. This work is based on research supported by the National Research Foundation of South Africa under grant IFR2011041500034.

References

Ashby N, Shahid-Saless B (1990) Geodetic precession or dragging of inertial frames? Phys Rev D 42:1118–1122

Ashby N (2002) Relativity and the global positioning system. Phys Today 55(5):41–47

Ashby N (2003) Relativity in the global positioning system. Living Rev Relativ, vol 6, p 1 [http://www.livingreviews.org/Articles/Volume6/2003-1ashby, online article: cited on 7 July 2010]

Ashby N (2004) The sagnac effect in the global positioning system. In: Rizzi G, Ruggiero ML (eds) Relativity in rotating frames, Kluwer Academic Publishers, Dordrecht, The Netherlands, pp 11–28

Ashby N (2005) Relativity in the global positioning system. In: Ashtekar A (ed) 100 years of relativity space-tine structure: Einstein and beyond. World Scientific Publishing Co, Singapore, pp 257–289

Ashby N (2006) Relativistic effects in the global positioning system. http://www.aapt.org/doorway/tgru/articles/Ashbyarticle.pdf, online article: cited on 1 May 2011

Ashby N, Nelson RA (2009) The global positioning system, relativity and extraterrestrial navigation, in relativity. In: lioner SA, Seidelmann PK, Soffel MH (eds) Fundamental astronomy: dynamics, reference frames, and data analysis proceedings IAU symposium no. 261, Cambridge University Press, pp 22–30

Bertotti B, Iess L, Tortora P (2003) A test of general relativity using radio links with the cassini spacecraft. Nature 425:374–376

Beutler G, Drewes H, Verdun A (2005a) The integrated global geodetic observing system (IGGOS) viewed from the perspective of history. J Geodyn 40(4–5):414–431

Beutler G, Mervart L, Verdun A (2005b) Methods of celestial mechanics: application to planetary system, geodynamics and satellite geodesy, vol II. Springer, Germany

Botai OJ, Combrinck L, Sivakumar V, Schuh H, Böhm J (2010) Extracting independent local oscillatory geophysical signals by geodetic tropospheric delay. In: IVS 2010 general meeting proceedings, pp 345–354. http://ivscc.gsfc.nasa.gov/publications/gm2010/botai.pdf

Burša M, Kenyon S, Kouba J, Šíma Z, Vatrt V, Vítek V, Vojtíšková M (2007) The geopotential value w_0 for specifying the relativistic atomic time scale and a global vertical reference system. J Geod 81:103–110

Carter WE, Robertson DS, MacKay JR (1985) Geodetic radio interferometric surveying: applications and results. J Geophys Res 90(B6):4577–4587

Cazenave A (1995) Geoid, topography and distribution of landforms, in global earth physics; a handbook of physical constants, American Geophysical Union

Ciufolini I (1986) Measurement of the lense-thirring drag on high-altitude laser- ranged artificial satellites. Phys Rev Lett 56:278–281

Ciufolini I, Wheeler JA (1995) Gravitation and inertia. Princeton University Press, Princeton

Ciufolini I, Lucchesi DM, Vespe F, Mandiello A (1996) Measurement of dragging of inertial frames and gravitomagnetic field using laser-ranged satellites, Il. Nuovo Cimento A 109:575–590

Ciufolini I, Pavlis E, Chieppa F, Fernandes-Vieira E, Pérez-Mercader J (1998) Test of general relativity and measurement of the lense-thirring effect with two earth satellites. Science 279:2100–2103

Ciufolini I, Pavlis EC (2004) A confirmation of the general relativistic prediction of the lense–thirring effect. Nature 431:958–960

Ciufolini I, Pavlis EC, Peron R (2006) Determination of frame-dragging using earth gravity models from CHAMP and GRACE. New Astron 11:527–550

Ciufolini I (2007) Gravitomagnetism, frame-dragging and lunar laser ranging, arXiv: 0704.3338v2 [gr-qc], 10 May 2007

Combrinck L (2008) Evaluation of PPN parameter gamma as a test of general relativity using SLR data. In: 16th international laser ranging workshop, (Poznan (PL) 13–17 Oct 2008), available at http://cddis.gsfc.nasa.gov/lw16/docs/papers/sci_6_Combrinck_p.pdf

Combrinck L (2010) Satellite laser ranging. In: Xu Guochang (ed) Sciences of Geodesy I, advances and future directions. Springer, Germany, pp 301–336

Combrinck L (2011) Testing general relativity theory through the estimation of PPN parameters γ and β using satellite laser ranging data. S Afr J Geol 114(3–4):549–560

Counselman CC, Kent SM, Knight CA, Shapiro II, Clark TA, Hinteregger HF, Rogers AEE, Whitney AR (1974) Solar gravitational deflection of radio waves measured by very-long-baseline interferometry. Phys Rev Lett 33(27):1621–1623

Crelinsten J (2006) Einstein's jury: the race to test relativity. Princeton University Press, Princeton

Cugusi L, Proverbio E (1977) Relativistic effects on the motion of the earth's satellites. In: Paper presented at the international symposium on satellite geodesy in Budapest from 28 June to 1 July 1977, J Geodesy, vol 51, pp 249–252

Cugusi L, Proverbio E (1978) Relativistic effects on the motion of earth's artificial satellites. Astron Astrophys 69:321–325

Damour T, Soffel M, Xu C (1994) General-relativistic celestial mechanics IV. Theory of satellite motion. Phys. Rev. D 49:618–635

de Sitter W (1916) On Einstein's theory of gravitation and its astronomical consequences. Mon Not Roy Astron Soc 77:155184

Drewes H (2007) science rationale of the global geodetic observing system (GGOS), in dynamic planet, monitoring and understanding a dynamic planet with geodetic and oceanographic tools. In: Tregoning P, Rizos C (eds) Planet earth. IAG Symposia, vol 130. Springer, pp 703–710

Dyson FW, Eddington AS, Davidson C (1920) A determination of the deflection of light by the sun's gravitational field, from observations made at the total eclipse of 29 May 1919. Phil Trans R Soc Lond A 220:291–333

Eddington AS (1923) The mathematical theory of relativity. Cambridge University Press, UK

Einstein A (1920) Relativity: the special and general theory. Henry Holt and Company, NY

Eubanks TM (1991) A consensus model for relativistic effects in geodetic VLBI. In: Eubanks TM (ed) Proceedings of the USNO workshop on relativistic models for use in space geodesy, pp 60–82

Eubanks TM, Matsakis DN, Martin JO, Archinal BA, McCarthy DD, Klioner SA, Shapiro S, Shapiro II (1997) Advances in solar system tests of gravity, American Physical Society, APS/AAPT Joint Meeting, 18–21 April, 1997 abstract #K11.05

Fomalont EB, Sramek RA (1975) A confirmation of Einstein's general theory of relativity by measuring the bending of microwave radiation in the gravitational field of the Sun. Astron J 199(3):749–755

Fomalont EB, Sramek RA (1976) Measurement of the solar gravitational deflection of radio waves in agreement with general relativity. Phys Rev Lett 36(25):1475–1478

Fomalont EB, Kopeikin SM (2003) The measurement of the light deflection from jupiter: experimental results. arXiv:astro-ph/0302294v2, pp 1–10

Heinkelmann R, Schuh H (2009) Very long baseline interferometry: accuracy limits and relativistic tests, In: Klioner SA, Seidelmann PK, Soffel MH (eds) Relativity in fundamental astronomy: dynamics, reference frames, and data analysis proceedings IAU symposium no. 261, Cambridge University Press

Hugentobler U (2008) Orbit Perturbations due to Relativistic Corrections. Unpublished notes available at ftp://maia.usno.navy.mil/conv2010/chapter10/add_info/

Huang C, Ries JC, Tapley BD, Watkins MM (1990) Relativistic effects for near-earth satellite orbit determination. Celest Mech Dyn Astron 48:167–185

Huang C, Liu L (1992) Analytical solutions to the four post-newtonian effects in a near-earth satellite orbit. Celest Mech Dyn Astron 53:293–307

Hoffman-Wellenhof B, Moritz H (2005) Physical geodesy. Springer-Verlag, Wien

Holdridge D (1967) An alternate expression for light time using general relativity, space programs, vol III. NASA, Washington, pp 2–4 (Summary 37–48)

Iorio L (2006) A critical analysis of a recent test of the lense–thirring effect with the LAGEOS satellites. J Geod 80:128–136

Iorio L (2007) The lense-thirring effect on orbits. In: Iorio L (ed) The measurement of gravitomagnetism. Nova Science Publishers Inc, NY, pp 73–86

Iorio L (2010a) On possible a priori "imprinting" of general relativity itself on the performed lense-thirring tests with LAGEOS satellites, communication and network, 2010, vol 2, pp 26–30. doi: 10.4236/cn.2010.21003 (Published Online February 2010 http://www.scirp.org/journal/cn)

Iorio L (2010b) Conservative evaluation of the uncertainty in the LAGEOS–LAGEOS II lense-thirring test. Cent Eur J Phys 8(1):25–32

Iorio L (2011) Effects of standard and modified gravity on interplanetary ranges. Int J Mod Phys D 20:181–232 (arXiv:1002.4585v5 [gr-qc])

Kaplan GH (2005) The IAU resolutions on astronomical reference systems time scales, and earth rotation models, explanation and implementation, United States Naval Observatory Circular No. 179, USNO, Washington

Kouba J (2004) Improved relativistic transformations in GPS. GPS Solutions 8:170–180

Klioner SA (1991) Proceedings of AGU chapman conference on geodetic VLBI: monitoring global change. In: Carter WE (ed) (NOAA Tech. Rep. NOS 137, NGS 49; Washington, DC: AGU), p 188

Klioner SA, Kopeikin SM (1992) Microarcsecond astrometry in space—relativistic effects and reduction of observations. Astron J 104:897–914

Klioner SA (2003) A practical relativistic model for microarcsecond astrometry in space. Astron J 125:1580–1597

Kopeikin SM (2001) Testing the relativistic effect of the propagation of gravity by very long baseline interferometry. Astrophys J 556:L1–L5

Kopeikin SM, Schäfer G (1999) Lorentz covariant theory of light propagation in gravitational fields of arbitrary-moving bodies. Phys Rev D 60:124002 (arXiv:gr-qc/9902030v5)

Lambert SB, Le Poncin-Lafitte C (2009) Determining the relativistic parameter γ using very long baseline interferometry. Astron Astrophys 499:331–335 (arXiv:0903.1615v1)

Larson KL, Ashby N, Hackman C, Bertiger W (2007) An assessment of relativistic effects for low earth orbiters: the GRACE satellites. Metrologia 44:484–490

Lebach DE, Corey BE, Shapiro II, Ratner MI, Webber JC, Rogers AEE, Davis JL, Herring TA (1995) Measurement of the solar gravitational deflection of radio waves using very-long-baseline interferometry. Phys Rev Lett 75(8):1439–1442

Lense J, Thirring H (1918) Uber die Einfluss der Eigenrotation der Zentralkorper auf die Bewegung der Planeten und Monde nach der Einsteinschen Gravitationstheorie. Phys Zeitschr, vol 19, p 156. English edition: Lense J, Thirring H (1984) Gen Relativ Gravitation (trans: Mashoon B, Hehl FW, Theiss DS (eds)), vol 16, p 711

Lindegren L (2009) Gaia: Astrometric performance and current status of the project, in: relativity in fundamental astronomy. In: Klioner SA, Seidelmann PK, Soffel MH Proceedings IAU symposium: dynamics, reference frames, and data analysis, proceedings IAU symposium no. 261, Cambridge University Press

Lucchesi DM (2003) LAGEOS II perigee shift and Schwarzschild gravitoelectric field. Phys Lett A 318:234–240

Lucchesi DM, Peron R (2010) Accurate measurement in the field of the earth of the general-relativistic precession of the LAGEOS II pericenter and new constraints on non-newtonian gravity. Phys Rev Lett 105:231103

MacMillan D, Petrachenko W, Niell A, Corey B, Behrend D, Schuh H (2011) VLBI2010: next generation VLBI System for geodesy and astrometry, Geophysical Research Abstracts, Vol. 13, EGU2011-13547, 2011, EGU General Assembly 2011

Margot J, Giorgini JD (2009) Probing general relativity with radar astrometry in the inner solar system. In: Klioner SA, Seidelmann PK, Soffel MH (eds) Relativity in fundamental astronomy: dynamics, reference frames, and data analysis, proceedings IAU symposium no. 261, Cambridge University Press

McCarthy DD, Petit G (2003) IERS conventions (2003), (IERS technical note; 32) Frankfurt am Maim: Verlag des Bundesamts für Kartographie und Geodäsie. Available at http://www.iers.org/nn_11216/IERS/EN/Publications/TechnicalNotes/tn32.html

Merkowitz SM (2010) Tests of gravity using lunar laser ranging. Living Rev Relativ, vol 13, p 7. Available at http://www.livingreviews.org/lrr-2010-7 (Online Article: cited on 12 Sep 2011)

Misner CW, Thorne KS, Wheeler JA (1973) Gravitation. W. H Freeman and Company, San Francisco

Moran J (1989) Introduction to VLBI. In: Felli M, Spencer RE (eds) Very long baseline interferometry, techniques and applications. Kluwer Academic Publishers, Dordrecht, pp 27–45

Moyer TD (2000) DESCANSO book series, deep space communications and navigation series. In: Yuen JH (ed) Formulation for observed and computed values of deep space network data types for navigation, vol 2

Montenbruck O, Gill E (2001) Satellite orbits: models, methods and applications. Springer-Verlag, Berlin

Müller J, Soffel M, Klioner SA (2008) Geodesy and relativity. J Geod 82:133–145

Murphy TW Jr, Adelberger EG, Battat JBR, Carey LN, Hoyle CD, LeBlanc P, Michelsen EL, Nordtvedt K, Orin AE, Strasburg CW, Stubss CW, Swanson HE, Williams E (2007) APOLLO: the apache point observatory lunar laser-ranging operation: instrument description and first detections. Publ Astron Soc Pac 120:20–37 2008 (arXiv:0710.0890v2)

Murphy T Jr (2009) Lunar ranging, gravitomagnetism, and APOLLO. Space Sci Rev 148:217–223

Newton I (1726) Principia: the mathematical principles of natural philosophy. English edition: (Motte A, Adee D), New York

Nelson RA, Ely TA (2006) Relativistic transformations for time synchronization and dissemination in the solar system. In: Proceedings of 38th annual precise time and time interval (PTTI) meeting. http://www.pttimeeting.org/archivemeetings/ptti2006.html, online article: cited on 9 May 2011

Neumann GA, Cavanaugh JF, Coyle DB, McGarry J, Smith DE, Sun X, Torrence M, Zagwodzki TW, Zuber MT (2006) Laser ranging at interplanetary distances. http://cddis.gsfc.nasa.gov/lw15/docs/papers/Laser Ranging at Interplanetary Distances.pdf, online article: cited on 11 Sept 2011. In: Proceedings of the 15th international workshop on laser ranging, Canberra, Australia

Pavlis DE, Poulose SG, Deng C, McCarthy JJ (2007) GEODYN II system documentation. SGT-Inc., Greenbelt, MD, contractor report

Perryman MAC, de Boer KS, Gilmore G, Høg E, Lattanzi MG, Lindegren L, Luri X, Mignard F, Pace O, de Zeeuw PT (2001) GAIA: composition, formation and evolution of the galaxy. Astron Astrophys 369:339–363

Petit G (2009) Relativity in the IERS conventions. In: Klioner SA, Seidelmann PK, and Soffel MH (eds) Relativity in fundamental astronomy: dynamics, reference frames, and data analysis. In: Proceedings IAU symposium no. 261, Cambridge University Press

Petit G, Luzum B (2010) IERS conventions (2010). In: Petit G, Luzum B (eds) (IERS Technical Note; No. 36). Bundesamts für Kartographie und Geodäsie, Frankfurt am Main. Available at http://www.iers.org/nn_11216/IERS/EN/Publications/TechnicalNotes/tn36.html

Pitjeva EV (2005) Relativistic effects and solar oblateness from radar observations of planets and spacecraft. Astron Lett 31:340–349
Robertson HP (1962) Relativity and cosmology. In: Deutsch AJ, Klemperer WB (eds) Space age astronomy. Academic, NY, pp 228–235
Robertson DS, Carter WE (1984) Relativistic deflection of radio signals in the solar gravitational field measured with VLBI. Nature 310:572–574
Robertson DS, Carter WE, Dillinger WH (1991) New measurement of the solar gravitational deflection of radio signals using VLBI. Nature 349:768–770
Ries JC, Huang C, Watkins MM (1988) Effect of general relativity on a near-earth satellite in the geocentric and barycentric reference frames. Phys Rev Lett 61:903–906
Riley WJ (2007) Handbook of frequency stability analysis. Hamilton Technical Services, USA, pp 22–23
Schuh H, Fellbaum M, Campbell J, Soffel M, Ruder H, Schneider M (1988) On the deflection of radio signals in the gravitational field of Jupiter. Phys Lett A 129:299–300
Schwarzschild K (1916) On the gravitational field of a mass point according to Einstein's theory. Sitzungsber Preuss AkadWiss, Phys Math Kl, vol 189, arXiv:physics/9905030v1, translated by S. Antoci and A. Loinger
Senior KL, Ray JR, Beard RL (2008) Characterization of periodic variations in the GPS satellite clocks. GPS Solut 12:211–225
Shapiro II (1964) Fourth test of general relativity. Phys Rev Lett 13:789–791
Shapiro II (1967) New method for the detection of light deflection by solar gravity. Science 157:806–808
Shapiro II, Pettengill GH, Ash ME, Stone ML, Smith WB, Ingalls RP, Brockelman RA (1968) Fourth test of general relativity: preliminary results. Phys Rev Lett 20:266
Shapiro SS, Davis JL, Lebach DE, Gregory JS (2004) Measurement of the solar gravitational deflection of radio waves using geodetic very-long-baseline interferometry data, 1979–1999. Phys Rev Lett 92(12):1–4 121101
Smith DE, Zuber MT, Xiaoli S, Neumann GA, Cavanaugh JF, McGarry JF, Zagwodzki TW (2006) Two-way laser link over interplanetary distance. Science 311:53
Soffel M, Klioner SA, Petit G, Wolf P, Kopeikin SM, Bretagnon P, Brumberg VA, Capitaine N, Damour T, Fukushima T, Guinot B, Huang T-Y, Lindegren L, Ma C, Nordtvedt K, Ries JC, Seidelmann PK, Vokrouhlicky′ D, Will CM, Xu C (2003) The IAU 2000 resolutions for astrometry, celestial mechanics, and metrology in the relativistic framework: explanatory supplement. Astron J 126:2687–2706
Sovers O, Fanselow J, Jacobs C (1998) Astrometry and geodesy with radio interferometry: experiments, models, results. Rev Mod Phys 70:1393–1454
Tapley BD, Schutz BE, Born GH (2004) Statistical orbit determination. Elsevier Academic Press, London
Turon C, O'Flaherty KS, Perryman MAC (eds) (2005) The three-dimensional universe with GAIA, ESA SP-576
Turyshev SG (2009) Experimental tests of general relativity: recent progress and future directions. Uspekhi Fizicheskikh Nauk, vol 179, p 3. arXiv:0809.3730 [gr-qc]
Turyshev SG, Farr W, Folkner WM, Girerd AR, Hemmti H, Murphy TW, Williams JG, Degnan JJ (2010) Advancing tests of relativistic gravity via laser ranging to Phobos. Experim Astron 28(2–3):209–249 (arXiv:1003.4961v2)
Vecchiato A, Lattanzi MG, Bucciarelli B, Crosta M, de Felice F, Gai M (2003) Testing general relativity by micro-arcsecond global astrometry. Astron Astrophys 399:337–342
Vallado DA (2001) Fundamentals of astrodynamics and applications, 2nd edn., Space technology libraryMicrocosm Press, California
Wei-qun Z, Chuan-fu L, Shuang-lin Y, Guan-zhong W, Yi-ping Z, Pei-hong Y, Jian Z (2001) A study and performance evaluation of hydrogen maser used in Chinese mobile VLBI stations. Chin Astron Astrophys 25(3):390–397
Will CM, Nordtvedt K Jr (1972) Conservation laws and preferred frames in relativistic gravity. I. Preferred-frame theories and an extended PPN formalism. Astrophys J 177:757–774

Will CM (2006) The confrontation between general relativity and experiment. Liv Rev Relativ. http://relativity.livingreviews.org/Articles/lrr-2006-3/, online article: cited on 12 Sep 2011, (arXiv:gr-qc/0510072v2)
Williams JG, Newhall XX, Dickey JO (1996) Relativity parameters determined from lunar laser ranging. Phys Rev D 53:6730–6739
Williams JG, Turyshev SG, Boggs DH (2004) Progress in lunar laser ranging tests of relativistic gravity. Phys Rev Lett 93:261101 (arXiv:gr-qc/0411113v2)
Williams JG, Turyshev SG, Boggs DH (2009) Lunar laser ranging tests of the equivalence principle with the earth and moon. Int J Mod Phys D 18:1129–1175 (arXiv:gr-qc/0507083v2)
Xu G (2007) GPS, theory, algorithms and applications, 2nd edn. Springer, Berlin
Zhang J, Zhang K, Grenfell R, Deakin R (2006) Short note: on the relativistic doppler effect for precise velocity determination using GPS. J Geod 80:104–110

Chapter 3
Global Terrestrial Reference Systems and Their Realizations

Detlef Angermann, Manuela Seitz and Hermann Drewes

1 Introduction

Geodetic reference systems are fundamental requisites for accurate and reliable geodetic results. Unambiguous reference systems are needed to refer the geodetic observations and estimated parameters to a unique global basis. Highly accurate, consistent and reliable realizations of the terrestrial reference systems are required for measuring and mapping the Earth's surface and its variations in time. These terrestrial reference frames are the basis for many practical applications, such as national and regional geodetic networks, engineering, precise navigation, geo-information systems, etc. as well as for scientific investigations in the Earth's system (e.g., tectonic plate motion, sea level change, seasonal and secular loading signals, atmosphere dynamics and Earth orientation excitation).

Historically, the establishment of global reference systems was first addressed in the late 1960s when space geodesy observations became available with the advent of artificial satellites. A decade later, the space technique of very long baseline interferometry (VLBI) was able to provide a direct link to extragalactical radio sources. While in the early days the measurement accuracy and its resolution was comparatively poor, today's space geodetic observation techniques have made enormous progress in precision and reliability. Today, space geodetic observation techniques allow one to determine geodetic parameters (e.g., station positions, Earth rotation) with a precision of a few millimetres (or even better). However, to exploit fully the potential of the space geodetic observations for investigations of various global and regional, short-term, seasonal and secular phenomena in the Earth's system, the reference frames must be realized with the highest accuracy,

D. Angermann (✉) · M. Seitz · H. Drewes
Deutsches Geodätisches Forschungsinstitut, Alfons-Goppel-Street 11,
80539 Munich, Germany
e-mail: angermann@dgfi.badw.de

G. Xu (ed.), *Sciences of Geodesy - II*, DOI: 10.1007/978-3-642-28000-9_3,

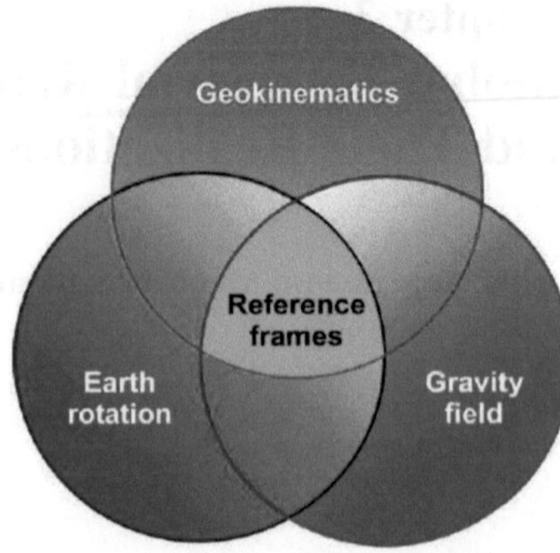

Fig. 3.1 Central role of reference frames (from Plag and Pearlman 2009)

spatial and temporal consistency and stability over decades. This is a prerequisite for the monitoring of physical processes of global change.

One of the difficulties for the definition and realization of global reference systems, however, is that the Earth itself is not a rigid body. With its atmosphere, oceans, ice cover, land surfaces, its interior and forces from external bodies, the Earth is a restless planet, which is subject to a large variety of dynamic processes. This makes the data analysis, modelling and interpretation of results more complicated and requires extreme efforts and care. On the other hand, observations and studies of the Earth system can only be improved if the underlying reference systems are realized with adequate quality.

The importance of reference systems and frames has long been recognised by many national and international organizations investing resources in the establishment and maintenance of reference frames. On the international level, the "Global Geodetic Observing System" (GGOS, Rothacher 2000), now a full component of the International Association of Geodesy (IAG), is the geodetic contribution to a comprehensive global observing system as it is presently set up by the "Group on Earth Observation" (GEO) in the form of the "Global Earth Observing System of Systems" (GEOSS) (Plag and Pearlman 2009). A key task and major goal of GGOS is the development of an observing system capable of measuring the Earth's variable shape, gravity field and rotation with an accuracy and consistency at the 1 mm level with high temporal and spatial resolution. As the linking element between these three pillars (see Fig. 3.1), the reference frames have to be established with the accuracy level specified by the GGOS requirements. The goal is to achieve a consistency and accuracy at the mm-level with a stability of 0.1 mm/year.

One prominent example for global change is the monitoring of global sea level rise, which is also an important driving factor for the accuracy requirements of geodetic reference frames. For almost 20 years satellite altimetry has provided accurate measurements for the determination of the current rate of sea level rise. The results of the computations performed at DGFI are shown in Fig. 3.2 (Bosch 2008).

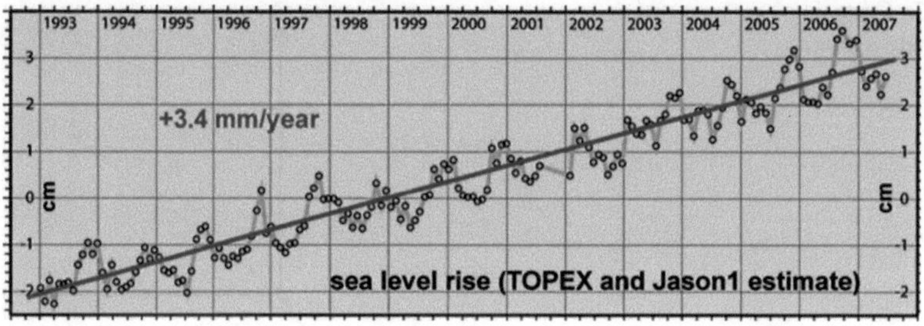

Fig. 3.2 Sea level rise obtained from multi-mission satellite altimetry using data from 1993 until 2007 (Bosch 2008)

The obtained rate of 3.4 mm/year is in good agreement with other studies and with the results published in the "Intergovernmental Panel on Climate Change (IPCC)" report (IPCC 2007). The present-day sea level change is much higher than the estimates for the twentieth century, which show a global average sea level rise of about 1.7 mm/year. This is probably an indication for an acceleration of this global change phenomenon. A precise and reliable observation of the present-day sea level change is of considerable interest because of its potential impact on human populations living in coastal regions and on islands as well as on the wider natural environment.

The estimated rates of global sea level rise from satellite altimetry strongly depend on the accuracy and long-term stability of the underlying reference frames. Any possible errors in station positions and velocities will have an impact on the orbit positions of the altimeter satellites, which then directly affects the results obtained for the global sea level change. A crucial factor is also the realization of the geodetic datum of the terrestrial reference frame (i.e., the origin and scale including their rates). Thus, a reliable monitoring of sea level change requires a high accuracy for geodetic reference frames with a high long-term stability over decades.

In geodesy, two fundamental reference systems, a celestial (space-fixed) and terrestrial reference system (earth-fixed) are required for the description of motion of astronomical bodies like the Earth, artificial satellites, or planets, as well as for the analysis of the space geodetic observations and the representation of results. In this chapter the focus is on global terrestrial reference systems and their realizations. After the introduction of some basic information, the authors give an overview about the terrestrial reference system and its definition. The next part is related to its realization, the terrestrial reference frame. In this context, the general concept, the history and the latest developments in the field of global reference frame realizations are addressed. Finally, the present status is discussed and challenges for future improvements are provided.

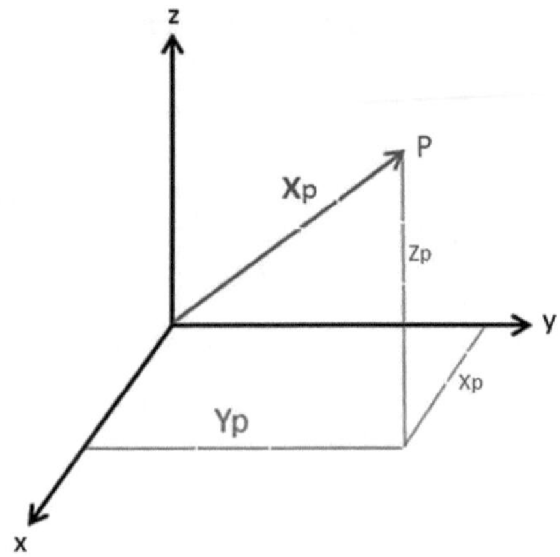

Fig. 3.3 Cartesian coordinate system

2 Basic Concepts and Fundamentals

Reference coordinate systems in satellite geodesy are global and geocentric in nature, because the satellite motion refers to the center of mass of the earth (geocenter). In a Cartesian coordinate system with the axes x, y, z the position of a point P is given by its position vector Xp (Fig. 3.3).

Conventionally, a terrestrial reference system is a spatial reference system co-rotating with the Earth. The origin of such a system is located in the geocenter, the orientation is equatorial (the z axis in the direction of the mean pole), and the scale is defined by an SI meter. The advantage of such a definition is that positions of points attached to the solid surface of the Earth have coordinates which undergo only small variation with time. These variations are mainly due to geophysical effects caused by various dynamic processes and forces from external bodies.

The International Terrestrial Reference System (ITRS) has been formally adopted and recommended for Earth Science studies by the International Union for Geodesy and Geophysics (IUGG) at its General Assembly in 2007 (IUGG 2007). For a unique relation between the reference system and coordinates of points, it is necessary to define constants, conventions, physical models and model parameters (e.g., Earth and ocean tide models, Love numbers). The ITRS is based on the conventions of the International Earth Rotation and Reference Systems Service (IERS), which serve as the necessary basis for the mathematical representation of geometric and physical quantities (Petit and Luzum 2010). Although, the coordinates in the ITRS are fully defined, they are not necessarily accessible.

To make the ITRS available to the users, a set of accessible parameters must be constructed to materialize the system (e.g., by a set of physical points with precisely determined coordinates in a Cartesian coordinate system). The ITRS is realized by a set of station coordinates and velocities for a global network at a given epoch, which constitutes a terrestrial reference frame. Thus, reference frames realize the reference system physically, i.e., by a solid materialisation of points, and mathematically, i.e., by the determination of parameters (e.g., geometric coordinates). The realization of the ITRS is called the International Terrestrial Reference Frame (ITRF).

The geodetic datum of the reference frame fixes unequivocally the relation between a frame and a system by allocating a set of "given" parameters, e.g., the coordinates of the origin of the system (x_0, y_0, z_0), the direction of the coordinate axes x, y, z, and the scale as a unit of length.

These concepts have been defined extensively by the astronomical and geodetic communities (e.g., Kovalevski et al. 1989; Boucher 2001; Drewes 2009a). In addition, some basic considerations and fundamentals related to reference systems, reference frames and the geodetic datum are given in Drewes (2009a). There it is pointed out that the definition of the reference system, the realization by a frame, and the allocation of the geodetic datum have to be strictly coherent. Reference systems, the geodetic datum, and reference frames form an order of hierarchy:

- The definition of a reference system must be completely unaffected by the realization of the reference frame and the geodetic datum, i.e., the realization of the system by the frame and the allocation of the datum must not change its definition.
- The definition of the datum has to be done by methods independent of the measurements of the reference frame, i.e., measurement errors or physical changes altering the observations of the frame must not affect the datum.
- The mathematical realization of the reference system has to be done by algorithms that keep the datum parameters fixed and follow strictly the principles defined by the reference system.

3 International Terrestrial Reference System

3.1 ITRS Definition

The International Terrestrial Reference System (ITRS), realized and maintained by the IERS, has been formally adopted and recommended for scientific and technological applications. According to IUGG Resolution 2 (IUGG 2007), the ITRS is a specific Geocentric Terrestrial Reference System (GTRS) for which the orientation is operationally maintained in continuity with past international agreements (i.e., BIH orientation).

The ITRS definition fulfils the following conditions (Petit and Luzum 2010):

- It is geocentric, the center of mass being defined for the whole Earth, including oceans and atmosphere.
- The unit length is the meter (SI). This scale is consistent with the TCG time coordinate for a geocentric local frame, in agreement with "International Astronomical Union" (IAU) and IUGG resolutions. This is obtained by appropriate relativistic modelling.

- Its orientation was initially given by the Bureau International de l'Heure (BIH) orientation of the BIH Terrestrial System (BTS) at epoch 1984.0.
- The time evolution of the orientation is ensured by using a no-net-rotation (NNR) condition with regard to horizontal tectonic motions over the whole Earth.

The ITRS is defined as geocentric, equatorial, and metric. This means that the coordinates given to the origin must be zero in order to refer to the geocenter: $x_0 = y_0 = z_0 = 0$. The orientation of the z axis is close to the (varying) Earth's rotation axis, and the x and y axes are in the equatorial plane. According to *The Metre Convention* (http://www.bipm.org/en/convention/) the scale is metric. Some aspects related to the datum definition are addressed below; more details are given in Drewes (2009a).

3.1.1 Origin

The geocentric origin can be realized by gravimetric parameters because the geocenter is defined as the Earth's center of mass (M = total mass of the Earth):

$$\begin{aligned} x_0 &= \iiint x \, \mathrm{dm}/M \\ y_0 &= \iiint y \, \mathrm{dm}/M \\ z_0 &= \iiint z \, \mathrm{dm}/M. \end{aligned} \tag{3.1}$$

The equations at (3.1) are related to the formulas below for the spherical harmonic coefficients of the Earth's gravity field (a = semi-major Earth axis):

$$\begin{aligned} C_{11} &= \iiint x \, \mathrm{dm}/\mathrm{a}\, M \\ S_{11} &= \iiint y \, \mathrm{dm}/\mathrm{a}\, M \\ C_{10} &= \iiint z \, \mathrm{dm}/\mathrm{a}\, M. \end{aligned} \tag{3.2}$$

This means that if in satellite positioning a gravity field model with $C_{11} = S_{11} = C_{10} = 0$ is used for the determination of the satellite orbits, a coordinate system with the origin $x_0 = y_0 = z_0 = 0$ (in the geocenter) is introduced. There is no degree of freedom for these three datum parameters. All satellite ephemerides are unequivocally and at any time estimated in the geocentric reference system. The geocentric ephemerides are related to the station coordinates of the terrestrial reference frame by distance measurements between satellites and ground stations, e.g., by Satellite Laser Ranging (SLR). They are also geocentric unless additional constraints are introduced, e.g., by fixing some terrestrial coordinates. Such constraints shall be avoided to conserve the datum given by the dynamics of the orbits. This holds true for all observation epochs: the origin is always in the geocenter, thus there cannot be a motion of the origin with respect to the geocenter or vice versa. The definition of the origin of the terrestrial reference frame is subject of various publications (e.g., Blewitt 2003; Dong et al. 2002; Wu et al. 2012).

3.1.2 Orientation

The orientation of the coordinate axes of the reference frame could, theoretically, also be defined by the Earth's gravity field, namely by the second degree and order spherical harmonic coefficients which are related to the orientation of the principal axis of inertia. This definition of the orientation is not used in practice because its determination is not as precise as for the origin, and the satellite orbits are not so sensitive w.r.t. its variations. Instead, the orientation of the ITRS is conventionally defined to be consistent with the orientation of the BTS at epoch 1984.0. Consistency between the sequent ITRF solutions is ensured by applying a NNR condition, that the actual realization is not rotated w.r.t. to the previous one. Thus the definition of the orientation is purely geometrical and arbitrary and not related to physical Earth parameters as is the case for origin and scale. That means that the orientation is not reproducible independently but depends on the previous ITRF realization. This is not satisfactory as errors and uncertainties in the realization of the orientation are propagated into the following ITRF solutions. An issue to be solved is the time evolution of the orientation from the reference epoch to all the other epochs. According to the ITRS definition, the time evolution is ensured by using an NNR-condition with regard to horizontal motions over the whole Earth. For the realization of the kinematic datum of the ITRF the geologic-geophysical model NNR-NUVEL-1A (Argus and Gordon 1991; DeMets et al. 1994) has been used (Altamimi et al. 2007).

3.1.3 Scale

The metric scale of the reference system is defined via the meter definition by the speed of light in vacuum. All the space techniques contain, in principle, information to realize the scale by fixing the speed of light. In the atmosphere, however, refraction effects caused by the ionosphere and troposphere have to be taken into account. As distances are determined by measurements of travel time, the electronic reference points (phase centers) have to be localized w.r.t. the geometric (materialized) reference points. Furthermore, it has to be investigated what other technique-specific biases and deficiencies regarding the modelling of the individual techniques may affect the realization of their scale. This is essential to assess the accuracy of the scale from space geodetic techniques.

3.2 Positions and Displacements of Reference Points

The definition of a conventional model to describe the station motions on a deformable Earth is complicated in nature. The reference points on the Earth's crust undergo a variety of motions: tidal displacements due to Earth's body tides and ocean tides (including ocean loading), polar motion, loading signals from atmospheric and hydrologic mass movements, tectonic motions, and other motions from internal mass movements. Certain reference points are also affected by volcanic effects or seismic events in the form of coseismic and postseismic displacements.

The ITRS is realized basically by a set of terrestrial points (reference points of tracking instruments or geodetic markers) referred to the solid Earth crust. According to the IERS Conventions (Petit and Luzum 2010), the station positions and displacements are described by coordinates of a reference point and constant velocities. This set of coordinates describes the secular evolution of the polyhedron over time. However, the application of such a linear model is not suitable for a description of the actual station motions on the Earth's crust, since these motions are to a large extent nonlinear caused by various geophysical effects. This is the reason that a so-called regularized position $x_R(t)$ is introduced by applying conventional corrections $\Delta x_i(t)$ in order to get a position with regular (nearly linear) time evolution. The general model connecting the instantaneous a priori position of a point on the Earth's crust at epoch t, $x(t)$, and a regularized position $x_R(t)$ is

$$x(t) = x_R(t) + \sum \Delta x_i(t). \tag{3.3}$$

In the conventional linear station motion model the regularized station position $x_R(t)$ at an epoch t is expressed as

$$x_R(t) = x_0 + v_0 * (t - t_0), \tag{3.4}$$

with the station position x_0 at a reference epoch t_0 and a constant velocity v_0. The numerical values x_0 and v_0 of a set of stations (reference points) constitute a specific realization of the ITRS. For some stations, it is necessary to consider several discrete linear segments in order to account for abrupt discontinuities in position (for example, due to earthquakes or changes in observing equipment).

From (3.3) it becomes obvious that the ITRS definition must include also well-defined conventions for the computation of the correction terms $\Delta x_i(t)$, which are given in the IERS Conventions (Petit and Luzum 2010). In the analysis of the space geodetic observations, these conventions shall be consistently applied by all analysis groups for the reduction of the space geodetic observations.

The displacements of reference points due to various geophysical phenomena and other effects is addressed in detail in Chap. 7 of the IERS Conventions 2010 (Petit and Luzum 2010). Three kinds of displacement are distinguished:

- Conventional displacements of reference points.
- Non-conventional displacements.
- Displacements of reference points of instruments.

The conventional displacements of reference markers on the crust include deformations of the solid Earth due to the body tides arising from the direct effect of the external tide-generating potential, displacement due to ocean tidal loading and to diurnal and semidiurnal atmospheric pressure loading, as well as those due to centrifugal perturbations caused by Earth rotation variations, including the pole tide. The displacements are described by geophysical models or gridded convolution results derived from geophysical models.

The second category includes non-conventional displacements of reference markers on the crust such as non-tidal motions associated with changing environmental loads (e.g., from atmosphere, ocean, continental water, post-glacial rebound), post seismic displacements, as well as other secular, episodic or seasonal variations caused, e.g., by mass redistributions within the Earth's system. The currently available models to account for these effects are usually less accurate, than those of the first category and also some deviations (up to several millimeters) between different models still exist. Thus, it is recommended by the IERS Conventions 2010 presently not to include such models in operational solutions for the generation of IERS products in order to get consistent results. However, the consequence of this recommendation is that the corresponding non-tidal loading displacements and other non-modelled effects will remain as signals in the geodetic time series results and will affect also the ITRF. To achieve further progress in this field, it is stimulated by the IERS that appropriate models should be made available to the user community through the IERS Global Geophysical Fluids Centre and its Special Bureaux, together with all necessary supporting information, implementation documentation, and software. These models should then be used and tested for data analyses and further studies.

The third category includes models for displacements of reference points of instruments which are to be considered when relating different instruments (e.g., SLR and VLBI) to each other or relating them to a physical marker. Typical examples are antenna phase center offsets and variations as well as thermal expansion of VLBI antennas. These effects are technique-dependent and the conventional models for these effects are kept and updated by the IAG services for the space techniques: the International GNSS Service (IGS) (Dow et al. 2009), the International Laser Ranging Service (ILRS) (Pearlman et al. 2007), the International VLBI Service for Geodesy and Astrometry (IVS) (Schlüter and Behrend 2007), and the International DORIS Service (IDS) (Willis et al. 2010). As some of these technique-specific effects depend on local environmental conditions, conventional models need to be based on reference values (e.g., for local temperature). A conventional model to determine the reference temperature that is recommended in the IERS Conventions 2010 is the global pressure and temperature (GPT) model (Böhm et al. 2007). The current version of the IERS Conventions includes also a model for VLBI antenna thermal deformation (Nothnagel 2008) which shall be used within the IVS, and a model for absolute phase center corrections for satellite and receiver antennas (Schmid et al. 2007) which has been applied by the IGS since November 2006.

Finally, a few remarks concerning the conventional approach and the application of correction models for station displacements shall be provided: If the station motions are represented by such a linear model as given in (3.4), in the ideal case, all nonlinear displacements of reference positions shall be captured by using conventional corrections $\Delta x_i(t)$. However, in practise this requirement cannot be fulfilled, since it is not possible to describe the irregular (nonlinear) station motions caused by various dynamic processes in the Earth system completely by conventional models. As a consequence, deviations of the nonlinear part of the real station motions from the conventional corrections will become visible as residuals in the reference point coordinate time series. This means, in other words, that the instantenous ITRF position of a reference point differs from the actual position estimated from space geodetic observations at a specific epoch. This can be considered as one of the limiting factors for the accuracy of present realizations of the terrestrial reference system. This issue is subject of ongoing discussions within the scientific community (see Sect. 6.3).

4 International Terrestrial Reference Frame

The IERS is responsible for the establishment and maintenance of the International Terrestrial Reference Frame (ITRF), a realization of the ITRS. The definition of the ITRS and the geophysical models to be used for its realization are specified in the IERS conventions (Petit and Luzum 2010). The ITRF is realized by the positions at a reference epoch and constant velocities for the IERS network stations derived from a combination of solutions. The contributing space techniques are VLBI, Satellite and Lunar Laser Ranging (SLR/LLR), Global Positioning System (GPS), and Doppler Orbitography and Radiopositioning Integrated by Satellite (DORIS).

Realizations of the ITRS are produced by the ITRS Centre hosted at the Institute Géographique National (IGN) in Paris, France. Within the re-organized IERS structure (since 2001), the ITRS Centre (formally called ITRS Terrestrial Reference Frame Section) is supplemented by ITRS Combination Centres which were included as additional IERS components to ensure redundancy for the ITRF computations and to allow for a decisive validation and quality control of the combination results. Three ITRS Combination Centres are established at Deutsches Geodätisches Forschungsinstitut (DGFI), IGN, and National Resources Canada (NRCan).

4.1 IERS Network

4.1.1 Definition of the IERS Network

In its first materialization which was the BTS 1984 (BTS84), the global network of geodetic reference stations was defined through all SLR, LLR, VLBI and Doppler/TRANSIT tracking instruments used by the various individual analysis centres contributing to the BIH (as the predecessor of the IERS) at that time (Boucher and Altamimi 1985). In the course of time, GPS stations from the IGS were added as well as the sites of the DORIS tracking network. Since its inception, the network

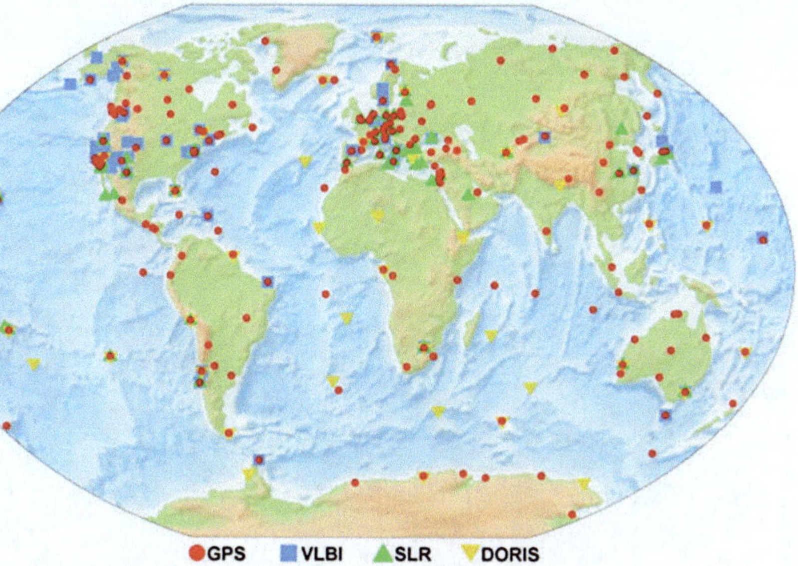

Fig. 3.4 Spatial distribution of technique-specific station networks

also included a selection of ground markers, specifically those used for mobile equipment and those currently included in local surveys performed for monitoring local eccentricities between instruments of co-location sites or for site stability checks.

Each point is currently identified by the attribution of a DOMES number. Historically, the DOMES numbering system was designed at the start of the MERIT campaign in the early 1980s in order to give an unambiguous identifier to all instrument reference points and markers involved in this MERIT project (Monitoring of Earth Rotation and Intercomparison of Techniques). This information was first published in a catalogue by the Bureau International de l'Heure (BIH) and entitled "Directory of MERIT Sites", hence DOMES (MERIT/COTES 1983). Since the official start of the IERS in 1988, the TRF section of the IERS Central Bureau continued this task for all ITRF contributing stations. Neighboring points are clustered into a site. The current rule is that all points which could be linked by a co-location survey (up to 30 km) should be included in a single site of the IERS network having a single DOMES site number.

Due to the progress within the space geodetic community and the huge effort of the technique-specific services, the IERS network was continuously expanded. Currently, about 40 VLBI-, 30 SLR-, 60 DORIS- and more than 300 GPS stations contribute to the IERS network (see Fig. 3.4).

Fig. 3.5 Geodetic Observatory Wettzell (Germany) with co-located VLBI, SLR, and GPS instruments. The observatory is jointly operated by the Bundesamt für Kartographie und Geodäsie (*BKG*) and the Forschungseinrichtung Satellitengeodäsie (*FESG*), Technische Universität München. Clearly visible are the recently installed twin telescopes, the 20 m radiotelescope and two laser ranging systems (Courtesy: A. Neidhardt, FESG)

4.1.2 Co-Location Sites

A co-location site is defined by the fact that two or more space geodesy instruments are occupying or have occupied simultaneously or sequentially at close locations. Typically, the distances between the different instruments are a few hundred meters, in some cases they are several kilometers. The connection between the reference points of different instruments shall be very precisely surveyed in three dimensions using classical surveys or GPS surveys. Today's accuracy requirement of 1 mm for these terrestrial reference vectors (geodetic local ties) is currently not achieved in most cases. It also has to be considered that there are different situations for the co-location sites such as simultaneous or non-simultaneous measurements and instruments using the same or different techniques. Usually, co-located points should belong to a unique IERS site. As an example, the Geodetic Observatory Wettzell (Bavarian Forest, Germany) is shown in Fig. 3.5.

Both the co-location sites and the geodetic local ties represent a key element in the ITRF computation. The ITRF relies strongly on the availability and quality of local ties in co-location sites as well as on a sufficient number and spatial distribution of these sites over the globe. Currently, there are about 90 co-location sites (see Fig. 3.6). These sites are equipped with (at least two) instruments of the four major observing techniques (VLBI, SLR, GPS, DORIS). Many of them are still

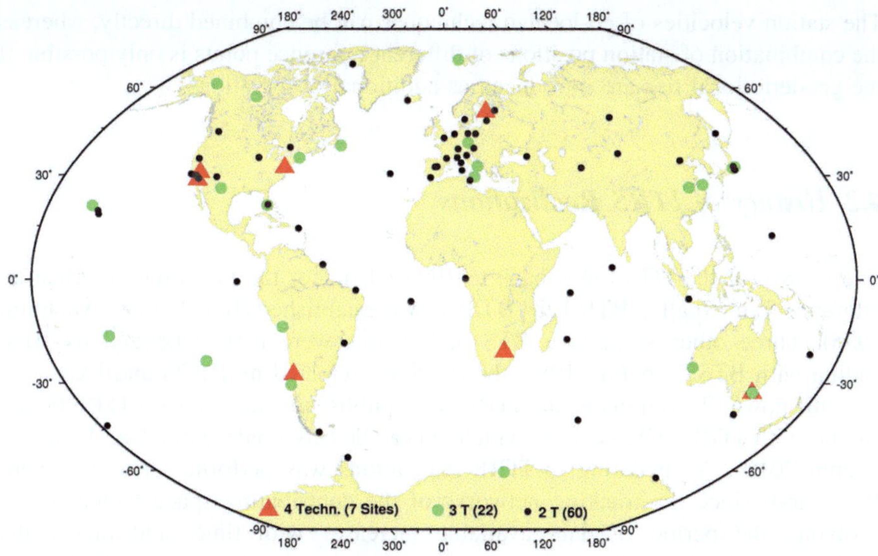

Fig. 3.6 Global distribution of ITRF2008 co-location sites

operating, but some others are mobile stations or stations that are not operating any more. For the ITRF2008 combination the co-location sites are distributed as follows: 60 sites with 2 techniques, 22 sites with 3 techniques, and 7 sites with 4 techniques. There are about ten co-location sites where no local tie information was available and several local ties show large discrepancies (>2 cm) w.r.t. the space geodetic estimates (e.g., Angermann et al. 2004; Krügel and Angermann 2007; Seitz et al. 2012).

It is obvious from Figs. 3.4 and 3.6 that the station networks and the co-location sites are rather inhomogeneously distributed. In particular, in the Southern hemisphere large regions with a sparse spatial distribution do exist. The majority of co-locations are between GPS and the other space techniques; there are only a few co-locations between VLBI and SLR. Thus, the GPS network plays a dominant role for the integration of the different techniques, which is rather problematic for the identification of the remaining technique-specific biases. Various local surveys have been performed during the last few years to determine the local tie vectors at co-location sites. The local tie vectors, which were used in the ITRF2008 combination, are available at http://itrf.ensg.ign.fr/local_surveys.php (Altamimi et al. 2011).

It shall be recalled that, without sufficient co-location sites and highly accurate local tie information, it is impossible to establish a unique and common reference frame for all major space techniques. Moreover, co-location sites are crucial for the comparison and validation of the individual space technique solutions to detect technique-specific biases. They provide the basis to combine parameters common to more than one technique, which strengthens the reference frame realizations.

The station velocities of co-located techniques can be combined directly, whereas the combination of station positions of different reference points is only possible if the geodetic local ties are introduced as additional observations.

4.2 History of ITRS Realizations

The history of the ITRF goes back to 1984 when, for the first time, a terrestrial reference frame (called BTS 1984 BTS84) was established (Boucher and Altamimi 1985). Three other successive BTS realizations were then generated by BIH, ending with BTS87, before 1988, the IERS was created by IUGG and IAU.

Until now, 12 versions of the ITRF were published, starting with ITRF88 and ending with ITRF2008, each of which superseded its predecessor (see Petit and Luzum 2010). An updating of ITRF realizations was performed (mostly every 1–3 years); since the tracking networks of the contributing space techniques is evolving, the period of data available increases with time, and finally the modelling and data analysis strategies as well as the combination methods were continuously improved in the last two decades.

The first realization was the ITRF88 followed by ITRF89 and ITRF90. These three realizations were computed by combining mean station positions (without station velocities) of VLBI, SLR, and LLR observing networks. At that time, it was recommended to adopt the AM0-2 model of Minster and Jordan (1978) for the velocities. Starting with ITRF91 and till ITRF93, combined velocity fields were estimated. The ITRF91 orientation rate was aligned to that of the NNR-NUVEL-1 model (Argus and Gordon 1991), and ITRF92–NNR-NUVEL-1A (DeMets et al. 1994), while ITRF93 was aligned to the IERS EOP series. Since ITRF94, full variance matrices of the individual solutions incorporated in the ITRF computations were used (Boucher et al. 1996). The ITRF96 was then aligned to ITRF94, and the ITRF97 to the ITRF96 using 14 parameter similarity transformations (Boucher et al. 1998, 1999). The ITRF2000 was intended to be a standard solution for geo-referencing and Earth science applications. Therefore, in addition to primary core stations observed by VLBI, SLR, LLR, GPS, and DORIS, the ITRF2000 was densified by regional GPS networks in Alaska, Antartica, Asia, Europe, North and South America, and the Pacific region (Altamini et al. 2002; Boucher et al. 2004).

Table 3.1 summarizes the number of stations and solutions for the ten ITRF realizations from ITRF88 till ITRF2000, compiled by the IERS Terrestrial Reference Frame Section, IGN, France. The ITRF network has improved with time in terms of the number of sites and co-locations as well as their distribution over the globe. GPS solutions have been included since 1991 and DORIS since 1994.

Since the re-organization of the IERS structure in 2001, the ITRS Centre, hosted at the Institute Geographique National (IGN), France, is supplemented by the ITRS Combination Centres (DGFI, IGN, NRCan). All these institutions (and also other groups) contributed to the development of refined combination

Table 3.1 Overview of the ITRF realizations compiled by the IERS terrestrial reference frame section, IGN France, from ITRF88 until ITRF2000; see http://lareg.ensg.ign.fr/ITRF

Realizations	# Stations	# Solutions				# Solutions
		VLBI	SLR	GPS	DORIS	Total
ITRF88	120	5	6	–	–	11
ITRF89	113	6	8	–	–	14
ITRF90	120	4	7	–	–	11
ITRF91	131	5	7	1	–	13
ITRF92	155	5	6	6	–	17
ITRF93	260	6	4	5	-	15
ITRF94	209	6	1	5	3	15
ITRF96	290	4	2	7	3	16
ITRF97	309	4	5	6	3	18
ITRF2000	477	3	9	6 + 8*	3	21 + 8*

Regional GPS solutions that were included in the ITRF2000 are represented by *

strategies for the terrestrial reference frame computations. One focus was on the investigation of the time evolution of the terrestrial reference frame by analyzing time series of station positions and geophysical parameters (origin, scale) obtained from VLBI, SLR, GPS, and DORIS solutions (Angermann et al. 2005; Meisel et al. 2005). The time series analysis allows one to detect discontinuities (e.g., those due to geophysical events such as earthquakes or instrumentation changes) that can be accounted for by piecewise-linear modeling. Furthermore, the investigation of the temporal behavior of the physical frame parameters (namely the origin and scale) has been performed, which provides insight into the individual space-geodetic solutions.

Based on these developments a completely refined strategy for the ITRF computations was defined by the ITRS Centre and the ITRS Combination Centres. This new strategy has then been applied, for the first time, to the ITRF2005 computations and was then also adopted to generate the ITRF2008. It is based on a time series of input solutions/normal equations with station positions and Earth orientation parameters (EOP) of the different space techniques, including the full variance covariance matrices. This enables one to account for discontinuities in the station positions by a piecewise linear parameterization and to detect other non-linear station motions (e.g., seasonal variations) as well as possible systematic effects in the time series. Furthermore, the EOP that were included in the ITRF2005 and ITRF2008 computations provide, as common parameters to all techniques, valuable information for the combination of the different space techniques (the so-called inter-technique combination).

The ITRF2005 and ITRF2008 computations also benefit from the re-organized IERS structure, since for both ITRS realizations two complete solutions were computed by the ITRS Combination Centres at IGN and DGFI; in addition, NRCan provided valuable input, in particular for GPS. The fact that IGN and DGFI use their own software packages and also apply different combination strategies allows for a decisive validation of the combination results, ensures an independent quality control, and provides the basis for an accuracy assessment of the ITRF.

In October 2006, the ITRF2005 was released by the ITRS Centre (Altamimi et al. 2007; Angermann et al. 2007, 2009). The ITRF2005 includes a time series of the space geodetic observations until the end of 2005. In November 2008, the IERS released a call for participation and solicited input data sets from the IERS Technique Centres (IGS, ILRS, IVS, IDS) and their related analysis centers for a new realization, the ITRF2008, to take full benefit of the latest developments. Since ITRF2005, 3 years of additional observation data have become available, new sites have been added to the ITRF network, possible changes in station motions need to be detected, and new geodetic local surveys have been performed. The models and standards for the data analysis of the different observation techniques have been continuously improved and unified to a much higher level as has been achieved for the ITRF2005.

In preparation for the ITRF2008, the IAG Services, together with their respective Analysis and Combination Centres, achieved significant improvements for the modeling and processing of the space geodetic observations. It was a remarkable effort to enable the capabilities for a self-consistent re-processing and per-technique combination of the VLBI, SLR, GPS, and DORIS observation time series covering (almost) the full history of observations. The data were processed by applying state-of-the-art models and parameterizations.

5 The Latest Realization, the ITRF2008

5.1 ITRF2008 Input Data

The input data sets for the ITRF2008 are time series of solutions or normal equations of station positions and EOP, including the full variance covariance matrices. The data were provided in the SINEX format by the corresponding Services of the IAG, namely the IGS, ILRS, IVS, and IDS. Remarkable progress has been achieved within these services regarding improved analysis strategies and re-processing capabilities of the long-observation time series. For the first time, for each of the contributing space-techniques, self-consistent reprocessed time series were generated, which cover time spans of more than 10 years for GPS and DORIS and about 25 years for SLR and VLBI. The homogeneously re-processed time series provide a fundamental basis for improved terrestrial reference frame computations.

The individual time series were combined per-technique by the four responsible techniques' combination centers, namely the NRCan for the IGS (Ferland 2010), the Institute for Geodesy and Geoinformation (IGG) of the University Bonn, Germany, for the IVS (Böckmann et al. 2010), the Agenzia Spaziale Italiana (ASI) for the ILRS (Bianco et al. 2000), and the Collecte Localisation par Satellite (CLS) in cooperation with the Centre National d'Etudes Spatiales (CNES), France, and NASA's Goddard Space Flight Center (GSFC), U.S.A., for the IDS (Valette et al. 2010; Willis et al. 2010).

The time resolution of the SINEX files is weekly for the satellite observations (GPS, SLR, DORIS) and daily (24 h session-wise) for VLBI. The VLBI data were provided in the form of unconstrained normal equations, and the data of the satellite techniques as solutions with loose or minimal constraints. This new type of input data provides the opportunity to analyze the time series of station positions and datum parameters such as the origin and scale of the contributing technique-specific networks. This enables one to account for discontinuities in the station positions and to detect other nonlinear station motions (e.g., seasonal variations) as well as possible systematic effects in the time series. Furthermore, the EOP provide as parameters common to all techniques valuable information for the combination of the different space techniques (the so-called inter-technique combination).

Compared to the ITRF2005 submissions, in particular the following improvements were achieved:

- New absolute phase center offsets and variation models for satellites and GPS receivers were used for the reprocessing of the GPS time series (Schmid et al. 2007). Furthermore a new tropospheric model (Böhm et al. 2006a, b) has been implemented by the IGS Analysis Centres.
- The consistently re-processed VLBI observations (back to 1980) account for the mean pole tide correction following the IERS Conventions (Petit and Luzum 2010). The new tropospheric model was also implemented by the IVS Analysis Centers and the thermal deformations of the VLBI antennas were modelled (Nothnagel 2008).
- The improved SLR time series (back to 1983) take into account new range bias values, a refined tropospheric modelling (Mendes and Pavlis 2004) as well as other station-dependent corrections (Pavlis et al. 2010).
- For the DORIS time series (back to 1993) the IDS achieved for the first time a self-consistent processing and per-technique combination of seven individual analysis center solutions. Furthermore, improved models for solar radiation pressure and for atmospheric drag estimation were implemented (Gobinddass et al. 2009).

Table 3.2 summarizes the major characteristics of the ITRF2008 input data. All the data files contain station positions and a set of EOP. The parameters and their time resolution provided by the individual techniques are given in Table 3.3.

In addition to these observation time series, the ITRF2008 input data comprise the geodetic local tie information, a key element for the integration of the different space techniques (inter-technique combination). All the local ties used in the ITRF2008 computation are provided in SINEX format and are available at the webpage of the ITRS Centre at http://itrf.ign.fr/local_surveys.php (Altamimi et al. 2011). The distribution of co-location sites is shown in Fig. 3.6.

Table 3.2 Input data sets for ITRF2008 (*TC*: techniques' combination centre, *NEQ*: constraint-free normal equation)

Technique	Service/TC	Data	Time period
GPS	IGS/NRCan	Weekly solutions	1997–2009
VLBI	IVS/IGG	24 h session NEQ	1980–2009
SLR	ILRS/ASI	Weekly solutions	1983–2009
DORIS	IDS/CLS-CNES-GSFC	Weekly solutions	1993–2009

Table 3.3 Parameter included in ITRF2008 input data

Parameter	GPS	VLBI	SLR_1	SLR_2	DORIS
Station positions	w	s	14d	w	w
Offsets of terrestrial pole	d	s	3d	d	d
Rates of terrestrial pole	d	s			
UT1–UTC		s			
LOD	d	s	3d	d	
Nutation offsets		s			

Resolutions are: *d* daily, *3d* three-daily, *s* 24h session-wise, *w* weekly, *14d* fortnightly. SLR_1 specifies the time interval 1980 until 1993, whereas SLR_2 denotes the period from 1993 to 2009

5.2 ITRF2008 Data Analysis

The ITRS Combination Centres at IGN and DGFI applied their preferred data analysis strategy by using their own combination software packages. The combination strategy of IGN is on the solution level by simultaneously estimating similarity transformation parameters with respect to the combined frame along with the adjustment of station positions, velocities, and EOP (Altamimi et al. 2007, 2011). The strategy applied at DGFI is based on the normal equation level and on the common adjustment of station positions, velocities, and EOP (Angermann et al. 2009; Seitz et al. 2012). A comparison of the data analysis and combination strategies of both ITRS Combination Centres is provided in Table 3.4.

Despite some differences between both strategies, the general procedure for the ITRF2008 computation is similar, which is based on two main parts:

- The accumulation (stacking) of the time series per technique to generate technique-specific solutions/normal equations.
- The combination of the per-technique solutions/normal equations.

In the following, the data analysis strategy applied at DGFI for the computation of the DTRF2008 is described. The computations have been performed with the DGFI Orbit and Geodetic Parameter Estimation Software (DOGS-CS) (Gerstl et al. 2000). Figure 3.7 gives an overview of the data flow and the data analysis procedure. Details on the methodology and the mathematical background are given in various publications (e.g., Angermann et al. 2004, 2007; Drewes et al. 2006; Seitz 2009; Seitz et al. 2012).

Table 3.4 Comparison of the combination strategies of IGN and DGFI

	IGN	DGFI
Software	CATREF	DOGS-CS
Solution name	ITRF2008	DTRF2008
Strategy	Solution level	Normal equation level
Time series combination	Stacking of minimum constrained solutions by applying 7 parameter transformations	Accumulation of normal equations
Inter-technique combination	Combination of per-technique solutions by applying 14 parameter transformations, IGN selected set of local ties	Accumulation of per-technique datum-free normal equations, without similarity transformations, DGFI selected set of local ties
Datum		
Origin	SLR	SLR
Scale	VLBI + SLR (weighted mean)	VLBI + SLR (weighted mean)
Rotation	3 NNR cond. w.r.t. ITRF2005	3 NNR cond. w.r.t. ITRF2005

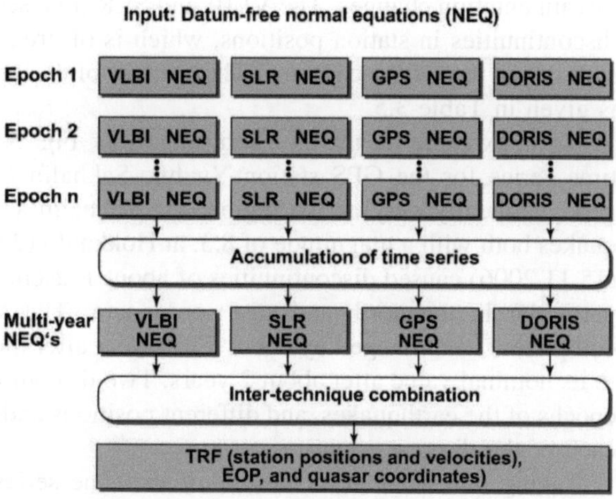

Fig. 3.7 Data flow and combination procedure for the DTRF2008 computation at the ITRS Combination Centre at DGFI

5.2.1 Accumulation of Time Series Per Technique

According to the combination methodology at DGFI, normal equations have to be generated from the ITRF2008 input data sets. For VLBI, the datum-free normal equations were used as they were provided by the IVS. In the case of SLR, normal equations were generated from the loosely constrained solutions. For GPS and DORIS the SINEX files were provided with minimum datum information. Since both techniques should not contribute to define the datum of the DTRF2008 solution, seven similarity transformation parameters were set up in the weekly normal equations. For each of the observation techniques stations with too few observations, which do not allow for the estimation of reliable station positions and velocities, were reduced.

Table 3.5 Number of stations and discontinuities per space technique in DTRF2008

Technique	# Stations	# Discontinuities
GPS	560	372
DORIS	132	48
SLR	124	30
VLBI	107	22

The resulting time series of epoch normal equations were analysed to detect nonlinear behavior in station position movement. In the first iteration, constant velocity parameters were set up in the epoch normal equations to parameterize linear station motions. The station positions of the weekly normal equations were transformed to positions at the reference epoch 2005.0 of the DTRF2008 solution. From the analysis of position time series discontinuities were identified in the station positions. The discontinuities are parameterized by setting up new position and velocity parameters for the corresponding stations. Most of the discontinuities were introduced for the GPS time series, which are in many cases caused by instrumentation changes. The VLBI and SLR time series are much less affected by discontinuities in station positions, which is of great advantage for the long-term stability of the reference frame. The number of discontinuities per space technique is given in Table 3.5.

As an example for the time series analysis, Fig. 3.8 shows the residual position time series for the GPS station Yuzhno-Sakhalin (YSSK), Russia, located in a geodynamically very active region, the Sakhalin seismic belt. Two large earthquakes both with a magnitude of 8.3, at Hokkaido (25.09.2003) and at Kuril Island (15.11.2006) caused discontinuities of about 1–2 cm in the position time series, in particular in the north and east component. The station velocity in the north component was changed by about 5 mm/year after the first earthquake, going back to its nominal value after about 2 years. Two discontinuities were introduced at the epochs of the earthquakes, and different positions and velocities were estimated for each of the three segments.

Taking into account the results of the time series analysis, the epoch normal equations were accumulated per technique. Technique-specific multi-year solutions were generated by adding minimum datum conditions w.r.t. the previous terrestrial reference frame realization, the ITRF2005. The time series analysis, the accumulation of the epoch normal equations, and the generation of multi-year solutions per technique were performed in an iterative procedure.

Based on the repeatability of the position time series, standard deviations of mean station positions were estimated, which were then used to estimate scaling factors for the technique-specific normal equations. This procedure was applied for each space technique by using a suitable subset of stations with long observation time spans. The resulting intra-technique solutions contain station positions, velocities, and daily EOP. For stations with discontinuities, separate positions and velocities were estimated for each segment. Statistical tests were applied to decide whether or not the estimated velocities for two different segments can be equated.

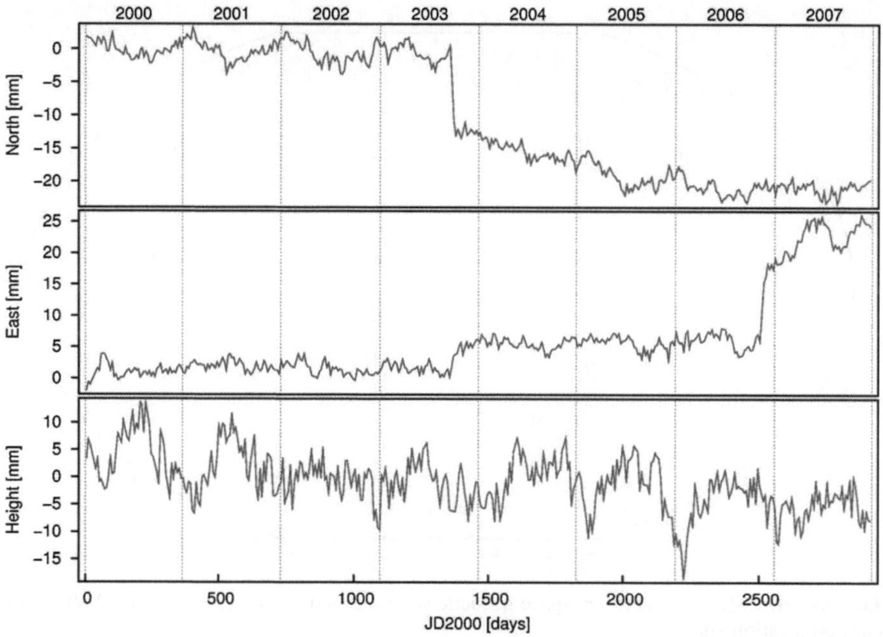

Fig. 3.8 Time series of station positions for GPS station yuzhno-sakhalin (*YSSK*), Russia

5.2.2 Inter-Technique Combination

Inputs for the combination of different techniques are the accumulated normal equations for DORIS, GPS, SLR and VLBI. The parameters include station positions, velocities and daily EOP. The connection of the different techniques' observations is given by local tie measurements between the instruments' reference points at co-location sites. These co-location sites (including the local tie vectors) are crucial elements for the inter-technique combination. They are essential to compare the individual space geodetic solutions and to integrate them into a common frame.

In a first step, the technique-specific solutions are compared with the local tie vectors at co-location sites. As an example, Fig. 3.9 displays the three-dimensional (3D) discrepancies between coordinate differences of the space geodetic solutions and the local tie vectors for 32 VLBI-GPS co-location sites. For some sites there is excellent agreement, but for a few others there are discrepancies of more than 2 cm. A similar result has been obtained for the comparison between GPS and SLR solutions and the corresponding local tie vectors at 30 co-location sites. The situation for a direct comparison between VLBI and SLR solutions is not satisfying, since there are only nine co-location sites between these two techniques and the observed discrepancies are mostly above 1 cm.

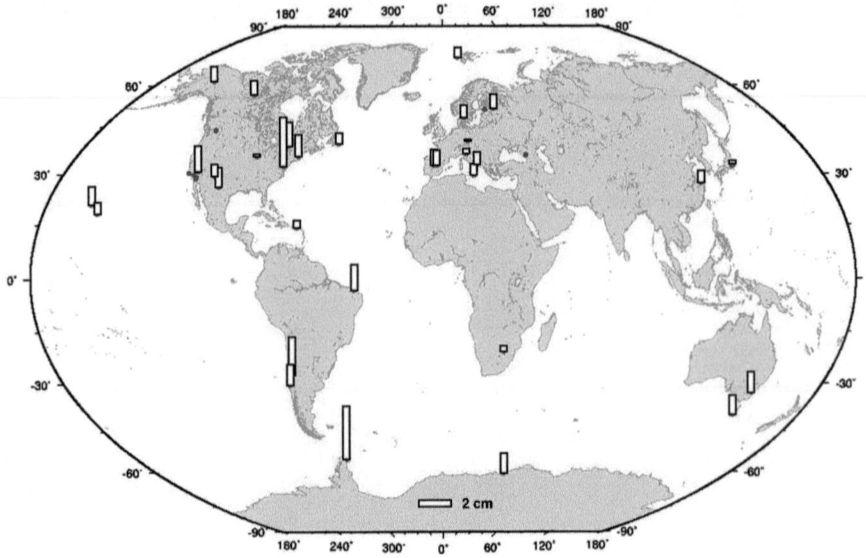

Fig. 3.9 3D differences between space geodetic solutions and local tie vectors for 32 VLBI and GPS co-location sites

An interpretation of the discrepancies between the terrestrial tie vectors and the coordinates differences obtained from the space geodetic solutions is rather difficult since various factors have to be considered. So, uncertainties in the measurements of the local ties, systematic effects in the solutions of the space techniques, uncertainties in the definition of the reference points of the instruments (e.g., antenna phase center), distance and height differences between co-located instruments (e.g., local effects, troposphere), discontinuities in station position, and velocity differences of co-located instruments may have an impact on the results.

The fact that the number and the spatial distribution of "high quality" co-location sites is not optimal and that there are several sites with rather large discrepancies between the space geodetic solutions and the local tie vectors can be considered as one of the major limiting factors for the inter-technique combination. Thus, the selection and handling of the local tie vectors is a crucial issue within the combination. The EOP estimates are used as a criterion to validate the selected local ties and to stabilize the inter-technique combination as additional "global ties". In the ideal case (without systematic errors) the EOP estimates must be (statistically) identical for all space techniques. The procedure, which has been applied at DGFI for the selection and handling of local ties, is described in, e.g., Krügel and Angermann (2007) and Seitz et al. (2012).

Other tasks of the inter-technique combination include the weighting of the different space techniques and the equating of station velocities at co-location sites. The relative weighting of the techniques is necessary, as their stochastic models are affected by inadequacies in the observation modelling. The station velocities of co-located instruments were estimated as separate parameters. The velocities were equated by applying conditions if the differences are statistically not significant. To generate the final DTRF2008 solution, minimum datum conditions have to be added to the combined normal equations, and the complete normal equation system has to be inverted. The origin (translations and their rates) is realized by the contributing SLR solutions and the scale and its time variation by SLR and VLBI. The orientation of the DTRF2008 is realized by NNR conditions w.r.t. ITRF2005. Altogether six condition equations are applied, three for the positions at the reference epoch 2005.0 and three for the velocities. The conditions were applied by using the position estimations of stable and globally distributed stations.

5.3 ITRF2008 Results

The ITRS Combination Centres at IGN and DGFI each computed a combined inter-technique solution for ITRF2008. The exchange of information (discontinuity lists, preliminary results, etc.) between both institutions contributed to the high quality of the final ITRF2008 results. The analysis and comparison of both solutions has shown that they are of the same high quality. There are only small discrepancies between both solutions, although they are based on different combination strategies and software packages.

It was decided by the ITRS Centre to release the ITRF2008 solution computed at IGN as the official ITRF2008 (Altamimi et al. 2011). The ITRF2008 network comprises 934 stations located at 580 sites, of which only 20% are located in the Southern hemisphere. All ITRF2008 data files and results are available at the ITRF web site: http://itrf.ign.fr/ITRF_solutions/2008/.

The ITRF2008 solution computed at DGFI is labeled as DTRF2008 (Seitz et al. 2012). The DGFI solution is available at the anonymous ftp server of DGFI at ftp://www.dgfi.badw.de/pub/DTRF2008.

The ITRF2008 results comprise station positions, station velocities, coordinates of the terrestrial pole as well as UT1-UTC and the Length of Day (LOD). The reference epoch of the station positions is 2005.0. Due to the fact, that quite a large number of discontinuities were introduced in the ITRF2008 computation, several stations get various solutions. This implies that the station positions and velocities are valid only for a certain time period, which has also been provided to the users.

Figure 3.10 shows the horizontal station velocities of the DTRF2008 solution computed at DGFI, which are dominated by plate tectonics. The station velocities are given in different colors for the different space-techniques. The majority of stations are located in the Northern hemisphere with clusters in Western Europe and North America. It shall also be noted that co-location sites may have different station velocities for co-located instruments if their estimated velocities differ significantly.

Figure 3.11 displays the station height velocities of the DTRF2008. Whereas for most of the stations, the height velocities are rather small (mostly below 1–2 mm/year), the stations located in Fennoscania show very nicely uplift rates in

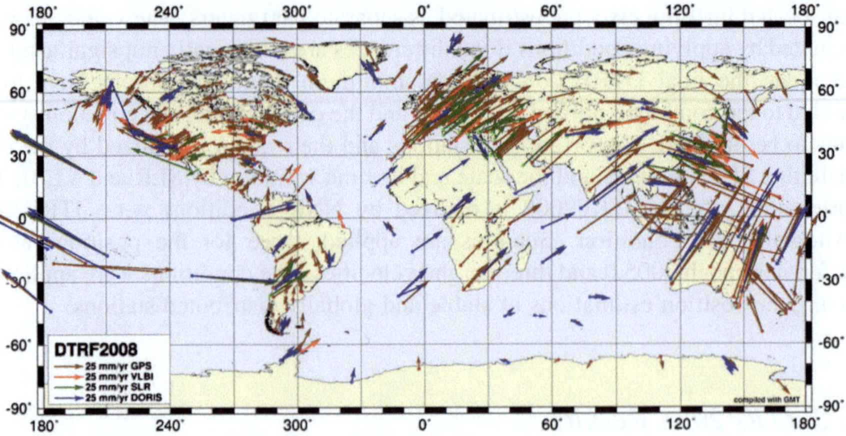

Fig. 3.10 Horizontal station velocities of the DTRF2008 solution computed at DGFI

the order of several mm/year, which are in good agreement with postglacial rebound models (e.g., Peltier 2004). The largest height velocity has been estimated for a GPS station in Bogota (Columbia), which is located on a building. Since this building is situated in a basin with soft ground it is subsiding at a rate of more than 5 cm/year. There are some other stations with rather large vertical velocities (>5 mm/year), which have to be interpreted with care, since local site effects and possible instrumentation effects may have an impact on the results.

As an additional result of the ITRF2008 computations, time series of station positions and datum parameters were obtained. These time series provide valuable information for detailed investigations of the combination results.

5.4 Comparison of ITRF2008 and DTRF2008

A comparison of the ITRF2008 and DTRF2008 ensures an independent quality control of the combination results and provides a valuable basis for an accuracy assessment of the ITRF. In principle, the station positions and velocities of both solutions can be compared directly without performing any transformation. However, it has to be considered that any possible datum differences between both solutions will directly enter into the comparison results. As a consequence it is not possible to distinguish between differences in the datum definition and discrepancies of the network geometry. Thus, the comparison of the ITRF2008 and DTRF2008 solutions was done by performing 14 parameter similar transformations (3 translations, 3 rotations, 1 scale, and their rates). These transformations were done separately for each space technique by using "high quality" and globally distributed core stations. As a result of these comparisons, two quality estimations were obtained for each technique-specific network: (1) the

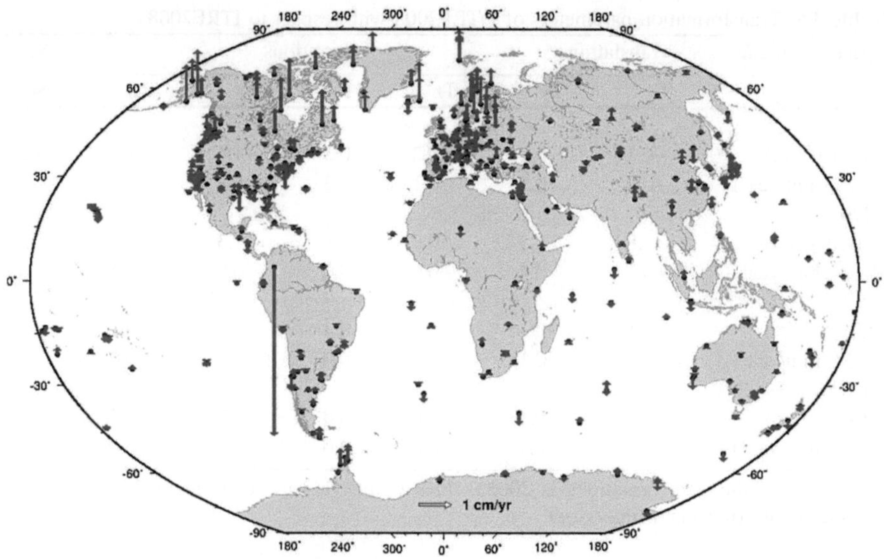

Fig. 3.11 Station height velocities of DTRF2008 solution

transformation parameters between the ITRF2008 and DTRF2008 as a measure for the accuracy of the datum definition (see Table 3.6) and (2) RMS differences for station positions and velocities as a measure for the accuracy of the network geometry (see Table 3.7) (Seitz et al. 2012).

The transformation parameters between ITRF2008 and DTRF2008 (given in Table 3.6) quantify the accuracy of the datum definition. The results indicate an accuracy of the datum parameters of about 5–6 mm for the positions (offsets at epoch 2005.0) and below 1 mm/year for the rates. These values represent the overall accuracy of the datum parameters (i.e., the squared sum of the individual transformation parameters) and reflect the accuracy of the ITRF2008 datum. The results indicate that clear differences exist between the different network parts. For SLR, which defines the ITRF2008 origin, the differences between both solutions are rather small, but the network parts of the other space-techniques show larger variations. This result is not surprising, since the relatively sparse distribution of co-location sites and the (partly) different selection and handling of local ties in the IGN and DFGI solutions certainly have an impact on the combination results. It can be concluded that the current distribution of co-location sites and the misfits between local ties and space-geodetic solutions do not allow for a realization of the ITRF2008 datum of better than 5 mm.

The RMS differences given in Table 3.7 reflect the agreement of the network geometries for the different space techniques. The best agreement has been achieved for VLBI, which can be explained by the fact that this space technique has the longest time span of almost 30 years of observations accompanied by only 22 discontinuities in total for all the 107 VLBI stations. In the case of GPS, the

Table 3.6 Transformation parameters of DTRF2008 with respect to ITRF2008

Space technique	Translations			Rotations			Scale
	Tx	Ty	Tz	Rx	Ry	Rz	Sc
SLR							
Offsets (mm)	−0.1	0.0	−0.3	0.5	−1.0	1.8	−2.0
Rates (mm/year)	−0.2	−0.5	0.1	0.3	0.4	0.4	0.1
VLBI							
Offsets (mm)	−1.8	1.3	−0.9	0.1	−1.3	5.3	2.1
Rates (mm/year)	0.4	0.4	−0.1	0.0	0.0	−0.1	−0.1
GPS							
Offsets (mm)	−1.1	0.1	−4.9	0.4	−1.3	0.1	−2.9
Rates (mm/year)	0.1	−0.1	0.0	0.0	0.1	0.0	0.0
DORIS							
Offsets (mm)	1.3	0.1	−3.0	0.0	−2.7	−3.3	3.2
Rates (mm/year)	−0.1	0.4	0.8	0.0	0.0	0.0	−0.1

The epoch of the transformations is 2005.0. Please note that the units for all transformation parameters are (mm) and (mm/year)

Table 3.7 RMS differences for station positions and velocities of DTRF2008 w.r.t. ITRF2008

	VLBI	SLR	GPS	DORIS
Positions (mm)	0.38	2.02	1.33	3.22
Velocities (mm/year)	0.09	0.82	0.19	0.98

The epoch of the transformations is 2005.0

large number of discontinuities and the shorter observation time period may be the major reasons for the larger discrepancies. The discrepancies for the SLR and DORIS network parts are larger compared to VLBI and GPS. They are in the range of 2–3 mm for the positions and 0.8–1.0 mm/year for the station velocities. The fact that the accuracy of the network geometries is better than for the datum parameter indicates that the datum realization is more critical in the ITRF computations. A key problem in this context is the relatively sparse distribution of high quality co-location sites and the observed misfits between local ties and space geodetic solutions. Progress in these fields is essential for further improvements of ITRF accuracy.

5.5 *Transformation Parameters from ITRF2008 to Past ITRF Realizations*

Table 3.8 lists the transformation parameters between ITRF2008 and their previous realizations. The values are extracted from the IERS Conventions 2010 (Petit and Luzum 2010) where the formulas for the 14 parameter similarity transformations are also given. It shall be noted that the transformation parameters

Table 3.8 Transformation parameters from ITRF2008 to past ITRF realizations

	Translations (mm)			Scale (ppb)	Rotations (mas)		
	Tx	Ty	Tz	Sc	Rx	Ry	Rz
ITRF2005	−2.0	−0.9	−4.7	0.94	0.00	0.00	0.00
Rates	0.3	0.0	0.0	0.00	0.00	0.00	0.00
ITRF2000	−1.9	−1.7	−10.5	1.34	0.00	0.00	0.00
Rates	0.1	0.1	−1.8	0.08	0.00	0.00	0.00
ITRF97	4.8	2.6	−33.2	2.92	0.00	0.00	0.06
Rates	0.1	−0.5	−3.2	0.09	0.00	0.00	0.02
ITRF96	4.8	2.6	−33.2	2.92	0.00	0.00	0.06
Rates	0.1	−0.5	−3.2	0.09	0.00	0.00	0.02
ITRF94	4.8	2.6	−33.2	2.92	0.00	0.00	0.06
Rates	0.1	−0.5	3.2	0.09	0.00	0.00	0.02
ITRF93	−24.0	2.4	−38.6	3.41	0.00	−1.48	−0.30
Rates	−2.8	−0.1	−2.4	0.09	0.00	−0.19	−0.07
ITRF92	12.8	4.6	−41.2	2.21	0.00	0.00	0.06
Rates	0.1	−0.5	−3.2	0.09	0.00	0.00	0.02
ITRF91	24.8	18.6	−47.2	3.61	0.00	0.00	0.06
Rates	0.1	−0.5	−3.2	0.09	0.00	0.00	0.02
ITRF90	22.8	14.6	−63.2	3.91	0.00	0.00	0.06
Rates	0.1	−0.5	−3.2	0.09	0.00	0.00	0.02
ITRF89	27.9	38.6	−101.2	7.31	0.00	0.00	0.06
Rates	0.1	−0.5	−3.2	0.09	0.00	0.00	0.02
ITRF88	22.8	2.6	−125.2	10.41	0.00	0.00	0.06
Rates	0.1	−0.5	−3.2	0.09	0.00	0.00	0.02

ppb refers to parts per billion (10^{-9}) and *mas* is the standard abbreviation for milliarcseconds. The units for rates are *per year*. The epoch for all transformations is 2000. 0
Source IERS conventions 2010

are adjusted values which depend on the weighting as well as on the number and distribution of the sites common to these frames. Therefore, using different subsets of common stations between two ITRF realizations to estimate transformation parameters would not necessarily yield values consistent with those displayed in Table 3.8. However, if reasonable station distributions are used for the transformations, the estimated parameters will only be changed slightly and the general outcome will not be affected.

The time series of the scale and translation parameters (in particular for the z-component) show a clear trend over time which may indicate some systematic effects in the series of ITRF realizations. This can be explained reasonably well by transferring imperfect modeled common network movements to the datum parameters (Drewes 2009a), which may enter into the transformation parameters between networks. The observed drifts between different ITRF realizations indicate that the datum definition is not stable over time.

In fact, the long-term stability of the terrestrial reference frame does not satisfy the accuracy requirements that have been specified for the Global Geodetic Observation System (GGOS). A prominent example for the monitoring of global change is the sea-level rise, which is in the order of 3 mm/year. For a reliable estimation of this rather small quantity the accuracy should be significantly better (i.e., by an order of magnitude). This means for the scale realization, which is directly correlated with station heights, that an accuracy of about 0.3 mm/year is required, which is equivalent to 0.5 ppb/year. The scale differences between the ITRF realizations indicate that this accuracy has not been achieved yet.

6 Discussion and Challenges for the Future

During the last two or three decades since space geodetic observations became available remarkable progress has been achieved in the field of global terrestrial reference frame computations. The establishment of the IAG Services for the space-techniques GPS, SLR, VLBI and DORIS and the tremendous effort of the participating organizations and institutions in their functions as station operators, data, analysis and combination centres contributes significantly to this success. The high accuracy level of today's space geodetic observations, the improvements of the station networks and the refinements of the modelling and data analysis procedures are a key pre-requisite for the high quality of the input data sets that were available for the latest realizations of the terrestrial reference frames.

Furthermore, the combination methodologies for the reference frame computations have also been significantly improved during the history of ITRF realizations. Starting with the combination of station positions (only) for the ITRF88, ITRF89 and ITRF90, the methodologies have been continuously improved. The current procedure is based on the time series combination of station positions and EOP based on self-consistently processed time series of VLBI, SLR, GPS and DORIS observations. Within the re-organized structure of the IERS, the ITRS Centre is supplemented by the ITRS Combination Centres. The fact, that IGN and DGFI computed each two fully combined solutions for the ITRF2005 and ITRF2008 provided a fundamental basis for the analysis of the terrestrial reference frame results and for a "quasi-independent" accuracy evaluation.

Although remarkable progress has been achieved in the fields mentioned above, there are still some limiting factors for the accuracy of terrestrial reference frame realizations and challenges for future improvements. The following issues are discussed in this section:

- IERS Network, co-location sites and local tie vectors (Sect. 6.1).
- Input data for the ITRF computations (Sect. 6.2).
- Nonlinear station motions (Sect. 6.3).
- Effect of large earthquakes (Sect. 6.4).
- Combination methodology and datum definition (Sect. 6.5).

Table 3.9 Three-dimensional differences (*mm*) between space geodetic solutions (DTRF2008) and local tie vectors

Co-locations	# Sites	<5 mm	5–10 mm	10–20 mm	>20 mm
GPS–VLBI	32	7	11	11	3
GPS–SLR	29	4	13	9	3
GPS–DORIS	40	2	12	16	10
VLBI–SLR	9	1	2	4	2
Total	110	14	38	40	18

Note that only those co-locations are considered which have a reasonable long time series (i.e., >2 years) of space geodetic observations

6.1 IERS Network, Co-Location Sites and Local Tie Vectors

One major limiting factor for the integration of the different space geodetic techniques, the inter-technique combination, is the rather inhomogeneous distribution of stations and the sparse distribution of "high quality" co-location sites. Table 3.9 gives an overview about the present status in this field.

The results clearly indicate that the discrepancies between the space geodetic solutions and the local tie vectors are too large (i.e., >1 cm) for many co-location sites. For about half of the co-locations, the discrepancies are above 1 cm. As already mentioned the interpretation of the discrepancies is difficult since various influence factors have to be considered. However, this topic is a key issue to improve the quality of the terrestrial reference frames and thus it shall be the subject of further investigations.

It is obvious that the long-term maintenance of co-location sites, their distribution, and the quality of the local tie measurements need to be improved. GGOS shall provide the frame to reinforce the international cooperation on co-location sites for long-term stability of the global reference frame. The goal specified in the GGOS2020 document (Plag and Pearlman 2009) is that a core network with about 40 globally well-distributed core stations shall be established. These stations shall co-locate the major geodetic observation techniques and a variety of additional sensors. In addition to the "classical" co-location on Earth, a challenge for the future would be the co-location of sensors in space.

6.2 Input Data for the ITRF Computations

The IAG Services, together with their respective Analysis and Combination Centers have achieved significant improvements for the modelling and processing of the space geodetic observations during the last years. However, there are still some deficiencies, e.g.:

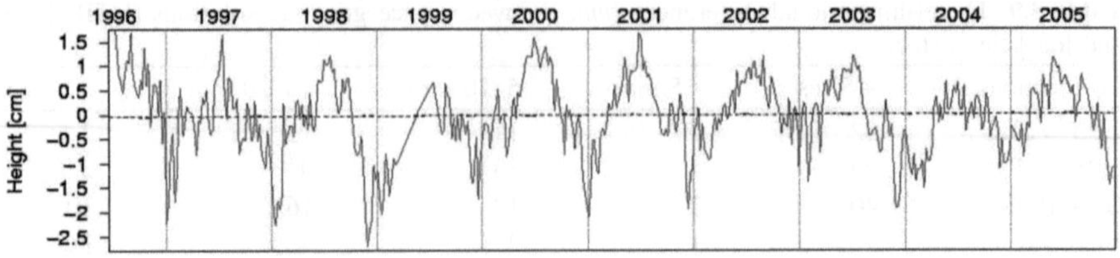

Fig. 3.12 Seasonal height variation for GPS station Irkutsk (*IRKT*), Siberia

- Most of the analysis centers have adopted and implemented the standards and models according to the most recent set of conventions, but most likely there are still some inconsistencies (e.g., different troposphere models, usage of different Earth's gravity field models, etc.). The adoption of exactly the same standards for the modelling and parameterization by all analysis centres is crucial to obtain really consistent results.
- Although capabilities for a self-consistent reprocessing of time series of station positions and EOP have been achieved, this is still a time consuming business (in particular for the large amount of GPS data) and a major limiting factor for a more frequent computation of subsequent ITRS realizations.
- Another issue is that most of the input data are provided as solutions with loose or minimum constraints which have to be removed from the input data. Even loose constraints may affect the combination results if they are applied manifold. The necessary inversion of the solutions may cause loss of precision by numerical effects (Drewes and Angermann 2003; Gerstl 2003). The IVS provided constraint-free normal equations for the ITRF2005/ITRF2008. DGFI also recommends the other services to provide such input data.
- Discontinuity tables for station positions need to be refined and must be fully consistent between techniques (e.g., for co-location sites affected by earthquakes) which has not yet been achieved. Discontinuities caused by instrumentation changes shall be avoided to the highest possible extent and of course they need to be recorded properly. It shall be kept in mind, that a large number of discontinuities weaken the long-term stability of the reference frame.

6.3 Nonlinear Station Motions

The combination of time series of station positions and EOP as carried out for the ITRF2005 and ITRF2008 computations provides valuable results to identify nonlinear effects in station motions. For many stations seasonal variations are visible, especially in the height component (see Fig. 3.12 as an example). This illustrates a shortcoming of the current ITRF approach, since the ITRF2008 station coordinates obtained with the linear model do not represent the "real" station positions. The differences between the actual position at a certain epoch and the

ITRF2008 coordinates can reach the level of up to 2 cm. As a consequence, the transformation of epoch solutions (e.g., campaign-style results) to the ITRF2008 is affected by these seasonal positions variations and this will lead to errors in the realized datum. The seasonal variations will also affect the velocity estimations, in particular for stations with relatively short observation time spans (i.e., <2 years). Thus, the current handling of nonlinear station motions is a major limiting factor for the accuracy of the terrestrial reference frame.

For an interpretation of the observed seasonal variations we recall the ITRS definition, which describes the station motions by a linear model for regularized positions $x_R(t)$ taking into account the conventional corrections $\Delta x_i(t)$ [see (3.3) and (3.4)]. As a consequence, any geophysical effects that are not considered in the conventional corrections will become visible as residuals in the position time series. Thus, the shape of the "residual" nonlinear station motions strongly depends on the models which are used for the reduction of the original observations. Part of the seasonal variations will disappear if, for example, atmospheric loading corrections are applied. As recommended in the IERS Conventions 2010, it was specified for the ITRF2008 input data, that all services shall not apply non-tidal loading corrections (e.g., atmospheric and hydrological loading). Thus, all these non-tidal loading effects on station positions are visible in the residual time series. The station height time series were compared with geophysical model results and it was found that a large part of the observed annual signals can be explained by non-tidal atmospheric and hydrologic loading variations (Seitz and Krügel 2009). However, as already mentioned in Sect. 3.2, the application of such geophysical models in operational processing is critical, and thus this issue is the subject of further studies. Another possibility is to extend the current linear model for the parameterization of station motions and to represent the seasonal variations mathematically by annual and semi-annual functions. However, the computation of a mean (averaged) annual motion is problematic, in particular if the seasonal variations are different over the observation time span. Furthermore, the additional parameters will affect the stability of the solutions, which is a particular problem for stations with relative short observation time spans. Thus, an appropriate handling of seasonal variations in station positions is a challenge for future ITRS realizations.

6.4 Effect of Large Earthquakes

The realizations of the ITRS consist of a set of site coordinates and velocities, i.e., a linear representation of site locations. It is known, however, that such a reference frame can be destroyed by a large earthquake within minutes over a very large region of the globe. Figure 3.13 shows the station displacements caused by the earthquake ($M = 8.8$) which occurred on 27 February 2010 in Chile.

The station CONZ (Conceptiŏn, Chile) was affected by a coseismic displacement of more than 3 m in a SW direction. More than 20 stations of the SIRGAS reference frame (Sànchez and Brunini 2009; Sànchez et al. 2012) were displaced by more than

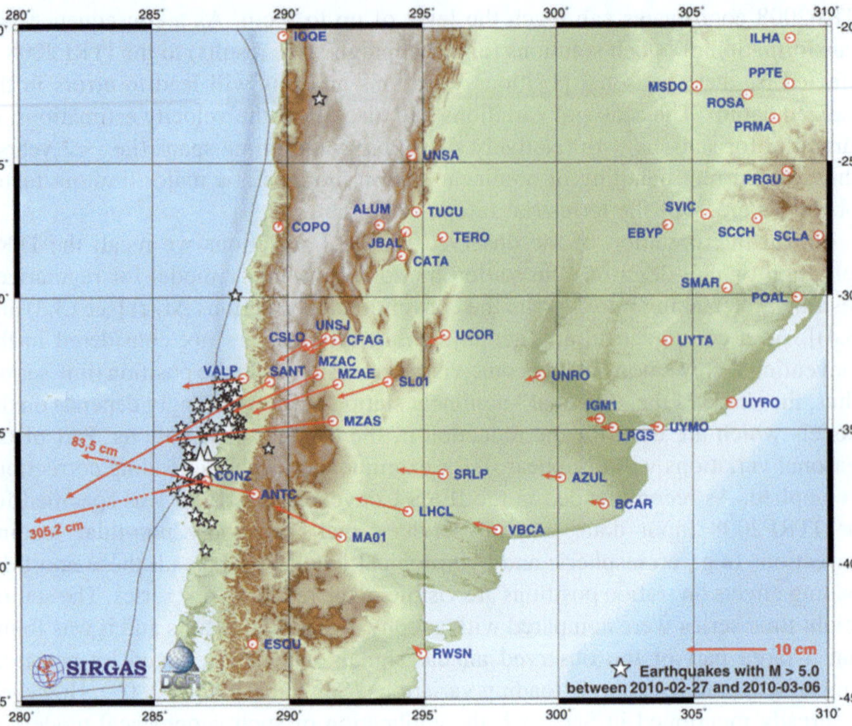

Fig. 3.13 Station displacements caused by the large ($M = 8.8$) earthquake on 27 February 2010 in Chile (Sànchez et al. 2012)

1.5 cm. Among them are also stations on the eastern site of the South American continent, at a distance of much more than 1,000 km from the epicentre. The stations are not only affected by a coseismic displacement—the station motions also differ significantly from the long-term velocities. Thus, it is evident, that after such a large earthquake, the ITRF coordinates and velocities of sites in a large region around the epicentre cannot be used as reference sites any longer. It is, therefore, an important goal to develop the necessary strategies and methods for providing terrestrial reference frame results much more frequently (e.g., as weekly or monthly epoch solutions) in addition to the "classical" multi-year reference frames.

6.5 Combination Methodology and Datum Definition

During the last few years remarkable progress has been achieved regarding the development of refined combination methods for the terrestrial reference frames, but further improvements are necessary to ensure that the results are fully consistent and to make optimal use of the strengths of the individual space techniques. Below several issues are addressed:

- Refined combination methods have to be developed to take into account the observed nonlinear station motions. As mentioned in Sect. 6.4 different possibilities for the handling of these nonlinear effects do exist in principle. However, the "optimal" strategy is not yet defined, and thus further studies are necessary in this field.
- Another important topic is the integration of the different space techniques. The current situation regarding the spatial distribution of co-location sites and the observed discrepancies between space geodetic solutions and geodetic local vectors is not satisfactory. Progress in this field is essential to achieve further improvements for the accuracy and consistency of the terrestrial reference frame results.
- Earlier it was shown, that after large earthquakes, the ITRF coordinates and velocities of the "classical" multi-year reference frames cannot be used in large regions. Therefore, there is a need to develop new methods for the combination of the different space techniques on an epoch basis (e.g., weekly or monthly).
- According to the ITRS definition, the origin of the terrestrial reference system shall be fixed in the geocenter, and coordinate changes caused by the station movements must go to the individual station coordinates and not into the datum. This is not ensured for the ITRF, since all common motions of the stations of the reference network are transformed into the similarity parameters (translation, orientation, expansion factor), and, thus violate the definition of the reference system.
- In current realizations of the terrestrial reference system, the orientation is aligned to that of the previous ITRF, and its rate is aligned, conventionally, to that of the geological model NNR-NUVEL-1A (e.g., DeMets et al. 1994). This geophysical model does not fulfil the NNR-condition of the ITRS definition because (1) it contains only 14 rigid plates (deformation zones are not included) and (2) the model reflects plate motions averaged over millions of years. Significant deviations from present-day motions are observed (e.g., Angermann et al. 1999; Altamimi et al. 2003). Thus, the kinematic datum shall be given by a present-day crustal motion and deformation model (e.g., APKIM2005; see Drewes 2009b).

References

Altamimi Z, Sillard P, Boucher C (2002) ITRF2000: a new release of the international terrestrial reference frame for earth sciene applications. J Geophys Res 107(B7):2214. doi:10.1029/2007JB000561

Altamimi Z, Sillard P, Boucher C (2003) The impact of a no-net-rotation condition on ITRF2000. Geophys Res Lett 30(2):1064. doi:10.1029/2992GL016270

Altamimi Z, Collilieux X, Legrand J, Garayt B, Boucher C (2007) ITRF2005: a new release of the international terrestrial reference frame based on time series of station positions and earth orientation parameters. J Geophys Res 112:B09401. doi:10.1029/2007/JB004949

Altamimi Z, Collilieux X, Métivie L (2011) ITRF2008: an improved solution of the international terrestrial reference frame. J Geod. doi:10.1007/s00190-011-0444-4

Angermann D, Klotz J, Reigber C (1999) Space-geodetic estimation of the Nazca-South America Euler vector, earth plant. Sci Lett 171:329–334

Angermann D, Drewes H, Krügel M, Meisel B, Gerstl M, Kelm R, Müller H, Seemüller W, Tesmer V (2004) ITRS combination center at DGFI: a terrestrial reference frame realization 2003, Deutsche Geodätische Kommission, Reihe B, Heft Nr. 313

Angermann D, Krügel M, Meisel B, Müller H, Tesmer V (2005) Time evolution of the terrestrial reference frame. In: Sanso F (ed) A window on the future of geodesy. IAG symposia, vol 128. Springer, Heidelberg, pp 3–8

Angermann D, Drewes H, Krügel M, Meisel B (2007) Advances in terrestrial reference frame computations. In: Tregoning P, Rizos C (eds) Dynamic planet, IAG symposia, vol 130. Springer, Heidelberg, pp 595–602

Angermann D, Drewes H, Gerstl M, Krügel M, Meisel B (2009) DGFI combination methodology for ITRF2005 computation. In: Drewes H (ed) Geodetic reference frames, IAG symposia, vol 134. Springer, Heidelberg, pp 11–16

Argus DF, Gordon RG (1991) No-net-rotation model of current plate velocities incorporating plate motion model NUVEL-1. Geophys Res Lett 18(11):2039–2042. doi:10.1029/91GL01532

Bianco G, Devoti R, Fermi M (2000) Investigation of the combination of space techniques. J Geodyn 30(3):337–353

Blewitt G (2003) Self-consistency in reference frames, geocenter definition, and surface loading of the solid earth. J Geophys Res 108(B2):2203. doi:10.1029/2002JB002082

Böckmann S, Artz T, Nothnagel A (2010) VLBI terrestrial reference frame contributions to ITRF2008. J Geod 84(3):201–219. doi:10.1007/s00190-009-0357-7

Böhm J, Niell A, Tregoning P, Schuh H (2006a) Global mapping function (GMF): a new empirical mapping function based on numerical weather model data. Geophys Res Lett 33: L07304. doi:10.1029/2005GL025546

Böhm J, Werl B, Schuh H (2006b) Troposphere mapping functions for GPS and very long baseline interferometry from european centre for medium-range weather forecasts operational analysis data. J Geophys Res Solid Earth 111:2406. doi:10.1029/2005/JB003629

Böhm J, Heinkelmann R, Schuh H (2007) Short-note: a global model of pressure and temperature for geodetic applications. J Geod doi:10.1007/s00190-007-0135-3

Bosch W (2008) Der Meeresspiegel—ansteigend und fast im Lot. Akademie Aktuell, 01/2008, 30–33, ISSN: 1436-753X

Boucher C, Altamimi Z (1985) Towards an improved realization of the BIH terrestrial reference frame. In: Mueller II (ed) The MERIT/COTES report on earth rotation and reference frames, vol 2. OSU/DGS, Columus

Boucher C, Altamimi Z, Feissel M, Sillard P (1996) Results and analysis of the ITRF94, IERS technical note 20, Observatoire de Paris, Paris. Available at http://www.iers.org/TN20

Boucher C, Altamimi Z, Sillard P (1998) Results and analysis of the ITRF96, IERS Technical Note 24, Observatoire de Paris, Paris. Available at http://www.iers.org/TN24

Boucher C, Altamimi Z, Sillard P (1999) The 1997 international terrestrial reference frame (ITRF97), IERS Technical Note 27, Observatoire de Paris, Paris. Available at http://www.iers.org/TN27

Boucher C (2001) Terrestrial coordinate systems and frames. In: Encyclopedia of Astronomy and Astrophysics, Version 1.0. Nature Publishing Group, Institute of Physisc publishing, Bristol

Boucher C, Altamimi Z, Sillard P, Feissel-Vernier M (2004) The ITRF2000, IERS Technical Note No. 31, Verlag des Bundesamtes für Kartographie und Geodäsie, Frankfurt am Main

DeMets C, Gordon RG, Argus DF, Stein S (1994) Effect of recent revisions to the geomagnetic time scale on estimates of current plate motions. Geophys Res Lett 21(20):2191–2194. doi:10.1029/94GL02118

Dong D, Yunck T, Heflin M (2002) Origin of the international terrestrial reference frame. J Geophys Res 108(B4):2200. doi:10.1029/2002JB002035

Dow JM, Neilan RE, Rizos C (2009) The international GNSS service in a changing landscape of global navigation satellite systems. J Geodesy 83(4–5):191–198. doi:10.1007/s00190-008-0300-3

Drewes H, Angermann D (2003) Remarks on some problems in the combination of station coordinate and velocity solutions, IERS Technical Note 30, Verlag des Bundesamtes für Kartographie und Geodäsie, Frankfurt am Main, pp 89–93

Drewes H, Angermann D, Gerstl M, Krügel M, Meisel B, Seemüller W (2006) Analysis and refined computations of the international terrestrial reference frame. In: Rothacher R, Rummel F, Schreiber B (eds) Observation of the earth system from space. Springer, Heidelberg

Drewes H (2009a) Reference systems, reference frames, and the geodetic datum—basic considerations. In: Sideris M (ed) Observing the changing Earth, IAG symposia, vol 133. Springer, Heidelberg, pp 3–10

Drewes H (2009b) The APKIM2005 as basis for a non-rotating ITRF. In: Drewes H (ed) Geodetic reference frames, IAG symposia, vol 134. Springer, Heidelberg, pp 95–99

Ferland R (2010) Description of IGS submission to ITRF 2008. Available at: http://itrf.ensg.ign.fr/ITRF_solutions/doc/IGSsubmission4ITRF2008.txt

Gerstl M, Kelm R, Müller H, Ehrnsperger W (2000) DOGSCS Kombination und Lösung großer Gleichungssysteme. Interner Bericht, DGFI, München

Gerstl M (2003) Numerical aspects on combination at the observation equation and normal equation level. IERS Technical Note 30, Verlag des Bundesamtes für Kartographie und Geodäsie, Frankfurt am Main, pp 89–93

Gobinddass ML, Willis P, de Viron O et al (2009) Improving DORIS geocenter time series using an empirical rescaling of solar radiation pressure models. Adv Space Res 44(11):1279–1287. doi:10.1016/j.asr.2009.08.004

IPCC AR4 WG1 (2007), In: Solomon S, Qin D, Manning M, Chen Z, Marquis M, Averyt KB, Tignor M, and Miller HL (eds) Climate Change 2007 The physical science basis, contribution of working group I to the forth assessment report of the intergovernmental panel on climate change, Cambridge University Press, ISBN: 978-0-0521-88009-1

IUGG (2007) International Union of Geodesy and Geophysics (IUGG) resolution number 2 of Perugia, IUGG General Assembly

Kovalevsky J, Mueller II, Kolaczek B (eds) (1989) Reference frames in astrometry and geophysics. Kluwer Academic Publisher, Dordrecht, p 474

Krügel M, Angermann D (2007) Frontiers in the combination of space geodetic techniques. In: Tregoning P, Rizos C (eds) Dynamic planet, IAG symposia, vol 130. Springer, Heidelberg, pp 158–165

Meisel B, Angermann D, Krügel M, Drewes H, Gerstl M, Kelm R, Müller H, Tesmer V (2005) Refined approaches for terrestrial reference frame computations. Adv Space Res 36(3):350–357

Mendes VB, Pavlis EC (2004) High-accuracy zenith delay prediction at optical wavelengts. Geophys Res Lett 31:14602. doi:10.1029/2004GL020308

Minster JB, Jordan TH (1978) Present-day plate motions. Geophys J Int 83(B11):5331–5354. doi:10.1029/JB083iB11p0331

MERIT/COTES joint working groups (1983) MERIT campaign: connection of reference frames, implementation plan

Nothnagel A (2008) Conventions on thermal expansion modelling of radio telescopes for geodetic and astrometric VLBI. J Geod 83(8):787–792. doi:10.1007/s00190-008-284-z

Pavlis E, Luceri C, Sciaretta C, Kelm R (2010) The ILRS contribution to ITRF2008. Available at: http://itrf.ensg.ign.fr/ITRF_solutions/doc/-ILRSSubmission4ITRF2008.pdf

Pearlman M, Noll C, Gurtner W, Noomen R (2007) The international laser ranging service and its support for GGOS, dynamic planet—monitoring and understanding a dynamic planet with geodetic and oceanographic tools. In: Rizos C, Tregoning P (eds) IAG Symposia 130, Springer, Heidelberg, pp 741–748, ISBN: 978-3-540-49349-5

Peltier WR (2004) Global glacial isostasy and the surface of the ice-age earth: the ICE-5G (VM2) model and GRACE. Annu Rev Earth Planet Sci 32:111–149

Petit G, Luzum B (2010) IERS Conventions (2010) IERS Technical Note 36, Verlag des Bundesamts für Kartographie und Geodäsie, Frankfurt am Main, http://tai.bipm.org/iers/conv2010

Plag H-P, Pearlman M (eds) (2009) The global geodetic observing system: meeting the requirements of a global society on a changing planet in 2020. Springer, Berlin

Rothacher M (2000) Towards an integrated global geodetic observing system, international association of geodesy symposia. In: Rummel R, Drewes H, Bosch W, Hornik H (eds) Towards an integrated global geodetic observing system (IGGOS), vol 120. Springer, New York, pp 41–52

Sànchez L, Brunini C (2009) Achievements and challenges of SIRGAS. In: Drewes H (ed) Geodetic Reference Frames, IAG Symposia, vol. 134, Springer, Berlin, pp 161–169. doi:10.1007/978-3-642-00860-3_42

Sánchez L, Seemüller W, Drewes H, Mateo L, González G, da Silva, Pampillón J, Martinez W, Cioce V, Cisneros D, Cimbaro S (2012) Long-term stability of the SIRGAS Reference Frame and episodic station movements caused by the seismic activity in the SIRGAS region. Proceedings of the IAG Symposium REFAG2010, Springer (accepted)

Schlüter W, Behrend D (2007) The International VLBI Service for geodesy and astrometry (IVS): current capabilities and future prospects. J Geodesy 81(6–8):379–387

Schmid R, Steigenberger P, Gendt G, Ge M, Rothacher M (2007) Generation of a consistent absolute phase center correction model for GPS receiver and satellite antennas. J Geod 81(12):781–798. doi:10.1007/s00190-007-0148-y

Seitz M (2009) Kombination geodätischer Raumbeobachtungsverfahren zur Realisierung eines terrestrischen Referenzsystems, Deutsche Geodätische Kommission, Reihe C, Heft Nr. 630, München

Seitz F, Krügel M (2009) Inverse model approach for vertical load deformations in consideration of crustal inhomogeneities. In: Drewes H (ed): Geodetic reference frames, IAG symposia, vol. 134, pp 23–29, Springer, Heidelberg. doi:10.1007/978-3-642-00860-3_4

Seitz M, Angermann D, Bloßfeld M, Drewes H, Gerstl M (2012) The DGFI Realization of the ITRS: DTRF2008. J Geod. doi:10.1007/s00190-012-0567-2

Valette JJ, Lemoine FG, Ferrage P, Yaya P, Altamimi Z, Willis P, Soudarin L (2010) IDS contribution to ITRF2008. In: Willis P (ed) DORIS: precise orbit determination and applications to the earth sciences, Advanced. Space Research. doi:10.1016/j.asr.2010.05.029

Willis P, Fagard H, Ferrage P, Lemoine FG, Noll CE, Noomen R, Otten M, Ries JC, Rothacher M, Soudarin L, Tavernier G, Valette JJ (2010) The International DORIS Service, toward majurity. In: Willis P (ed) DORIS: scientific applications in geodesy and geophysics. Adv Space Res 45(12):1408–1420. doi:1016/j.asr.2009.11.018

Wu X, Ray J, van Dam T (2012) Geocenter motion and its geodetic and geophysical implications. J Geod 58:44–61

Chapter 4
Photogrammetry

Paula Redweik

1 Introduction

1.1 Definition and Short History

The term Photogrammetry, meaning at its origin measuring on photographs, has been defined as the science and technique of interpreting and evaluating the form, dimension and position of objects by analyzing and measuring images of them. The result of photogrammetric procedures is a precise three-dimensional geometric reconstruction of the object that can be orthogonally projected onto a plane (normally horizontally or vertically) at a certain scale or visualized in a perspective static or dynamic representation in a computer monitor for further evaluation. Due to the fact that the measuring does not occur directly on the object, photogrammetry is considered a remote sensing technique. Historically, this last term appeared about 90 years after Meydenbauer had invented in 1867 the German word *Photogrammetrie* to name his method of measuring on photographs for architectural surveys (Grimm 2007).

Although the frontier between photogrammetry and remote sensing became more and more permeable, there is a general tendency of applying the first term to the methods involving photographs while the second is reserved for those dealing with images from the Earth surface captured by artificial satellites, including non visible radiation sensors.

Two branches with partial independent development can be considered in photogrammetry: terrestrial and aerial including space. Terrestrial photogrammetry includes all the projects in which the camera has somehow a physical contact with

P. Redweik (✉)
Department of Geographic Engineering, Geophysics and Energy, Faculty of Sciences, University of Lisbon (FCUL), 1749-016 Campo Grande, Lisboa, Portugal
e-mail: pmredweik@fc.ul.pt

G. Xu (ed.), *Sciences of Geodesy - II*, DOI: 10.1007/978-3-642-28000-9_4,

the Earth surface, while aerial photogrammetry deals with photographs taken by cameras on aerial platforms (airplanes, helicopters, unmanned aerial vehicles (UAV), balloons, kites, space shuttles, orbital stations or even early satellites).

According to Konecny (2002) the history of photogrammetry can be divided into four relevant phases, successive in time, having nevertheless co-existed during one or two decades in the transition:

- From 1850 to 1900: plane table photogrammetry. Elevations or plans of the object are compiled through graphical methods applied to measurements of angles and directions in convergent photos.
- From 1901 onwards: analogical photogrammetry. Stereoscopy is used in optical and mechanical stereoplotters to reconstruct the object from stereo pairs.
- From 1953 onwards: analytical photogrammetry. Mathematical models are used to translate the geometric relations between photos and object.
- From 1972 onwards: digital photogrammetry. The process is carried out on digital photographs allowing application of the mathematical models from analytical photogrammetry together with digital image processing operators.

Origins of photogrammetry might be located at the time of the Italian Renaissance (fifteenth and sixteenth centuries), as painters first used geometric rules, based on the observation of the surrounding world, in order to represent three-dimensional scenes in a two-dimensional plane surface (wall or canvas), producing a depth sensation in the observer. Those were the rules of linear perspective, probably invented in 1420 by the Italian architect Brunnelleschi and applied by several other painters such as Piero de la Francesca, Leonardo da Vinci and Albrecht Dürer.

The objective of photogrammetry is somehow to perform the reverse operation. A photograph is already a perspective representation of a scene in a 2D plane surface. From two (or more) different perspectives of the same scene it is possible to reconstruct geometrically a spatial scene (3D).

Pioneers of this technique were the French colonel Aimé Laussedat and the already mentioned German engineer Albrecht Meydenbauer. While the former compiled maps from photos obtained by a camera in a kite or in other high situated platforms (hill tops and roofs), the latter used photogrammetry for architectural surveys and later also for topographic plans. Worth mentioning is the, to a large extent still existing, archive of metric photographs that Meydenbauer founded in 1885 containing high quality metric images of several monuments in Germany and in the near east. Glass plates obtained with Meydenbauer's self built metric cameras, presenting unusual original dimensions up to 40×40 cm^2, are still part of the archive (Albertz 2008).

Several inventions in the late nineteenth and beginning of the twentieth centuries represented technical milestones for photogrammetry: the floating mark (1892 by F. Stolze) for 3D measuring in stereoscopic models, the stereocomparator (1901 by C. Pulfrich), the stereoautograph (1908 by E. von Orel) and the double projector (1917 by M. Gasser) were the foundation for countless plotting instruments developed in the twentieth century, turning photogrammetry into a productive science and technique. The last mentioned was the first photogrammetric instrument to be designed for vertical photography. As an ideal platform for the photographic camera in the acquisition of near vertical photos, the invention of the airplane by the Wright brothers in 1903 was quickly adopted for cartography after less successful trials on platforms that included balloons, kites and even pigeons. The invention of the serial-photo aerial camera (1915 by O. Messter) allowed aerial photo coverage for systematic mapping. In the 1930s aerial stereo photogrammetry became the principal method for map production.

The invention of the electronic computer (1941 by K. Zuse) and its rapid development permitted the implementation of analytical solutions that had been proposed for a long time principally by S. Finsterwalder, T. von Scheimpflug and O. von Gruber. Their complexity had inhibited its adoption until then. Algorithms for orientation and triangulation have been improved and the most precise photogrammetric instruments ever made were developed: analytical plotters. Analytic calibration of terrestrial cameras as well as solutions using series of convergent photos in terrestrial photogrammetry were implemented, enlarging the scope of applications and increasing the popularity of photogrammetry in other fields of application.

The step to the digital era was a short one with great consequences. The success of digital image processing techniques, developed to be first applied to satellite Earth images (first computer readable Earth images sent in 1972 by Landsat-1), and its automation potential attracted photogrammetry researchers. In the 1980s film aerial photographs began to be scanned so that the same digital processing methods could be applied. Since photos could be handled in a digital format, the necessity for mechanical and optical components for photogrammetric instruments no longer existed. Except for a scanner, necessary for the analog–digital transformation of the photos, a computer was all that was needed to perform photogrammetric operations. These included not only stereo plotting but also orthorectification. Software packages based on the programs already applied in analytical plotters soon included both options. This development turned out to be very revolutionary from an economic point of view. Producers of scanners and scanning services were quick in coming. Traditional producers of photogrammetric instruments (stereoplotters and orthorectifiers) saw the competition increase from plain software producers. Small size companies emerged, no longer needing a great investment in expensive instruments to participate in projects. Traditional photogrammetry producers lost their monopoly in the market.

The digital photogrammetric process could now be partially automated, shifting the emphasis from the acquisition to the quality control. Since 2000, when the first digital aerial cameras were presented to the photogrammetric community at the Amsterdam Congress of the International Society for Photogrammetry and Remote Sensing (ISPRS), aerial photos can be acquired directly in digital format. It became possible to acquire simultaneously in one flight panchromatic, colour and near-infrared photos. There was no need to scan the photos any more. There was also no longer a need for photographic laboratories. Digital format prevailed from image acquisition to the end products.

1.2 Applications and Limitations

The scope of applications for photogrammetry is large, the principal fields being the following:

- Mapping and acquisition of geoinformation
- Documentation
- Monument preservation and architecture
- Aerial, terrestrial and underwater archaeology
- Monitoring earth surface and building deformations
- Civil engineering studies
- Automobile, aeronautical and nautical industries
- Dental, orthopaedic medicine and biomechanics
- Forensic applications

From all fields, mapping and geoinformation acquisition is the most important application in terms of work volume. In fact, since the 1930s either new topographic maps or updates of existing maps are executed by means of photogrammetry. Typical scales produced reach from 1:1,000 to 1:25,000. Greater map scales are mostly required only for limited areas being surveyed by topographic methods. Smaller scales can be produced by means of cartographic generalization.

Several facets of monument preservation activity take advantage of photogrammetric techniques: from the photographic archive of architecturally or historically important buildings, of façades along a street or even of a whole town district, to the individual survey of a monument for restoration.

Archaeologists use photogrammetry not only for detecting sites from the air but also for mapping and documenting different stages of diggings and to survey underwater sites.

For deformation detection and evaluation on the Earth surface (due to mining activities or underground works) as well as on buildings photogrammetric monitoring has the advantage of delivering not only geometric but also pictorial information of the object in each of the observation epochs. This can be of interest for the interpretation of results.

Photogrammetry can also be applied to studies of civil engineering, for instance in laboratory tests of dam and bridge prototypes and for real dam monitoring (in conjunction with laser scanning).

Industry as well as medicine turns to photogrammetry due to the fact that it is a non invasive technique. Automobile, aeronautical and nautical industries apply it in the monitoring of articulated robots and in the quality control of products. In crash tests photogrammetry is also applied in order to evaluate deformations in particular parts of the tested car.

Forensic applications involve documentation and evaluation of traffic accidents and crime scenes, identification of intruders by means of surveillance video cameras and association of a crime weapon to the injuries or damages caused.

Comparing with other methods for measuring and evaluating objects, photogrammetry offers several advantages such as:

- Relative short on-site work that includes the time of image acquisition (in the case where they are not acquired by fixed (video) cameras) and the time for an eventual ground control acquisition. The principal part of the work is done at the office.
- The prime material, the photographic image, is a dense register of information of the object.
- Once the image is acquired, the information can be recovered any time (a day, a month, a year, 50 years, a century after) as long as the image is preserved.
- Moving phenomena can be evaluated.
- It is a non invasive technique.
- It allows one to acquire information about unreachable objects, due either to distance or to a dangerous environment.

Nevertheless some limitations are also present in this technique. In fact, only what can be seen on the photograph can also be evaluated. A proper measurement of an object can be inhibited by deep shadows or occlusion by neighbouring objects.

The majority of terrestrial photogrammetry projects are close-range applications with objectives other than mapping. Refer to Luhmann et al. (2006) for consolidation of this matter. The following sections will be focused on aerial photogrammetry as a technique to produce geoinformation.

2 Image Acquisition

2.1 Aerial Cameras

Aerial photogrammetry is based on photographs of a part of the Earth surface taken by high precision cameras specially built to be mounted in an aerial platform. Evaluation of aerial photographs can be focused on interpretation or measuring/mapping. For interpretation of details, images with high geometric and radiometric resolution are needed. These can be obtained by a camera in a vertical or oblique axis position depending on the aim of the operation. Geometric properties of the camera and its lenses are not a priority for this type of evaluation. For measuring, instead, a high geometric resolution and stable geometry of the camera, as well as a near distortion free objective, are required. Radiometric resolution is not a priority, although a higher radiometric resolution improves differentiation of objects.

For mapping purposes, vertical aerial photos are used. Oblique aerial photos have also been used in recent years for 3D urban modelling.

The cameras described in the next sections are metric cameras designed for photogrammetry.

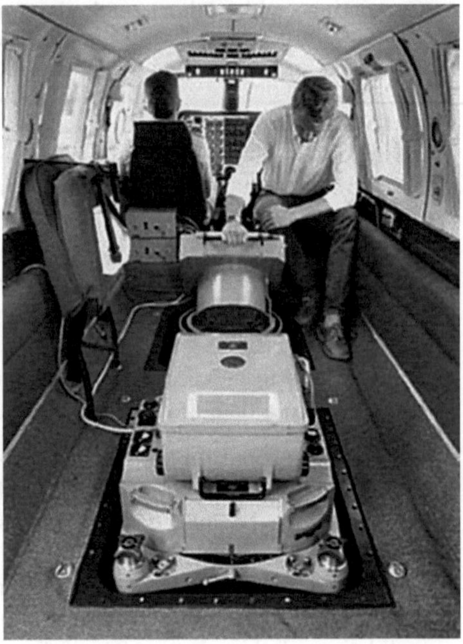

Fig. 4.1 Analogue cameras inside an aircraft

2.1.1 Analogue Cameras

Analogue aerial cameras capture the images on film. Some of the older cameras also used photographic glass plates as sensor. Although the production of analogue cameras has been stopped since digital aerial cameras have been adopted by the photogrammetric community, analogue cameras prevailed for more than 80 years and several are still operational today.

An aerial camera is normally designed to be mounted over an aperture in the lower fuselage of an airplane with its objective looking downwards (Fig. 4.1). For special missions, cameras have been mounted in a space shuttle, an orbital station and in early satellites.

The analogue camera is composed of the following principal components (Fig. 4.2):

- A camera mount
- A lens cone
- A film magazine
- A motor
- Instruments for navigation and camera control

The camera mount is the interface to the aircraft and can be gyro-stabilized. In its center there is a circular opening where the lens cone is inserted. On the focal plane of the objective there is a frame containing the fiducial marks. These were four in number in the older models of analogue cameras, located either in the corners or in the middles of the sides. Since the 1990s the newer models showed

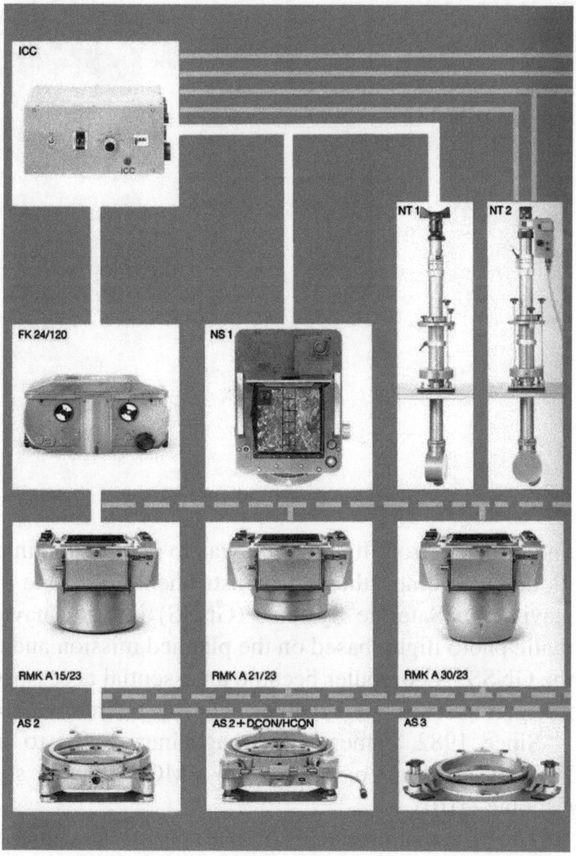

Fig. 4.2 Components of an analogue aerial camera Zeiss RMK-A. From *bottom* to *top*: *line 1* several camera mounts; *line 2* lens cones for different focal length (15, 21 and 30 cm); *line 3* film magazine, overlap controller, navigation periscopes; *line 4* motor

eight fiducial marks in corners and sides (RMK-TOP, RC30). Furthermore, some models presented coded fiducial marks in order to allow their automatic detection and identification during the later processing. The film magazine contained two film spools and was placed over the lens cone so that a portion of the unwrapped film could occupy the focal plane. A vacuum device ensured the film was held flat on the focal plane during exposure. Analogue cameras had several instruments for camera control and navigation help. Online monitoring of the course, allowing corrections of effects caused by aircraft drift and relevant changes in ground relief, was normally assured by an overlap controller and a navigation telescope. The overlap controller consisted of a device mounted parallel to the camera over an aperture through which the navigator could see the ground passing by under the airplane. A moving chain could be seen in the field of view sliding with a velocity according to the chosen overlap between photos along the strip. If the ground and the chain moved asynchronously the overlap value, which controlled the elapsed time between exposures, had to be adjusted. The navigation telescope was also mounted parallel to the camera axis over an aperture and had a field of view of 90°. Its axis was normally looking forwards with an inclination of 40–50°

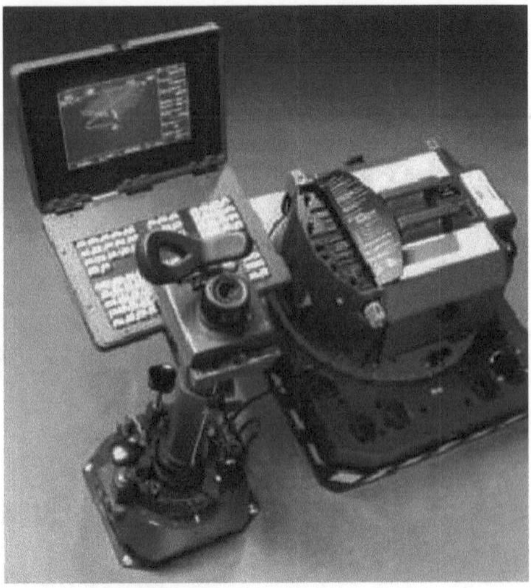

Fig. 4.3 Analogue aerial camera Leica RC30 with navigation periscope and computer with navigation software (Leica Geosystems)

under the horizon. Its function was to help maintain the course along a strip. Later models combined these two instruments into one and the integration of Global Navigation Satellite Systems (GNSS) into the navigation allowed a more automatic photo flight, based on the planned mission and on the position information of the GNSS. A computer became an essential accessory of the aerial camera and the navigation telescope became a control instrument (Fig. 4.3).

Since 1982, camera film magazines began to incorporate a mechanism for Forward Motion Compensation—FMC, allowing still sharper aerial photographs (Hobbie 2010).

The principal producers of analogue aerial cameras in Europe were the German companies Carl Zeiss Jena and Zeiss Oberkochen (both later Zeiss-Intergraph) and the Swiss company Wild Heerbrugg (later Leica Geosystems). More than 4,000 exemplars (2,500 Zeiss cameras and 1,600 Wild cameras) were sold for civil applications all around the world (Hobbie 2010; Fricker et al. 2003). An even greater number were produced for military applications, especially during World War II. In the United States of America, the principal producer of aerial cameras was the Fairchild Camera and Instruments Corporation. By the 1930s the Fairchild K-17 camera used the proportions 6 inches focal length to 9×9 square inches photo format. This soon became a standard for aerial mapping cameras: wide angle objective (15 cm) and $23 \times 23\ cm^2$ format. European cameras earlier than 1956 used rather smaller formats ($13 \times 13\ cm^2$, $13 \times 18\ cm^2$, $18 \times 18\ cm^2$).

2.1.2 Digital Cameras

Analogue aerial cameras, together with the aerial photographic film industry, reached a performance level that became standard for the development of digital cameras. The first large format digital aerial cameras that could compete with the analogue ones for mapping purposes were presented to the photogrammetric community at the ISPRS Congress in Amsterdam in 2000: the Z/I-Imaging DMC and the Leica ADS. Just a few years later digital cameras had already conquered the market. UltraCamD from Vexcel (now Microsoft) was the third camera of this type to appear.

Medium format digital cameras also began to respond to the needs for aerial photos for instance in association with airborne LIDAR (Leica ALS with camera Leica RCD105 and Optech ALTM with camera DiMAC Ultralight). Medium format system cameras are also being developed by Intergraph Z/I Imaging (RMK-D) and Vexcel Imaging (UltraCamL).

Smaller cameras equipped with Bayer pattern have also found a market for smaller projects (e.g. DSS Applanix, IGI DigiCam, DLR-3 K), for example traffic monitoring, remote sensing, forestry, corridor mapping and orthoimage generation. Nearly every year new concepts and models are being presented to the photogrammetric world.

Digital aerial cameras, like common digital cameras, do not need film for keeping the images. These are captured in electronic photosensors placed in the focal plane of an objective and are saved in an electromagnetic memory unit. The photosensors used so far are CCD. The size limitation for feasible and economic CCD array arrangements up to 2010 was less than the usual image format in aerial photogrammetry. This fact drove researchers toward solutions to achieve a performance comparable to the last generation of analogue aerial cameras in terms of format and resolution. Two different concepts were adopted in this development: the linear array sensor and the matrix sensor. Unlike analogue cameras for mapping purposes, digital ones were designed to capture panchromatic, colour and near infrared images in the same flight and with the same camera. This fact contributed significantly to the success of digital cameras.

Linear Array Sensor Cameras

The concept of the linear array sensor camera had its origin in the three line sensors developed by Deutsches Zentrum für Luft und Raumfahrt (DLR) as MOMS camera used in 1993 in the Spacelab and in 1996 in the Priroda Spacestation, as WAOSS for the 1994 MARS Mission and as HRSC for the 1996 Mars mission (Albertz et al. 1992). The objective of this sensor was to obtain three different perspectives from the same trajectory, allowing automatic generation of a digital model of the ground and the following generation of maps, height models and orthoimages. Adapted to Earth reality, this concept led to an aerial camera with one objective and a set of linear array sensors located in the focal plane, occupying several parallel positions oriented transverse to the flight direction. Basically three arrays are responsible for the panchromatic image acquisition: one for the nadir perspective, one for the forward and one for the backward perspectives. The Leica Airborne Digital Scanner is a system of this kind for cartographic purposes as well as the Jenoptik JAS-150 and the Wehrli 3-DAS-1. The Leica

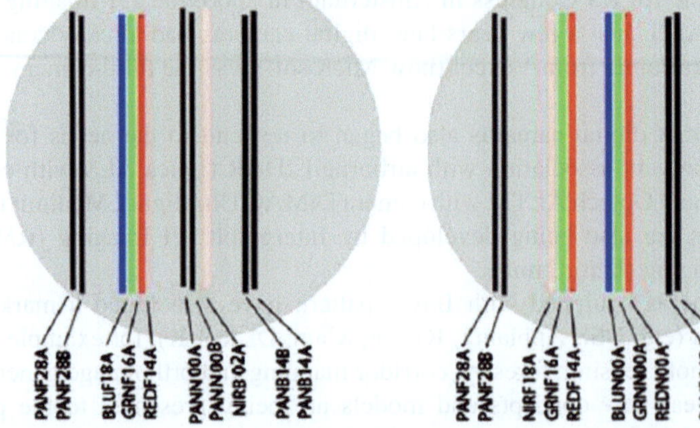

Fig. 4.4 Some variants for line array disposition on the focal plane and line designation of the ADS 80. *Left*: the basic variant (adapted from Leica Geosystems)

ADS80 has a focal length of approximately 62 mm and a field of view of 64° across track. Each sensor array has 12,000 pixels with $6.5 \times 6.5\ \mu m^2$ pixel dimension. The position and quantity of sensor arrays on the focal plane can vary according to the version and purpose of the actual camera. Each panchromatic channel can consist of one or two lines shifted by half a pixel (staggered arrays). Additional single arrays are responsible for the multi-spectral (MS) channels (Red + Green + Blue + Near Infra Red). The basic variant of ADS presents ten lines with the disposition presented in Fig. 4.4 (left).

The images are acquired in a push-broom technique along the flight line. Each array captures an image line of a determined perspective from a particular aircraft position and attitude at a particular time.

The original scenes are built by composition of adjacent image lines. The delivered images are rectified after correction for attitude changes of the sensor. These are recovered from position and orientation data acquired during the flight. Therefore a high quality Inertial Measuring Unit (IMU) is integrated in the camera. There are as many original scenes as sensor arrays, each of them having the format of a long strip. The forward and the backward looking perspectives build different angles with the nadir looking perspective (28.4° and 14.2°) at the projection centre (Fig. 4.5). This allows for several stereoscopic evaluation alternatives (backw. + forw., backw. + nadir, nadir + forw.). A tetrachroid filter (Fig. 4.6) is used for splitting the arriving multispectral radiation from an elementary area on the ground in their red, green, blue and near Infra Red components, allowing each of them to be captured by a different sensor array while guaranteeing the co-registering of the different channels (Fig. 4.7).

The photogrammetric processing of ADS images is not compatible with the established mapping production workflow, needing extra software solutions. At the

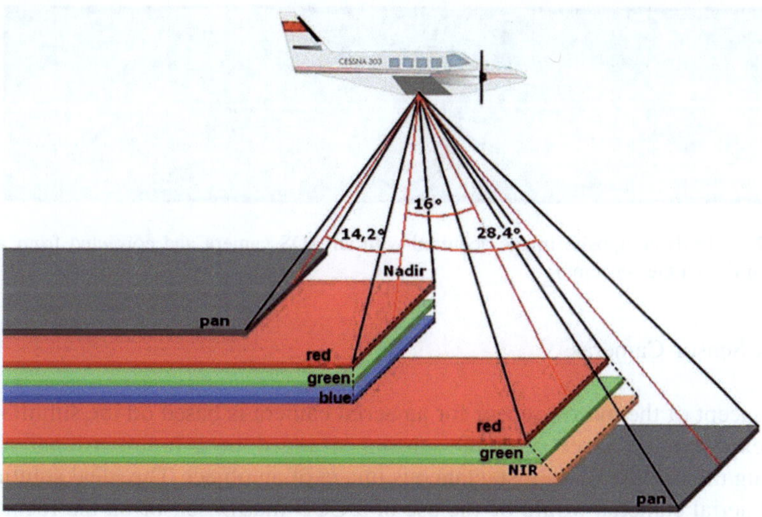

Fig. 4.5 Angles on the projection centre between each perspective and the nadir perspective planes of the ADS 80 (adapted from Leica Geosystems)

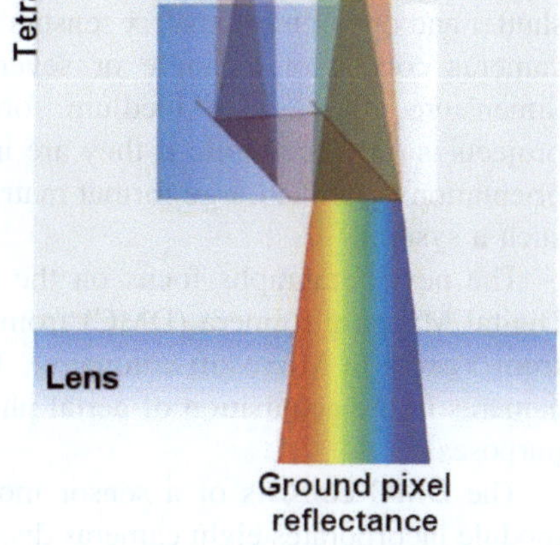

Fig. 4.6 Tetrachroid filter used in the ADS camera (Leica Geosystems)

beginning of the digital camera era this might have been an obstacle for the ADS to become widespread. Meanwhile most of the photogrammetric software providers offer solutions for the processing of line scanner images.

Fig. 4.7 A RGB composite image acquired with an ADS camera and corrected from attitude variation (Leica Geosystems)

Matrix Sensor Cameras

The concept of the matrix sensor for an aerial camera is based on the simultaneous light exposure of a set of CCD sensors arranged in a bidimensional matrix, allowing the capture of an instantaneous image of an object. The ideal solution for digital aerial cameras would be the use of a CCD matrix sensor on the focal plane with at least the same information content as analogue aerial cameras (area $23 \times 23\ cm^2$). This solution was not achievable at the launch of large format digital matrix cameras. In order to achieve an image area acceptable for cartographic purposes, camera developers had to search for innovative solutions using these relatively small CCD matrixes. Basically, the solutions found had already been applied in an analogue version in the early years of photogrammetric mapping as a trial to increase the model area (McCurdy et al. 1944). Instead of one, multiple cameras had to be assembled and the multiple images obtained had to be fused in the end into a bigger one that could then be processed. Large format digital aerial cameras are in fact several cameras each with its own objective and shutter and one or more matrix sensors in its own focal plane. Also medium format cameras consist of a single or several cameras. Due to their reduced image dimensions, the use of medium format cameras for small photogrammetric projects is only economic if they are integrated with GNSS/IMU positioning and orientation systems. Large format matrix cameras, however, are independent from such a system.

The next paragraphs focus on the two dominating cameras of this type: the Digital Mapping Camera (DMC) from Intergraph Z/I Imaging and the UltraCam from Vexcel (a Microsoft company). They have nearly totally replaced analogue cameras in the acquisition of aerial photographs for mapping and remote sensing purposes.

The DMC consists of a sensor module and mass memory units. The sensor module incorporates eight cameras disposed as shown in Fig. 4.8. Each of the four central cameras contains a 7,168 × 4,096 pixels CCD chip with 12 μm pixel dimension and an objective with 120 mm focal length and maximum aperture f/4. These cameras capture the high resolution sub-images for the panchromatic channel. The four peripheral cameras are the multispectral cameras (MS). They have a 3,072 × 2,048 pixels CCD chip with 12 μm pixel dimension and an

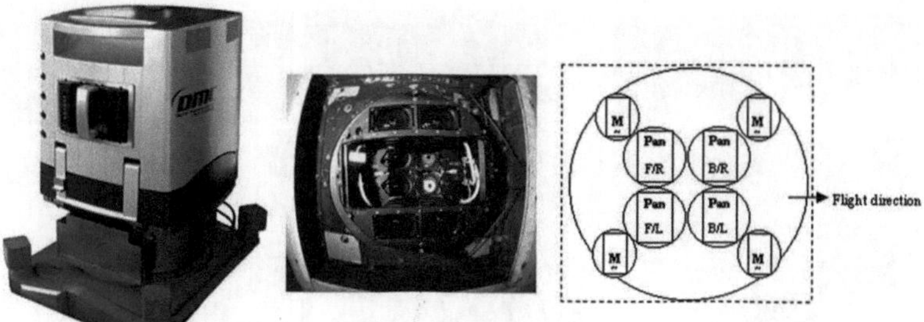

Fig. 4.8 DMC: (*left*) sensor module and solid state disk for mass memory; (*centre*) sensor head showing the eight objectives; (*right*) camera disposition (Intergraph)

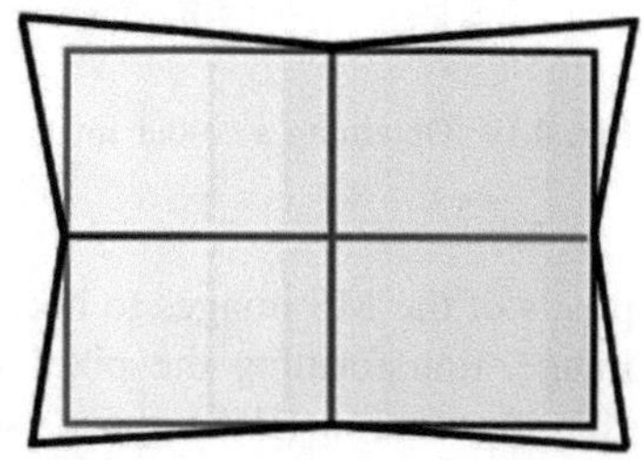

Fig. 4.9 Areas covered by the panchromatic mosaic (*black*) and by each MS camera (*outer grey rectangle*)

objective with 25 mm focal length, maximum aperture f/4. Colour filters ensure that each of the cameras captures a distinct portion of the electromagnetic spectrum (R, G, B and NIR). The MS images have less resolution than the panchromatic ones.

The four panchromatic cameras are assembled with their axis slightly tilted so that the four images have a small overlap. The four MS cameras are instead mounted in a vertical configuration. Each MS camera captures the whole area covered by the panchromatic mosaic (Fig. 4.9). A final high resolution panchromatic image with 13,824 × 7,680 pixels is built from the four original panchromatic sub-images in a post-processing step. This includes a platform calibration (determination of the 3D orientation of all the cameras in relation to an internal reference system through bundle adjustment using common points in the overlapping areas of the four images) after consideration of the individual camera calibration parameters (interior orientation) and a projection of the original image pixels to a virtual plane. The resulting virtual image is a distortion free central perspective of the object as taken from a single objective camera with 120 mm focal length (Cramer 2004). The pixel dimension is 12 μm square and the shorter side of the image is parallel to the flight direction. The colour and near infrared channels are also projected onto the same virtual plane.

The technique of pan-sharpening is then applied to obtain RGB, near infrared (NIR) and false colour infrared (CIR) images with the same geometric resolution as the panchromatic image. Pan-sharpening consists, for example, of dividing the

Fig. 4.10 Obtaining a colour image (*right*) through pan-sharpening (Intergraph)

pixels of the MS images to become the same pixel dimension as the panchromatic image, transforming the pixel values of the RGB colour space to the Intensity, Hue, Saturation (IHS) space, substituting the I value in each pixel through the corresponding pan value and retransforming in the RGB space (Fig. 4.10).

The generation of the final virtual image is independent of navigation data (GNSS/IMU). The frame rate of the DMC, the time needed for saving the images after each exposure, is 2.1 s. This camera uses Time Delay and Integration (TDI) sensors for compensating for the aircraft's movement along the flight line during exposure, allowing a longer exposure time. It works like an electronic FMC. Short frame rates and TDI allow for small ground sample distances (GSD) and sharp images even from lower altitudes.

The UltraCam large format digital camera from the Austrian company Vexcel Imaging, a Microsoft Company since 2006, is also based on multiple cameras with several matrix CCDs. Launched onto the market in 2003, this camera has rapidly evolved from the first presented model UltraCam$_D$ to the latest UltraCam Xp WA. The evolution goes through nearly all the components, involving the sizes of the CCD matrix, of the sensors, of the virtual image and the type of objectives and respective focal length. However, the basic concept has been maintained. UltraCam presents eight individual cameras with parallel aligned axes. Four of them, aligned in the flying direction, capture panchromatic images and the other four are provided with optical filters for MS image capture (Fig. 4.11). The pan cameras all have the same field of view but the sensors are located in distinct positions in the focal plane. In the foremost camera there is one CCD matrix sensor. The second camera contains two, the third (the master cone) contains four and the last one again has two CCD matrix sensors. Since the projection centres of the cameras are shifted, Vexcel developed the concept of syntopic imagery for the acquisition of the panchromatic channel. The individual shutters are triggered with certain time

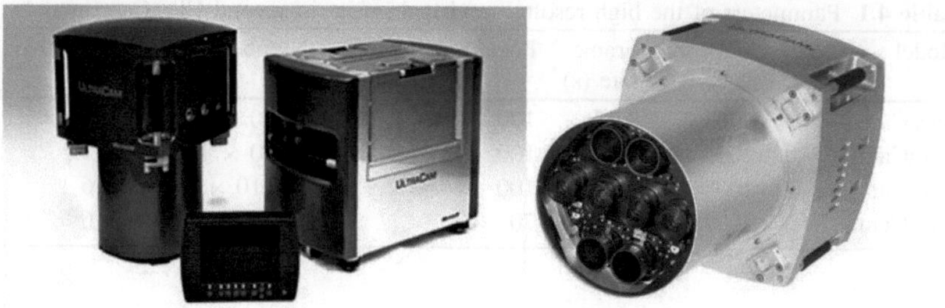

Fig. 4.11 Large format camera UltraCam from Microsoft Vexcel Imaging (Vexcel)

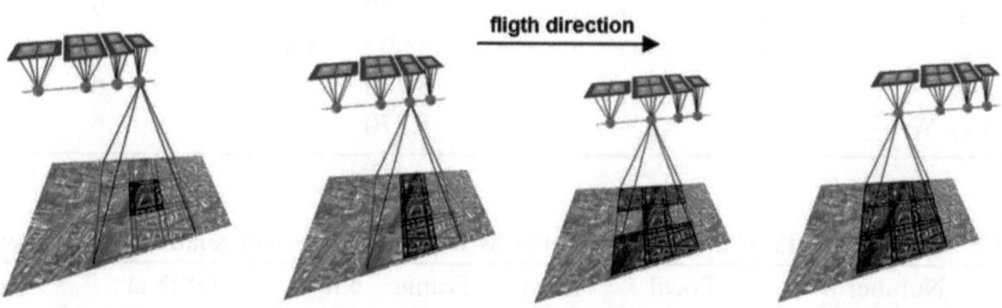

Fig. 4.12 Syntopic image acquisition with UltraCam large format camera (Vexcel)

delays depending on the flying speed, so that all panchromatic images are actually captured from the same space position of the projection centres (Fig. 4.12). This process has the disadvantage of making the camera sensitive to strong vibrations.

The final panchromatic virtual image, built by stitching the nine original sub-images on the frame defined by the four matrixes in the master cone, is one large format central perspective. For this operation, the interior orientation parameters of each camera head (focal length, principal point, radial distortions) and for each CCD matrix (shift, scale, shear and perspective distortions), determined by calibration in the laboratory, first have to be applied to the sub-images. These, which have a small overlap, are then used to calculate the eventual tilts between the camera axes and to define the relative orientation of the four camera heads through bundle adjustment. The final panchromatic image is theoretically distortion free (Leberl and Gruber 2005). The geometric problems of merging the nine sub-matrixes have been improved by so-called monolithic stitching, transforming the nine sub-matrixes into the homogenous green image with lower resolution (Ladstätter et al. 2010). As this monolithic stitching is contradictory to the syntopic mode, the synchronous acquisition mode of all cameras is preferred.

Table 4.1 indicates the image parameters for several models of the UltraCam large format camera (Leberl and Gruber 2005; Wiechert and Gruber 2009). The colour channels are composed for obtaining lower resolution colour or colour infrared (CIR) images (Table 4.2). High resolution RGB and CIR images are obtained by pan-sharpening after projecting the composite images of the several channels into the same frame of the high-resolution panchromatic image. TDI is

Table 4.1 Parameters of the high resolution virtual image in several UltraCam models

Model	Phase-in	Frame rate (s)	Focal length (mm)	Number of pixels	Pixel size (μm)
$UltraCam_D$	2003	0.75	100	7,500 × 11,500	9
$UltraCam_X$	2006	1.35	100	9,420 × 14,430	7.2
$UltraCam_{Xp}$	2008	2.0	100	11,310 × 17,310	6
$UltraCam_{Xp}$ WA	2009	2.0	70	11,310 × 17,310	6

Table 4.2 Parameters of RGB and CIR composite images in several UltraCam models

Model	Focal length (mm)	Number of pixels	Pixel size (μm)
$UltraCam_D$	28	3,680 × 2,400	9
$UltraCam_X$	33	4,992 × 3,328	7.2
$UltraCam_{Xp}$	33	5,770 × 3,770	6
$UltraCam_{Xp}$ WA	23	5,770 × 3,770	6

Table 4.3 Technical data of the DMC II versions with base to height relation for 60% end lap l

Camera	Number of pixels	Focal length (mm)	Pixel size (μm)	Frame rate (s)	b/h ($l = 60\%$)	GSD at h = 1,000 m (cm)	Relation pan/MS
DMC II 140	12,096 × 11,200	92	7.2	2	0.35	7.8	1: 2.0
DMC II 230	15,104 × 14,400	92	5.6	1.7	0.35	6.1	1: 2.5
DMC II 250	17,216 × 14,656	112	5.6	1.7	0.29	5.0	1: 3.2

also used in the acquisition of the original sub-images in order to compensate for the aircraft's forward motion, improving image sharpness. As smaller sensors need longer exposure, the frame rate, the minimum time interval between two frames, has been growing as the pixel dimension has been getting smaller (Table 4.1).

The evolution by digital aerial cameras has been very fast. Since all the above-mentioned digital cameras are built in a modular concept it is easy to replace their obsolete components by newly developed ones. Numbers from 2008 (Gruber 2008) reveal a total of 244 digital aerial cameras sold around the world, sharing ADS40 25%, DMC 30% and UltraCam 42% of the market.

With the now available large format CCD matrixes from DALSA the situation has changed. Z/I Imaging developed the DMC II equipped with just one large format panchromatic CCD matrix and four cameras for RGB and NIR. The large size homogenous panchromatic CCD-matrix has the advantage of very small systematic image errors (Jacobsen 2011) (Table 4.3).

The high number of pixels used by the DMC II accommodate more information than analogue aerial photos. A comparison of the details which could be extracted for topographic mapping from DMC, UltraCAM and ADS40 images as well as scanned aerial photos having different GSD show that only $8{,}520^2$ pixels are required for the information contents of scanned aerial photos compared with original digital images, when these are not degraded by lower effective resolution (Jacobsen 2009). The better geometric performance of digital cameras compared to analogue photos has been demonstrated by the camera test of the German Society of Photogrammetry, Remote Sensing and Geoinformation (Jacobsen et al. 2010).

2.2 Planning the Photo Flight

The flight mission, specially performed for the acquisition of a set of aerial photos that can be processed photogrammetrically, must be carefully planned according to the objectives of the coverage and the restrictions to the flight. Nowadays several camera producers offer software that supports the mission planning. Most of them go further and also support the photo mission based on what has been previously planned in terms of course and overlaps. This requires satellite positioning supported navigation and connections with receiver and camera.

Items to define at the beginning of the planning include:

- Area to survey, its location, dimensions and spatial disposition (two dimensional or one dimensional extent)
- Photo scale or GSD
- Pretended overlap
- Aerial camera/objective to use
- Aircraft
- Epoch of the year and time of day for the photo flight

The area to survey is usually defined in digital format. In the case of a two-dimensional extent, the boundary of the area can be delivered to the planner as a closed polygon (e.g. the boundary of a county) or as a set of coordinates of the vertexes. In the case of a one-dimensional region the axis of the region must be delivered as an open polygon or a set of coordinates of its vertexes, as well as the width for each side of the axis that should be surveyed. The first case applies to standard mapping projects while the second concerns special surveying projects, known as corridor mapping along rivers, coastline and terrain where highways or railways, for instance, exist or are to be constructed. The flight plan shall consider straight strips over the region to survey. All the necessary plane turns should be specified outside. While one-dimensional regions are covered by strips with straight axis whose orientation varies, two-dimensional regions are covered by parallel strips (Fig. 4.13).

The photo scale, in a planning context, is the relation between focal length and flying height. In a processing context, however, it is a factor for both measurement accuracy and the amount of information of the object that can be obtained from the image. Since digital images began to be photogrammetrically processed the

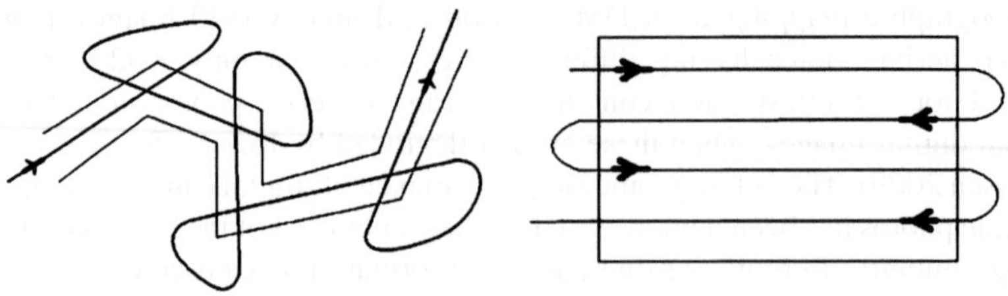

Fig. 4.13 Flying course for one-dimensional (*left*) and for two-dimensional (*right*) regions

Table 4.4 Typical overlaps in photogrammetric coverages

Objective of the coverage	% End lap	% Side lap	Notes
Stereo plotting/measuring/correlating	60	20–30	
Triangulation (high precision)	60	30–60	Plus cross strips
Orthoimage production (existing DEM)			
Urban	60–80	60–80	
Rural	30	30	
Photointerpretation			
Features	30–60	20	
Terrain	60	20	

concept of GSD, meaning the distance on the ground between the centres of two adjacent pixels, began to be applied instead of the photo scale. Like the photo scale, the GSD is also normally not constant along the image area, varying due to photo axis inclination and ground relief. Assuming a square pixel, a mean GSD can be computed from the image pixel (linear) size multiplied by the denominator of the mean photo scale.

Photogrammetric flight missions are planned in order to obtain the coverage of a region composed by strips of vertical photos with a certain overlap between successive photos. In the case of several strips these should also partially overlap. In this way, the same object on the ground will generally be pictured in two or more aerial images. The overlap is expressed as a percentage of the side dimension of the photo, along (end lap l) and across (side lap q) the flight direction. The objective of the end lap is pairwise stereoscopic plotting, measuring and correlating while the side lap normally has the function of avoiding gaps between strips. Nevertheless, both end and side lap are important for good geometric stability in the triangulation of the whole block of strips. Typical overlaps for different types of projects are indicated in Table 4.4.

Table 4.5 gathers the formulas used to calculate some parameters of the photo flight. The meaning of the variables is illustrated in Fig. 4.14. Although the example shown refers to stereoscopic coverage, the formulas are valid for all other cases. These formulas are also used in the elaboration of a budget for the flying mission by setting a standard price for a flying hour and for one image.

Table 4.5 Photo flight parameters

Photo scale	$E_f = 1/m_f = f/h$
Photo side on the ground	
Along flight direction	$S_1 = s_1 \cdot m_f$
Across flight direction	$S_2 = s_2 \cdot m_f$
Photo basis	$b = B/m_f$
Flying height above ground	$h = f \cdot m_f$
Absolute flying height	$Z_0 = h + Z$
End lap %	$\ell = (1 - B/S_1)\ 100$
Side lap %	$q = (1 - A/S_2)\ 100$
Basis length for ℓ% end lap	$B = S_1\ (1 - \ell/100)$
Strip axes distance for q% side lap	$A = S_2\ (1 - q/100)$
Area covered by a photo on the ground	$A_f = S_1 \cdot S_2 = s_1 \cdot m_f \cdot s_2 \cdot m_f$
Area covered by a stereo model	$A_m = (S_1 - B)\ S_2$
Area of model overlapping	$A_{sm} = (S_1 - 2B)\ S_2$
New area covered by a model from the second strip on	$\mathrm{An} = A \cdot B$
Number of stereo models per strip	$\mathrm{Nm} = \mathrm{int}((L/B) + 1)$
Number of photos per strip	$\mathrm{Nf} = \mathrm{Nm} + 1$
Number of strips in the block	$\mathrm{Ns} = \mathrm{int}((Q/A) + 1)$
Time interval between photos	$\Delta t\ [\mathrm{s}] = B[\mathrm{m}]/v[\mathrm{m/s}]$

Note int (x) = integer part of x

Variables in Table 4.5:

A Distance between strip axes
B Real basis: distance between successive exposure points
F Calibrated focal length
s1 Image side along the flight direction
s2 Image side across the flight direction
h Height above the ground
Z Terrain height
Zo Plane's absolute height
v Plane's mean speed during photo mission
L Strip length
Q Block width

An end lap of 60% guarantees total stereoscopy of the strips and a theoretical constant overlap of 20% between triplets (three successive photos) along the strip except for the first and last pair of photos. For better homogeneity of the results, the mission should be planned so that those pairs lie outside the project area.

For high precision triangulation, the overlap combination 60/60 plus cross strips also with 60/60 (main strips) and 60/30 (cross strips) is preferred, increasing the number of photos where each object point appears and can be measured.

For the production of orthoimages (Sect. 4.3), since the contents of the image will be transferred to the final product, the radial distortion due to objects' height can be a problem. If a digital terrain model is applied in the orthorectification, high objects will appear distorted, the top of the object in a false planimetric position

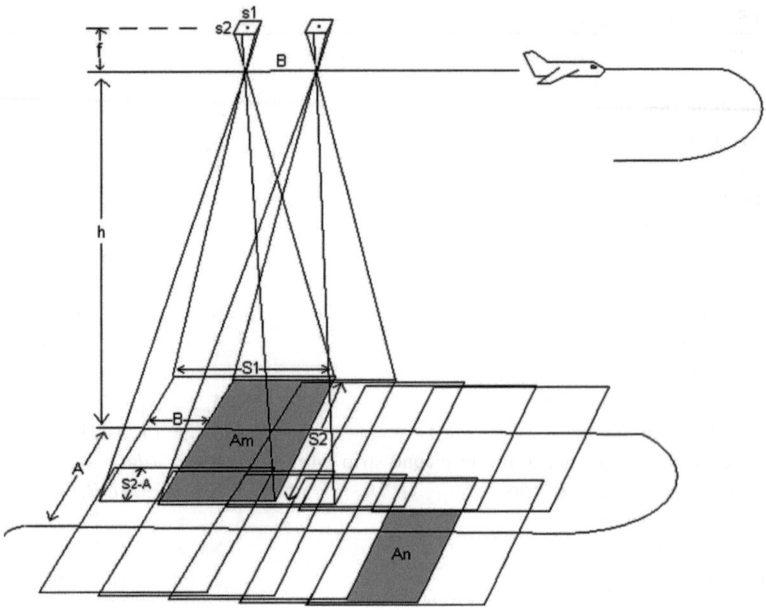

Fig. 4.14 Photo flight scheme

and vertical planes, like building façades that should be invisible, appear projected on the ground causing an undesired leaning buildings effect and occlusions on the orthoimage. If a digital surface model is used instead, occlusions caused by the radial distortion of higher objects in the aerial photo turn into gaps of radiometric information for the rectification operation (Fig. 4.15). Therefore, aerial coverages for orthoimage projects should present as little radial distortion as possible. In rural environments where high buildings are not so common, a 30/30 or even lesser overlap would be sufficient since stereoscopy is not necessary for orthoimage production. Such a coverage could only be used in a case where no aerial triangulation is required for the determination of the image orientation, which seldom happens.

In urban environments, the end lap should be larger to allow a view to the streets and to make it possible to limit the orthorectification to just a central square of the photo where radial distortion is tolerable. Since the radial distortion Δr of a point depends on:

- the radial distance r of the point to the centre of the vertical aerial photo.
- the height difference Δh between the point and the projection of the photo centre on the ground (actually the projection of the principal point on the ground).
- the flying height above ground, h_0.

Fig. 4.15 Radial distorted high buildings in aerial photo (*left*) and gaps (in black) corresponding to occlusions in orthoimage (*right*) (Oda et al. 2004)

the side dimension of the central square to rectify, *sqs*, can vary according to (4.1) where r_{max} is the maximum radial distance corresponding to the tolerable radial distortion Δr_{tol} affecting objects with the maximum height difference Δh_{max} referred to the ground principal point.

$$\begin{aligned} sqs &= r_{\max}\sqrt{2} \\ r_{\max} &= \Delta r_{\mathrm{tol}}\cdot\frac{h_0}{\Delta h_{\max}} = \Delta r_{\mathrm{tol}}\cdot\frac{f\cdot m_f}{\Delta h_{\max}} \end{aligned} \tag{4.1}$$

The square side, multiplied by the scale denominator, defines, in this case, the distance B to fly between successive exposures. The distance A between adjacent strip axes should also be defined by *sqs* in the same way, so that there are no gaps between adjacent central squares (Fig. 4.16).

The overlaps for coverages intended to be photointerpreted depend on the objective of the interpretation. For landforms interpretation stereoscopy is indispensable, requiring a 60/20 overlap, while, for projects related to feature counting and ground cover classification, stereoscopy can be often dispensed and a 30/20 overlap is enough if no aerial triangulation is required.

For all measuring projects, although requiring more photographs, a 30% side lap is preferable to 20%. The effective area, meaning the area to be processed in each photo reaching from its centre to half the overlap to all adjacent photos (Fig. 4.17) is smaller in the first case, reducing the effects of radial distortion due to relief and object's height.

The aerial camera to be used must be chosen according to the mission's objective. The option between analogue or digital camera tends to disappear but it still exists. Analogue cameras have alternative objectives with several focal lengths allowing coverages of the same scale from different heights and of several scales from the same height. Digital cameras often have just one focal length. For a smaller GSD (bigger scale), cameras can reach their limits when the time to save the

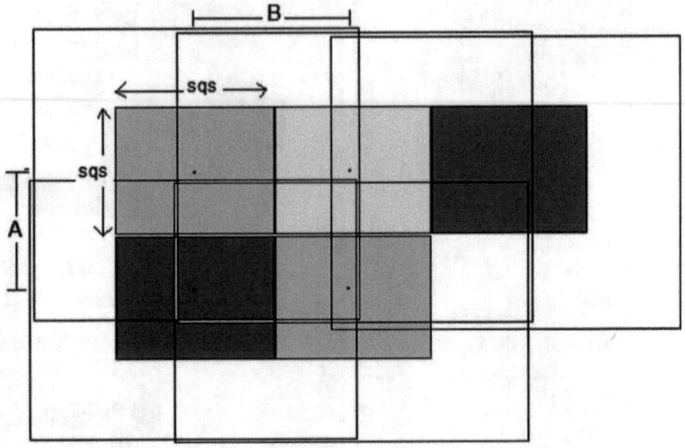

Fig. 4.16 Disposition of aerial photos for central squares orthorectification

Fig. 4.17 Theoretical effective area (*grey region*) of an aerial photograph in an intermediate strip

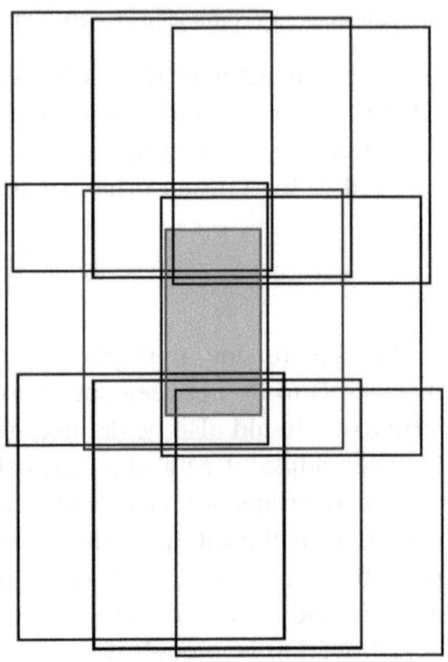

captured images is longer than the required time interval between exposures for the pretended end lap. As derived from (4.1), the focal length f influences the radial distortion due to relief or the object's height, Δr, in the opposite way. The longer the focal length, the smaller the distortion. That is an aspect to consider, especially for projects covering high mountains and deep valleys or urban regions with high buildings. Flying height and pretended GSD also play a role in this matter.

The choice of an aircraft at this stage of the photo flight planning is related to the size of the project. If there are alternatives, a plane should be chosen according to its autonomy (flying time without need to refuel), minimum flying speed and maximum flying height. Double engine planes are preferable.

To guarantee both good luminosity of the photographs and a reduction of shadow sizes, photo flights should normally be done when the sky is clear and the sun is high above the horizon. A generally adopted threshold for the sun angle above the horizon is 30° but because of higher sensitivity for digital cameras this may be reduced to 20°. The daily temporal window, during which a photo flight can be made having a sun angle greater than 30°, varies with the season and decreases with the latitude of the region. For a latitude greater than 73.43° N or S the sun never reaches that height. For those lands (Antarctica, some islands in North Canada, Greenland and a small part of Siberia) that threshold cannot be maintained. As trees can cover important details photo flights should be done when foliage is reduced to a minimum, which is just before the leaves sprout in deciduous trees in early spring. For some projects of photo interpretation, shadows can be an advantage, helping to identify objects. In that case, the photo flight should be done with a lower sun angle. Digital camera images possess a higher dynamic range than analogue images (12 bit instead of 8 bit), allowing visibility and measuring of details in shadow areas. Consequently, flight temporal windows can be enlarged for digital images.

For planning objectives it is assumed at first that the terrain is horizontal and the photos vertical (vertical axis). Knowledge about the terrain relief, normally through map information, must be used to refine the planning.

The output of flight mission planning is a flying map in digital format. It consists of the planned course of the plane as a set of oriented straight lines containing the planned location of the exposure points over a geographic basis (Fig. 4.18). Some flight mission supporting software produce interfaces to virtual globes available on the internet, allowing visualization of the planned image footprints on the ground, which is represented as a 3D terrain model with realistic texture, and the detection of gaps due to the influence of relief in the overlaps (Fig. 4.19). The accuracy of the footprint projection depends on the quality of the DTM used by the mapping program in the region of interest.

For mountainous areas the accuracy and information content has to be fulfilled at the lowest ground height, while the overlap has to be planned considering the highest ground elevation. A photo flight parallel to the contour lines is preferred.

3 Image Georeferencing

3.1 Coordinate Systems in Photogrammetry

The main objective of photogrammetry, no matter to which field of application it is applied, is to measure quantities on images or stereo models and transform them in corresponding quantities in the photographed object. To perform the transformation from image into object space and reverse, coordinate systems in each space (image space, model space, object space) are required. The most important system

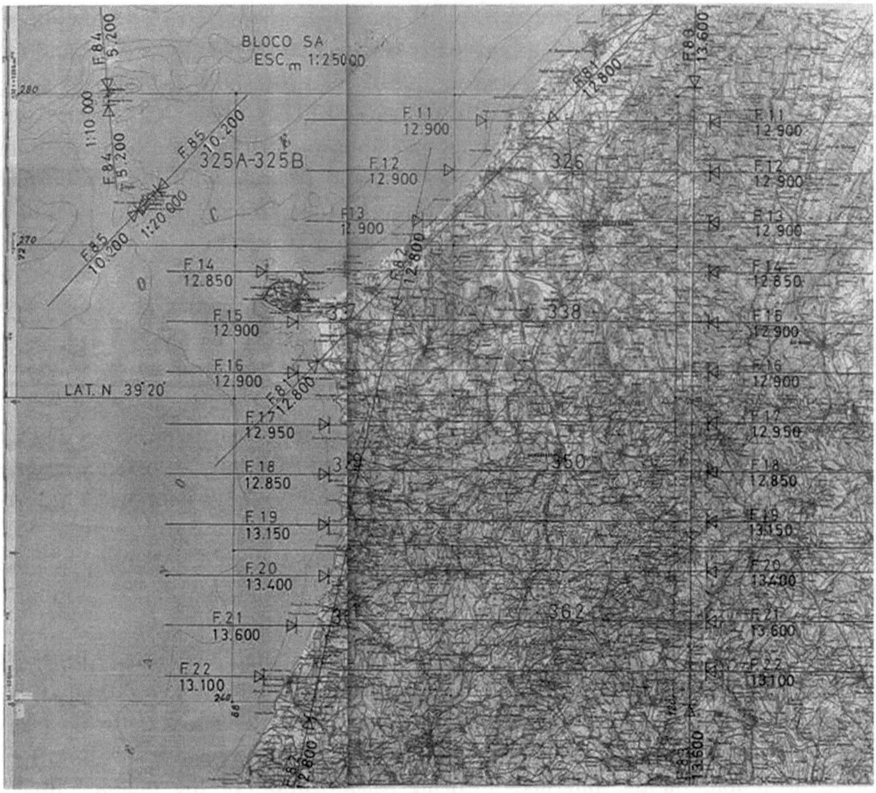

Fig. 4.18 Flying map showing strips, start and end exposure points and flying height in feet

in the image space is the so-called image coordinate system or photo coordinate system. Its origin is located in the projection centre with focal length *f*, also called principal distance, as z-value. The axes form a right handed system. The perpendicular projected origin in the image plane, parallel to the xy plane, is represented by the principal point of auto collimation (PPA). For analogue photos this point is related to the fiducial marks by the camera calibration. Fiducial marks are images of references that exist in the focal plane of an analogue camera. The coordinates of the PPA referred to the fiducial centre (FC) are normally designated by (x_o,y_o). Instead of the principal point of auto collimation, the adjusted principal point of best symmetry (PPS) is normally used. For digital cameras the principal point is expressed by its pixel address.

Digital images, scanned or captured with digital cameras, have an associated pixel coordinate system. It is normally left-handed Cartesian with the origin in the centre or corner of the upper left pixel (0,0). The integer coordinates of a pixel indicate the row and column of its position on the image. For scanned analogue images the relation between image and camera is determined by the fiducial marks. For original digital images there are no fiducial marks. The relation to the camera is obtained through the

Fig. 4.19 Projection of the flight mission on Google Earth

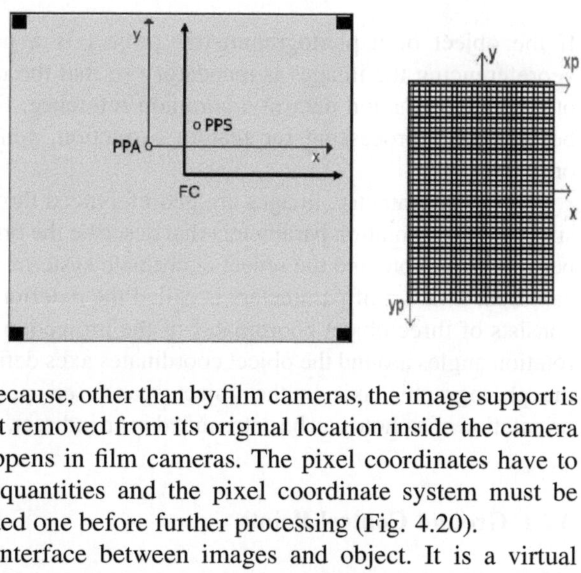

Fig. 4.20 Fiducial (*dashed*) image (*grey*) and pixel (*black*) coordinate systems in analogue and digital images

pixel positions in the image because, other than by film cameras, the image support is fixed to the camera and is not removed from its original location inside the camera after each exposure as it happens in film cameras. The pixel coordinates have to be transformed into metric quantities and the pixel coordinate system must be transformed into a right handed one before further processing (Fig. 4.20).

The model space is an interface between images and object. It is a virtual Cartesian 3D space that can be perceived by stereoscopic observation of a stereo pair although it doesn't physically exist. In analogue plotting instruments a model space could be materialized through mechanical axes of the instrument and 3D model coordinates could be measured instead of photo coordinates. The origin of this system is arbitrary as well as the scale. Most of the photogrammetric problems are solved today without resorting to this intermediate space, but some software still refers to model coordinates in a stage between relative and absolute orientation of stereo pairs.

In most cases the national coordinate system is used as the object coordinate system. It is not exactly a Cartesian and right-handed system, so the geometric difference to a Cartesian coordinate system has to be respected, for example, by Earth curvature correction. In some cases local coordinate systems tangential to the Earth ellipsoid are used. The following scheme shows the sequence of the transformations needed. For details see Seeber (2003) and Krakiwsky (1973).

Photogrammetric object (ground) coordinates (X, Y, Z)
= topocentric coordinates (X, Y, Z)

⬇ ⬆

Geocentric rectangular coordinates $(X, Y, Z)_G$

⬇ ⬆

Elipsoidic curvilinear coordinates (φ, λ, η)

⬇ ⬆

Map coordinates (E, N) + orthometric height (H)

3.2 *Indirect Georeferencing*

If the object of a photogrammetric project is a portion of the Earth surface, georeferencing the images is mandatory so that the results can be integrated with other geoinformation data in a common reference. Georeferencing must be done before image processing for feature extraction, generation of terrain models or orthoimages.

In photogrammetry, images are georeferenced through the determination of the six exterior orientation parameters that describe the original spatial relation existing between the photo and the object coordinate systems at the moment the image was captured. This set of parameters is called the exterior orientation of the image and consists of three object coordinates of the projection centre (X_0, Y_0, Z_0) and three rotation angles around the object coordinates axes defining the spatial orientation of the photographic axis in the object space, (ω, φ, κ) (Fig. 4.21). The exterior orientation of a photo can be determined in an indirect, a semi-direct and a direct way.

3.2.1 Ground Control Points

The indirect georeferencing needs information about the object, normally in the form of a set of ground control points (GCP). These are points in the object space whose object coordinates are known. In close-range photogrammetry it is also common to use other types of geometric information on the object, such as distances and geometric constraints for object lines, for instance parallelism, perpendicularity, verticality and horizontality.

There are two types of ground control points: natural and targeted. Natural points are conspicuous points on the ground or on built objects where the

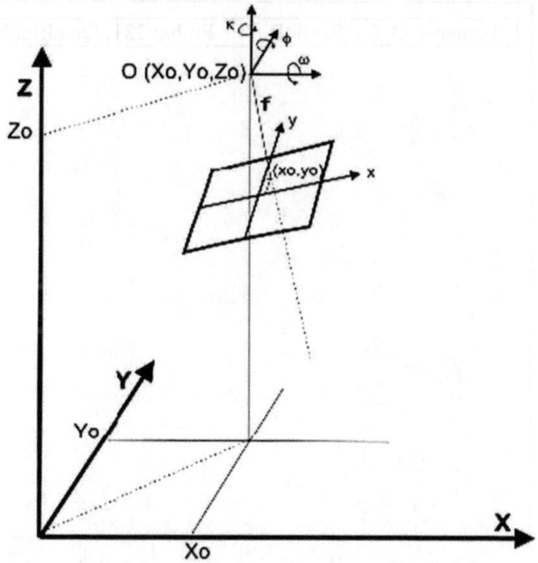

Fig. 4.21 Elements of interior and exterior orientation

assignment between image and ground situation should be most certain, such as corners of sidewalks or walls, crossroads, etc. Fig. 4.22). Symmetric objects should be preferred. Targeted control points have to be signalized on the ground before the photo flight takes place. Their location must be planned based on the flying map and they should be set in near horizontal places with no occluding objects in the immediate neighbourhood in order to be seen from the air in several photos, that is, from several perspectives. They consist of a central regular geometric figure (circle, square or equilateral triangle) and two to four identification strips, depending the number on the available place in loco (Fig. 4.23). They are marked with canvas, plastic film or paint in a contrasting colour to the surrounding ground, white, yellow or red being the preferred colours (Fig. 4.24). The centre of the central figure is actually the GCP to be coordinated.

Normally the whole target is built as a function of just one parameter which is the diameter of the central circle, the side of the central square or triangle, the dimension of which depends on the pretended image scale in order to be easily identifiable and measurable.

GCP coordinates are determined through geodetic positioning techniques. Whenever possible, satellite positioning systems are used. The uncertainty associated with the GCP coordinates should be less than the GSD in planimetry and less than the mean flying height above ground multiplied by 10^{-4} for altimetry.

Targeted GCPs allow higher accuracy of the georeferencing but the targeting operation is an expensive one. Independent of the kind of GCP, the surveyor that coordinates them must elaborate a localization scheme of each point and its surroundings so that it can be identified and measured in the aerial image. Current coordinating missions often revert to internet mapping programs or virtual globes

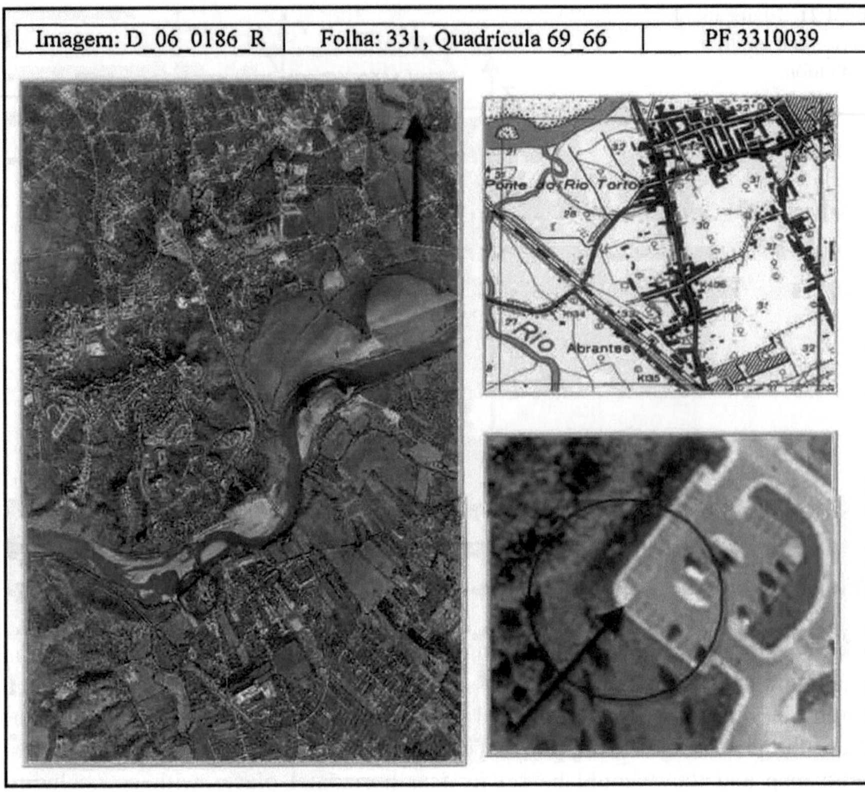

Fig. 4.22 Location of a natural GCP

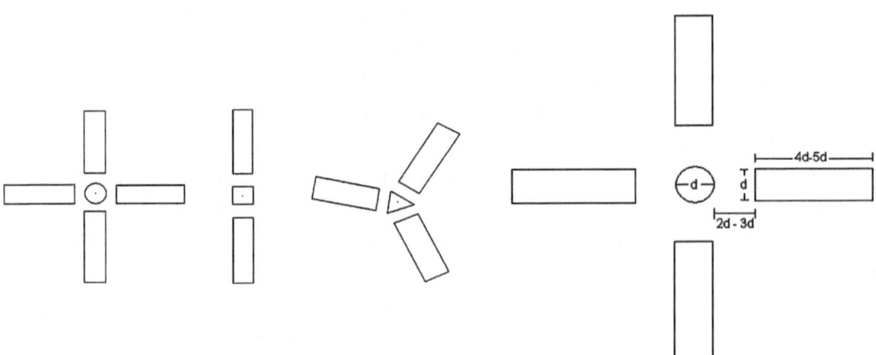

Fig. 4.23 Shapes for targeted GCPs and construction rules

to indicate the location of the coordinated points (Fig. 4.25). For greater regions a small data base is organized including the GCPs, their coordinates and pictorial information in the form of map and image extracts.

Fig. 4.24 Targeting operation

Fig. 4.25 GCP management using Google Earth

3.2.2 Spatial Resection

Indirect georeferencing is based on the analytical relation between image coordinates and object coordinates expressed by the collinearity (4.2, 4.3) valid for a central projection:

$$\begin{cases} x = x_0 - f \dfrac{r11(X-X_0)+r21(Y-Y_0)+r31(Z-Z_0)}{r13(X-X_0)+r23(Y-Y_0)+r33(Z-Z_0)} \\ y = y_0 - f \dfrac{r12(X-X_0)+r22(Y-Y_0)+r32(Z-Z_0)}{r13(X-X_0)+r23(Y-Y_0)+r33(Z-Z_0)} \end{cases} \quad (4.2)$$

$$\begin{cases} X = X_0 + (Z - Z_0)\frac{r11(x-x_0)+r12(y-y_0)-r13.f}{r31(x-x_0)+r32(y-y_0)-r33.f} \\ Y = Y_0 + (Z - Z_0)\frac{r21(x-x_0)+r22(y-y_0)-r23.f}{r31(x-x_0)+r32(y-y_0)-r33.f} \end{cases} \tag{4.3}$$

In (4.2) and (4.3):

- x, y are the image coordinates of a point
- X, Y, Z are the object coordinates of the same point
- x_0, y_0, f are interior orientation parameters (offset of the principal point to the fiducial centre or to the image centre and principal distance)
- X_0, Y_0, Z_0 are the object coordinates of the projection centre (three parameters of the exterior orientation)
- r_{ij} are the elements of the spatial rotation matrix existing between image and object coordinate systems (functions of ω, φ and κ, the remaining three parameters of the exterior orientation)

The image coordinates may be improved for systematic image errors, lens distortion and other geometric effects, adapting the mathematical model to the physical one.

The linearization of (4.2) yields a pair of linear observation equations relating observed image coordinates to small variations (corrections) of all the parameters involved in the collinearity conditions (4.4).

$$\begin{aligned} v_x = {} & \mathrm{d}x_0 + \left(\frac{\partial x}{\partial c}\right)^0 \mathrm{d}c + \left(\frac{\partial x}{\partial \omega}\right)^0 \mathrm{d}\omega + \left(\frac{\partial x}{\partial \phi}\right)^0 \mathrm{d}\varphi + \left(\frac{\partial x}{\partial \kappa}\right)^0 \mathrm{d}\kappa + \left(\frac{\partial x}{\partial X_0}\right)^0 \mathrm{d}X_0 \\ & + \left(\frac{\partial x}{\partial Y_0}\right)^0 \mathrm{d}Y_0 + \left(\frac{\partial x}{\partial Z_0}\right)^0 \mathrm{d}Z_0 + \left(\frac{\partial x}{\partial X}\right)^0 \mathrm{d}X + \left(\frac{\partial x}{\partial Y}\right)^0 \mathrm{d}Y + \left(\frac{\partial x}{\partial Z}\right)^0 \mathrm{d}Z + (x^0 - \bar{x}) \\ v_y = {} & \mathrm{d}y_0 + \left(\frac{\partial y}{\partial c}\right)^0 \mathrm{d}c + \left(\frac{\partial y}{\partial \omega}\right)^0 \mathrm{d}\omega + \left(\frac{\partial y}{\partial \phi}\right)^0 \mathrm{d}\varphi + \left(\frac{\partial y}{\partial \kappa}\right)^0 \mathrm{d}\kappa + \left(\frac{\partial y}{\partial X_0}\right)^0 \mathrm{d}X_0 \\ & + \left(\frac{\partial y}{\partial Y_0}\right)^0 \mathrm{d}Y_0 + \left(\frac{\partial y}{\partial Z_0}\right)^0 \mathrm{d}Z_0 + \left(\frac{\partial y}{\partial X}\right)^0 \mathrm{d}X + \left(\frac{\partial y}{\partial Y}\right)^0 \mathrm{d}Y + \left(\frac{\partial y}{\partial Z}\right)^0 \mathrm{d}Z + (y^0 - \bar{y}) \end{aligned} \tag{4.4}$$

In (4.4):

- v_x,v_y are the corrections to the image coordinates
- $d\,u$ is the correction to the generic parameter u
- $(\partial x/\partial u)^0$ is the derivative of the first equation of (4.2) relating to the parameter u calculated for approximate values of the parameters
- $(\partial y/\partial u)^0$ is the derivative of the second equation of (4.2) relating to the parameter u calculated for approximate values of the parameters
- $\mathrm{d}x_0$, $\mathrm{d}y_0$ are corrections to the approximate offset values of the principal point
- $\bar{x}, \bar{y}$—are the observed (measured) image coordinates
- x^0, y^0 are image coordinates calculated by (4.2) using approximate parameters

Refer to Kraus (1993) or Luhmann et al. (2006) for the explicit formulas of the derivatives.

In the case of a frame image whose interior orientation is known, at least three non-collinear GCPs that appear on the image are needed to determine the six parameters of exterior orientation. For each GCP a pair of observation equations can be formulated where the terms in dx_0, dy_0, dc, dX, dY and dZ are all null, only remaining the terms of the exterior orientation parameters. Derivatives are calculated with approximate values for those parameters. Three GCPs yield six equations with six unknowns which are the corrections to the approximate values. The problem, called spatial resection, is solved iteratively.

In the case where the interior orientation parameters of the image are also unknown, after defining an approximate value for f and considering x_0, y_0 equal zero, the observation (4.4) will have nine unknowns (corrections to the approximate parameters: six of exterior and three of interior orientations) which can be solved iteratively with the information of at least five non-collinear GCPs. With normal aerial images based on GCPs located approximately in a plane, the interior orientation parameters cannot be determined if no projection centre coordinates are available.

3.2.3 Triangulation

Photogrammetric triangulation consists of consistently determining the spatial orientation of a set of images, based on imaging geometry of the set (not just of one image) and ground information, using a reduced number of well distributed GCPs, instead of three per photo as required by spatial resection. Two basic mathematical models are alternatively applied to perform a photogrammetric triangulation: the spatial similarity transformation (4.5) and the above-mentioned collinearity (4.2, 4.3). The triangulation can be solved strip- or blockwise depending on the consideration of the photogrammetric unit as being a strip or a whole block of models or images.

The spatial similarity transformation is used in the historical so-called independent models adjustment.

$$X = X_{0m} + mRx_m \tag{4.5}$$

In (4.5):

- X denotes the object coordinates of a point
- x_m denotes the model coordinates of the same point
- X_{0m} denotes the object coordinates of the origin of the model coordinates system
- m denotes the scale factor existing between model and object coordinate systems
- R denotes the spatial rotation matrix between model and object coordinate systems, function of three rotation angles Ω, Φ, K

This method was often applied in the era of analogue instruments.

The collinearity equations are the basis of the so-called bundle adjustment. In this method there is no need to recover the relative orientation, since the observations are photo coordinates and not model coordinates. The method consists on

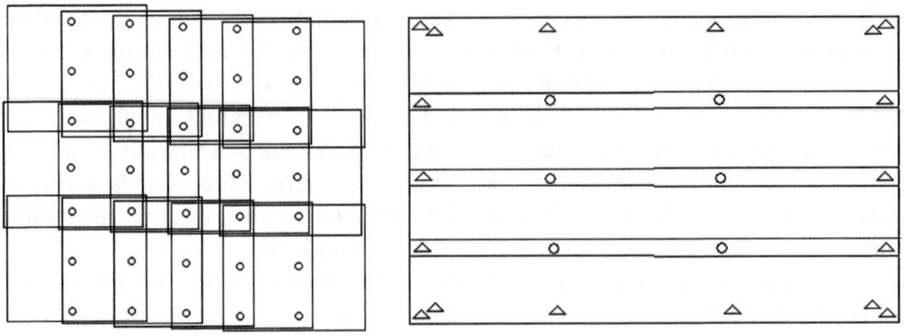

Fig. 4.26 Classical distribution of block triangulation points. *Left* tie points (*circles*). *Right* GCPs (*triangles*—planimetry and height, *circles*—just height). Distances between GCPs along the strip: three to four bases

measuring photo coordinates of GCPs as well as of tie points between images in any overlapping area (Fig. 4.26). Observation equations like (4.4) are formulated for all measured points in every photo they appear in. The discrepancy between the real image geometry, being the same along the block, and the perspective geometry can be determined by a set of additional parameters. A least squares adjustment yields the six exterior orientation parameters of each of the photos, georeferencing them. Furthermore, the three object coordinates of each measured point and the additional parameters related to the camera are also delivered as a result. This type of triangulation is the most flexible because it can be applied to any coverage, it can include images from several cameras and it can be used to calibrate a camera. It is the most widely applied method in analytical/digital photogrammetry.

The measurement of tie points for triangulation can be substantially simplified through an automatic detection of homologous points in neighbouring images. This operation is based on several digital image processing techniques and is optimized in digital photogrammetric software for the kind of textures appearing in the images. Template matching and feature based matching are the general concepts underlying those techniques. Most automatic triangulation routines use pyramids composed of a sequence of images with decreasing geometric resolution, built by resampling of the original full resolution aerial image. Feature based matching algorithms are applied to the highest pyramid level of neighbouring images in order to detect similar structures. The candidates' locations are transferred to the immediate lower level in the pyramid, corresponding to images of which the geometric resolution is higher. Candidates' matching is again tested and new candidates are searched for. The process is repeated down to the lowest level of the pyramid corresponding to the original image pair. Here, patches around the final candidates are searched with template matching algorithms in order to locate one pair of homologous points within the patch. A least squares adjustment refines the homologous positions in the sub-pixel domain. The detection of tie points can be spread all over the image area or be limited just to the positions of the classical distribution shown in Fig. 4.26, left. The strategic regions shown there in each

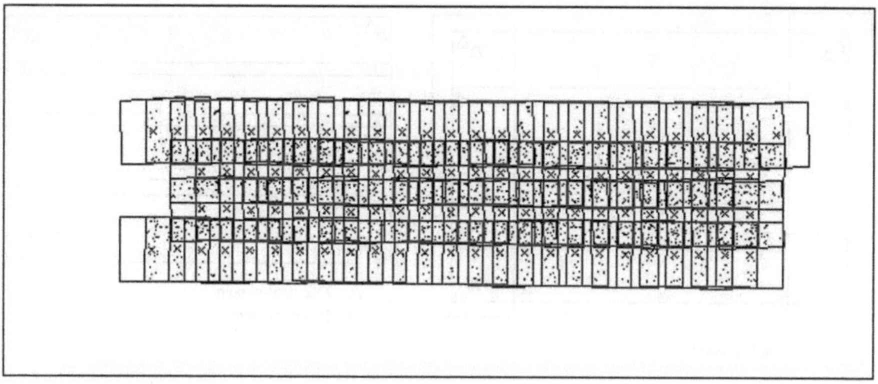

Fig. 4.27 Triangulated block with automatic determined tie points

model correspond to the location of the six von Gruber points. They represent the preferential location of homologous point pairs for the determination of the relative orientation between images. In the case of automatic aerial triangulation quite a higher number of tie points is preferred as shown in Fig. 4.27.

3.3 Semi-Direct Georeferencing

While indirect georeferencing is based exclusively on ground control points, semi-direct georeferencing takes advantage of other positioning technologies. As the spatial component of the Global Positioning System (GPS), a navigation satellite system maintained by the government of the United States, became operational in the 1990s, efforts were made by photogrammetry scientists to use this technique for the direct determination of the position of the camera projection centre. By relative kinematic GNSS positioning, the object coordinates X_0, Y_0, Z_0 of the projection centres can be determined. Each measured projection centre becomes a ground control point. Having so many spatial GCPs, theoretically no more control on the ground would be needed. In fact, the quantity required is significantly reduced, which represents a great economy for photogrammetry projects. Today, GNSS is standard for any flying mission.

A photo flight for triangulation with GNSS requires an airplane equipped with a GNSS receiver and an aerial camera with an exposure moment register. Furthermore, a second GNSS receiver has to be placed on the ground within a radius of 50 km to the surveying area. In this way a relative kinematic positioning of the receiver on board can be made, based on carrier phase measurements, with a relative accuracy of up to ±5 cm. A permanent GNSS ground network, if dense enough, can also be used for this purpose, dispensing from the receiver on the ground. The trajectory of the aircraft during the photo flight can be determined in post-processing with an absolute accuracy better than 0.5 m.

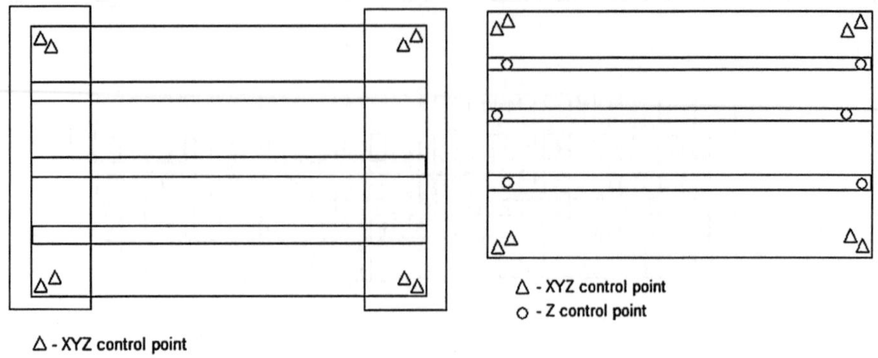

Fig. 4.28 Ground control configuration for blocks with (*left*) or without (*right*) crossing strips

The flight/navigation plan should preview broader turns between strips in order to avoid signal interruptions to some satellites due to strong wing inclination. Signal interruptions require a re-initialization of the system not always possible to perform on a flight, causing positioning errors in the subsequent trajectory. The flight plan should preferably include additional crossing strips covering the beginnings and ends of the standard strips to allow a determination of systematic errors in the GNSS positioning. Crossing strips should also be repeated every 30 bases when standard strips are too long. The experience showed that, although very reduced, some GCPs are still needed. Recommended configurations are shown in Fig. 4.28 (Jacobsen 2004).

Triangulation with GNSS data includes an additional set of observation equations for each projection centre observed. These equations, three per image, relate the object coordinates of the projection centre with the post-processed geodetic coordinates of the centre of the GNSS antenna at the exposure moment of the image, interpolated from the trajectory. In analogue cameras this projection centre is the exterior node of the objective. In digital cameras it is a virtual point determined during the process of image mosaicing.

$$\begin{bmatrix} v_X^{GNSS} \\ v_Y^{GNSS} \\ v_Z^{GNSS} \end{bmatrix} = \begin{bmatrix} X_{0j} \\ Y_{0j} \\ Z_{0j} \end{bmatrix} - \begin{bmatrix} X_j^{GNSS} \\ Y_j^{GNSS} \\ Z_j^{GNSS} \end{bmatrix} + R\omega\varphi\kappa . \begin{bmatrix} dx \\ dy \\ dz \end{bmatrix} + \left(\begin{bmatrix} ax \\ ay \\ az \end{bmatrix} + \begin{bmatrix} bx \\ by \\ bz \end{bmatrix} .(t - t_0) \right)_k \tag{4.6}$$

In (4.6):

- (X_{0j}, Y_{0j}, Z_{0j}) are the object coordinates of the projection centre of photo j
- $(X_j^{GNSS}, Y_j^{GNSS}, Z_j^{GNSS})$ are the object coordinates of the GNSS antenna at the exposure moment of photo j
- (dx, dy, dz) is the offset existing between antenna and projection centre in the aircraft, given in the camera coordinate system (usually directly measured)

- $R_{\omega\varphi\kappa}$ is the rotation matrix between camera (image) and object coordinate system (equals plane attitude if the camera is fixed to the plane)
- (*ax, ay, az*) and (*bx, by, bz*) are considered constant in each strip *k* and are the constant offset and the linear drift of the GNSS positioning, the latter depending on time interval $(t - t_0)$

As GNSS coordinates refer to WGS84 and GCP coordinates refer to a local datum, both sets have to be transformed to a common coordinate system following the scheme shown in Sect. 3.1 before they can be integrated in the photogrammetric adjustment. Geoids undulation must also be considered for height conversion.

3.4 Direct Georeferencing

Direct georeferencing of aerial images can be achieved by applying a device on the aircraft that integrates GNSS positioning with inertial positioning. An Inertial Measuring Unit (IMU) is a device composed of accelerometers and gyroscopes that measure linear and angular accelerations of a moving body along and around three perpendicular axes. Attached to an aerial camera during a photo flight, those accelerations, after double integration, deliver not only the position coordinates for the exposure moment but also the attitude angles in relation to an initial position of the camera. Converted to a convenient coordinate system, those data result in the exterior orientation parameters of the camera for each exposure moment.

Both positioning systems, GNSS and IMU, have their pros and cons. While a good absolute accuracy is possible using GNSS techniques, IMU is known by its good relative accuracy. GNSS is more or less stable with time while IMU shows significant drift with time. GNSS positioning fails when not enough satellites (minimum of four) are being received. IMU is independent of any other system. GNSS needs to be re-initialized after each interruption in order to solve ambiguities. IMU works continuously, having a high data delivering rate, while GNSS's delivering rate is not so high. IMU has the above-mentioned advantage of also delivering orientation angles of the body it is attached to.

An integration of the two systems has several advantages regarding the robustness of the positioning system. IMU can be initialized at the same rate as GNSS delivers absolute positions, significantly reducing the effects of drift on the IMU results. GNSS can be reinitialized after a lack of signal with the position delivered by IMU. The trajectory of the plane during a photo mission can thus be reliably determined.

GNSS/IMU systems on the market are capable of determining positions with an accuracy of 5–30 cm and orientation angles with an accuracy of 0.004–0.01°. They are indispensable for airborne laser scanning (LIDAR) missions, which require the most precise systems, and for digital cameras with linear array sensors. For photogrammetry projects, the use of a GNSS/IMU system allows direct

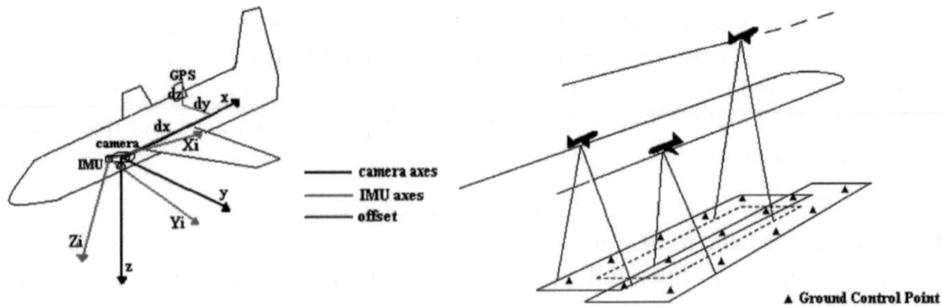

Fig. 4.29 Calibration parameters (*left*) and calibration flight (*right*)

georeferencing, meaning acquiring the six exterior orientation parameters of all photos of a coverage without GCPs (except for the ground station of the second GNSS receiver) and without triangulation. A theoretical reduction of global project costs along with the feasibility of 'unorthodox' projects, such as irregular blocks, inaccessible terrain, regions of poor contrast where matching algorithms fail (dense forests, sand, water bodies) and small format images, made the use of GNSS/IMU very attractive. Nevertheless, the system has to be often calibrated because the calibration may not be stable for long periods of time.

In order to achieve reliable georeferencing it is recommended to fly a figure eight shaped initialization curve before the exposures, which has to be repeated in the case of long strips, and to photograph over a calibration field before, during or after the actual project flight. The configuration of the calibration flight should be as shown in Fig. 4.29: two strips flown in opposite directions over a field covered with targeted GCPs in the classical distribution shown in Fig. 4.26, right.

This configuration will allow us to determine the shift between projection centre and IMU centre as well as the misalignment of the IMU axes relative to the image/camera coordinate system (Fig. 4.29, left). Corrections to the GNSS antenna offset can also be determined. These calibration parameters are estimated from the differences between observed (by GNSS/IMU) and calculated (by triangulation) exterior orientation parameters. If the calibration flight cannot be done at the same scale as the project flight, a third strip at a different height is necessary in order to model the influence of height in the interior orientation parameters (Redweik and Jacobsen 2007). The calibration area can be located inside or outside the project area. Most flying companies maintain a calibration field in the neighbourhood of the airport where their airplanes take off and land. If the calibration zone is far away from the project area, one must take geoids undulation and map projection scale variations into account.

The actual project flight can have an arbitrary strip configuration and can include coastal regions, water bodies, forests, sand deserts, etc. There is no need for ground control except for the GNSS receiver station when a GNSS permanent network cannot be used. The strips should be started two bases before the area to be covered for stabilization of the IMU after a turn.

The GNSS/IMU delivered data for the project flight have to be corrected by the calibration parameters in order to result in exterior orientation parameters for each photo. Again the transformation for a local datum has to be considered.

3.4.1 Integrated Sensor Orientation

Since the IMU is not so accurate in the determination of the attitude angle corresponding to k (yaw or heading) as for the other two angles, problems may occur in the stereo visualization of pairs and consequently in the 3D feature extraction in direct georeferenced images. The quality of the georeference by means of GNSS/IMU can be improved if the exterior parameters are adjusted in a bundle triangulation without ground control points, in order to achieve a minimum of discrepancies in a set of well distributed tie points. This operation is called integrated sensor orientation and is particularly useful for mapping projects.

Due to miscalibration or lack of knowledge about the geoid in certain regions, the direct georeference data are not accurate enough for projects requiring a higher accuracy. It is always recommended to test the reliability of the data for the purpose in mind by means of a set of check points on the ground.

3.4.2 Georeferencing of Linear Array Images

Linear array images cannot be triangulated in the same way as frame images. As already mentioned in Linear Array Sensor Cameras, these images are captured by a camera with several linear arrays in the focal plane of the unique objective. Taking the example of ADS80, JAS-150 or Wehrli 3-DAS-1, the cameras have a high quality IMU associated with it that delivers attitudes at a high rate. The GNSS receiver delivers positions at a lower rate. The post processed trajectory is interpolated for sensor position and attitude at intervals of the sampling rate. This would correspond to the capture time interval between two successive lines in a scene, meaning there are exterior orientation parameters available for each image line. The georeference supported on GNSS/IMU data proved to be good enough for some applications, but more accurate mapping applications need more accuracy than the system can afford. A triangulation based on ground control points is needed not only for a better georeference but also to calibrate system components (sensor lines + GNSS + IMU).

Original scenes, called level 0, are first transformed into level 1 scenes by means of the GNSS/IMU derived exterior orientations for each line. They are rectified to a plane. In the rectified images it is possible to detect homologous points between the three scenes (backwards, nadir and forwards) automatically. For the triangulation, the scene coordinates of each point (line and sample) have to be back transformed to the coordinates in level 0, the original image. Ground control points are interactively measured on all the scenes. Orientation fixes are then determined. These are discrete points of the trajectory, identified in the timeline, to which six exterior orientation parameters of the sensor correspond.

The triangulation aims the adjustment of the exterior orientation parameters in the orientation fixes. In the first versions of the triangulation software for linear array images, orientation fixes should be equally spaced in time and their distance should be equivalent to the short base. The short base corresponds to the ground distance between backward and nadir ground lines (corresponding to the smaller aperture angle in flight direction at the projection centre in Fig. 4.5). This configuration assures the best quality in the results. Recent versions of the algorithm allow for different spaced orientation fixes along the flight lines according to the pretended project accuracy.

$$\begin{cases} x_{ij} = -f\frac{r11_j(X_i-X_{0j})+r21_j(Y_i-Y_{0j})+r31_j(Z_i-Z_{0j})}{r13_j(X_i-X_{0j})+r23_j(Y_i-Y_{0j})+r33_j(Z_i-Z_{0j})} \\ y_{ij} = -f\frac{r12_j(X_i-X_{0j})+r22_j(Y_i-Y_{0j})+r32_j(Z_i-Z_{0j})}{r13_j(X_i-X_{0j})+r23_j(Y_i-Y_{0j})+r33_j(Z_i-Z_{0j})} \end{cases} \tag{4.7}$$

$$\begin{aligned} X_{0j} &= a_j X_{0_k} + (1-a_j)X_{0_{k+1}} - \delta X_{0j} \\ Y_{0j} &= a_j Y_{0_k} + (1-a_j)Y_{0_{k+1}} - \delta Y_{0j} \\ &\dots \\ k_j &= a_j k_k + (1-a_j)k_{_{k+1}} - \delta k_j \end{aligned} \tag{4.8}$$

$$a_j = \frac{t_{k+1} - t_j}{t_{k+1} - t_k}$$

$$\begin{aligned} \delta X_{0j} &= a_j X_{0_k}^{\text{GPS}} + (1-a_j)X_{0_{k+1}}^{\text{GPS}} - X_{0_j}^{\text{GPS}} \\ &\dots \\ \delta k_j &= a_j k_{_k}^{\text{IMU}} + (1-a_j)k_{_{k+1}}^{\text{IMU}} - k_{_j}^{\text{IMU}} \end{aligned} \tag{4.9}$$

The mathematical model applied is the relation between image coordinates and ground coordinates, and is based on collinearity equations expressed as a function of the orientation parameters at the two neighbouring fixes and the time interval between the actual line and those fixes (4.7–4.9). For details see Hinsken et al. (2002); Gruen and Zhang (2003) and Hinsken and Boehrer (2010).

In (4.7–4.9):

- x_{ij}, y_{ij} denote the scene coordinates of a point i captured at the time point t_j
- X_i, Y_i, Z_i the corresponding ground coordinates
- $r_{lm\ j}$ the elements of the rotation matrix for the time point t_j
- f the principal distance
- the indexes k and $k + 1$ denote parameters corresponding to the neighbouring fixes before and after the time point t_j
- a is an interpolation coefficient
- the superscript indexes *GPS* and *IMU* indicate the source of the data

As in frame images, triangulation tie points are needed between the three scenes of each strip and between neighbouring strips.

4 Image Processing

Several kinds of geoinformation can be produced today by photogrammetric processing of aerial images. One can divide them between vector and raster geoinformation if the criteria respect the format under which they are saved. Another classification will be 2D, 2.5D or 3D products according to the spatial dimension of the extracted information. Non-geometric information can also be obtained by photo interpretation, so that a division can be made between geometric/graphic and semantic information. The principal photogrammetric products are line maps, orthoimages, digital terrain/elevation models and 3D urban models.

Photogrammetric image processing is done today at digital photogrammetric workstations. These stations are essentially powerful computers where the photogrammetric software runs and include some hardware to enable stereoscopic viewing on the monitor for an interactive evaluation.

Line maps (vector, 2D + 1D) have been the traditional photogrammetric product since the 1920s and are still being produced, though with newer methods, instruments and objectives. Line maps have the advantage of being a very concise abstract representation of the Earth surface. Orthoimages (raster, 2D) are another very popular product because they are faster and less expensive to produce than line maps, having the information content of an image and the geometry of a map. They present much more visual information than a line map but have the disadvantage of being neither filtered nor interpreted, requiring extra effort from the user when searching for information. Orthoimage production, for average precision, is fully automatic once a digital elevation model is provided. Digital elevation models (vector or raster, 2.5D) began to be just a means to an end, consisting of surface covering altimetry necessary for orthoimage production. Now they constitute an independent product required for several other applications, such as visualization of 3D extending phenomena. More recent products are 3D city models (vector and/or raster, 3D) whose geometric components can essentially be extracted by aerial and terrestrial photogrammetry. Apart from the visual products previous referred to, photogrammetry can also be applied for measuring the object by means of georeferenced vertical or oblique aerial images, for instance for cadastral purposes.

The photogrammetric methods used for geographic data generation for cartography or GIS (Geographic Information Systems) are essentially two: stereoplotting and orthorectification. Multiple-image plotting, a technique involving measurements in coverages of convergent images, has been optimized for close-range applications and has also started to be introduced in aerial applications for geoinformation extraction in the domain of city modelling (3D-GIS).

4.1 Stereoplotting

Stereoplotting is a feature extraction method that requires a floating mark and stereoscopic visualization of the common part of a pair of images, taken with approximately parallel axes. Several stereoscopy methods are available in digital photogrammetric workstations. The most advantageous are the method of alternating images visualized with active shutter glasses (Fig. 4.30) and the method of polarized images visualized with passive polarized filter glasses. Some workstations also provide other stereo view options like anaglyph mode or split screen. A floating mark is a device with a geometric regular shape, materialized in various ways and visible in the 3D space. It allows the measurement of 3D object coordinates of every visible point in the stereo model once the exterior orientations of the two images or the absolute orientation of the stereo model are recovered. By means of the floating mark points, lines, and polygons can be measured to define the geometric details of the objects existing on the Earth surface and visible in both images. The relief can also be extracted from the stereo model using the floating mark, in the form either of contour lines or of a dense set of isolated height points (Fig. 4.31).

Which objects are to be extracted and how they have to be defined, geometrically and semantically, depend on the purpose of the collected features and has to be predefined in the form of a catalogue before the plotting operation. In the case of a feature collection for an official line map, the catalogues are defined by a land or national cartographic authority. Most existing catalogues contain long lists of feature codes and their corresponding graphic attributes (e.g. line type, line weight, symbol), without any other intelligence associated, i.e. non-graphic attributes. Those catalogues were designed with the final aim of a graphic representation of the information. The query potential of graphic data structures is very limited, turning their conversion into an object relational GIS database rather difficult. More recently, following the trend that photogrammetric products are not a final product but input data for GIS, there is a strong demand to integrate directly elements being photogrammetrically collected in a GIS compatible database format. These so-called GIS shapefiles contain graphic and non-graphic attributes as well as topology (relationship to neighbouring features). Solutions are also being provided to convert already acquired graphic data into a compatible structure for GIS.

4.2 Three-Dimensional Modelling

4.2.1 DTM and DSM

Relief on the Earth surface can be modelled in several ways:

- By a discrete set of height points in a regular or irregular distribution over the surface
- By contour lines with constant height and variable orientation in the horizontal plane

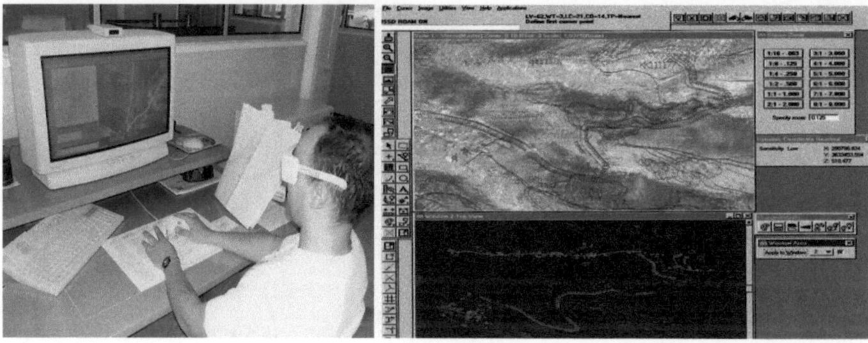

Fig. 4.30 Intergraph digital photogrammetric workstation with shutter glasses (*left*) and working environment (*right*)

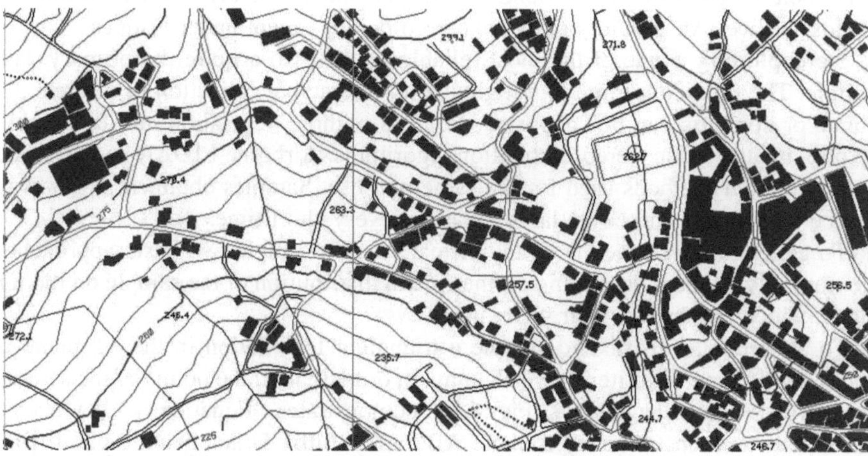

Fig. 4.31 Stereoplotted information for a line map at scale 1:5000

- By profiles with constant orientation in the horizontal plane and variable height
- By a function $Z = f(X,Y)$ yielding one height value for each point (X,Y)

The function $Z = f(X,Y)$ approximates the real relief and it is generally called a digital elevation model (DEM). It is a 2D surface immersed in a 3D space and must be estimated from a sample of real height points. The denser the sample, the better the approximation. On the other hand, a dense sample often becomes a huge amount of data, difficult to process and to manage once it is processed. So a compromise has to be found. Two types of digital elevation models are normally considered: digital terrain model (DTM), when only the ground is modelled and digital surface model (DSM), when built objects and trees are included in the height model (Fig. 4.32). Therefore, differences between a DTM and a DSM are greater in urban areas. A normalized DSM (nDSM) is obtained by subtracting

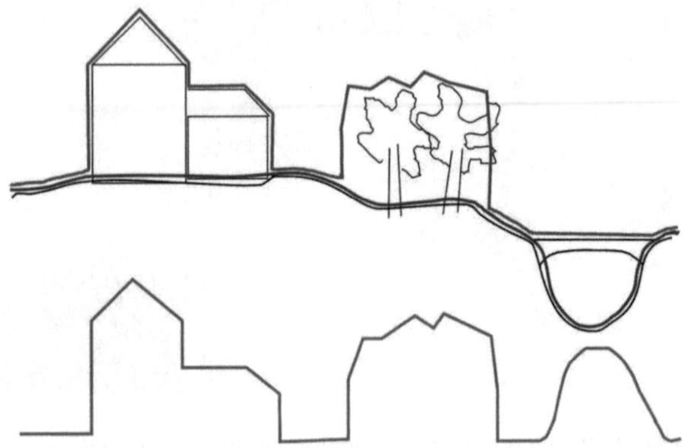

Fig. 4.32 Digital Elevation Models: DSM (*black*), DTM (*dashed*), nDSM (*bottom*)

the DTM from the DSM. In this product only built objects and trees are shown with their height from the ground to the top, independently from ellipsoids or geoids surface. All ground points have zero height except under bridges.

Stereoplotting is one of the most reliable methods to acquire the primary data sample. The time-consuming but high quality interactive collection of discrete 3D points or contours can be supported, in areas without vegetation and favourable texture, by automatic stereo-correlation algorithms. Airborne LIDAR and airborne radar are other methods used for the same purpose. Samples obtained in this way can be very dense and cover large areas in a very short time.

To generate the surface from the primary height sample two basic approaches are normally adopted depending on the spatial distribution of the sample on the *XY* plane, which can be irregular or regular.

To an irregular distributed sample a triangular tessellation in the *XY* plane is applied following the criteria of Delaunay, in order that all the points of the sample are vertexes of triangles, as Fig. 4.33 demonstrates. Equation (4.10) shows one condition each Delaunay triangle ABC must fulfil. D is a generic point from the sample different from A, B and C. This corresponds to the condition that every other point of the sample lies outside the circumcircle of ABC.

$$\begin{vmatrix} X_A - X_D & Y_A - Y_D & (X_A^2 - X_D^2) + (Y_A^2 - Y_D^2) \\ X_B - X_D & Y_B - Y_D & (X_B^2 - X_D^2) + (Y_B^2 - Y_D^2) \\ X_C - X_D & Y_C - Y_D & (X_C^2 - X_D^2) + (Y_C^2 - Y_D^2) \end{vmatrix} \leq 0 \qquad (4.10)$$

This is the so-called Triangulated Irregular Network (TIN) model. The height for a generic point (X,Y) is interpolated on the corresponding triangular plane facet.

If the sample point distribution is regular in the *XY* plane, like a matrix, two sorts of models can be generated by a square tessellation of the plane: one with square cells of constant height centred on each sample point, the other with curved surfaces limited by squares on the *XY* plane, whose vertexes are the sample height points. The height of a generic point (X,Y) is the height of the corresponding cell in the first case. In the second model the height must be interpolated from the heights of the

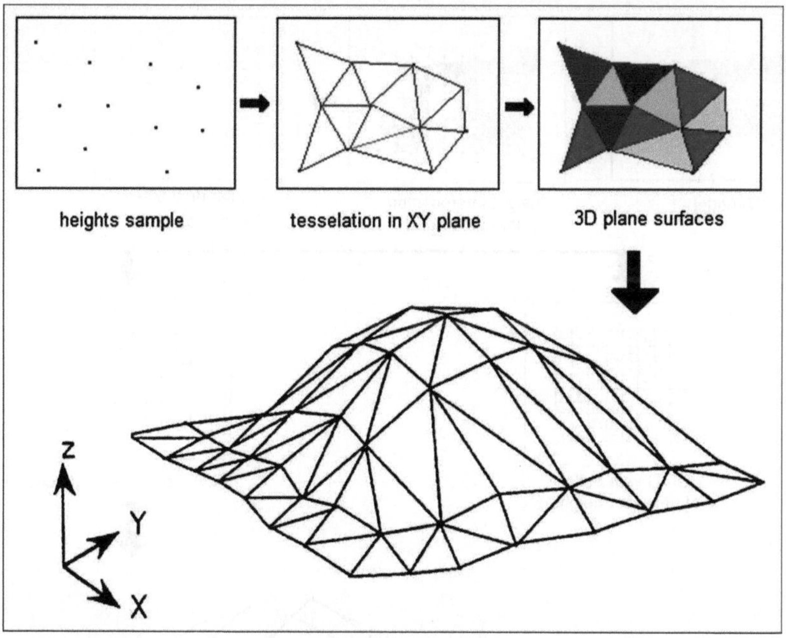

Fig. 4.33 Digital elevation model generated by a TIN

vertexes of the corresponding cell or also of neighbouring cells. Both models are classified as GRID models in most literature sources. From an irregular distributed primary sample it is also possible to build a GRID model, creating first a TIN and then interpolating the heights for the points of a regular sample. This produces a secondary sample over which the GRID model is built. The nodes of the grid do not correspond with original heights any more (Fig. 4.34).

As the Earth relief presents discontinuities like stream lines, ridges, cliff top lines, etc., these are normally included in the height sample as breaklines that the model algorithm shall preserve. Stereoplotting is a very efficient technique for the accurate acquisition of terrain breaklines (Fig. 4.35). TIN models can easily integrate breaklines in the primary sample, while in GRID models a variable mesh density, being higher in the neighbourhood of breaklines, improves the terrain model.

Airborne LIDAR samples can be classified in order to separate ground and non-ground elevations, so that it is possible to produce both a DTM and a DSM from the same initial sample. This is particularly useful in forest regions, where the laser impulse that presents a flying height dependent footprint diameter is reflected partially from the foliage and partially from the ground itself, generating multiple returns for the same impulse.

In the generation of a DSM from a sample, it must be considered that every vertical surface of the object is a discontinuity of the height surface. Therefore, buildings' boundaries, bridges and other vertical limits of objects have to be included as breaklines in the primary sample (Fig. 4.36).

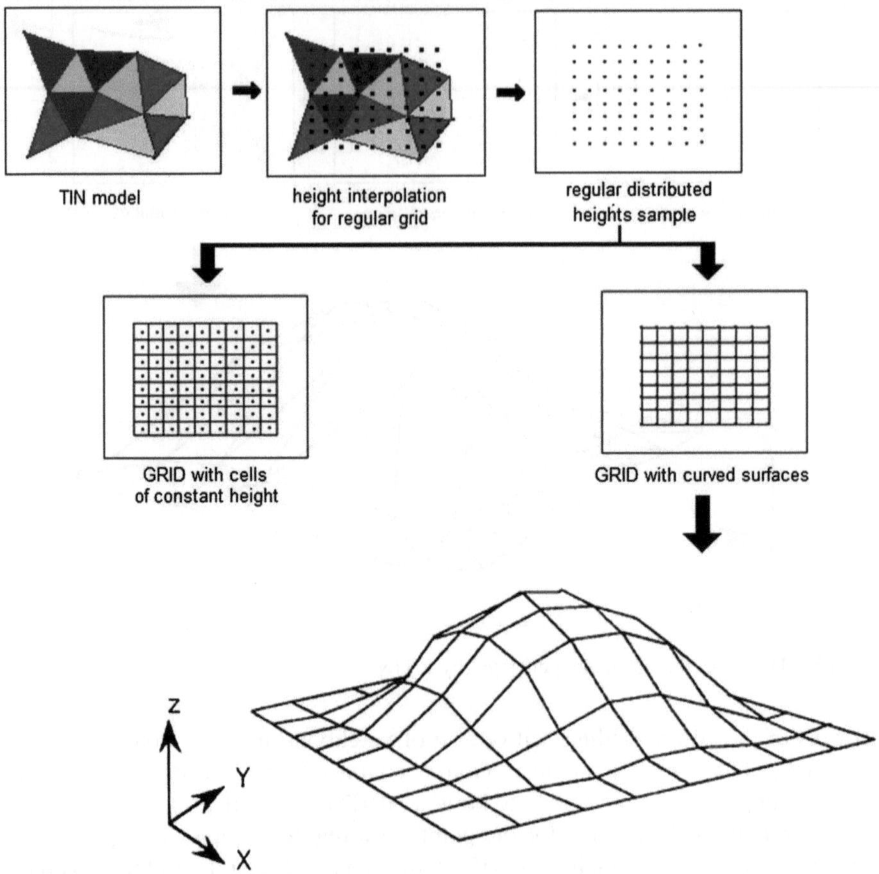

Fig. 4.34 Generation of GRID models

4.2.2 3D City Models

Three dimensional city models are a virtual 3D copy of the real world focused on the description and visualization of cities and urban centres. Apart from the attractive appearance and immediate visual interpretation potential of such models by a normal user, with a lifetime acquaintance with a 3D environment and naturally able to recognise objects in 3D, such models are useful in georeferenced applications in fields where the third dimension is essential for a suitable analysis. Examples of such fields are the following:

- Mapping of buildings and their heights
- Simulation of new buildings and developments
- Updating and keeping cadastral data
- Pervious and impervious surface inventories

Fig. 4.35 Perspective representation of a DTM obtained from stereoplotted data

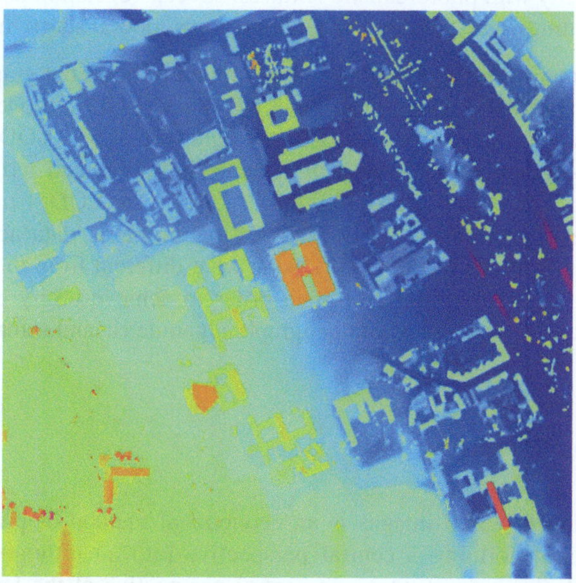

Fig. 4.36 Chromostereoscopic representation of a DSM obtained from LIDAR data

- Tree registers
- Disaster management
- Security and intelligence
- Line-of-sight and view shed analysis

- Sound propagation and noise control
- Radio network planning
- Tourism

A 3D city model describes geospatial objects existing on an urban environment (buildings, streets, vegetation, city furniture, etc.) allying their positional, geometrical and graphical components to topological relations and semantic attributes.

Three-dimensional city models can have different scales and levels of detail according to the objective. A standardization of representation and semantic rules has been developed over the last few years within the Open Geospatial Consortium (OGC) in order to allow that 3D city models, built in different environments, can be connected, analysed and updated (Döllner et al. 2006). According to the initiators of this proceeding, 3D city models can be classified in five levels of detail (LOD) as follows:

- LOD 0 a 2.5-D digital surface model with or without an orthoimage or a map draped on it
- LOD 1 block representation of buildings with prismatic solids and flat roofs
- LOD 2 buildings with differentiated roof structures and surfaces
- LOD 3 buildings showing architectural details (windows, balconies, dormers, etc.) eventually with photorealistic textures, streets, vegetation and city furniture
- LOD 4 buildings with interior structures (rooms, doors, stairs, furniture, etc.)

Photogrammetric methods are normally applied to build the graphic/geometric component of LOD 0 models. Data for LOD 1 and LOD 2 models can also be produced by stereo plotting or from LIDAR data (Fig. 4.37). LOD 3 can involve terrestrial photogrammetric surveys of relevant buildings and terrestrial LIDAR surveys of characteristic objects (Fig. 4.38). Details for LOD 4 models are normally produced by means of CAD. The photogrammetric data acquisition must follow the specifications for a 3D GIS, somewhat different from the specifications for a graphic line map or a 2D GIS, in order to achieve a city describing data structure that can be queried across geographic, geometric and semantic attributes.

4.3 Orthorectification

The rectification of aerial images is an economical method of producing geoinformation by transforming a central perspective into an orthogonal projection. A frame aerial image constitutes one central perspective of the terrain, while by linear array images each line of each scene is a different central perspective. The result of the orthorectification, in both cases, is an image with homogeneous scale, where objects lie in the correct planimetric position, and with the same information density of the original aerial image. This image is denominated orthoimage and is geometrically a map like product. The already presented collinearity (4.2) are, once again, the mathematical model for the transformation. Orthoimages are generated for each aerial image of a coverage in an automatic process provided

Fig. 4.37 LOD 1–3D city model: University of Lisbon—Part of the Campus

Fig. 4.38 Details of a LOD 3—3D city model of Lisbon (Branco 2008)

that a DTM already exists. For an orthoimage map sheet with a predefined rectangular terrain area, normally a mosaic of orthoimages must be built, which requires a subsequent radiometric adjustment in the transition areas. For a paper support output, a map grid is normally superimposed onto the image, as well as some text information. Altimetry can also be superimposed in the form of contour lines derived from the DTM. For the sake of legibility, the set of contour lines is reduced. Figure 4.39 illustrates the required sequence of analytical transformations for the orthoimage generation. The output image pixels are associated with the object coordinate system. A pixel (X, Y), still without information (colour), is projected in the DTM, which yields a Z value. The terrain point (X, Y, Z) is then transformed by means of collinearity equations to the input aerial image whose interior and exterior orientations are known. From the neighbouring pixels of the calculated image position (x, y), an interpolated grey value g_{int} is calculated and then attributed to the initial pixel (X, Y) of the output image. This procedure that

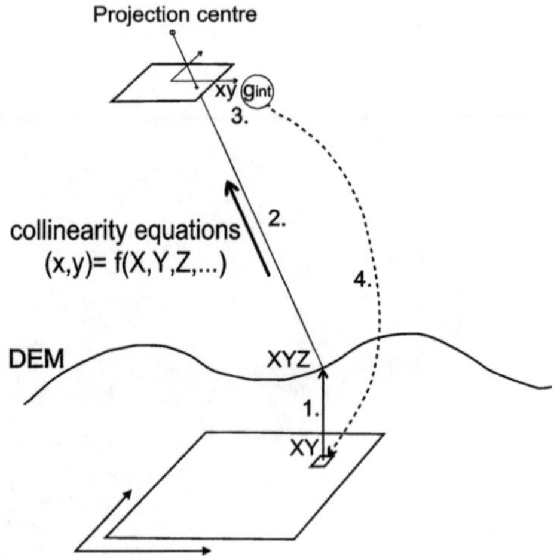

Fig. 4.39 Orthoimage generation

begins and ends in the output image guarantees the absence of gaps in the final image.

4.3.1 True Orthos

The quality of an orthoimage depends on the resolution of the aerial images that are at its origin, on the accuracy of the orientation parameters and on the quality of the digital elevation model used in the rectification algorithm. Normal orthoimages are produced using a DTM, correcting the images for radial distortion caused by relief at ground level. In urban areas, however, an orthoimage produced in this way seldom yields a satisfying product, because elevated objects may appear more or less leaning, depending on the position relative to the image centre they had in the original image. The top of the objects is not imaged at the correct planimetric position, only the ground. Furthermore, there is no ground information ‘under’ the leaning buildings. Wider objective angles and lower flying heights increase the leaning buildings effect.

The use of a high precision DSM where all the elevated objects are modelled, allow the production of an orthoimage where the top and the base of vertical objects coincide and the planimetric position of every pixel is correct. Such DSM can be obtained by stereoplotting or by combination of this with LIDAR data for the height information.

It remains a question as to where to get the lost radiometric information after the rectification with a DSM, meaning the objects that were occluded by higher objects. A solution normally adopted is to fly with a higher overlapping, as already mentioned in Sect. 2.2.

The images obtained capture the elevated objects from several different perspectives and the leaning effect occurs in different radial directions, showing in each picture a different part of the ground (or other occluded objects) in the direct

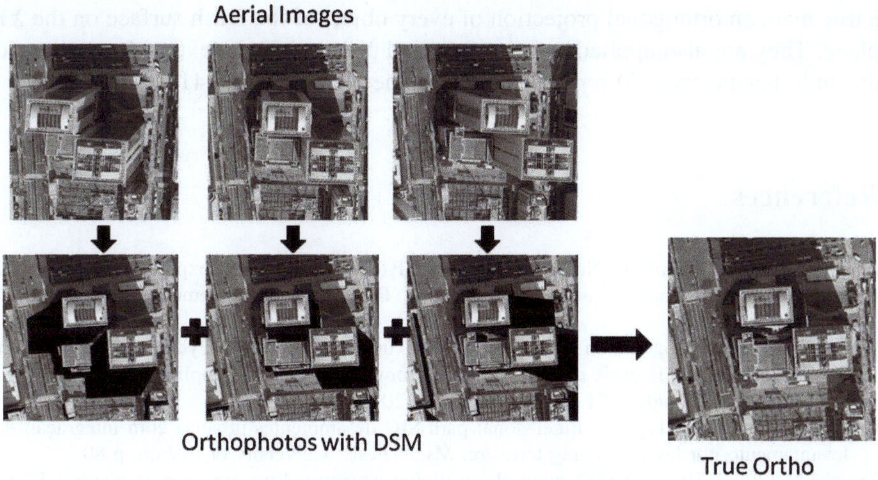

Fig. 4.40 Generation of true orthos (adapted from Oda et al. 2004)

Fig. 4.41 Perspective representation of a DTM with draped orthoimage (Porto Santo Island)

neighbourhood of the elevated object. The orthoimages generated with DSM from neighbouring images show gaps at different locations. Because every pixel in every orthoimage is in its correct planimetric position, a mosaic can be generated by simply filling the gaps with patches of radiometric information from the other rectified pictures and adjusting the radiometry of the final product (Fig. 4.40).

Several image processing algorithms are being implemented in order to automate the production of true orthoimages. Nevertheless, because of the high precision DSM required and the eventual intensive editing operations, it is a much more expensive product than the normal orthoimage. A true ortho is geometrically a true map: an orthogonal projection of every object of the Earth surface on the *XY* plane. They are also applied as georeferenced photo realistic texture for static and dynamic perspective 3D representations of the terrain (Fig. 4.41).

References

Albertz J, Ebner H, Heipke C, Neukum G, Scholten F (1992) The camera experiments HRSC and WAOSS on the mars 94 mission. Int Arch Photogrammetry Remote Sens Comm 1 29(B1):130–137

Albertz J (2008) Photographien als historische Dokumente–Albrecht Meydenbauer als Pionier der Denkmalpflege. Historic maps and imagery for modern scientific applications, Bern. URL: http://maps.unibe.ch/data/X1.pdf. Accessed Jan 2010

Branco P (2008) Modelação tridimensional para SIG de ambientes urbanos com integração de levantamento por laser-scanning terrestre. Msc. Thesis, University of Lisbon, p 80

Cramer M (2004) The EuroSDR network on digital camera calibration—report phase 1 URL: http://www.ifp.uni-stuttgart.de/eurosdr/euroSDR-Phase1-report.pdf. Acessed May 2010

Döllner J, Kolbe TH, Liecke F, Sgouros T, Teichmann K (2006) The virtual 3d city model of Berlin—managing, integrating and communicating complex urban information. In: Proceedings of the 25th international symposium on Urban Data Management UDMS 2006 Aalborg, Denmark URL: www.citygml.org/1535/. Accessed May 2010

Fricker P, Chapuis A, Hughes D, Traversari E, Schreiber, P, Schapira F (2003) Entwicklung der Photogrammetrie in der Schweiz URL: http://www.wild-heerbrugg.com/photogrammetry.htm. Accessed March 2010

Grimm A (2007) The origin of the term photogrammetry. URL: http://www.ifi.uni.stuttgart.de/publications/phowo07/080Grimm.pdf. Accessed Jan 2010

Gruber M (2008) Vexcel imaging photogrammetry products. Vexcel Imaging/Microsoft Photogrammetry User Meeting, Tokyo, Sept 2008

Gruen A, Zhang L (2003) Sensor modeling for aerial triangulation with three-line-scanner (tls) imagery. J Photogrammetrie Fernerkundung Geoinf 2(2003):85–98

Hinsken L, Miller S, Tempelmann U, Uebbing R, Walker S (2002) Triangulation of LH systems'ADS40 imagery using ORIMA GPS/IMU. Int Arch Photogrammetry Remote Sens Spatial Inf Sci 34(Part 3A): 156–162

Hinsken L, Boehrer N (2010) Automation in quality analysis of triangulation results from ADS images, European Calibration and Orientation Workshop, Barcelona, 2010

Hobbie D (2010) Die Entwicklung Photogrammetrischer Verfahren und Instrumente bei Carl Zeiss in Oberkochen. Deutsche Geodätische Komission bei der Bayrischen Akademie der Wissenschaften, Reihe E—Geschichte der Geodäsie. Nr. 30

Jacobsen K (2004) Combined adjustment with kinematic GPS-data and/or IMU-data. IMU misalignment. User Manual of Program System BLUH

Jacobsen K (2009) Potential of large format digital aerial cameras. Map World Forum, Hyderabad, GIS Development; http://www.gisdevelopment.net/technology/emerging/mwf09_Karsten_AerialCameras.htm. Accessed Mar 2010

Jacobsen K, Cramer M, Ladstädter R, Ressl C, Spreckels V (2010) DGPF project: evaluation of digital photogrammetric camera systems geometric performance. Photogrammetrie Fernerkundung Geoinfor 2:83–97

Jacobsen K (2011) Geometric property of large format digital camera DMC II 140. Photogrammetrie Fernerkundung Geoinfor 2011(2):71–79

Konecny G (2002) Geoinformation: remote sensing, photogrammetry and geographic information systems, 2nd edn, CRC Press, Boca Raton, p 280

Krakiwsky EJ (1973) Conformal map projections in geodesy. lecture notes 37, Department of Geodesy and Geomatics Engineering, University of New Brunswick
Kraus K (1993) Photogrammetry. Fundamentals and standard processes, vol. 1. Ferdinand Dümmels Verlag, Bonn, p 402
Ladstädter R, Gruber M, Wiechert A (2010) Monolithic stitching: one sensor geometry for multiple sensor camera. ASPRS 2010 annual conference San Diego.
Leberl F, Gruber M (2003) Flying the new large format digital aerial camera Ultracam. Photogrammetric Week'03, Dieter Fritsch (ed), Wichman. pp 67–76
Leberl F, Gruber M (2005) *ULTRACAM-D:* understanding some noteworthy capabilities. Photogrammetric Week'05, Dieter Fritsch (ed.), Wichman, pp 57–68
Luhmann T, Robson S, Kyle S, Harley I (2006) Close range photogrammetry: principles, techniques and applications. Whittles Publishing, Scotland, p 528
McCurdy PG, Woodward LA, Davidson JI, Wilson RM, Ask RE (eds) (1944) Manual of photogrammetry (Preliminary edn). American Society of Photogrammetry, Pitman Publishing Corporation, p 819
Oda K, Lu W, Uchida O, Doihara T (2004) *Triangle*-based visibility analysis and true orthoimage generation. In: Proceedings of the ISPRS congress Istambul, vol 35, pp 623–628, URL: http://www.isprs.org/proceedings/XXXV/congress/comm3/comm3.aspx
Redweik P, Jacobsen K (2007) Handling uncalibrated GPS/IMU data for medium scale mapping. Photogrammetrie, Fernerkundung, Geoinfor, Schweizerbart Science Publishers 2(2007): 99–108
Seeber G (2003) Satellite geodesy, 2nd edn. Walter de Gruyter, Berlin, p 589
Wiechert A, Gruber M (2009) Vexcel imaging gmbh—innovating in photogrammetry*:* UltraCamXp, UltraCamLp and UltraMap. Phtogrammetric Week'09 URL: www.ifp.uni-stuttgart.de/publications/phowo09/040Wiechert.pdf. Accessed May 2010

Chapter 5
Regional Gravity Field Modeling: Theory and Practical Results

Heiner Denker

1 Introduction

Geodesy, with its three core areas positioning and reference systems, Earth rotation determination, and gravity field modeling, is striving for a relative accuracy of at least 10^{-9} for all relevant quantities, and to a great extent this goal has already been reached (10^{-9} corresponds to about 6 mm relative to the Earth's radius and $10^{-8}\,\mathrm{ms}^{-2} = 1\,\mu\mathrm{Gal}$ in terms of gravity). Regarding gravity field modeling, the highest accuracy demands are from geodesy, especially Global Navigation Satellite System (GNSS) positioning, oceanography, and geophysics. In this context, the geoid and quasigeoid are of major interest; e.g., these quantities are required for the transformation between the purely geometric GNSS (ellipsoidal) heights and gravity field related heights as well as for the modeling of the (mean) dynamic ocean topography (DOT), requiring accuracies at the level of about 1 cm or even below. In this way, the importance of geoid and quasigeoid modeling has increased considerably—also for economic reasons—and as early as 1982 Torge (1982) postulated a "renaissance of the geoid."

Over the past decades, significant progress has been achieved in the collection of high-resolution gravity and terrain data, computing and modeling techniques, as well as the operational availability of satellite data from several altimetry and gravity field missions. Of special interest are the results from the CHAllenging Minisatellite Payload (CHAMP, active from 2000 to 2010), Gravity Recovery And Climate Experiment (GRACE, operational since 2002), and Gravity field and steady-state Ocean Circulation Explorer (GOCE, launched in 2009) missions; while the CHAMP and GRACE missions already delivered the long wavelength geoid

H. Denker (✉)
Institut für Erdmessung (IfE), Leibniz Universität Hannover (LUH),
Schneiderberg 50, 30167 Hannover, Germany
e-mail: denker@ife.uni-hannover.de

G. Xu (ed.), *Sciences of Geodesy - II*, DOI: 10.1007/978-3-642-28000-9_5,

with an accuracy of about 1 cm up to a resolution (half wavelength) of 650 and 200 km, respectively, the GOCE mission is targeting at an accuracy of 1–2 cm for the geoid and 1 mGal for gravity, both at a resolution of approximately 100 km. In addition, the GRACE mission allows the determination of the long wavelength geoid (up to about 350 km resolution) with an accuracy of 1 mm on a monthly basis, from which time variations of the geoid can be deduced. However, due to the required satellite altitudes of at least a few 100 km (below this level, satellite orbits become unstable due to air drag, etc.), the pure satellite gravity field solutions alone can only provide long wavelength gravity field models associated with an omission error (gravity field components not included in the model) of several decimeters regarding the geoid. Consequently, even in the future, only a combination of the satellite gravity fields with high-resolution terrestrial (mainly gravity and terrain) data can provide the complete geoid spectrum (all wavelengths) with an accuracy of 1 cm or even better. In this context, the satellite and terrestrial data ideally complement each other, with the satellite data providing accurately the long wavelength gravity field structures, while the terrestrial data sets, with potential weaknesses in the large-scale accuracy and coverage, mainly contribute the short wavelength features.

Altogether, there is considerable interest in high-resolution regional gravity field modeling, i.e., on a provincial to national and continental scale (several 1,000 km), especially with regard to the geoid and quasigeoid. At present, the combination of terrestrial data sets with up-to-date satellite gravity field models allows the calculation of geoid and quasigeoid models with accuracies of a few centimeters, provided that high-resolution and high-quality terrestrial and satellite data are utilized; furthermore, in view of the GOCE mission, the accuracy may be improved to the level of about 1 cm in the near future, being close to the general accuracy goal of 10^{-9} in geodesy.

This chapter specifically refers to the experiences gained at the Institut für Erdmessung (IfE), Leibniz Universität Hannover (LUH), Germany, within the field of gravity field modeling, especially the calculation of the geoid and quasigeoid. IfE has a long tradition in local and regional geoid and quasigeoid determinations, starting even before the (GNSS) Global Positioning System (GPS) era, when at the beginning of the 1970s a test network was set up in the Harz mountains in Northern Germany, this being the classic geoid research area of Helmert (Torge 1977). Then, with the advent of the GPS, studies on geoid and quasigeoid modeling with centimeter accuracy were intensified, and in a small test area near Hannover, Germany, it could be shown for the first time that an agreement between gravimetric and GPS/leveling results at the centimeter level is in fact possible (Denker and Wenzel 1987). These computations were subsequently extended to larger regions, covering Lower Saxony, Germany (Denker 1988), and the whole of Germany (Denker 1989). Based on these experiences, IfE proposed to perform corresponding computations for the whole of Europe, and this task has been supported by the International Association of Geodesy (IAG) since 1990; IfE served as the computing center within the IAG Geoid Sub-Commission for Europe from 1990 to 2003, from 2003 (when the new IAG structure was implemented) to 2011 the work was supported in the form of an IAG Commission 2 Project "CP2.1—European Gravity and Geoid Project (EGGP)," and since 2011 this task has continued as IAG Sub-Commission 2.4a "Gravity and Geoid in Europe." Major results of this IAG enterprise are the high-resolution European geoid and quasigeoid models EGG1997 (Denker and Torge 1998) and EGG2007/2008 (see Sect. 4).

After providing the motivation for preparing this chapter, the necessary fundamentals of physical geodesy are described in Sect. 2, including reference systems, basic gravity field properties, the geoid and height systems, the normal gravity field, as well as some remarks about temporal gravity field variations, tidal systems, and atmospheric effects. The intention of this section is to provide all the basics needed for high-precision gravity field modeling with as few approximations as possible. Section 3 describes the methodology of gravity field modeling, where the disturbing potential is the primary quantity of interest. The emphasis is on the spatial gravity field description related to quantities defined at the Earth's surface, such as the disturbing potential as well as the height and gravity anomalies, which require no assumptions about the Earth's interior gravity field (in contrast to the geoid). After giving an overview on geodetic boundary value problems, the linearization of the boundary conditions (observation equations) is discussed, aiming at the rigorous implementation of a high-degree geopotential model as a reference field. Then the spherical and constant radius approximations are introduced, leading to the classical Poisson, Hotine, Stokes, and other integral formulas. Afterwards, the solution of Molodensky's boundary value problem (related to the Earth's surface) and Stokes's boundary value problem (related to the geoid) are outlined. In addition, the spectral combination technique, least-squares collocation, and astronomical leveling are mentioned. Although most of these modeling techniques can be utilized globally, they are primarily used at regional (provincial to national and continental) scale in connection with the remove–compute–restore (RCR) technique and topographic reductions as well as a global geopotential model, which is described at the end of Sect. 3. Finally, Sect. 4 gives some practical results related to the European geoid and quasigeoid calculations carried out at IfE, starting with a discussion of the data requirements, then an outline of the European gravity and geoid project and the collected gravity and terrain data sets is given, followed by an overview on the development and evaluation of the European Gravimetric (Quasi)Geoid model EGG2008.

2 Fundamentals of Physical Geodesy

2.1 Reference Systems

The definition and realization of reference systems has become a major part of geodesy. Length, mass, and time are basic quantities used in geodesy, the units being meter (m), kilogram (kg), and second (s), respectively, as defined through the

International System of Units (SI), see BIPM (2006). Furthermore, fundamental constants (e.g., the gravitational constant) are regularly updated and recommended by the Committee on Data for Science and Technology (CODATA); the latest set of constants originates from 2006 (Mohr et al. 2008).

For the modeling of the Earth's gravity field, global and local reference systems are needed. In this context, a terrestrial reference system (TRS), also denoted as Earth-fixed (global) reference system, is of vital importance. A TRS is a spatial reference system co-rotating with the Earth in its diurnal motion in space, in which points at the solid Earth's surface undergo only small variations with time (e.g., due to geophysical effects related to tectonics or tides). With regard to the terminology, it is fundamental to distinguish between a "reference system," which is based on theoretical considerations or conventions, and its realization, the "reference frame," to which users have access. The International Earth Rotation and Reference Systems Service (IERS) is in charge of defining, realizing and promoting the International Terrestrial and Celestial Reference System (ITRS and ICRS, respectively), including the necessary transformations; the use of these reference systems is recommended by the International Astronomical Union (IAU) and the International Union of Geodesy and Geophysics (IUGG).

The ITRS origin is at the center of mass of the whole Earth including oceans and atmosphere (geocenter), the unit of length is the meter (SI), the orientation is equatorial and initially given by the Bureau International de l'Heure (BIH) terrestrial system at epoch 1984.0, and the time evolution of the orientation is ensured by using a no-net-rotation condition with regard to the horizontal tectonic motions over the whole Earth. Accordingly, the *Z*-axis is directed towards the IERS reference pole (i.e., the mean terrestrial North Pole), the axes *X* and *Y* span the equatorial plane, with the *X*-axis being defined by the IERS reference meridian (Greenwich), such that the coordinate triplet *X*, *Y*, *Z* forms a right-handed Cartesian system. The instantaneous North Pole (more precisely, the Celestial Intermediate Pole, CIP, which is defined conventionally by the IERS precession and nutation models) deviates from the IERS reference pole by the effect of polar motion (described by rectangular coordinates x_P, y_P). The ITRS is materialized by the International Terrestrial Reference Frame (ITRF), consisting of the three-dimensional positions and velocities of stations observed by space geodetic techniques, where the positions are regularized in the sense that high-frequency time variations (mainly geophysical ones) are removed by conventional corrections. The most recent realization of the ITRS is the ITRF2008 with the reference epoch 2005.0. The accuracy of the geocentric positions (*X*, *Y*, *Z*) is at the level of 1 cm or better. For further details including Earth orientation parameters (EOPs) and transformations see IERS (2010), Kovalevsky and Seidelmann (2004), and Angermann et al. (2012). A corresponding TRS is the World Geodetic System 1984 (WGS84) which is intended to be as closely coincident as possible with the ITRS; it is maintained by the National Geospatial-Intelligence Agency (NGA), U.S.A., for use with the NAVSTAR Global Positioning System (GPS). The latest realization of WGS84 (i.e., the terrestrial reference frame, TRF) is denoted as "Reference Frame G1150;" it agrees with the ITRF at the level of 1 cm.

Moreover, the WGS84 definitions also include the parameters of a level ellipsoid (see Sect. 2.5); for further details see NIMA (1997 and 2002).

Customarily, owing to the Earth's general shape (approximately spherical with a slight flattening at the poles), ellipsoidal geographic coordinates (ellipsoidal latitude, longitude, and height, φ, λ, h, also known as geodetic coordinates), based on an ellipsoid of revolution (ellipse rotating about its minor (polar) axis), are employed in many geodetic applications; regarding the ITRF solutions; IERS (2010) recommends the Geodetic Reference System 1980 (GRS80) ellipsoid (for further details see Sect. 2.5). In addition, spherical coordinates (polar distance or spherical colatitude, spherical longitude, radius, θ, λ, r; the spherical and ellipsoidal longitudes are identical) are of great significance in gravity field modeling. After introducing a reference ellipsoid (e.g., by the geometrical parameters a = semimajor axis and e = first eccentricity), the following relation holds:

$$\mathbf{X} = \begin{pmatrix} X \\ Y \\ Z \end{pmatrix} = \begin{pmatrix} (N+h)\cos\varphi\cos\lambda \\ (N+h)\cos\varphi\sin\lambda \\ ((1-e^2)N+h)\sin\varphi \end{pmatrix} = \begin{pmatrix} r\sin\theta\cos\lambda \\ r\sin\theta\sin\lambda \\ r\cos\theta \end{pmatrix}. \tag{5.1}$$

N is the prime vertical radius of curvature of the reference ellipsoid, which can be computed as

$$N = \frac{a}{\sqrt{1-e^2\sin^2\varphi}}. \tag{5.2}$$

The inversion of the system (5.1) is straightforward for the spherical case and can be carried out iteratively for the ellipsoidal case (Torge 2001); in addition, formula (5.1) can be used to transform ellipsoidal into spherical coordinates (e.g., needed in connection with spherical harmonic expansions; see Sects. 2.2 and 3.3) and vice versa.

In this context, it should be noted that historically many ellipsoidal systems and the associated Cartesian systems (i.e., national reference frames) were non-geocentric. This is mainly due to the orientation of the classical geodetic networks by astronomical observations, which provide only direction information, but no direct access to the geocenter. In this case, the non-geocentric X_{ng}, Y_{ng}, Z_{ng} system can be transformed to the X, Y, Z system by a three-dimensional similarity transformation (Torge 2001):

$$\mathbf{X} = \mathbf{X}_0 + (1+m)\,\mathbf{R}(\varepsilon_{X_{ng}}, \varepsilon_{Y_{ng}}, \varepsilon_{Z_{ng}})\,\mathbf{X}_{ng} \tag{5.3}$$

with

$$\mathbf{X}_0 = \begin{pmatrix} X_0 \\ Y_0 \\ Z_0 \end{pmatrix}, \quad \mathbf{R}(\varepsilon_{X_{ng}}, \varepsilon_{Y_{ng}}, \varepsilon_{Z_{ng}}) = \begin{pmatrix} 1 & \varepsilon_{Z_{ng}} & -\varepsilon_{Y_{ng}} \\ -\varepsilon_{Z_{ng}} & 1 & \varepsilon_{X_{ng}} \\ \varepsilon_{Y_{ng}} & -\varepsilon_{X_{ng}} & 1 \end{pmatrix}, \tag{5.4}$$

where $\mathbf{X}_0$ is a translation vector with the coordinates of the origin of the X_{ng}, Y_{ng}, Z_{ng} system with respect to the geocenter, m is a (small) scale correction, $\mathbf{R}$ is a rotation matrix with three (small) Eulerian angles, and $\mathbf{X}_{ng}$ is the coordinate vector in the non-geocentric system. It should be noted that the transformation model (5.3) is a linearized formula which is sufficient due to the small size of the rotation angles; furthermore, the sign conventions are according to Torge (2001). The geodetic datum describes the orientation of any geodetic system with respect to the global geocentric system (seven transformation parameters of (5.3)) and also includes the parameters of the reference ellipsoid employed. Numerous examples of national (non-geocentric) geodetic datums can be found in NIMA (1997) and Torge (2001); the translation parameters can reach several hundreds of meters, the scale corrections may be in the order of 10^{-5}, the rotation angles are usually quite small at the level of 1 arc second or below (due to the orientation by astronomical observations), and miscellaneous reference ellipsoids were employed for national geodetic surveys, carried out since the nineteenth century.

After introducing ellipsoidal coordinates in the vectors $\mathbf{X}$ and $\mathbf{X}_{ng}$ in (5.3), while also allowing a change of ellipsoid parameters, the following transformation formula is obtained for the ellipsoidal heights:

$$h = h_{ng} + \cos\varphi\cos\lambda\, X_0 + \cos\varphi\sin\lambda\, Y_0 + \sin\varphi\, Z_0 - \Delta a + a\sin^2\varphi\,\Delta f, \quad (5.5)$$

with the ellipsoidal heights h and h_{ng} in the geocentric and non-geocentric system, respectively, and the changes in the ellipsoid parameters $\Delta a = a - a_{ng}$ (semi-major axis) and $\Delta f = f - f_{ng}$ (flattening). The above formula is based on spherical approximations (terms of the order of the flattening $o(f)$ are neglected), where the rotation angles have no effect. The translation parameters (X_0, Y_0, Z_0) can be converted into changes in the ellipsoidal coordinates of a fundamental station, and hence be interpreted as height shift and tilts in north–south and east–west direction of the respective ellipsoid surfaces. Formula (5.5) is applied, e.g., for the transformation of geoid and quasigeoid heights. Corresponding formulas for the transformation of latitudes and longitudes (φ_{ng}, $\lambda_{ng} \rightarrow \varphi$, λ) can be found in Torge (2001); in addition, more precise transformation formulas without the usual spherical approximations are given in Heck (2003). Finally, it is assumed in the following that all coordinates refer directly (e.g., through the application of space geodetic techniques) or have been converted (from national geodetic datums) consistently to an IERS reference frame (e.g., ITRF 2008, epoch 2005.0).

Because most geodetic and astronomical observations refer to the Earth's gravity field by orientating observation instruments along the local vertical (through levels or plummets), local coordinate systems related to the Earth's gravity field are introduced. These local astronomical (Cartesian) systems have their origin at the observation point P, the z-axis points toward the zenith (tangent of the plumb line, outer normal of the level surface), while the x-axis (north) and y-axis (east) span the horizontal plane, which is tangential to the level surface at P. The local astronomical system (x, y, z) is left-handed. As the direction of the plumb line (local vertical) with respect to the global geocentric system is given by the astronomical latitude Φ and longitude Λ (see Fig. 5.1), the coordinate vector

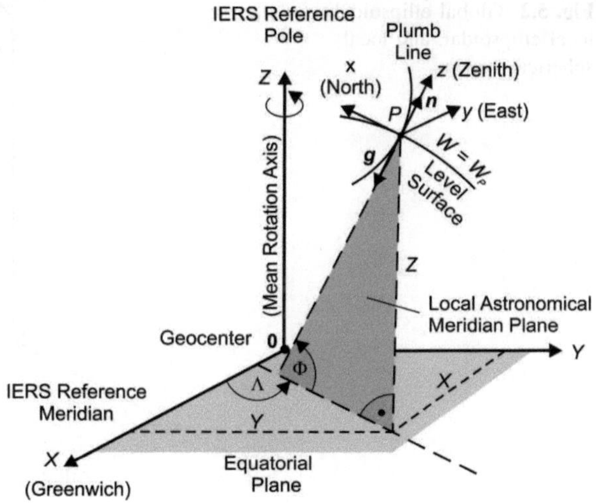

Fig. 5.1 Earth-fixed global geocentric and local astronomical system

x in the local astronomical system can be transformed into a coordinate difference vector $\mathbf{\Delta X}$ in the global geocentric system by

$$\mathbf{\Delta X} = \mathbf{A_x x} \tag{5.6}$$

with

$$\mathbf{A_x} = \mathbf{A_x}(\Phi, \Lambda) = \begin{pmatrix} -\sin\Phi\cos\Lambda & -\sin\Lambda & \cos\Phi\cos\Lambda \\ -\sin\Phi\sin\Lambda & \cos\Lambda & \cos\Phi\sin\Lambda \\ \cos\Phi & 0 & \sin\Phi \end{pmatrix}. \tag{5.7}$$

The transformation matrix $\mathbf{A_x}$ is orthogonal, thus

$$\mathbf{A_x^{-1}} = \mathbf{A_x^T}. \tag{5.8}$$

Accordingly, local ellipsoidal and spherical (Cartesian) coordinate systems can be introduced as needed, for example in connection with the rigorous calculation of parameters related to the actual, normal, or anomalous gravity field (see Sects. 2.3, 2.5, and 3.2). Regarding the local ellipsoidal system (x^e, y^e, z^e), the z^e-axis points toward the ellipsoidal zenith (ellipsoidal normal), with the x^e-axis (north) and y^e-axis (east) being perpendicular to it. On the other hand, in the local spherical system (x^s, y^s, z^s), the z^s-axis points radially outwards (along the geocentric radius vector), with the axes x^s, y^s (north and east) again being perpendicular to it (see Fig. 5.2). Analogous to (5.6), the transformation to the global system is given by

$$\mathbf{\Delta X} = \mathbf{A_{x^e} x^e}, \quad \mathbf{A_{x^e}} = \mathbf{A_{x^e}}(\varphi, \lambda), \tag{5.9}$$

$$\mathbf{\Delta X} = \mathbf{A_{x^s} x^s}, \quad \mathbf{A_{x^s}} = \mathbf{A_{x^s}}(90^\circ - \theta, \lambda). \tag{5.10}$$

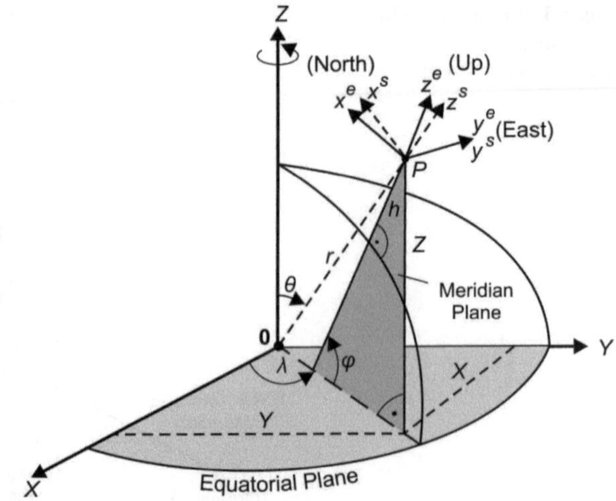

Fig. 5.2 Global ellipsoidal, local ellipsoidal, and local spherical system

The above transformation matrices $\mathbf{A}_{\mathbf{x}^e}$ and $\mathbf{A}_{\mathbf{x}^s}$ are defined according to (5.7); the only difference is that the parameters Φ, Λ are replaced by φ, λ and $90° - \theta$, λ, respectively. The transformation matrices also relate to the coordinate basis vectors of the corresponding Cartesian coordinate systems, e.g., if $\mathbf{e}_\mathbf{X}$, $\mathbf{e}_\mathbf{Y}$, $\mathbf{e}_\mathbf{Z}$ are unit vectors along the coordinate axes X, Y, Z, and $\mathbf{e}_{\mathbf{x}^s}, \mathbf{e}_{\mathbf{y}^s}, \mathbf{e}_{\mathbf{z}^s}$ are the corresponding vectors along the axes x^s, y^s, z^s, the following relation holds (see also Wenzel 1985):

$$\begin{pmatrix} \mathbf{e}_\mathbf{X} \\ \mathbf{e}_\mathbf{Y} \\ \mathbf{e}_\mathbf{Z} \end{pmatrix} = \mathbf{A}_{\mathbf{x}^s} \begin{pmatrix} \mathbf{e}_{\mathbf{x}^s} \\ \mathbf{e}_{\mathbf{y}^s} \\ \mathbf{e}_{\mathbf{z}^s} \end{pmatrix} = \begin{pmatrix} -\cos\theta\cos\lambda & -\sin\lambda & \sin\theta\cos\lambda \\ -\cos\theta\sin\lambda & \cos\lambda & \sin\theta\sin\lambda \\ \sin\theta & 0 & \cos\theta \end{pmatrix} \begin{pmatrix} \mathbf{e}_{\mathbf{x}^s} \\ \mathbf{e}_{\mathbf{y}^s} \\ \mathbf{e}_{\mathbf{z}^s} \end{pmatrix}. \quad (5.11)$$

Moreover, the transformation formulas can also be combined to transform coordinates and basis vectors from one local system to another.

2.2 *Newton's Law of Gravitation and Potential*

According to Newton's law of gravitation, two point masses m_1 and m_2 attract each other with gravitational (attractive) force which is directly proportional to the product of their masses and inversely proportional to the square of the distance l between them. The gravitational force is directed from either point mass to the other and applies equally to one mass as the other. The vector form of Newton's law is given by

$$\mathbf{F}_\mathbf{b} = -G\frac{m_1 m_2}{l^2}\frac{\mathbf{l}}{l}, \quad (5.12)$$

where $\mathbf{l}/l$ is a unit vector pointing from m_1 to m_2, $\mathbf{F}_\mathbf{b}$ is the gravitational force vector attached to m_2 and pointing to m_1, and G is known as Newton's gravitational constant. G can be determined by experiment and the current best value (recommended by CODATA 2006; Mohr et al. 2008) is

$$G = (6.67428 \pm 0.00067) \times 10^{-11}\ \mathrm{m^3\,kg^{-1}\,s^{-2}}. \tag{5.13}$$

The SI unit of force is m kg s^{-2} with the special name "newton" and symbol "N,"—see BIPM (2006). In this context, gravitation is exclusively based on Newton's classical formulation; for a discussion of some aspects related to Einstein's theory of general relativity consult Kovalevsky and Seidelmann (2004), IERS (2010), Jekeli (2009), or Müller et al. (2008).

A gravitational acceleration (also termed gravitation) can be ascribed to the gravitational force $\mathbf{F_b}$, which represents the acceleration that one mass undergoes due to the gravitational attraction of the other. From (5.12) it follows, for the attracted point P (after dropping the indices),

$$\mathbf{b} = -G\frac{m}{l^2}\frac{\mathbf{l}}{l}, \tag{5.14}$$

where m is the attracting mass, $\mathbf{l} = \mathbf{r} - \mathbf{r}'$, with $\mathbf{r}$ and $\mathbf{r}'$ being the position vectors of the attracted point P and the source point P', respectively. By the law of superposition, the gravitational acceleration of an extended body like the Earth can be computed as the vector sum of the accelerations generated by the individual point masses (or mass elements), yielding

$$\mathbf{b} = \mathbf{b}(\mathbf{r}) = -G\iiint_{\text{Earth}} \frac{\mathbf{r}-\mathbf{r}'}{|\mathbf{r}-\mathbf{r}'|^3}\,\mathrm{d}m = -G\iiint_{\text{Earth}} \frac{\mathbf{r}-\mathbf{r}'}{|\mathbf{r}-\mathbf{r}'|^3}\,\rho\,\mathrm{d}v\ ,\quad \rho = \rho(\mathbf{r}')\ , \tag{5.15}$$

where dm is the differential mass element, ρ is the volume density (unit kg m^{-3}), and dv is the volume element. The SI unit of acceleration is ms^{-2} (BIPM 2006). However, the non-SI unit gal is still used frequently in geodesy and geophysics, and it is also listed in BIPM (2006) under "non-SI units accepted for use with the SI, Table 9" (name of unit: gal; symbol of unit: Gal):

$$1\ \mathrm{Gal} = 1\,\mathrm{cm\,s^{-2}} = 0.01\,\mathrm{m\,s^{-2}},\quad 1\mathrm{K\ Gal} = 10\,\mathrm{m\,s^{-2}},\quad 1\ \mathrm{mGal} = 10^{-5}\,\mathrm{m\,s^{-2}},\quad 1\mu\mathrm{Gal} = 10^{-8}\,\mathrm{m\,s^{-2}}. \tag{5.16}$$

The gravitational acceleration vectors $\mathbf{b}$ form a conservative vector field, also known as potential field. A conservative vector field is a vector field which is the gradient of a scalar potential function. It has the important property that line integrals from one point to another are path independent and, conversely, path independence is equivalent to the vector field being conservative. Conservative vector fields are also irrotational or non-vortical, meaning that (in three dimensions) they have vanishing curl; the converse of this property (i.e., fields with vanishing curl are conservative) is also true if the domain is simply connected (Kellog 1953).

The gravitational acceleration vector **b** can be represented as the gradient of the gravitational potential *V*:

$$\mathbf{b} = \operatorname{grad} V\,. \tag{5.17}$$

The gradient vector has the properties that it points in the direction of greatest change of the potential function *V*, its magnitude equals the rate of change with respect to the distance in this direction, and it is everywhere normal to a surface of constant potential *V*. Furthermore, the directional derivative of *V* with respect to the distance in a particular direction (specified by a vector **d**) is the projection of grad*V* onto that direction.

For a point mass *m*, see (5.14), the gravitational potential is given by

$$V = \frac{Gm}{l}, \quad \text{with} \quad \lim_{l\to\infty} V = 0\,, \tag{5.18}$$

and correspondingly for the Earth, see (5.15), the potential is obtained by

$$V = V(\mathbf{r}) = G \iiint\limits_{\text{Earth}} \frac{\mathrm{d}m}{l} = G \iiint\limits_{\text{Earth}} \frac{\rho\,\mathrm{d}v}{l}, \qquad \text{with} \quad \lim_{l\to\infty} V = 0\,, \tag{5.19}$$

where in both cases the latter condition implies that the potential is regular at infinity. It can easily be shown that the acceleration vectors in (5.14) and (5.15) result from (5.18) and (5.19) by applying the gradient operator, respectively. In this context, it should be noted that the potential *V* is defined with a positive sign in geodesy, in contrast to physics, where it is usually defined with opposite sign (conceptually closer to potential energy; Jekeli 2009).

According to potential theory, the gravitational effect of concentric homogeneous mass shells is equal to the effect of the entire mass being concentrated at the center of mass of the object. This property is useful for approximating the effect of celestial bodies at larger distances or the computation of atmospheric effects (see Sect. 2.6). On the other hand, this relates to the inverse problem of potential theory; the inverse problem (determination of the masses from the potential), in contrast to the direct problem (determination of the potential from the masses), has no unique solution, because, in general, there are infinitely many mass distributions possible, which are in accordance with a given exterior potential function.

The gravitational potential at a point *P* indicates the work done by gravitation in order to move the unit mass from infinity ($V = 0$) to *P* (Sigl 1985; Torge 2001). The unit of the potential is $\mathrm{m^2s^{-2}}$. If the density structure and geometry of the entire Earth were known, (5.19) would permit the calculation of the gravitational potential and its functionals. In reality, of course, this information is not available with sufficient accuracy, e.g., densities are known with only two or three significant digits, and global Earth models merely consider radial density structures. Therefore, the determination of the exterior potential field can be solved only by measurements performed at or above the Earth's surface (boundary value problems; see Sects. 3.1–3.5); for a comprehensive presentation of this and further specialized topics of potential theory (e.g., existence and uniqueness theorems, or the classical integral theorems of Gauss, Green, and Stokes, which are of great importance in physical geodesy), reference should be made to the textbooks of Kellog (1953), MacMillan (1958), and Sigl (1985); furthermore, a concise overview is given in Jekeli (2009).

The gravitational (volume) potential V according to (5.19) and its first derivatives are continuous and bounded everywhere; this holds even in the case that the evaluation point P is on the bounding surface or inside the mass distributions (case $l = 0$, weak singularity; see Jekeli 2009; Torge 2001). The second derivatives of V satisfy under certain conditions on the mass density ρ (so-called Hölder conditions; see Kellogg 1953; Heck 1997) the following partial differential equation, known as Poisson's equation:

$$\Delta V = \frac{\partial^2 V}{\partial x^2} + \frac{\partial^2 V}{\partial y^2} + \frac{\partial^2 V}{\partial z^2} = V_{xx} + V_{yy} + V_{zz} = -4\pi G\rho\,, \qquad (5.20)$$

where Δ is the Laplace operator, and x, y, z are coordinates in any Cartesian system. Consequently, the second derivatives of V exhibit discontinuities where abrupt changes of the mass density ρ occur. A special case of the above equation applies for those regions where the density vanishes (i.e., in free space); then Poisson's equation turns into the Laplace equation:

$$\Delta V = 0\,. \qquad (5.21)$$

The Laplace operator Δ may also be defined as the divergence of a gradient field, i.e.,

$$\Delta V = \text{div}\,(\text{grad}\,V)\,. \qquad (5.22)$$

As the divergence operator represents the (gravitational) flux generated per unit volume at each point of the field, (5.22) emphasizes that the sources of the gravitational field are the masses, i.e., the divergence of the field is zero in free space and non-zero inside the masses. A vector field with constant zero divergence is also called solenoidal (or incompressible).

The solutions of the Laplace equation are known as harmonic functions, which are important in many fields of mathematics and physics, such as potential fields related to gravitation, electrostatics, magnetics, etc. For instance, every Newtonian potential is a harmonic function in free space, and conversely, every harmonic function can be represented as a Newtonian potential of a mass distribution (Jekeli 2009). Formally, (5.21) represents a partial differential equation of second order for V, which holds in the exterior space of the Earth (the atmosphere, etc., are neglected for the moment). Like any differential equation, a complete solution is obtained only with the application of boundary conditions (conditions which the solution must satisfy at the boundary of the region, i.e., the Earth's surface; Jekeli 2009).

Laplace's equation (5.21) can be solved by introducing an appropriate coordinate system. In geodesy, the solution based on spherical polar coordinates $(\theta,\ \lambda,\ r)$ is of great significance. A solution of $\Delta V = 0$, rewritten in spherical coordinates, can be found by the method of separation of variables, where the solution of V is postulated as $V(\theta, \lambda, r) = f(\theta)\, g(\lambda)\, h(r)$. The general solution can be written as

$$V(\theta,\lambda,r) = \frac{GM}{a}\sum_{n=0}^{\infty}\left(\frac{a}{r}\right)^{n+1}\sum_{m=0}^{n}(C_{nm}\cos m\lambda + S_{nm}\sin m\lambda)P_{nm}(\cos\theta)\,, \quad (5.23)$$

where n, m are integers denoted as degree and order of the expansion, GM is the geocentric gravitational constant (gravitational constant G times the mass of the Earth M), a is in the first instance an arbitrary constant, but is typically set equal to the semimajor axis of a reference ellipsoid, $P_{nm}(\cos\theta)$ are the associated Legendre functions of the first kind, and C_{nm}, S_{nm} are the spherical harmonic coefficients (also denoted as Stokes's constants). In particular in satellite geodesy, sometimes the following conventions are used:

$$\begin{aligned} J_{nm} &= -C_{nm}, \quad K_{nm} = -S_{nm}, && \text{for } n \neq 0 \text{ or } m \neq 0\,,\\ J_{00} &= C_{00} = 1, && \text{for } n = 0 \text{ and } m = 0\,. \end{aligned} \quad (5.24)$$

For the case $m = 0$, the index m is usually dropped, leading to the coefficients J_n and the Legendre polynomials $P_n(\cos\theta)$.

The (unitless) spherical harmonic coefficients C_{nm}, S_{nm} represent mass integrals, as the spherical harmonic expansion (5.23) is just another way of expressing the volume integral over the Earth's masses in (5.19). Furthermore, the low degree coefficients have a simple physical interpretation. The coefficient $C_{00} = 1$ leads to the zero degree term GM/r of the gravitational potential, which represents the effect of a point mass, or equivalently a radially layered spherical Earth. The degree one terms are associated with the coordinates of the Earth's center of mass; they are forced to zero if the coordinate system is geocentric. The terms of degree two are connected with the moments and products of inertia (see Torge 2001). Regarding the magnitude of the harmonic coefficients, C_{20} ($J_2 = -C_{20}$ is also known as the dynamical form factor, characterizing the Earth's flattening) is more than three orders of magnitude smaller than the central term, and the remaining coefficients contribute again at least two to three orders of magnitude less than C_{20}, indicating that the bulk of the potential can be described by an ellipsoidal model.

For numerical reasons, it is convenient to introduce the so-called fully normalized associated Legendre functions and corresponding spherical harmonic coefficients:

$$\begin{aligned} &\overline{P}_{nm}(\cos\theta) = F_{nm}\,P_{nm}(\cos\theta)\,, \quad \begin{Bmatrix}\overline{C}_{nm}\\ \overline{S}_{nm}\end{Bmatrix} = \frac{1}{F_{nm}}\begin{Bmatrix}C_{nm}\\ S_{nm}\end{Bmatrix}\,,\\ &F_{nm} = \sqrt{k(2n+1)\frac{(n-m)!}{(n+m)!}} \quad \text{with} \quad k = \begin{cases}1 & \text{for } m = 0\\ 2 & \text{for } m \neq 0\end{cases}. \end{aligned} \quad (5.25)$$

With (5.25), the spherical harmonic expansion (5.23) can be written compactly as

$$V(\theta, \lambda, r) = \sum_{n=0}^{\infty} \left(\frac{a}{r}\right)^{n+1} \sum_{m=-n}^{n} \overline{V}_{nm} \overline{Y}_{nm}(\theta, \lambda), \tag{5.26}$$

with

$$\overline{Y}_{nm}(\theta, \lambda) = \overline{P}_{n|m|}(\cos\theta) \begin{cases} \cos m\lambda & \text{for } m \geq 0 \\ \sin|m|\lambda & \text{for } m < 0 \end{cases}, \quad \overline{V}_{nm} = \frac{GM}{a} \begin{cases} \overline{C}_{nm} & \text{for } m \geq 0 \\ \overline{S}_{nm} & \text{for } m < 0 \end{cases}. \tag{5.27}$$

The functions $r^{-(n+1)}\overline{Y}_{nm}(\theta, \lambda)$ in (5.26) are called solid spherical harmonics, while $\overline{Y}_{nm}(\theta, \lambda)$ are called Laplace's surface spherical harmonics, fulfilling the orthogonality relations

$$\frac{1}{4\pi} \iint_{\sigma} \overline{Y}_{nm}(\theta, \lambda) \overline{Y}_{n'm'}(\theta, \lambda)\, d\sigma = \begin{cases} 1 & \text{for } n = n' \text{ and } m = m' \\ 0 & \text{otherwise} \end{cases}, \tag{5.28}$$

where σ is the unit sphere, and $d\sigma$ is the corresponding surface element. In this context, the spherical harmonic expansion may also be regarded as a complete system of orthogonal basis functions (eigenfunctions), with the coefficients being the corresponding eigenvalues, which can be interpreted as the spectrum on the sphere (Jekeli 2009).

The infinite spherical harmonic series (5.23), or equivalently (5.26), converges uniformly for all $r > R_c$, where R_c is the radius of the sphere that encloses all terrestrial masses (the so-called Brillouin sphere), while the convergence below this sphere down to the Earth's surface (i.e., in free space) has been a subject of controversy in the literature and is still not fully solved. However, due to the theorem of Runge-Krarup, any regular harmonic function can be approximated arbitrarily well by a spherical harmonic series in the mass-free space, and hence convergence of the series can be assumed there for all practical applications, including truncated spherical harmonic series. For further details on this topic, see, e.g., Moritz (1980) or Jekeli (1983, 2009). Besides the spherical harmonic series expansion, which is of outstanding importance in geodesy, other solutions of the Laplace equation (5.21) also exist for specific coordinate types. Of some relevance are the ellipsoidal harmonics, which are based on elliptical coordinates (β, λ, u; β = reduced latitude; λ = ellipsoidal longitude; u = semiminor axis of a confocal ellipsoid; see Heiskanen and Moritz 1967). Ellipsoidal harmonics are used, e.g., in connection with the (ellipsoidal) normal gravity field (Heiskanen and Moritz 1967) or for intermediate results within the development of high-degree global geopotential models (Pavlis et al. 2008); the elliptical coordinates, β, λ, u, should not be confused with the ellipsoidal geographic coordinates, φ, λ, h, which do not admit a separation of variables solution of the Laplace equation (Grafarend 1988, Jekeli 2009).

2.3 The Earth's Gravity Field

A body rotating with the Earth experiences the resultant of the gravitational force, $\mathbf{F_b}$, and the centrifugal force, $\mathbf{F_c}$, due to the Earth's rotation, while an artificial satellite, not rotating with the Earth, is affected only by gravitation. Regarding the centrifugal force, $\mathbf{F_c}$, again an acceleration (acting on a unit mass) can be ascribed to it, which is directed outwards and perpendicular to the rotation axis. Based on the Earth-fixed reference system (*X*, *Y*, *Z*; see Sect. 2.1), the centrifugal acceleration is given by

$$\mathbf{z} = \mathbf{z}(\mathbf{p}) = \omega^2 \mathbf{p}, \quad \mathbf{p}^{\mathrm{T}} = (X \quad Y \quad 0), \quad p = |\mathbf{p}| = \sqrt{X^2 + Y^2} \,, \tag{5.29}$$

where ω is the angular velocity, and $\mathbf{p}$ is the distance vector from the rotation axis (Z), with $\mathbf{p}$ and $\mathbf{z}$ having the same direction.

The centrifugal acceleration vectors $\mathbf{z}$ also form a conservative vector field and hence can be represented as the gradient of a potential function Z. With

$$\mathbf{z} = \operatorname{grad} Z \,, \tag{5.30}$$

the centrifugal potential Z becomes

$$Z = Z(p) = \frac{\omega^2}{2} p^2 \,. \tag{5.31}$$

Applying the Laplace operator on Z yields

$$\Delta Z = 2\,\omega^2 \,, \tag{5.32}$$

i.e., the centrifugal potential Z is not harmonic, as opposed to V.

The gravity acceleration (or gravity) vector $\mathbf{g}$ is the resultant of the gravitation $\mathbf{b}$ and the centrifugal acceleration $\mathbf{z}$:

$$\mathbf{g} = \mathbf{b} + \mathbf{z} \,. \tag{5.33}$$

The force of gravity $\mathbf{F_g}$ is obtained by multiplying $\mathbf{g}$ by the mass m of the attracted object, i.e., $\mathbf{F_g} = m\,\mathbf{g}$. The direction of $\mathbf{g}$ is the direction of the plumb line (vertical), the magnitude g is called the gravity intensity (or often just gravity; see Torge 2001). In this context, time variations are not considered here, assuming that they are taken into account by appropriate reductions (see Sect. 2.6).

Finally, with

$$\mathbf{g} = \operatorname{grad} W = \mathbf{b} + \mathbf{z} = \operatorname{grad} V + \operatorname{grad} Z \,, \tag{5.34}$$

the gravity potential W of the Earth is given by

$$W = W(\mathbf{r}) = V + Z = G \iiint_{\text{Earth}} \frac{dm}{l} \; + \; \frac{\omega^2}{2} p^2 \,. \tag{5.35}$$

From (5.20) and (5.32), the generalized Poisson equation is obtained:

$$\Delta W = -4\pi G\rho + 2\,\omega^2, \tag{5.36}$$

which reduces to the generalized Laplace equation in free space:

$$\Delta W = 2\,\omega^2. \tag{5.37}$$

The gravity vector **g** and accordingly the gradient operator can be expressed in various coordinate systems. Regarding the global Cartesian Earth-fixed coordinate system *X*, *Y*, *Z*, the following representations are common:

$$\mathbf{g} = \operatorname{grad} W = \begin{pmatrix} \dfrac{\partial W}{\partial X} \\ \dfrac{\partial W}{\partial Y} \\ \dfrac{\partial W}{\partial Z} \end{pmatrix} = \begin{pmatrix} W_X \\ W_Y \\ W_Z \end{pmatrix} = [\mathbf{W_X}] = W_X\mathbf{e_X} + W_Y\mathbf{e_Y} + W_Z\mathbf{e_Z}, \tag{5.38}$$

where the gradient vector is first defined as a column vector, and second written by means of the unit vectors $\mathbf{e_X}$, $\mathbf{e_Y}$, $\mathbf{e_Z}$, pointing along the coordinate axes *X*, *Y*, *Z*, respectively. In the case where spherical coordinates θ, λ, r are employed (e.g., in connection with a spherical harmonic expansion of the gravitational potential *V*), it is convenient to represent **g** with respect to the local spherical system (x^s, y^s, z^s; see Sect. 2.1):

$$\mathbf{g} = \operatorname{grad} W = \begin{pmatrix} \dfrac{\partial W}{\partial x^s} \\ \dfrac{\partial W}{\partial y^s} \\ \dfrac{\partial W}{\partial z^s} \end{pmatrix} = \begin{pmatrix} W_{x^s} \\ W_{y^s} \\ W_{z^s} \end{pmatrix} = [\mathbf{W_{x^s}}] = W_{x^s}\mathbf{e_{x^s}} + W_{y^s}\mathbf{e_{y^s}} + W_{z^s}\mathbf{e_{z^s}}, \tag{5.39}$$

where $\mathbf{e_{x^s}}$, $\mathbf{e_{y^s}}$, $\mathbf{e_{z^s}}$ are again unit vectors pointing along the local coordinate axes x^s, y^s, z^s, respectively. The derivatives of *W* with respect to the local spherical system can be obtained by using the chain rule for differentiation (Tscherning 1976a, b), e.g.,

$$\frac{\partial}{\partial x^s} = \frac{\partial\theta}{\partial x^s}\frac{\partial}{\partial\theta} + \frac{\partial\lambda}{\partial x^s}\frac{\partial}{\partial\lambda} + \frac{\partial r}{\partial x^s}\frac{\partial}{\partial r}, \quad \text{and} \quad \frac{\partial}{\partial y^s}, \frac{\partial}{\partial z^s} \quad \text{accordingly,} \tag{5.40}$$

which, given (5.1) and (5.10), leads to

$$[\mathbf{W_{x^s}}] = \left(-\tfrac{1}{r}W_\theta \quad \tfrac{1}{r\sin\theta}W_\lambda \quad W_r\right)^{\mathbf{T}}, \tag{5.41}$$

with W_θ being the derivative with respect to θ, etc. Then with (5.10) or (5.11), the transformation to the Earth-fixed system yields

$$[\mathbf{W_X}] = \mathbf{A_{x^s}}[\mathbf{W_{x^s}}]\,. \tag{5.42}$$

On the other hand, in the Earth-fixed system, the components of the gravity vector **g** can also be expressed by the astronomical latitude and longitude Φ, Λ (plumb line parameters; see Fig. 5.1):

$$\mathbf{g} = -g\mathbf{n} = -g \begin{pmatrix} \cos\Phi \cos\Lambda \\ \cos\Phi \sin\Lambda \\ \sin\Phi \end{pmatrix}, \tag{5.43}$$

which together with (5.38) yields

$$\begin{aligned} &\sin\Phi = -\frac{1}{g} W_Z, && \cos\Phi = \frac{1}{g}\sqrt{W_X^2 + W_Y^2}, \\ &\sin\Lambda = -\frac{W_Y}{g\cos\Phi} = -\frac{W_Y}{\sqrt{W_X^2 + W_Y^2}}, && \cos\Lambda = -\frac{W_X}{g\cos\Phi} = -\frac{W_X}{\sqrt{W_X^2 + W_Y^2}}, \\ &g = \sqrt{W_X^2 + W_Y^2 + W_Z^2}\,. \end{aligned} \tag{5.44}$$

Φ, Λ, and $W(X, Y, Z)$ form the so-called natural coordinates (Torge 2001). A surface of constant gravity potential W is designated as equipotential, level, or geopotential surface, and the gravity vector $\mathbf{g}$ is everywhere normal to it.

Following the line of thought used for deriving (5.41), the second derivatives of the gravity potential W with respect to the local spherical system are given by (Moritz 1971; Tscherning 1976a)

$$\begin{aligned} [\mathbf{W}_{\mathbf{x}^s\mathbf{x}^s}] &= \operatorname{grad}\mathbf{g} = \operatorname{grad}(\operatorname{grad} W) = \begin{pmatrix} W_{x^sx^s} & W_{x^sy^s} & W_{x^sz^s} \\ W_{y^sx^s} & W_{y^sy^s} & W_{y^sz^s} \\ W_{z^sx^s} & W_{z^sy^s} & W_{z^sz^s} \end{pmatrix} \\ &= \begin{pmatrix} \underline{\frac{1}{r}W_r + \frac{1}{r^2}W_{\theta\theta}} & \frac{1}{r^2\sin\theta}(\cot\theta\, W_\lambda - W_{\theta\lambda}) & \frac{1}{r^2}W_\theta - \frac{1}{r}W_{\theta r} \\ \cdots & \underline{\frac{\cot\theta}{r^2}W_\theta + \frac{1}{r}W_r + \frac{1}{r^2\sin^2\theta}W_{\lambda\lambda}} & \frac{1}{r\sin\theta}\left(-\frac{1}{r}W_\lambda + W_{\lambda r}\right) \\ \cdots & \cdots & \underline{W_{rr}} \end{pmatrix}, \end{aligned} \tag{5.45}$$

where, e.g., $W_{\theta\theta}$ is the second derivative with respect to θ, etc. The corresponding matrix with respect to the Earth-fixed system (X, Y, Z) is given by

$$[\mathbf{W}_{\mathbf{XX}}] = \mathbf{A}_{\mathbf{x}^s}[\mathbf{W}_{\mathbf{x}^s\mathbf{x}^s}](\mathbf{A}_{\mathbf{x}^s})^{\mathrm{T}}\,. \tag{5.46}$$

The matrix of the second derivatives of W, e.g., in the form (5.45) or (5.46), is also denoted as gravity gradient tensor, Eötvös tensor, or Marussi tensor; it can be expressed by means of curvature parameters, which completely describe the geometry (curvature) of the level surfaces and the plumb lines (Marussi 1985). The gravity gradient tensor includes only five independent elements; the matrix is symmetric (because of curl $\mathbf{g}$ = curl grad $W = 0$, i.e., $W_{xy} = W_{yx}$, etc.) and the trace must fulfil the conditions (5.36) or (5.37), respectively.

The approach of computing, in the first instance the (first and second) derivatives with respect to the local spherical system, followed by a transformation to the Earth-fixed system, involves no approximations (Wenzel 1985) and is particularly suitable in connection with spherical harmonic expansions, e.g., the high-degree geopotential model EGM2008 (Pavlis et al. 2008). It should also be noted that (5.41) and (5.42) for the first derivatives as well as (5.45) and (5.46) for the second derivatives can be utilized accordingly for each of the components of *W*, i.e., *V* and *Z*; moreover, they may be used in connection with the disturbing potential and its functionals (see Sect. 3.2). With regard to the presently active satellite mission GOCE (ESA 1999; Rummel et al. 2011), the second derivatives of *V* are of special interest. Finally, the results in the Earth-fixed system can be further transformed to any other Cartesian system of interest.

At this point it is emphasized that the primary goal of physical geodesy is the determination of the gravity potential *W* as a function of position; if $W(\mathbf{r})$ were known, then all parameters of interest could be derived from it, including the gravity vector $\mathbf{g}$ (direction parameters, Φ, Λ, and magnitude, g, see (5.44)), the curvature of the level surfaces and plumb lines (depending on the second derivatives of *W*, e.g., (5.46)), as well as the form of the equipotential surfaces (by solving the equation $W(X, Y, Z) = \text{const.}$). However, as mentioned above, the gravity potential *W* cannot be computed directly based on (5.35) due to the insufficient knowledge about the density structure of the entire Earth; instead, the determination of the exterior potential field must be solved by measurements performed at or above the Earth's surface (see Sect. 3).

2.4 The Geoid and Heights

The geoid is of great importance in geodesy, oceanography, geophysics, and other Earth sciences, serving as a reference surface for heights over the continents as well as for the dynamic ocean topography (DOT). The geoid was introduced by C.F. Gauss as an equipotential surface of the Earth's gravity field, coinciding with the mean sea level (MSL) of the idealized oceans (i.e., homogeneous water masses at rest, subject only to the time-invariable force of gravity; see Torge 2001). The basis of this definition, i.e., the geoid being a selected equipotential surface of the Earth's gravity field (with $W = W_0$), is of fundamental importance and is still useful today. Given the gravity potential value W_0, the equation of the geoid is

$$W = W(\mathbf{r}) = W_0 \,. \tag{5.47}$$

As discussed in the previous section, the geoid is a closed and continuous surface, even inside the Earth; however, inside the masses, the curvature of the geoid may exhibit discontinuities where abrupt density changes occur. The vertical distance between the geoid and a given reference ellipsoid is denoted as the geoid height or geoid undulation.

The geoid is conceptionally chosen to approximate (in some mathematical sense) the mean ocean surface (Rapp 1995). However, mean sea level (MSL), derived by averaging the instantaneous sea surface over a sufficiently long time span (e.g., at least 1 year), may also vary over longer time spans (for instance, due to the global secular sea level rise of about 1–2 mm/year, observed over the past 100 years; Torge 2001), and furthermore it does not coincide with a level surface due to the forcing of the oceans by winds, atmospheric pressure, and buoyancy (as a result of density differences associated with corresponding temperature and salinity differences) in combination with gravity and the Earth's rotation. The deviation of MSL from a best fitting equipotential surface (geoid) is denoted as the (mean) dynamic ocean topography DOT. The DOT reaches maximum values of about ±2 m (Rapp and Balasubramania 1992; Bosch and Savcenko 2010) and is of vital importance for oceanographers, as it allows the derivation of the absolute circulation of the oceans (Wunsch and Gaposchkin 1980; Condi and Wunsch 2004).

Accordingly, a refined definition of the geoid is needed, which could be based on a (global) minimum condition with regard to the deviation of MSL from a best fitting level surface, and which should also consider that MSL is not constant in time. In principle, two options exist: either the geoid definition has to refer to a certain epoch of MSL, or a time-dependent geoid linked to the respective MSL could be introduced. However, corresponding to the general geodetic practice of reducing time-variable quantities to a quasi-static state (see also Sect. 2.6), mainly the first option is feasible (e.g., for employing the geoid as a height reference surface). Nevertheless, the gravity potential of a best fitting level surface of MSL at a given point in time may be expressed by the (static) geoid potential W_0 (associated with a certain reference epoch) and a linear change with time in first approximation (e.g., completely corresponding to the ITRF station positions and velocities). Finally, a refined geoid definition must also include specifications regarding the treatment of the notable permanent tide effects (see Sect. 2.6). For further details on the definition and realization of the geoid as well as the W_0 aspect see Heck (2004) and Heck and Rummel (1990).

The numerical value for the geoid potential W_0 can in principle be deduced from the MSL spatial positions with respect to the (global) Earth-fixed reference system and a geopotential model such as EGM2008 (Pavlis et al. 2008), possibly supplemented by an oceanographic or geodetic model of the mean DOT; the (absolute) gravity potential W is derived for the MSL (or MSL minus MDOT) points from the geopotential model ($W = V + Z$; (5.26) and (5.31), presupposing that V is regular at infinity), and finally some averaging procedure is applied to reduce random effects. The MSL positions can be taken from satellite altimetry (available within the latitude band ±86°; see Andersen et al. 2010), providing directly the sea surface height (SSH), or from tide gauge and Global Navigation Satellite System (GNSS) observations (see Fig. 5.3). The first approach based on satellite altimetry gives access to almost the entire ocean domain and was applied by Bursa et al. (2002), while the latter approach is restricted to the existing tide gauge stations at the coasts. Furthermore, it is noted that a numerical value for W_0 (based on the satellite altimetry approach; Bursa et al. 2002) is provided in the IERS conventions (IERS 2010):

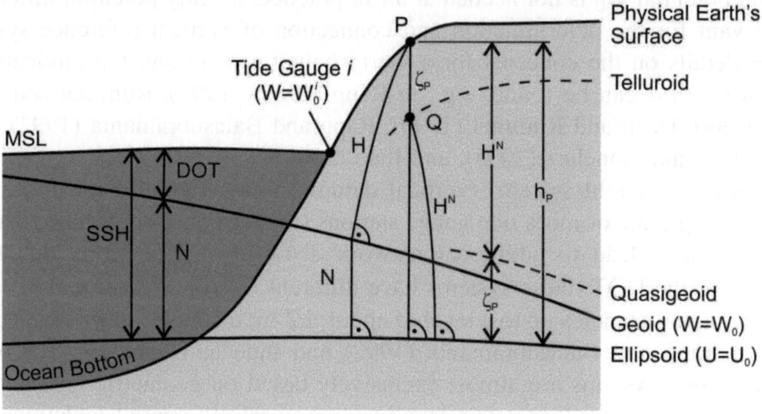

Fig. 5.3 Geoid, quasigeoid, heights, continental topography, mean sea level (MSL), and dynamic ocean topography (DOT)

$$W_0 = 62{,}636{,}856.0\ \mathrm{m^2s^{-2}}\,. \tag{5.48}$$

However, Sanchez (2008) published W_0 values, differing by about 2–3 $\mathrm{m^2s^{-2}}$ from the above value (corresponds to a vertical distance of about 2–3 dm).

After defining the geoid by a (conventional) value W_0, it is still necessary to realize it, i.e., to find the position of the geoid in space, or equivalently to find for a given point P at the Earth's surface the vertical distance between P and the geoid, which corresponds to determining the potential difference $W_0 - W_P$ (see Fig. 5.3). The most promising procedure is to use GNSS stations or networks together with a global geopotential model (e.g., EGM2008), ideally supplemented by local gravity observations; based on the ITRF positions for the GNSS sites, the corresponding (absolute) gravity potential (W_P) can be computed (from the global model, possibly plus local gravity data), yielding then the potential difference $W_0 - W_P$ and hence the position of the geoid. Another option for realizing the geoid is to use tide gauge stations (or connected leveling stations) together with a DOT model (see Fig. 5.3). Naturally, several stations should be employed in either approach to average out random effects.

The definition of a world height system and the W_0 topic have been discussed for more than 25 years; within the International Association of Geodesy (IAG), this subject has been treated recently within the Inter-Commission Project 1.2, Vertical Reference Frames (Ihde 2009) and is now continued within the GGOS (Global Geodetic Observing System) project. Corresponding to the IERS approach, most likely, there will be an ideal "vertical reference system" (VRS) with corresponding conventions (including a conventional W_0 value), and a realization, the "vertical reference frame" (VRF). In this context, Heck (2004) emphasizes that the absolute gravity potential (including W_0) is dependent on the assumption of regularity at infinity, see (5.19), and that the numerical value of the absolute potential W_0 is not needed at all in practice, as only potential differences are relevant for the determination and connection of vertical reference systems. Further details on the concepts for a world height system and the unification of national systems can be found, e.g., in Rapp (1983a, 1995), Rummel and Teunissen (1988), Heck and Rummel (1990), Rapp and Balasubramania (1992), Heck (2004), Ihde and Sanchez (2005), and Ihde (2009).

Historically, height systems (vertical datums) were related to mean sea level (MSL) through one or more tide gauge stations (the introduction of more than one tide gauge may lead to additional network distortions). Therefore, due to the existence of the DOT, these systems have different reference surfaces ($W = W_0^i$), implying inconsistencies up to a level of about ± 2 m; examples of this kind can be found in Rapp and Balasubramania (1992) and Ihde and Sanchez (2005). The existing height systems are almost exclusively based on geometric leveling (also called spirit leveling). Geometric leveling is a quasi-differential technique, providing height differences δn (backsight minus foresight reading) with respect to the local astronomical system (see Fig. 5.1). Over longer distances, the non-parallelism of the level surfaces cannot be neglected, as it results in a path dependence of the results. This problem can be overcome by introducing potential differences, which are path independent because the gravity field is conservative (see Sects. 2.2 and 2.3). With

$$\mathrm{d}W = \frac{\partial W}{\partial x}\mathrm{d}x + \frac{\partial W}{\partial y}\mathrm{d}y + \frac{\partial W}{\partial z}\mathrm{d}z = \mathrm{grad}W\,\mathbf{ds} = \mathbf{g}\,\mathbf{ds} = -g\,\mathrm{d}n\ , \tag{5.49}$$

$\mathrm{d}n$ being the distance along the outer normal of the level surface (zenith), the geopotential number C is defined as

$$C = W_0 - W_P = -\int_{P_0}^{P} \mathrm{d}W = \int_{P_0}^{P} g\,\mathrm{d}n\ , \tag{5.50}$$

where P_0 is an arbitrary point on the geoid (height reference surface) and P is a point on the Earth's surface. Thus, in addition to the leveling results ($\mathrm{d}n$), gravity observations (g) are also needed along the path between P_0 and P (with regard to the required spacing and accuracy of the gravity points, see Torge 2001). Historically, the geopotential numbers were referred to local reference surfaces (W_0^i), but henceforth no distinction is made between W_0^i (local reference surface) and W_0 (geoid).

The geopotential numbers are ideal quantities for describing the direction of water flow, i.e., water flows from points with higher geopotential numbers C to points with lower values. However, the geopotential numbers have the unit $\mathrm{m^2s^{-2}}$ (or $10\ \mathrm{m^2s^{-2}} = 1$ kGal m $= 1$ gpu), and are thus somewhat inconvenient in disciplines like civil engineering, etc. Therefore, a conversion to metric heights is desirable, which can be achieved by dividing the C values by an appropriate gravity value. Widely used are the orthometric heights (e.g., U.S.A., Canada,

Austria, and Switzerland) and normal heights (e.g., Germany, and many other European countries), which also play an important role in gravity field modeling due to the strong height dependence of various gravity field quantities (Sect. 3).

The orthometric height H is defined as the distance between the surface point P and the geoid, measured along the curved plumb line (see Fig. 5.3), which explains the common understanding as "height above sea level" (Torge 2001). The orthometric height can be derived from (5.50) by expanding the right side by H and integrating along the plumb line from the geoid to the surface point P:

$$H = \frac{C}{\bar{g}}, \quad \bar{g} = \frac{1}{H}\int_0^H g \,\mathrm{d}H, \tag{5.51}$$

where $\bar{g}$ is the mean gravity along the plumb line (inside the Earth). As $\bar{g}$ cannot be observed directly (besides some stations with borehole gravity data; Strange 1982), hypotheses about the interior gravity field are necessary, which is one of the main drawbacks of the orthometric heights. Assuming a constant density of the topographic masses (2,670 kgm^{-3}) as well as a flat topography (so-called Poincaré-Prey reduction) leads to

$$\bar{g}_{[\mathrm{ms}^{-2}]} = g_{P\,[\mathrm{ms}^{-2}]} + 0.424 \times 10^{-6} H_{[\mathrm{m}]} \,, \tag{5.52}$$

where g_P is the gravity value at the surface point P. The heights based on the mean gravity estimate (5.52) are denoted as Helmert-orthometric heights H^H (they are used, e.g., for the North American Vertical Datum, NAVD88; Zilkoski et al. 1995). For a discussion on refined procedures for the computation of $\bar{g}$, see Marti and Schlatter (2001), Flury and Rummel (2009), or Sjöberg (2010). Lastly, it is noted that points with equal orthometric heights are normally associated with slightly different level surfaces, which is due to the non-parallelism of the level surfaces.

In order to avoid the hypotheses about the Earth's interior gravity field, the normal heights H^N were introduced by Molodensky (e.g., Molodenskii et al. 1962). The normal height is defined as

$$H^N = \frac{C}{\bar{\gamma}} \,, \quad \bar{\gamma} = \frac{1}{H^N}\int_0^{H^N} \gamma \,\mathrm{d}H^N, \tag{5.53}$$

where $\bar{\gamma}$ is a mean normal gravity value along the normal plumb line, and γ is the magnitude of the normal gravity vector (for further details see next section). Consequently, the normal height H^N is measured along the slightly curved normal plumb line (Heiskanen and Moritz 1967; Torge 2001); it is in the first instance defined as the elevation of the telluroid above the ellipsoid, but can also be considered as the elevation of the surface point P above the quasigeoid (for details see Sect. 3.2 and Fig. 5.3). The quasigeoid is not a level surface and has no physical interpretation. Hence, the concept of the normal height and quasigeoid is less illustrative than that of the orthometric height and geoid, respectively, but it has the significant advantage that it is exclusively based on quantities of the Earth's exterior gravity field, avoiding any hypotheses about the interior field. Furthermore, it is noted that Heck (2003) defines the normal height in a conceptually different way as the ellipsoidal height (measured along the straight ellipsoidal normal) of the telluroid; in this case, (5.53) has to be adapted by computing the corresponding mean normal gravity value along the ellipsoidal normal, using the normal gravity component in the direction of the ellipsoidal normal instead of the absolute value of normal gravity itself. However, due the small length difference between the ellipsoidal normal and the normal plumb line (see below), the concept from Heck (2003) and the classical formulation can be considered as equivalent for all practical applications.

Another option is the so-called dynamic height H^{dyn}, which is defined as

$$H^{\mathrm{dyn}} = \frac{C}{\gamma_0^{45}}, \tag{5.54}$$

where γ_0^{45} is a constant normal gravity value, usually at the ellipsoid surface at 45° latitude. As the dynamic heights differ from the geopotential numbers only by a constant factor, points with the same H^{dyn} are located on the same level surface. As the dynamic heights have no geometric interpretation, and because the corrections to the raw leveling are quite large, they are not very widely used in practice.

Historically, when no gravity values were available for the computation of the geopotential numbers C, normal gravity values were used in some cases, resulting in the so-called normal-orthometric heights H^{NO} (e.g., still in use in Australia, Featherstone et al. 2011, or employed formerly in Germany, Heck 2003); these heights can be regarded as an approximation to the normal heights (Wolf 1974), but have the significant disadvantage of being path-dependent, in contrast to the above defined heights H, H^N, and H^{dyn}. Therefore, the H^{NO} are unsuitable for a modern height system.

It is also worth mentioning that the raw leveling results along lines (Δn) can be converted directly into corresponding height differences (ΔH, ΔH^N, ΔH^{dyn}) by the orthometric, normal, and dynamic corrections, respectively (Torge 2001; Heck 2003). Moreover, although the precision of geometric leveling is rather high (standard deviation for a 1 km traverse about 0.2–1.0 mm), it is important to keep in mind, that geometric leveling is a differential technique and hence susceptible to systematic errors; examples are the differences between the second and third geodetic leveling in Great Britain (about 0.2 m in north–south direction over about 1,000 km distance; Kelsey 1972), corresponding differences between the old and new leveling in France (about 0.25 m from the Mediterranean Sea to the North Sea, also mainly in north–south direction, distance about 900 km; Rebischung et al. 2008), as well as inconsistencies of more than 1 m across Canada and the U.S.A. (differences between different levelings and with respect to an accurate geoid; Véronneau et al. 2006; Smith et al. 2010). Regarding Canada and the U.S.A., this led to the decision to abandon geometric leveling completely and to use GNSS techniques together with a so-called "geoid based vertical datum," which shall be introduced by 2013 (Canada) and 2021 (U.S.A.), respectively (Smith et al. 2010).

In recent years, some authors (e.g., Steinberg and Papo 1998; Kumar 2005) became proponents of purely (geometric) ellipsoidal height systems, which neglect the effect of gravity. However, these are considered as unsuitable for any application involving fluid flow, among others (Vaníček 1998). However, the approach of using the GNSS technique and a geoid based vertical datum, as initiated in Canada and the U.S.A., appears to be a good alternative to avoid the time-consuming and expensive geometric leveling, especially in view of the now possible geoid and quasigeoid accuracies (see Sect. 4).

The geoid and quasigeoid serve as the zero height surfaces (vertical datum) for the orthometric and normal heights, respectively. With regard to the ellipsoidal heights h (from GNSS observations), the following relation holds:

$$h = H + N = H^N + \zeta\,, \tag{5.55}$$

where N is the geoid height or geoid undulation, and ζ is the quasigeoid height or height anomaly (see also Fig. 5.3). The above equation neglects that in the strict sense the relevant quantities are measured along slightly different lines in space. The ellipsoidal height (h) of a point in space is measured along the straight ellipsoidal normal, while the corresponding normal height (H^N) is measured along the slightly curved normal plumb line (see Fig. 5.3); the length difference between both paths can be estimated from the curvature of the normal plumb line, yielding less than 10^{-7} m for a point 10 km above the ellipsoid, which can be safely neglected. For the orthometric heights, the corresponding length difference can be roughly estimated by means of the deflection of the vertical (the angle between the actual plumb line and the ellipsoidal normal; see Fig. 5.3), resulting in an effect of about 0.4 mm for a station height of 10 km and an extreme deflection of the vertical of 1′. Hence, (5.55) is accurate at the millimeter level for all practical cases; another possibility would be to work with the corresponding potentials instead of the heights.

A transformation between the orthometric and normal heights, or geoid and quasigeoid heights, is possible by combining (5.51), (5.53), and (5.55), giving

$$H^N - H = N - \zeta = \frac{\bar{g} - \bar{\gamma}}{\bar{\gamma}} H = \frac{\bar{g} - \bar{\gamma}}{\bar{g}} H^N \approx \frac{\Delta g_B}{\bar{\gamma}} H\,. \tag{5.56}$$

The difference $H^N - H$ or $N - \zeta$ is mainly depending on the station height as well as $\bar{g} - \bar{\gamma}$, which is approximately the (simple) Bouguer anomaly Δg_B (Heiskanen and Moritz 1967; Torge 2001). In this context, it is noted that the Bouguer approximation in (5.56) is virtually exact in connection with the Helmert-orthometric heights (Forsberg and Tscherning 1997). The magnitude of the difference $H^N - H$ or $N - \zeta$ can reach several centimeters to about 1 dm in low mountain ranges, about 3–5 dm (or even more) in the high mountains such as the European Alps or Rocky Mountains, and about 3 m in the Himalayan Mountains (Rapp 1997; Marti and Schlatter 2001; Tenzer et al. 2005; Flury and Rummel 2009). On the oceans, the geoid and quasigeoid practically coincide (Torge 2001), as the effect of the DOT is only marginal.

Finally, regarding the orthometric heights, the various procedures in use for the computation of the mean gravity value $\bar{g}$ (e.g., (5.52) or more sophisticated methods) may lead to substantially different results (e.g., Marti and Schlatter 2001; Flury and Rummel 2009); therefore, it is essential to ensure that the heights H and the corresponding geoid undulations N are consistent such that (5.55) is satisfied.

2.5 The Normal Gravity Field

The normal gravity field is introduced as an approximation of the Earth's gravity field. On the one hand it should provide a reasonably good agreement with the real field, since it is used for the linearization of the observation equations, and on the other hand it should be simple to compute, as well as useful for other disciplines (Torge 2001). Based on these considerations, the level ellipsoid (or so-called Somigliana-Pizetti normal field) is almost exclusively used; another argument may also be the utilization of an ellipsoid for station coordinates (see Sect. 2.1). However, today, with the availability of very accurate satellite gravity field models (e.g., from the GRACE and GOCE missions), it is also worth considering the employment of a complete spherical harmonic expansion up to some maximum degree $n_{\max}$.

Corresponding to the gravity potential W, the normal gravity potential U is introduced as the sum of the normal gravitational potential V^N and the normal centrifugal potential Z^N:

$$U = V^N + Z^N \; . \tag{5.57}$$

The associated normal gravity vector is given by

$$\boldsymbol{\gamma} = \operatorname{grad} U \; . \tag{5.58}$$

The direction of $\boldsymbol{\gamma}$ is the direction of the normal plumb line, the magnitude γ is the normal gravity (intensity).

The normal gravity field of the level ellipsoid solely depends on four parameters. These can be two parameters describing the size and shape of the ellipsoid (for example, the semimajor axis, a_N, and flattening, f_N), the Earth's rotation rate, ω_N, and the total mass of the Earth, M_N. Numerical values for such parameters are recommended from time to time by the IUGG, IAG, etc. The latest set of constants was recommended by the IUGG and IAG in 1979 at the XVIIth General Assembly of the IUGG in Canberra (e.g., Moritz 2000), known as the Geodetic Reference System 1980 (GRS80), with the four defining parameters being a_N, J_2^N, ω_N, and GM_N. An updated (current best) set of parameters is also provided in IERS (2010). Besides GRS80, the WGS84 level ellipsoid is frequently used (NIMA 1997). It is defined by the geometrical parameters a_N and f_N and the physical parameters GM_N and ω_N; apart from the significantly different GM_N values of GRS80 and WGS84, the a_N and

ω_N parameters are identical, while the flattening parameters show only marginal differences (corresponding to 3×10^{-5} m with respect to the semiminor axis).

All parameters related to the normal gravity field of the level ellipsoid can be computed by closed formulas based on ellipsoidal harmonics (Heiskanen and Moritz 1967). However, the normal gravitational potential of the level ellipsoid can also be expanded in a rapidly converging spherical harmonic series; due to the symmetry with respect to the rotational axis as well as the equator, only the even zonal coefficients are non-zero, and an expansion up to degree 10 is fully sufficient. The spherical harmonic series approach, proposed by Tscherning (1976a), is well suited for the computation of U and its first and second derivatives, gives accurate results everywhere in space (including satellite positions), is easy to use in connection with high-degree Earth gravity field models such as EGM2008, and can also be generalized to more complicated normal gravity fields, e.g., based on a complete spherical harmonic gravitational model up to some degree $n_{\max}$.

Considering a complete spherical harmonic expansion of the normal gravitational potential (up to degree $n_{\max}$) as well as the centrifugal potential (with $p = r \sin\theta$) according to (5.26) and (5.31), respectively, yields

$$U(\theta,\lambda,r) = V^N + Z^N = \sum_{n=0}^{n_{\max}} \left(\frac{a_N}{r}\right)^{n+1} \sum_{m=-n}^{n} \overline{V}^N_{nm} \overline{Y}_{nm}(\theta,\lambda) + \frac{\omega_N^2}{2} r^2 \sin^2\theta, \quad (5.59)$$

with

$$\overline{V}^N_{nm} = \frac{GM_N}{a_N} \begin{cases} \overline{C}^N_{nm} & \text{for } m \geq 0 \\ \overline{S}^N_{nm} & \text{for } m < 0 \end{cases}. \quad (5.60)$$

The first and second derivatives of U with respect to the spherical coordinates θ, λ, r can be derived easily from the above equation (require the derivatives of the associated Legendre functions), which can then be used to compute the derivatives with respect to the local spherical system (x^s, y^s, z^s) in analogy to (5.41) and (5.45). Regarding the normal gravity vector, the transformation to the Earth-fixed system yields

$$\boldsymbol{\gamma} = [\mathbf{U_X}] = \mathbf{A_{x^s}} [\mathbf{U_{x^s}}] = -\gamma \begin{pmatrix} \cos\varphi^\gamma \cos\lambda^\gamma \\ \cos\varphi^\gamma \sin\lambda^\gamma \\ \sin\varphi^\gamma \end{pmatrix}, \quad (5.61)$$

where $\varphi^\gamma, \lambda^\gamma$, describing the direction of the normal gravity vector (in correspondence with (5.43)), as well as γ can be computed by applying (5.44) accordingly. This procedure is exact, involves no approximations, and works everywhere in space. In the case of the level ellipsoid, φ^γ, λ^γ, are identical with the normal latitude and longitude, φ^N, λ^N, with $\lambda = \lambda^N$ (the normal plumb line of the level ellipsoid is only slightly curved in the meridian plane).

The first and second derivatives of U can be expressed as well with respect to a local coordinate system oriented at the normal gravity vector. Corresponding to the local ellipsoidal system (see Sect. 2.1), a Cartesian system (x^γ, y^γ, z^γ) is introduced, where the z^γ-axis points in the opposite direction to that of γ, and the axes x^γ, y^γ are pointing north and east, respectively. The local system (x^γ, y^γ, z^γ) and the local ellipsoidal system (x^e, y^e, z^e) deviate by the curvature of the normal plumb line. In the local system (x^γ, y^γ, z^γ), the horizontal components (x^γ, y^γ) of the normal gravity vector are zero and the vertical component (z^γ) is equal to the negative value of γ. With the transformation matrix $\mathbf{A}_{\mathbf{x}^\gamma} = \mathbf{A}_{\mathbf{x}^\gamma}(\varphi^\gamma, \lambda^\gamma)$, defined analog to (5.7), the second derivatives of U with respect to the local system (x^γ, y^γ, z^γ) are given by (Tscherning 1976a)

$$[\mathbf{U}_{\mathbf{x}^\gamma\mathbf{x}^\gamma}] = (\mathbf{A}_{\mathbf{x}^\gamma})^{\mathrm{T}} \mathbf{A}_{\mathbf{x}^s} [\mathbf{U}_{\mathbf{x}^s\mathbf{x}^s}] (\mathbf{A}_{\mathbf{x}^s})^{\mathrm{T}} \mathbf{A}_{\mathbf{x}^\gamma} \,. \tag{5.62}$$

Equations (5.61) and (5.62) were programmed and tested for the level ellipsoid (with $n_{\max} = 10$); the agreement with the results from closed formulas on the basis of ellipsoidal harmonics (Heiskanen and Moritz 1967) was better than 10^{-5} m^2 s^{-2} for the potential U and 10^{-11} m s^{-2} in γ (everywhere in space; the results are based on 8 byte variables and the remaining differences are mainly due to rounding errors, etc.). Regarding the level ellipsoid, further testing is possible, as the components of the matrix (5.62) must fulfil the following conditions (Tscherning 1976a):

$$\left.\begin{aligned} &U_{x^\gamma y^\gamma} = U_{y^\gamma z^\gamma} = 0 \\ &U_{x^\gamma x^\gamma} + U_{y^\gamma y^\gamma} + U_{z^\gamma z^\gamma} - 2\omega_N^2 = 0 \end{aligned}\right\} \quad \text{everywhere in space,}$$

$$U_{x^\gamma x^\gamma} = -\frac{\gamma}{M}\,, \quad U_{y^\gamma y^\gamma} = -\frac{\gamma}{N} \Big\} \quad \text{at the ellipsoid surface.} \tag{5.63}$$

The first set of conditions results from the fact the plumb lines of the level ellipsoid are only curved in the meridian plane, the second condition is based on the generalized Laplace equation, and third, at the surface of the level ellipsoid, the second derivatives of U in the direction of x^γ, y^γ are associated with the principal radii of curvature of the ellipsoid (M: meridian; N: prime vertical).

Traditionally, near the Earth's surface, the normal gravity of the level ellipsoid is computed from a Taylor series expansion with respect to the ellipsoidal height:

$$\gamma = \gamma_0 + \left(\frac{\partial \gamma}{\partial h}\right)_0 h + \frac{1}{2}\left(\frac{\partial^2 \gamma}{\partial h^2}\right)_0 h^2 + \frac{1}{6}\left(\frac{\partial^3 \gamma}{\partial h^3}\right)_0 h^3 + \frac{1}{24}\left(\frac{\partial^4 \gamma}{\partial h^4}\right)_0 h^4 + \cdots \quad . \tag{5.64}$$

The normal gravity at the level ellipsoid is given by the (rigorous) formula of Somigliana (e.g., Torge 2001):

$$\gamma_0 = \frac{a_N \gamma_a \cos^2\varphi + b_N \gamma_b \sin^2\varphi}{\sqrt{a_N^2 \cos^2\varphi + b_N^2 \sin^2\varphi}} = \gamma_a \frac{1 + k \sin^2\varphi}{\sqrt{1 - e_N^2 \sin^2\varphi}} \quad \text{with } k = \frac{b_N \gamma_b}{a_N \gamma_a} - 1\,, \tag{5.65}$$

where a_N, b_N are the semimajor and semiminor axes of the ellipsoid, e_N is the first eccentricity, and γ_a, γ_b are the normal gravity values at the equator and pole, respectively. The computation of the partial derivatives in (5.64) was investigated in detail by Wenzel (1989), suggesting that the first derivative be computed from the Bruns equation and the higher derivatives be taken from a spherical harmonic expansion of U based on J_2^N only. This procedure can be slightly improved by also considering J_4^N in the second derivative term, yielding

$$\begin{aligned}
\left(\frac{\partial\gamma}{\partial h}\right)_0 &= -\gamma_0\left(\frac{1}{M}+\frac{1}{N}\right) - 2\omega_N^2 \\
&= -\frac{\gamma_0}{a_N\,(1-e_N^2)}(1-e_N^2\sin^2\varphi)^{1/2}(2-e_N^2(1+\sin^2\varphi)) - 2\omega_N^2, \\
\left(\frac{\partial^2\gamma}{\partial h^2}\right)_0 &= \frac{GM_N}{r^4}\left(6-60\left(\frac{a_N}{r}\right)^2 J_2^N P_2(\cos\theta) - 210\left(\frac{a_N}{r}\right)^4 J_4^N P_4(\cos\theta)\right), \\
\left(\frac{\partial^3\gamma}{\partial h^3}\right)_0 &= \frac{GM_N}{r^5}\left(-24+360\left(\frac{a_N}{r}\right)^2 J_2^N P_2(\cos\theta)\right), \\
\left(\frac{\partial^4\gamma}{\partial h^4}\right)_0 &= 120\frac{GM_N}{r^6}.
\end{aligned} \tag{5.66}$$

Equations (5.64)–(5.66) were also programmed, and the results were compared with those from closed formulas based on ellipsoidal harmonics; the differences were always below 2×10^{-9} m s^{-2} (0.2 μGal) for stations up to $h = 10$ km, with the fourth order term being insignificant (max. 3×10^{-10} m s$^{-2} = 0.03$ μGal). Furthermore, if the derivative terms in (5.64) are computed by expansions with respect to the flattening f_N (see Heiskanen and Moritz 1967), terms up to the second order of f_N, $o(f_N^2)$, are needed for the first vertical derivative of γ (or better apply the Bruns equation, which is rigorous), terms up to $o(f_N)$ should be used for the second derivative, while the third derivative can be based on a spherical approximation, yielding an accuracy of about 1×10^{-8} m s$^{-2} = 1$ μGal. On the other hand, formulas considering only terms up to $o(f_N)$ in the first vertical derivative of γ and a spherical approximation of the second derivative (as found frequently in textbooks, e.g., Heiskanen and Moritz 1967; Torge 2001) may result in errors with a magnitude of about 1×10^{-6} m s$^{-2} = 100$ μGal (again for heights up to 10 km), which is insufficient.

Finally, the Taylor series (5.64) opens a simple way to compute the mean normal gravity value

$$\overline{\gamma} = \frac{1}{h}\int_0^h \gamma\, dh = \gamma_0 + \left(\frac{\partial\gamma}{\partial h}\right)_0 \frac{h}{2} + \left(\frac{\partial^2\gamma}{\partial h^2}\right)_0 \frac{h^2}{6} + \left(\frac{\partial^3\gamma}{\partial h^3}\right)_0 \frac{h^3}{24} + \left(\frac{\partial^4\gamma}{\partial h^4}\right)_0 \frac{h^4}{120} + \ldots, \tag{5.67}$$

as needed in connection with normal heights.

2.6 Temporal Gravity Field Variations and the Atmosphere

The Earth's body and gravity field undergo changes at different temporal and spatial scales, which has been described in detail, e.g., by Torge (1989). The acquisition, analysis, description, and interpretation of such changes are treated in the field of geodynamics. The largest temporal variations are due to tidal effects (mainly moon and sun, but also planets), leading to (mostly) periodical deformations of the Earth's crust with maximum amplitudes of about 30 cm and corresponding gravity changes with amplitudes up to about 200 μGal (roughly 10^{-7} g); for details see Torge (1989, 2001) and Timmen (2010). Generally, gravimetric measurements are reduced for the effect of tides, atmospheric mass redistributions (mainly by simple admittance functions), and Earth rotation variations (Torge et al. 1987; Torge 1989; Timmen 2010). In general, the Earth's gravity field varies with time due to mass redistributions in the geosphere, atmosphere, hydrosphere (including the cryosphere), and biosphere; selected examples related to hydrology (ground water) and postglacial rebound (Fennoscandia) are presented in Timmen (2010). According to Torge (2001), gravity changes due to mass redistributions do not exceed the order of 10^{-9}–10^{-8} g.

Since 2002, the US–German GRACE satellite mission is providing the Earth's time-variable and static gravity field globally with unprecedented temporal and spatial resolution, which has greatly improved the understanding of mass redistributions in the atmosphere, oceans, water reservoirs, and cryosphere. An overview on the GRACE mission and early results is given in Tapley et al. (2004a, b), while reviews of recent GRACE results can be found in Wahr (2009) as well as Cazenave and Chen (2010). The GRACE results are mainly provided as monthly global spherical harmonic models, which can then be employed for studying periodic and secular variations of the Earth's gravity field; for this purpose, the GRACE models are usually restricted to a ground resolution of about 300–400 km, because the GRACE errors become larger with increasing resolution (e.g., Tapley et al. 2004a; Wahr 2009). The largest-amplitude signals are related to water storage variability on land (Wahr 2009); they have mainly annual periods with amplitudes up to about 10 mm and 10 μGal in terms of geoid and gravity, respectively. Figure 5.4 shows the linear (secular) trend of the geoid as derived from a sequence of 103 monthly GRACE solutions (Release 04) from GFZ (Helmholtz-Centre Potsdam—German Research Centre for Geosciences, GFZ); the GRACE data, covering the time span August 2002–September 2011, were smoothed by applying a Gaussian filter with a radius of 400 km. The largest geoid trends are related to the ice losses in Greenland (–2 mm/year), Alaska, and Antarctica, as well as water storage changes in the Amazon region, but also the post glacial rebound signals over North America (about +1 mm/year) and Fennoscandia (about 0.5 mm/year) are clearly visible.

In the following it is assumed that temporal variations of the Earth's body and gravity field have been taken into account by appropriate reductions or have been averaged out over sufficiently long time periods (this includes the station coordinates as realized, e.g., by the ITRF solutions). In this context, the largest

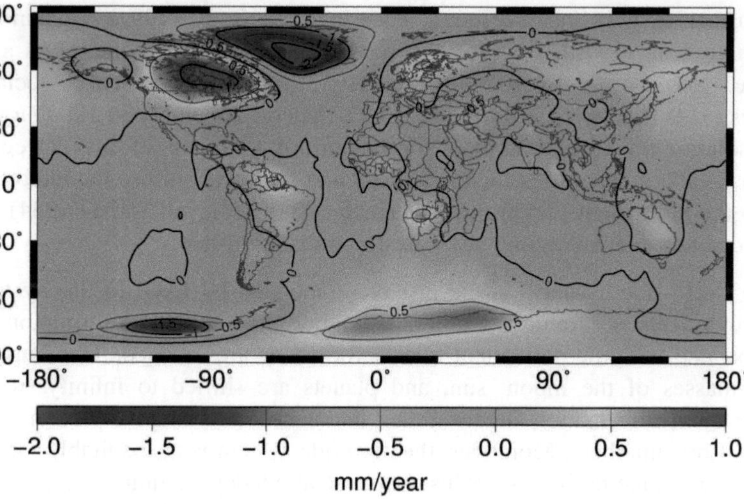

Fig. 5.4 Linear (secular) geoid trends derived from a sequence of 103 monthly GRACE solutions (Release 04) from GFZ, covering the time span August 2002 to September 2011, after Gaussian filtering with a radius of 400 km

variations are due to tidal effects, which can be reduced relatively easily with sufficient accuracy (e.g., Timmen 2010). On the other hand, effects like ground water changes or postglacial rebound are more difficult to handle (e.g., Timmen 2010). In general, the reduction of all relevant data sets to a certain epoch is an appropriate solution to reach a (quasi) static state; considering the postglacial rebound signal in Fennoscandia as an example, this can be done by means of models (e.g., the existing uplift model of Ågren and Svensson 2007) or observation time series (e.g., from GRACE).

The tidal attraction acts in a direct and indirect way. The direct (or gravitational) attraction deforms the elastic Earth, which causes an (additional) indirect change of the gravitational potential (deformation potential). While the direct effects can be computed easily by astronomical tidal theory, the calculation of the indirect effects requires knowledge about the Earth's elastic parameters (primarily Love and Shida numbers). Regarding the tidal reductions of geodetic parameters (e.g., potential, gravity, station coordinates, physical heights), both the direct and indirect effects contain time-dependent (periodic) parts as well as time-independent (permanent, zero frequency) parts; the computation of the latter portion of the indirect (deformation) effects requires the fluid (secular) Love numbers, which differ substantially from the standard (second degree) elastic values and are unobservable (cannot be determined experimentally). In this context, a long-lasting and still ongoing discussion relates to the handling of the notable permanent parts of the tidal potential (see below), e.g., documented by the publication of Ekman (1996) with the title "the permanent problem of the permanent tide." A comprehensive treatise of the permanent tide subject is given in Ekman (1989a, b), and further reviews can be found, e.g., in Rapp et al. (1991), Ekman (1996), Poutanen et al. (1996), Heck (2004), or Mäkinen and Ihde (2009). As a consequence, based on the report from Rapp (1983b), the International Association of Geodesy (IAG) adopted at the IAG/IUGG General Assembly in Hamburg, 1983, the resolution no. 16, stating that "for the uniform treatment of tidal corrections to various geodetic quantities such as gravity and station positions, the indirect effect due to the permanent yielding of the Earth be not removed" (IAG 1984).

Altogether, the following cases are to be distinguished:

- The "zero tide system" is the one recommended by IAG. In this system, the direct effects are removed completely, but the indirect deformation effects associated with the permanent tidal deformation are retained. This implies that the masses of the moon, sun, and planets are shifted to infinity, while the permanent deformation effects are left untouched, avoiding the problem with the fluid Love numbers. Moreover, the zero tide system is also suitable for solving boundary value problems (BVPs) in physical geodesy, requiring that no masses exist external to the boundary surface (harmonicity condition; see Sect. 3).
- In the "mean tide system," only the periodic tidal effects are removed, but the permanent parts (both direct and indirect) are retained. Thus, the mean values reflect the mean situation (shape) in the presence of the moon, sun, and planets, which is the natural system in connection with oceanography and satellite altimetry observations as well as station positions, noting that the mean and zero crust (station positions) are identical (both include the permanent deformation effects). On the other hand, the mean tide system has the disadvantage that it is not free of external masses (i.e., unsuitable for solving BVPs in physical geodesy); to overcome this problem, Zeman (1987) suggested modifying the normal gravity potential.
- The "tide-free system" (or non-tidal system) is aiming at the removal of all tidal effects (periodic and permanent direct and indirect effects). In this case, the required (unobservable) fluid Love numbers have to be replaced by conventional values; therefore, this system is also denoted sometimes as "conventionally tide-free."

Further discussions on advantages and disadvantages of each of the three concepts can be found in the references mentioned above. IAG has recommended the zero tide system, oceanographic applications may require the mean tide system, while the positioning domain (including the ITRF solutions; see IERS 2010) mainly uses the non-tidal system, thus not following the IAG recommendations.

Therefore, since different applications are usually associated with particular tidal systems, transformation formulas are needed for the conversion from one system to another. Considering only the dominating degree two terms of the tidal potential, each of the three tidal systems is directly associated with a corresponding $\overline{C}_{20}$ potential coefficient. Following IERS (2010), a quantity

$$\Delta\overline{C}_{20}^{\text{perm}} = A_0 H_0 k_{20} = -1.39141 \times 10^{-8} k_{20} \tag{5.68}$$

is introduced; then with the potential Love number k_{20}, the following relations hold:

$$\overline{C}_{20}^{nt} = \overline{C}_{20}^{zt} - \Delta\overline{C}_{20}^{\text{perm}}, \tag{5.69}$$

$$\overline{C}_{20}^{mt} = \overline{C}_{20}^{zt} + \Delta\overline{C}_{20}^{\text{perm}} / k_{20}, \tag{5.70}$$

where the superscripts *nt*, *zt*, and *mt* stand for non-tidal (tide-free), zero tide, and mean tide system, respectively. Starting with the zero tide coefficient $\overline{C}_{20}^{zt} = -484.16948 \times 10^{-6}$ (epoch J2000.0) and $k_{20} = 0.29525$ from IERS (2010), the above equations give the non-tidal and mean tide coefficients $\overline{C}_{20}^{nt} = -484.16537 \times 10^{-6}$ and $\overline{C}_{20}^{mt} = -484.18339 \times 10^{-6}$, respectively. Corresponding transformation formulas for geoid undulations can be obtained on the basis of the above coefficients and the spherical harmonic expansion (5.26) with $r = a$, $\varphi \approx 90° - \theta$, and a mean (normal) gravity value:

$$N_{[m]}^{nt} = N_{[m]}^{zt} + 0.0879_{[m]} \sin^2\varphi - 0.0293_{[m]}, \tag{5.71}$$

$$N_{[m]}^{mt} = N_{[m]}^{zt} - 0.2977_{[m]} \sin^2\varphi + 0.0992_{[m]}. \tag{5.72}$$

Further transformation formulas for other quantities such as gravity and heights can be found, e.g., in Ekman (1989a) or Mäkinen and Ihde (2009); the derivation of refined transformation formulas is described in Ihde et al. (2008).

To a first approximation, the atmospheric effect can be computed by using spherical approximations and a radially layered spherical density model, ignoring the topography. Then for a point *P* with radius *r* above the reference sphere (with radius *R*) the gravitational attraction can be split into an exterior and interior part, associated with the masses below and above the point *P*, respectively. Now it is well known from potential theory that the potential inside a spherical shell is constant, and thus the attraction is zero. Since the exterior gravitational field of concentric homogeneous mass shells is equal to the effect of the entire mass being concentrated at the center of mass of the object, the effect of the atmosphere on gravity is given by

$$g^A = \frac{G\,m(r)}{r^2}, \tag{5.73}$$

where $m(r)$ is the mass of all atmospheric layers below *P*. Introducing the total mass of the atmosphere, M_A, and $m(r) = M_A - M(r)$, yields

$$g^A = \frac{G M_A}{r^2} - \frac{G M(r)}{r^2} = \frac{G M_A}{r^2} - \delta g^A, \tag{5.74}$$

where $M(r)$ is the atmospheric mass above P. Accordingly, the gravitational potential of the atmosphere is given by

$$V^A = \frac{GM_A}{r} + G\int_r^{\infty} \frac{GM(r)}{r^2}\mathrm{d}r = \frac{GM_A}{r} - \delta V^A, \tag{5.75}$$

$$M(r) = 4\pi \int_r^{\infty} \rho_A(r)\, r^2 \mathrm{d}r\ . \tag{5.76}$$

The respective first terms on the right side of (5.74) and (5.75) represent the effect of a point mass (M_A); they are included in the normal potential (GM_N includes the mass of the solid Earth and atmosphere). On the other hand, the second terms on the right side represent the non-harmonic contributions; they are denoted as the atmospheric gravity and potential corrections δg^A and δV^A, respectively. The sign convention is here in accordance with IAG (1970), and Moritz (2000); the sign is defined such that the corrections have to be added to the observed quantities (i.e., the non-harmonic atmospheric contribution is reduced); then after gravity field modeling, the correction terms may be subtracted again from the final results to be consistent with what is being observed within the atmosphere. The correction terms can be tabulated easily based on an atmospheric model; examples are the values recommended by IAG (1970), which are based on ellipsoidal density models suggested by Ecker and Mittermayer (1969), or those computed by Wenzel (1985), using spherical approximations. The potential effect, δV^A, has a maximum value of only about 0.06 $\mathrm{m}^2/\mathrm{s}^2$ (at the reference sphere with radius R), and is commonly neglected. The atmospheric gravity corrections, δg^A, can be approximated by the following formula derived by Wenzel (1985):

$$\delta g^A_{[\mathrm{mGal}]} = 0.874 - 9.9 \times 10^{-5} h_{[\mathrm{m}]} + 3.5625 \times 10^{-9} h^2_{[\mathrm{m}]}\ . \tag{5.77}$$

The above formula is based on spherical theory and applicable for heights up to about 8 km; the results differ by not more than 0.005 mGal from the values recommended by IAG (1970), considering ellipsoidal density models. Both quantities, the atmospheric gravity and potential corrections δg^A and δV^A, depend on the masses above a given station P, and hence go to zero for large radii (elevations); therefore, they need not be considered at satellite altitude. Further details on atmospheric effects can be found in Ecker and Mittermayer (1969), Rummel and Rapp (1976), Christodoulidis (1979), Moritz (1980), as well as Andersen et al. (1975), Andersen (1976), and Sjöberg and Nahavandchi (2000), also considering topographic information. The accuracy of the atmospheric potential based on the simple spherical model without topography may be estimated as about 0.1 $\mathrm{m}^2/\mathrm{s}^2$ (Christodoulidis 1979; Denker 1988). In the future, improvements of the atmospheric correction scheme may be necessary (see also Forsberg 2010), as implemented already in connection with the GRACE gravity field mission (Flechtner et al. 2010) or absolute gravimetry (Gitlein and Timmen 2006).

3 Gravity Field Modeling

3.1 Geodetic Boundary Value Problems

In mathematics, within the field of differential equations, a boundary value problem (BVP) is a differential equation together with a set of additional constraints, called the boundary conditions, which the solution to the BVP must satisfy. BVPs arise in several branches of physics, engineering, etc., in connection with any differential equation (e.g., Morse and Feshbach 1953). To be useful in applications, a BVP should be well-posed, i.e., a unique solution should exist with respect to the given input. Much theoretical work in the field of partial differential equations is devoted to proving that boundary value problems arising from scientific and engineering applications are in fact well-posed.

Potential theory may be defined as the study of potential functions related to conservative vector fields (or potential fields). With regard to gravitation, solutions of the Laplace and Poisson differential equations are sought (divergence-free and divergence-involving problems). Sometimes, potential theory is also defined exclusively as the study of harmonic functions, i.e., the solution of Laplace's equation. In this context, a BVP consists of finding a harmonic function V in the space outside of the closed (star-shaped) boundary surface, which fulfils the boundary conditions and is regular at infinity. Commonly, three types of BVPs are distinguished (Sigl 1985; Jekeli 2009):

- BVP of the first kind, also known as the Dirichlet problem. Solve for the potential function V in the exterior space, given its values on the boundary surface.
- BVP of the second kind, also known as the Neumann problem. Solve for V in the exterior space, given its normal derivatives on the boundary surface (derivatives in the direction of the surface normal).
- BVP of the third kind, also known as the mixed BVP or Robin problem. Solve for V in the exterior space, given a linear combination of V and its normal derivative on the boundary surface.

In addition, the category of oblique derivative problems can be introduced, related to the cases where the derivatives of the potential function are not given in the direction of the boundary surface normal. Furthermore, interior and exterior problems can be distinguished, related to the space interior and exterior to the boundary. However, in geodesy, the exterior BVPs are of prime importance.

Geodetic boundary value problems (GBVPs) may be considered as the combined determination of the Earth's figure and gravity field from geodetic observations (at the Earth's surface or its exterior). Besides the traditional terrestrial geodetic observations, such as potential (differences), gravity, and astronomical latitudes and longitudes, new types of boundary data become available from satellite techniques, etc., involving new types of GBVPs (e.g., mixed and overdetermined GBVPs). A comprehensive overview on GBVPs is given in Sansò (1995) as well as Heck (1997), where the latter publication specifically highlights that the primary unknown, to be solved in the framework of the GBVPs, is the exterior gravity potential W outside the boundary surface. In this connection, W can only be computed indirectly from the boundary data by solving a GBVP, while a direct computation is not possible, mainly due to insufficient knowledge about the Earth's density structures. However, once the potential function W is known, all relevant quantities can be derived from it (see Sect. 2.3). The basic assumptions in the following are that the Earth behaves like a rigid and non-deformable body, uniformly rotating about a body-fixed axis (Moritz 1980; Heck 1997); i.e., all time-variable effects have to be taken into account by appropriate reductions in order to reach a quasi-static state (e.g., by referring all quantities to a given epoch; see Sect. 2.6).

Depending on the type of boundary data as well as the type and number of unknown functions to be solved from geodetic observational data, several GBVP formulations can be distinguished (Heck 1997). At first, a subdivision into fixed and free GBVPs is appropriate, involving the assumption of a known or unknown boundary surface, respectively. Hence, fixed GBVPs are always associated with a completely known boundary surface (e.g., fixed by coordinate vectors $\mathbf{X}$ derived from GNSS techniques), and therefore the only remaining unknown is the potential function W; this leads to the fixed gravimetric GBVP when employing gravity observations (provide the magnitude of the gravity vector) as the most important boundary data. On the other hand, regarding the free GBVPs, the information on the geometry of the boundary is either incomplete or missing entirely. When employing again gravity observations as boundary data, this results in Molodensky's boundary value problem (Molodenskii et al. 1962), i.e., the classical free gravimetric GBVP, which can be further subdivided into the vectorial free GBVP (astronomical variant of Molodensky's problem), where the position of the boundary is completely unknown (in total four unknowns; i.e., three coordinates in $\mathbf{X}$ and the potential W), and the scalar free GBVP (geodetic variant of Molodensky's problem), where the horizontal positions are known (e.g., gravity points with given ellipsoidal latitudes and longitudes), resulting in only two unknowns, one for the vertical coordinate (e.g., the ellipsoidal height) and a second one for the potential W (Heck 1997). The latter case can be considered as quite close to the hitherto applied geodetic practice, where, e.g., the horizontal coordinates of gravity stations were traditionally based on geodetic networks, mostly allowing a transformation to the Earth-fixed system with sufficient accuracy.

3.2 Linearization of the Boundary Conditions

The most important boundary data are gravity observations, carried out at (or near) the Earth's surface. Two essential cases can be distinguished with respect to the available station coordinates. The first case is related to the scalar-free GBVP and corresponds to the more traditional geodetic practice, where the vertical coordinates (physical heights) are mainly derived from geometric leveling, while the horizontal station coordinates (ellipsoidal latitudes and longitudes) are usually based on corresponding national horizontal control networks. In this context, it is assumed that the ellipsoidal coordinates are finally referred to the Earth-fixed system, either by adequate transformations from the national networks or directly by GNSS observations. Hence, in the scalar free GBVP, the horizontal coordinates are known, but the vertical spatial positions (ellipsoidal heights) of the observation sites (boundary) are unknown. The second case is related to the (scalar) fixed GBVP, where the geometry of the boundary is assumed to be completely known. This corresponds to the modern geodetic practice, where in many cases GNSS techniques are employed, giving directly the entire position vector $\mathbf{X}$ of all observation sites (boundary) with respect to the Earth-fixed system (either coordinates X, Y, Z or φ, λ, h). However, also with GNSS techniques, the vertical accuracy is never as good as the horizontal accuracy, and, in addition, some effort is required to get an accurate connection to the ITRF.

The boundary conditions for various geodetic boundary value problem (GBVP) formulations are in general nonlinear. This means that the relevant observations (boundary data) depend in a nonlinear way on the unknown gravity potential function W; they can be considered as nonlinear functionals of W, see (5.44). As no mathematical tools exist for solving nonlinear GBVPs (Heck 1997), the boundary conditions (observation equations) must be linearized. For this purpose, a known reference potential must be introduced, and, in addition, a known reference surface has to be adopted in the case of the free GBVPs. Regarding the reference potential, traditionally the level ellipsoid is used, but today one of the highly accurate satellite models from the recent satellite missions GRACE and GOCE can also be employed. The question of the reference surface is related to the definition of telluroid; for details on different telluroid mappings (Molodensky, Hirvonen, isozenithal, Marussi, and gravimetric telluroid) see Heck (1986). With respect to both the reference potential and the reference surface, it is important that the approximate values are sufficiently close to the real situation, such that a one-step solution is sufficient, or a convergent iteration process can be constructed (e.g., Rummel 1988; Heck 1997).

In the first instance, the normal gravity field of the level ellipsoid is employed for approximating the gravity potential. Moreover, it is supposed that at least the horizontal positions (ellipsoidal latitudes and longitudes) and in the case of fixed GBVPs also the ellipsoidal heights, referring to the Earth-fixed system, are known for the observation sites; this is considered as realistic with respect to today's geodetic practice. In addition, without going into detail about different telluroid mappings (Grafarend 1978a; Heck 1986), the reference surface or telluroid is defined according to Molodensky (Heck 1986, 1997):

$$\varphi_Q = \varphi_P \,, \quad \lambda_Q = \lambda_P \,, \quad U_Q = U_0 - C = U_0 - (W_0 - W_P) \,. \tag{5.78}$$

The above equations associate each point P at the Earth's surface with a corresponding telluroid point Q, serving then as a known linearization point. The first two conditions fix the horizontal position of the telluroid point Q, requiring that the surface point P and the telluroid point Q are located on the same ellipsoidal normal, while the third condition defines the vertical position of Q based on the (observable) geopotential number C and the reference potential U_O, which is usually identified with the constant potential of the surface of the level ellipsoid; in principle, a known value of W_O could also be employed, but this option is not pursued any further here. Regarding most of the existing geopotential numbers, these were historically referred to a fundamental datum point P_0^i with a corresponding local reference surface W_0^i (with the exact numerical value being typically unknown); in the following, no distinction is made between a local vertical datum (W_0^i) and the global case (with W_O related to the geoid; for further details see Sect. 2.4). Furthermore, the reference potentials (W_O and W_0^i) may be considered as additional unknowns in the solution of the GBVPs, which must be counterbalanced by additional observations (GNSS and leveling); for details see Rummel and Teunissen (1988) or Heck and Rummel (1990). Finally it is also worth mentioning that the above telluroid definition according to Molodensky and the definition from Hirvonen, where the point Q is put on the same normal plumb line as P, are practically equivalent (Heck 1986); especially for the vertical component, the difference between both telluroid definitions is completely negligible due to the very small curvature of the normal plumb lines.

If the normal potential U is associated with the level ellipsoid, then the ellipsoidal height of the telluroid point Q, defined according to (5.78), is virtually identical with the normal height H^N (see Fig. 5.5); this results from (5.53) with $C = U_0 - U_Q$ and the fact that the ellipsoidal height (h) and the normal height (H^N) of Q differ by less than 10^{-7} m for heights up to 10 km (see Sect. 2.4). In addition, the position anomaly vector, defined as the difference of the position vectors of P and Q, respectively, has zero horizontal components with respect to the local ellipsoidal system in Q, while the vertical component is the height anomaly ζ_P (see Fig. 5.5).

The disturbing (or anomalous) potential is defined for an arbitrary point P in space by

$$T_P = W_P - U_P \,. \tag{5.79}$$

Assuming that the centrifugal parts in W and U are identical, the disturbing potential T may also be expressed as, see (5.35), and (5.57),

$$T_P = V_P - V_P^N \,, \tag{5.80}$$

and hence T is harmonic outside the Earth's surface and regular at infinity, see (5.19) and (5.21):

$$\Delta T = 0, \qquad \lim_{l \to \infty} T = 0 \,. \tag{5.81}$$

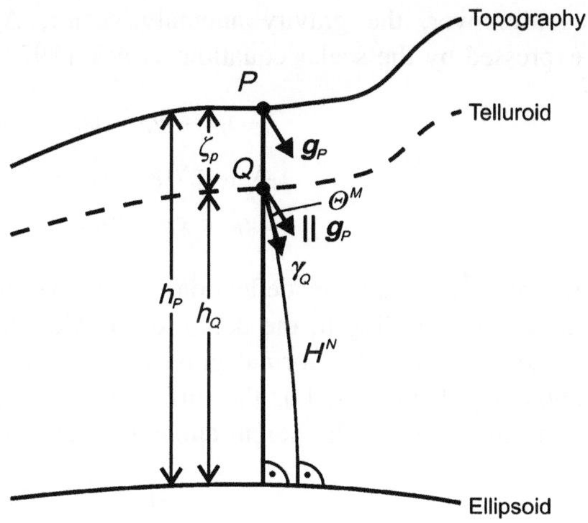

Fig. 5.5 Earth's surface, telluroid, ellipsoid, actual and normal gravity, deflection of the vertical according to Molodensky

With regard to the solution of the free GBVPs, the domain of harmonicity of T has to be extended down to the surface of the (known) telluroid, which causes a small problem from the theoretical side (Heck 1997).

Corresponding to the disturbing potential T, the gravity disturbance vector is defined as

$$\boldsymbol{\delta g}_P = \mathbf{g}_P - \boldsymbol{\gamma}_P = \text{grad } W_P - \text{grad } U_P = \text{grad } T_P \, , \tag{5.82}$$

while the scalar gravity disturbance is given by

$$\delta g_P = g_P - \gamma_P \, , \tag{5.83}$$

noting that the term "disturbance" is always used for one-point functions related to the same point in space (Grafarend 1978a; Heck 1997). Besides the observations in $\mathbf{g}_P$ (Φ, Λ, g), the computation of the gravity disturbance vector $\boldsymbol{\delta g}_P$ requires the spatial coordinates of P with respect to the Earth-fixed system (e.g., from GNSS) in order to be able to compute $\boldsymbol{\gamma}_P$.

Corresponding to (5.79) and (5.82) or (5.83), the potential anomaly

$$\Delta W_P = W_P - U_Q = T_P + (U_P - U_Q) \, , \tag{5.84}$$

and the gravity anomaly vector

$$\boldsymbol{\Delta g}_P = \mathbf{g}_P - \boldsymbol{\gamma}_Q = \boldsymbol{\delta g}_P + (\boldsymbol{\gamma}_P - \boldsymbol{\gamma}_Q) = \text{grad } T_P + (\text{grad } U_P - \text{grad } U_Q) \, , \tag{5.85}$$

are introduced. Thus the "anomalies" are two-point functions related to the surface point P and the telluroid point Q (Grafarend 1978a; Heck 1997). The computation of the normal gravity vector $\boldsymbol{\gamma}_Q$ requires the spatial position vector of Q, e.g., its ellipsoidal latitude, longitude, and height, the latter being virtually identical with the normal height H^N (see above and Sect. 2.4), derived from geometric leveling through the geopotential number C.

Moreover, the gravity anomaly vector, $\Delta \mathbf{g}_P$, according to (5.85), can be expressed by the scalar equations (Heck 1997)

$$\begin{aligned} \Delta\varphi_P^{\gamma} &= \Phi_P - \varphi_Q^{\gamma} = \xi_P^M , \\ \Delta\lambda_P^{\gamma} &= \Lambda_P - \lambda_Q^{\gamma} = \eta_P^M / \cos\varphi_Q^{\gamma} , \\ \Delta g_P &= g_P - \gamma_Q , \end{aligned} \tag{5.86}$$

where ξ_P^M, η_P^M are the deflections of the vertical in north–south and east–west direction according to the definition of Molodensky (angles between the gravity vector at P and the normal gravity vector at Q), and Δg_P is the scalar gravity anomaly, also denoted as the surface free-air gravity anomaly.

In this context, the height anomaly is also a two-point function, given by

$$\zeta_P = h_P - h_Q. \tag{5.87}$$

Considering now the Molodensky telluroid, the combination of (5.78) and (5.84) yields $\Delta W_P = W_0 - U_0$, while a Taylor development of the normal potential U around the telluroid point Q gives

$$U_P = U_Q + \left(\frac{\partial U}{\partial h}\right)_Q \zeta_P + \ldots = U_Q - \gamma_Q \zeta_P + \ldots \quad . \tag{5.88}$$

Finally, by inserting both expressions into (5.84), Bruns's formula is obtained as

$$\zeta_P = \frac{T_P}{\gamma_Q} - \frac{W_0 - U_0}{\gamma_Q}. \tag{5.89}$$

In the above formula, the second term is neglected in many cases, thus assuming that the condition $W_0 = U_0$ holds. Furthermore, (5.89) is based only on the first term of the series expansion (5.88), i.e., a linear approximation, neglecting the nonlinear terms. In addition, it is noted that within the framework of linear approximations, e.g., the quantities T, ζ, δg, Δg, etc., are considered as small of first order, while products of such terms are small, of second order, and thus negligible, leading to $T_Q \approx T_P$, $\mathrm{grad}T_Q \approx \mathrm{grad}T_P$, etc. (Moritz 1980). Consequently, if the point P is unknown, which applies to the free GBVPs, T_P cannot be evaluated and has to be replaced by T_Q in (5.89). However, if the position of the point P in space is known, e.g., associated with the fixed GBVPs, a virtually rigorous version of formula (5.89), neglecting only the slightly different directions of the ellipsoidal normal and the normal plumb line, can be derived by expressing $U_P - U_Q = -\bar{\gamma}_{QP}\zeta_P$, corresponding to (5.53), where $\bar{\gamma}_{QP}$ is the mean normal gravity value along the line from Q to P (e.g., Wenzel 1985), giving

$$\zeta_P = \frac{T_P}{\bar{\gamma}_{QP}} - \frac{W_0 - U_0}{\bar{\gamma}_{QP}} \quad \text{with} \quad \bar{\gamma}_{QP} = \frac{1}{h_P - h_Q}\int_{h_Q}^{h_P} \gamma \,\mathrm{d}h. \tag{5.90}$$

Given the point P in space, the use of (5.89) instead of (5.90) leads to maximum errors in the height anomalies of about a few millimeters (see also Wenzel 1985), and hence, as the mean normal gravity value $\bar{\gamma}_{QP}$ is easy to compute, in the simplest case as ½($\gamma_Q + \gamma_P$), (5.90) should be preferred (noting that it is not consistent with the concept of linear approximations explained above).

The linearization of the nonlinear boundary conditions (observation equations) related to the scalar fixed (δg_P), the vectorial free ($\mathbf{\Delta g}_P$), and the scalar free GBVP (Δg_P) is treated in detail in Heck (1997) and Seitz (1997), also investigating the nonlinear terms and the resulting linearization errors (the nonlinearities arise from the free boundary as well as the use of the norm operator to compute the vector lengths, e.g., (5.83) and (5.86)). A rigorous linearization of the boundary conditions is not intended here. Therefore, considering only linear approximations (see above) of (5.89), (5.83), and (5.86) results in

$$\zeta_P = \frac{T}{\gamma} - \frac{W_0 - U_0}{\gamma}, \tag{5.91}$$

$$\delta g_P = -\frac{\partial T}{\partial z^e} = -\frac{\partial T}{\partial h}, \tag{5.92}$$

$$\xi_P^M = -\frac{1}{\gamma}\frac{\partial T}{\partial x^e} = -\frac{1}{\gamma\,(M+h)}\frac{\partial T}{\partial \varphi}, \tag{5.93}$$

$$\eta_P^M = -\frac{1}{\gamma}\frac{\partial T}{\partial y^e} = -\frac{1}{\gamma\,(N+h)\cos\varphi}\frac{\partial T}{\partial \lambda}, \tag{5.94}$$

$$\Delta g_P = -\frac{\partial T}{\partial h} + \frac{1}{\gamma}\frac{\partial \gamma}{\partial h}T - \frac{1}{\gamma}\frac{\partial \gamma}{\partial h}(W_0 - U_0). \tag{5.95}$$

where the right sides of the above formulas have to be evaluated at the boundary surface, i.e., the telluroid in the case of the free GBVPs. The negative sign in the vertical deflection components follows from the sign conventions for the height anomalies and the vertical deflections (Torge 2001). For the derivation of (5.95), a Taylor series expansion of $\gamma_P - \gamma_Q$, analogous to (5.88), is used; the equation is also denoted as the fundamental equation of physical geodesy. Furthermore, the introduction of spherical approximations in the above formula system, i.e., the omission of terms of the order $o(f)$ with $f \approx 1/300$ (often called ellipsoidal effects), not affecting (5.91), gives

$$\delta g_P = -\frac{\partial T}{\partial r}, \tag{5.96}$$

$$\xi_P^M = -\frac{1}{\gamma r}\frac{\partial T}{\partial \varphi}, \tag{5.97}$$

$$\eta_P^M = -\frac{1}{\gamma r\cos\varphi}\frac{\partial T}{\partial \lambda}, \tag{5.98}$$

$$\Delta g_P = -\frac{\partial T}{\partial r} - \frac{2}{r}T + \frac{2}{r}(W_0 - U_0), \tag{5.99}$$

where $\partial\gamma/\partial h = \partial\gamma/\partial r = -2\gamma/r$ (spherical approximation) is utilized. Moreover, the subscripts P and Q are dropped on the right sides of (5.91)–(5.99), noting again that the linearized boundary conditions hold on the (known) boundary surface, which is the Earth's surface for the fixed GBVPs and the telluroid for the free GBVPs, respectively; hence, the linearization process is associated with a transformation of the free GBVPs into fixed ones, with the telluroid then serving as the (known) boundary. In this context, it has to be stressed that the boundary conditions in spherical approximation also still relate to the Earth's surface (fixed GBVPs) or the telluroid (free GBVPs); in other words, spherical approximations include only the omission of ellipsoidal terms, but do not imply that the boundary is replaced by a sphere (see also Moritz 1980 and Heck 1997). In addition, planar approximations, also neglecting terms of the order $o(h/R)$ associated with a mean Earth radius R, may be used for very local applications (e.g., Moritz 1980), but are not discussed here. Considering the linearized boundary conditions (5.91)–(5.99), it is clear that the derivatives of T are in general not normal to the (known) boundary surface (Earth's surface or telluroid, respectively), leading to the so-called oblique derivative BVPs (Heck 1997; Sideris 2011a). Furthermore, in the boundary condition for the gravity disturbance the radial derivative of T appears, while in the corresponding equation for the gravity anomaly a linear combination of the radial derivative of T and T itself occurs, similar to the second (Neumann) and third (Robin) BVP of potential theory, involving normal derivatives.

So far it has been assumed that the level ellipsoid is used for the linearization of the observation equations, implying that the height anomalies, i.e., the separation between the Earth's surface and the telluroid, following closely the Earth's surface, reach maximum values of about 100 m with an RMS (root-mean-square) value of roughly 30 m. Thus the use of spherical approximations, i.e., the omission of terms of the order $o(f)$ with $f \approx 1/300$, may in certain cases lead to significant errors at the milligal and decimeter level in the derived gravity and height anomalies, respectively (Heck 1997; Hipkin 2004). Obviously, the non-spherical and nonlinear terms cannot be neglected in precise gravity field modeling, and thus have to be considered by appropriate reductions (see, e.g., Heck 1991), especially in view of the present accuracy requirements for regional and global computations (e.g., the GRACE and GOCE satellite missions), aiming at accuracies at the millimeter to centimeter level for height anomalies.

On the other hand, the effect of the linearization and spherical approximation errors can be substantially reduced by introducing a higher degree reference field, e.g., a complete spherical harmonic model derived from the satellite missions GRACE (e.g., Mayer-Gürr et al. 2010; Kurtenbach et al. 2009) and GOCE (e.g., Pail et al. 2011) or the combined model EGM2008 (Pavlis et al. 2008), extended by a centrifugal component. In this context, it is important to stress that the satellite-only models are inherently unaffected by any spherical approximations, etc., while EGM2008 is derived on the basis of ellipsoidal harmonics and thus also hardly affected by such effects. In addition, it is essential that the gravity field parameters derived from the high-degree models are computed rigorously without any (spherical, etc.) approximations, as outlined in Sect. 2.3, such that the residuals with respect to the global model are virtually exact. Now, if the normal gravity field of the level ellipsoid is replaced by a global geopotential model complete to degree and order 100, 200, and 360, the residual height anomalies reduce to about 1.0, 0.4, and 0.2 m RMS, with corresponding maximum values of about 18, 11, and 4 m, respectively. Hence, a high-degree reference field leads to a much better approximation of the real situation, with the separation between the Earth's surface and the corresponding telluroid reducing by about one to two orders of magnitude, as compared to the linearization with respect to the level ellipsoid. Accordingly, the effect of linear and spherical approximations decreases in the modeling of the (rigorously derived) residual gravity field parameters, such that height anomalies may be deduced with accuracies at the centimeter to a few millimeters level; this is also supported by the numerical investigations in Heck (1997) and Seitz (1997). Further insight into this problem can be gained as well through closed-loop simulations with synthetic data (see, e.g., Wolf 2008).

The rigorous implementation of a high-degree geopotential model as a reference field in the linearization process must strictly follow the procedure described above for the level ellipsoid; the geopotential model is associated with a gravity potential

$$W^M = V^M + Z^M \,, \tag{5.100}$$

based on a spherical harmonic expansion of the gravitational part V^M as well as a centrifugal component Z^M. Correspondingly, the gravity vector is defined as

$$\mathbf{g}^M = \text{grad}\, W^M = \text{grad}\, V^M + \text{grad}\, Z^M \,. \tag{5.101}$$

The notation is chosen here in line with the real gravity potential of the Earth and its functionals, because the currently available high-degree geopotential models allow a quite good approximation of the real gravity field. Now, within the linearization process, W^M and $\mathbf{g}^M$ have to take the place of U and γ, respectively. Hence, the disturbance quantities related to a single point P in space with given coordinates can be computed directly, while the anomaly quantities require the definition of a telluroid associated with the geopotential model. In accordance with (5.78) it follows that

$$\varphi_{Q*} = \varphi_P \,, \quad \lambda_{Q*} = \lambda_P \,, \quad W^M_{Q*} = U_Q = U_0 - C = U_0 - (W_0 - W_P), \tag{5.102}$$

where $Q*$ is the telluroid point in conjunction with the geopotential model. The advantage of employing a high-degree reference field instead of the level ellipsoid is that the point $Q*$ is much closer to P than Q (see above as well as Fig. 5.6). Again it is assumed that the ellipsoidal latitude and longitude of P and thus $Q*$ are given, while the ellipsoidal height of P is unknown and that of $Q*$ can in principle be computed iteratively, starting with $h_{Q*\,(0)} = h_Q = H^N$ and $W^M_{Q*\,(0)}(\varphi_{Q*}, \lambda_{Q*}, h_{Q*\,(0)})$, while then the equation

$$h_{Q*\,(i)} = h_{Q*\,(i-1)} + \frac{1}{g^M_{Q*\,(i-1)}}\left(W^M_{Q*\,(i-1)}(\varphi_{Q*}, \lambda_{Q*}, h_{Q*\,(i-1)}) - W^M_{Q*}\right), \quad i = 1, \ldots, \infty\,, \tag{5.103}$$

has to be iterated until convergence, with the nominal potential value being $W^M_{Q*} = U_Q = U_0 - C$, which is deduced from leveling and a conventional U_0 value. The time-consuming part in this iteration process is the calculation of the potential values W^M, requiring the evaluation of the spherical harmonic expansion; however, the convergence of (5.103) is very good, e.g., regarding a geopotential model complete to degree and order 360, the maximum error of h_{Q*} was found to be less than 1×10^{-5} m after only two iterations ($i = 2$). Furthermore, it is noted that the geopotential model gravity value $g^M_{Q*\,(i-1)}$ in (5.103) can be replaced by practically any constant (normal) gravity value γ, as the term in parentheses on the right side of (5.103) goes to zero within the iteration process.

After all, it is convenient to express also W^M as the sum of the potential of the level ellipsoid U and a corresponding disturbing potential T^M. Considering (5.79), this leads directly to the disturbing potential, associated with the high-degree geopotential model

$$T^*_P = W_P - W^M_P = (U_P + T_P) - (U_P + T^M_P) = T_P - T^M_P\,, \tag{5.104}$$

and the corresponding height anomaly (of P with respect to $Q*$; see Fig. 5.6)

$$\zeta^*_P = h_P - h_{Q*} = (h_P - h_Q) - (h_{Q*} - h_Q) = \zeta_P - \zeta^M_{Q*}\,, \tag{5.105}$$

which, following the line of thought used to derive (5.89), can be expressed in linear approximation as

$$\zeta^*_P = \frac{T^*_P}{g^M_{Q*}} - \frac{W_0 - U_0}{g^M_{Q*}}\,, \tag{5.106}$$

and accordingly ζ^M_{Q*} is given by

$$\zeta^M_{Q*} = \frac{T^M_{Q*}}{\gamma_Q}\,, \tag{5.107}$$

noting that more rigorous versions of the above two equations can be obtained analogous to (5.90) by introducing corresponding mean gravity values $\bar{g}^M_{Q*P}$ and $\bar{\gamma}_{QQ*}$, respectively.

In the same way, the gravity disturbance and anomaly vectors with respect to the geopotential model are given by

Fig. 5.6 Telluroid associated with level ellipsoid and high-degree reference geopotential model

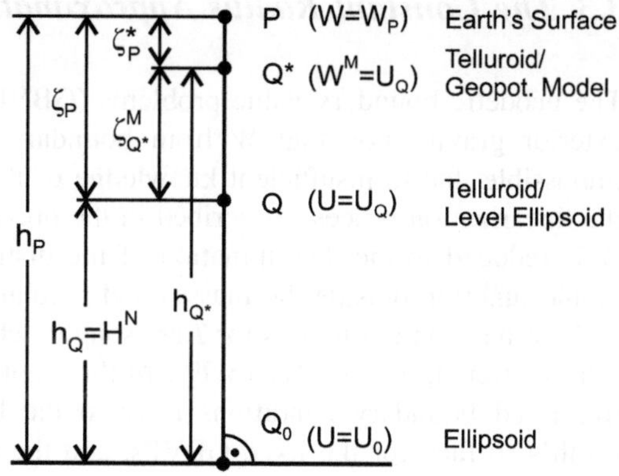

$$\boldsymbol{\delta g}_P^* = \mathbf{g}_P - \mathbf{g}_P^M = (\gamma_P + \boldsymbol{\delta g}_P) - (\gamma_P + \boldsymbol{\delta g}_P^M) = \boldsymbol{\delta g}_P - \boldsymbol{\delta g}_P^M \, , \tag{5.108}$$

$$\boldsymbol{\Delta g}_P^* = \mathbf{g}_P - \mathbf{g}_{Q*}^M = (\gamma_Q + \boldsymbol{\Delta g}_P) - (\gamma_Q + \boldsymbol{\Delta g}_{Q*}^M) = \boldsymbol{\Delta g}_P - \boldsymbol{\Delta g}_{Q*}^M \, . \tag{5.109}$$

Consequently, the rigorous linearization with respect to a high-degree geopotential model leads to residual quantities, which closely correspond to those used in the well-known remove-compute-restore (RCR) technique, in which topographic (or mass) information is additionally taken into account (see Sect. 3.9). In this context, it is pointed out again that the parameters from the geopotential model should be derived without (spherical, etc.) approximations at the appropriate positions in space, and it is also noted that the concepts introduced in Sect. 2.3, computing first the potential derivatives with respect to a local spherical system, followed by a transformation to the desired target system, can be applied as well for the anomalous gravity field quantities. Now for the case that only normal heights exist, the telluroid point $Q*$ can be computed with sufficient accuracy (see above) after only two iterations of (5.103); in practice this means that the geopotential model has to be evaluated only at the two heights $h_{Q*\,(0)} = h_Q = H^N$ and $h_{Q*\,(1)} = H^N + \zeta_{Q*\,(0)}$, yielding T_Q and $T_{Q*} \approx T_{Q*\,(1)}$ in Q and $Q*$, respectively. With regard to a geopotential model complete to degree and order 360, the maximum differences between T_Q and T_{Q*} reach about 0.25 $\mathrm{m^2s^{-2}}$, while $T_{Q*\,(1)}$ is accurate to better than 1×10^{-4} $\mathrm{m^2s^{-2}}$, corresponding to about 0.025 m and 1×10^{-5} m in terms of height anomalies. The maximum differences between T_Q and T_{Q*} are considered as significant, and hence the geopotential model should be evaluated at the appropriate positions in space. This was also pointed out by Tscherning (2004), mentioning that in this way "it is possible to come close to making no approximation at all."

3.3 The Constant Radius Approximation

The geodetic boundary value problems (GBVPs) aim at the determination of the exterior gravity potential W from boundary data, as a direct computation is impossible due to insufficient knowledge of the Earth's density structure. Within the linearization process, described in the previous section, the task of computing W is reduced to the determination of the disturbing potential T, which is a harmonic function outside the masses and regular at infinity, see (5.81).

Now harmonic solutions for T are sought, which satisfy the boundary conditions (observations), e.g., (5.91)–(5.99). In this context, it is pointed out again that the linearized boundary conditions refer to the known boundary surface (i.e., the Earth's surface for the fixed GBVPs, and the telluroid for the free GBVPs), and the derivatives of T, appearing in the boundary conditions, are generally not normal to the boundary surface (oblique derivative GBVPs).

However, even the linearized GBVPs based on spherical approximations, e.g., (5.96)–(5.99), do not permit rigorous analytical solutions in closed form as long as the boundary is not spherically shaped (Heck 1997). Therefore, Rummel (1988) and Heck (1997) discuss iterative solutions, where terms due to the Earth's flattening and topography of order $o(f)$ and $o(h/a)$ are involved, respectively. The omission of all non-spherical terms (i.e., assuming a spherical boundary) leads to the so-called constant radius approximations, which also contribute the dominating terms in iterative GBVP solution schemes.

In the following, several well-known formulas are derived on the basis of the constant radius approximation. In the first instance, as the disturbing potential T is a harmonic function, it may be expanded in spherical harmonics based on (5.80), (5.26), and (5.59), resulting in

$$T(\theta,\lambda,r) = \sum_{n=0}^{\infty}\left(\frac{a}{r}\right)^{n+1}\sum_{m=-n}^{n}\overline{T}_{nm}\overline{Y}_{nm}(\theta,\lambda) = \sum_{n=0}^{\infty}\left(\frac{a}{r}\right)^{n+1}T_n(\theta,\lambda,a), \qquad (5.110)$$

with

$$\overline{T}_{nm} = \frac{GM}{a}\left(\begin{Bmatrix}\overline{C}_{nm}\\ \overline{S}_{nm}\end{Bmatrix} - \frac{GM_N}{GM}\left(\frac{a_N}{a}\right)^n\begin{Bmatrix}\overline{C}^N_{nm}\\ \overline{S}^N_{nm}\end{Bmatrix}\right) \quad \text{for} \quad \begin{Bmatrix}m\geq 0\\ m<0\end{Bmatrix}, \qquad (5.111)$$

where the different GM and a values in V and V^N lead to a rescaling of the coefficients of the normal potential, noting that (5.110) and (5.111) may also be expressed with respect to a_N and GM_N, whatever is more convenient. The summation in the above equation starts at degree $n = 0$ to account for possible differences in the GM and GM_N quantities related to V and V^N, respectively. Another option would be to compute T directly as the difference of V and V^N, requiring no rescaling of any coefficients. Furthermore, T_n (θ, λ, a) are the Laplace surface harmonics of degree n, referring to the radius $r = a$.

Assuming now that T is given on a sphere with radius $r = a$, (5.110) can be inverted easily by applying the orthogonality relations (5.28), yielding

$$\overline{T}_{nm} = \frac{1}{4\pi} \iint_{\sigma} T(\theta', \lambda', a)\, \overline{Y}_{nm}(\theta, \lambda)\, d\sigma \,. \tag{5.112}$$

Utilizing this result as well as the decomposition formula for the Legendre polynomials gives directly

$$T_n(\theta, \lambda, a) = \frac{2n+1}{4\pi} \iint_{\sigma} T(\theta', \lambda', a) P_n(\cos\psi)\, d\sigma \,, \tag{5.113}$$

where ψ is the spherical distance between the two points $P\,(\theta, \lambda, a)$ and $P'\,(\theta', \lambda', a)$ with

$$\cos\psi = \cos\theta \cos\theta' + \sin\theta \sin\theta' \cos(\lambda - \lambda') \,. \tag{5.114}$$

Correspondingly, the gravity disturbance may be expressed in spherical approximation according to (5.96) as

$$\delta g(\theta, \lambda, r) = -\frac{\partial T}{\partial r} = \sum_{n=0}^{\infty} \left(\frac{a}{r}\right)^{n+2} \frac{n+1}{a} T_n(\theta, \lambda, a) = \sum_{n=0}^{\infty} \left(\frac{a}{r}\right)^{n+2} \delta g_n(\theta, \lambda, a) \,. \tag{5.115}$$

The above approach can be applied to other gravity field parameters as well and leads to the so-called Meissl scheme, which in its extended form, based on tensor spherical harmonics, is also applicable to the first and second horizontal derivatives of T (Rummel and van Gelderen 1995; Rummel 1997). Thus in spherical approximation, the following simple spectral relations (eigenvalue expressions) hold:

$$\delta g_n(\theta, \lambda, a) = \frac{n+1}{a} T_n(\theta, \lambda, a), \tag{5.116}$$

$$\Delta g_n(\theta, \lambda, a) = \begin{cases} -\frac{1}{a} T_0(\theta, \lambda, a) + \frac{2}{a}(W_0 - U_0) & \text{for } n = 0 \\ \frac{n-1}{a} T_n(\theta, \lambda, a) & \text{for } n > 0 \end{cases}, \tag{5.117}$$

$$T_n(\theta, \lambda, r) = \left(\frac{a}{r}\right)^{n+1} T_n(\theta, \lambda, a), \tag{5.118}$$

$$\delta g_n(\theta, \lambda, r) = \left(\frac{a}{r}\right)^{n+2} \delta g_n(\theta, \lambda, a), \tag{5.119}$$

$$\Delta g_n(\theta,\lambda,r) = \left(\frac{a}{r}\right)^{n+2} \Delta g_n(\theta,\lambda,a). \tag{5.120}$$

Basically, the spherical harmonic expansion (5.110) together with (5.112) represents a solution to the Dirichlet BVP for a spherical boundary. Now, the spherical harmonic series has its advantages mainly in global analyses and the spectral interpretation, while integral formulas (Green's function representations) are better suited for local applications; this is easily recognized by the fact that a change in one boundary value leads to changes in all spherical harmonic coefficients, while a change in a remote boundary value affects the local field computed by integral formulas only marginally, as the integration kernels essentially depend on the inverse distance or higher powers thereof (Jekeli 2009).

Combining (5.110) and (5.113) and assuming that the disturbing potential T is given on a sphere with radius $r = R$ (instead of $r = a$ used above) leads to the well-known Poisson integral (Heiskanen and Moritz 1967; Jekeli 2009)

$$T(\theta,\lambda,r) = \frac{1}{4\pi} \iint_\sigma T(\theta',\lambda',R)\, U(\psi,r)\, \mathrm{d}\sigma \tag{5.121}$$

with the Poisson kernel

$$\begin{aligned} U(\psi,r) &= \sum_{n=0}^{\infty} \left(\frac{R}{r}\right)^{n+1} (2n+1)\, P_n(\cos\psi) \\ &= \frac{R(r^2 - R^2)}{l^3} \end{aligned} \tag{5.122}$$

and the spatial distance

$$l = \sqrt{r^2 + R^2 - 2rR\cos\psi}\ . \tag{5.123}$$

The above equations allow the upward continuation of any harmonic function from a sphere with radius R to another sphere with radius $r > R$, e.g., they can be directly applied to the (spatial) function $r\,\Delta g$, which is harmonic, while Δg itself is not a harmonic function (for details see Heiskanen and Moritz 1967).

Furthermore, utilizing (5.113) correspondingly for the derivation of the gravity disturbance surface harmonics and inserting this result into (5.116) and (5.110) yields the Hotine integral (Hotine 1969; Heck 1997)

$$T(\theta,\lambda,r) = \frac{R}{4\pi} \iint_\sigma \delta g\,(\theta',\lambda',R)\, H(\psi,r)\, \mathrm{d}\sigma \tag{5.124}$$

with the Hotine kernel

$$\begin{aligned} H(\psi, r) &= \sum_{n=0}^{\infty} \left(\frac{R}{r}\right)^{n+1} \frac{2n+1}{n+1} P_n(\cos\psi) \\ &= \frac{2R}{l} - \ln\frac{l+R-r\cos\psi}{r(1-\cos\psi)} . \end{aligned} \tag{5.125}$$

It should be noted that the summation in (5.122) and (5.125) starts at degree $n = 0$, and thus the complete spectrum including the degrees zero and one can be determined from the boundary data. However, with regard to gravity anomalies, the situation is different, as (5.117) has to be employed accordingly, where the factor $(n - 1)$ appears; i.e., the gravity anomalies have no first degree harmonics and hence the corresponding disturbing potential harmonics cannot be determined from gravity anomalies. This leads to a non-unique solution of the BVP, and therefore additional constraints are necessary to achieve a unique solution for T. The first-degree harmonic coefficients are proportional to the center-of-mass coordinates and can be enforced to zero with an appropriate definition of the coordinate system (for further details see Rummel 1995 and Heck 1997). Corresponding to (5.124) and (5.125), the complete solution for the disturbing potential based on gravity anomalies is given by

$$T(\theta, \lambda, r) = \sum_{n=0}^{1} \left(\frac{R}{r}\right)^{n+1} T_n(\theta, \lambda, R) + \frac{R}{4\pi} \iint_{\sigma} \Delta g\,(\theta', \lambda', R)\, S(\psi, r)\, \mathrm{d}\sigma, \tag{5.126}$$

with the (extended) Stokes kernel

$$\begin{aligned} S(\psi, r) &= \sum_{n=2}^{\infty} \left(\frac{R}{r}\right)^{n+1} \frac{2n+1}{n-1} P_n(\cos\psi) \\ &= \frac{2R}{l} + \frac{R}{r} - 3\frac{Rl}{r^2} - \frac{R^2}{r^2}\cos\psi \left(5 + 3\ln\frac{l+r-R\cos\psi}{2r}\right). \end{aligned} \tag{5.127}$$

In the above equations, the zero-degree and first-degree terms of T are handled separately, because the Stokes integral, the right term in (5.126), conventionally excludes these components (e.g., Heiskanen and Moritz 1967; Jekeli 2009); this is also evident from the Stokes kernel (5.127), where the summation only starts at degree $n = 2$. Furthermore, according to (5.110) and (5.111), the zero-degree term of T in (5.126) can be expressed as

$$T_0(\theta, \lambda, R) = \frac{GM - GM_N}{R}, \tag{5.128}$$

and the corresponding zero-degree term of ζ follows from Bruns's formula as

$$\zeta_0(\theta, \lambda, R) = \frac{1}{\gamma_0} T_0(\theta, \lambda, R) - \frac{1}{\gamma_0}(W_0 - U_0). \tag{5.129}$$

In addition, it is noted that (5.124) and (5.125) solve the Neumann BVP, while (5.126)–(5.128) solve the Robin BVP, provided the boundary is a sphere. Furthermore, all integral kernels (Green's functions) introduced above, have singularities at $\psi = 0$, when the computation point is on the sphere with radius R. Therefore, the inner zone contributions have to be evaluated separately by expanding the boundary data in a Taylor series and performing the integration term by term (Heiskanen and Moritz 1967; Bian 1997; Torge 2001).

For the Stokes integral (with $r = R$), the inner zone contribution can be computed by approximating $S(\psi) = S(\psi, R) = 2/\psi = 2\,R/s$, with s being the planar distance (the first term in (5.127) dominates for $\psi \to 0$); then integration over a spherical cap with radius s_0 gives the inner zone contribution

$$T_i = \Delta g_P\, s_0 + \ldots, \tag{5.130}$$

while an integration over a rectangular area (see also Haagmans et al. 1993) results in

$$\begin{aligned} T_i &= \Delta g_P \frac{1}{2\pi} \int_{-\Delta x/2}^{+\Delta x/2} \int_{-\Delta y/2}^{+\Delta y/2} \frac{1}{\sqrt{x^2 + y^2}} \mathrm{d}x\, \mathrm{d}y + \ldots \\ &= \Delta g_P \frac{1}{2\pi} \left(\Delta x \ln \frac{d + \Delta y}{d - \Delta y} + \Delta y \ln \frac{d + \Delta x}{d - \Delta x} \right) \quad \text{with} \quad d = \sqrt{\Delta x^2 + \Delta y^2}, \end{aligned} \tag{5.131}$$

where Δx, Δy are the side lengths (in meters) of the rectangle in x (north–south) and y (east–west) direction, respectively. If the area sizes of the spherical cap and the rectangle are chosen to be identical ($\pi s_0^2 = \Delta x \Delta y$), (5.131) always gives slightly larger inner zone contributions than (5.130); the differences are at the few percent level, depending on the ratio of $\Delta x/\Delta y$ (for further details see Haagmans et al. 1993).

Finally, based on the spherical and constant radius approximations, virtually any gravity field quantity can be obtained in space from any other quantity on a spherical boundary (Jekeli 2009). Further examples are the Vening-Meinesz integral formulas, which allow the computation of deflections of the vertical from gravity anomalies by applying (5.97) and (5.98) to the Stokes integral, as well as the inverse Stokes, Hotine, and Vening-Meinesz integral formulas, which are all presented, e.g., in Jekeli (2009).

3.4 Solutions to Molodensky's Boundary Value Problem

In the previous section, different GBVP solutions are derived based on the spherical and constant radius approximation, assuming that the boundary is a sphere. Now, more complicated boundaries require additional corrections or iterative solution schemes. The scalar free GBVP, formulated first by Molodensky (e.g., Molodenskii et al. 1962), is based on gravity observations at the Earth's surface. Within the linearization process, the scalar free GBVP is transformed into a fixed one by approximating the Earth's surface by means of the telluroid, serving then as the (known) boundary surface, to which the boundary conditions as well as the boundary data (i.e., the gravity anomalies) refer (see Sect. 3.2). Furthermore, the scalar free GBVP in linear and spherical approximation is often called the simple Molodensky problem (Moritz 1980; Heck 1997).

Molodensky's problem can be solved in various ways; detailed derivations can be found in Moritz (1980). An efficient solution, avoiding integral equations, is provided by the method of analytical continuation (Moritz 1980; Sideris 1987, 2011a). In this method, the gravity anomalies Δg (at any point on the telluroid) are reduced to the normal level surface passing through the given computation point P (point level); the resulting anomalies are denoted as $\Delta g'$ and can be obtained by

$$\Delta g' = \sum_{n=0}^{\infty} g_n\,, \quad g_n = -\sum_{j=1}^{n} (H^N - H_P^N)^j L_j(g_{n-j})\,, \quad \text{starting with} \quad g_0 = \Delta g\,. \tag{5.132}$$

Thus the g_n terms are evaluated recursively based on the vertical derivative operator L_j with

$$L_j = \frac{1}{j!}\frac{\partial^j}{\partial h^j} = \frac{1}{j}\frac{1}{(j-1)!}\frac{\partial^{j-1}}{\partial h^{j-1}}\frac{\partial}{\partial h} = \frac{1}{j}L_{j-1}L_1 = \frac{1}{j!}L^j = \frac{1}{j!}LLL\ldots L \quad (j \text{ times})\,. \tag{5.133}$$

In the above equation, the abbreviation $L = L_1$ is introduced, and in planar approximation ($R \to \infty$) L becomes the surface operator

$$L(f) = L_1(f) = \frac{R^2}{2\pi}\iint_\sigma \frac{f - f_P}{l_0^3}\,\mathrm{d}\sigma\,, \quad l_0 = 2R\sin\frac{\psi}{2}\,. \tag{5.134}$$

Now, the analytically continued gravity anomalies $\Delta g'$ refer to the normal level surface passing through the computation point P, and hence the disturbing potential T at P can be obtained by applying Stokes's integral operator $\mathbf{S}$ (right term of (5.126)) to $\Delta g'$, yielding

$$T = \mathbf{S}(\Delta g') = \mathbf{S}(\Delta g) + \sum_{n=1}^{\infty} \mathbf{S}(g_n) = \sum_{n=0}^{\infty} T_n\,. \tag{5.135}$$

Thus, the main contribution to the Molodensky solution is provided by the Stokes term, while the further so-called Molodensky terms consider that the data are not given on a level surface. Moreover, the zero-degree and first-degree terms of T are omitted in the above equation, but the notes given for the Stokes integral apply to the Molodensky solution as well (see previous section). Finally, it should be noted that the Molodensky terms T_n and the Laplace surface harmonics of T have nothing in common.

Regarding the above computation procedure, it is important to realize that the g_n terms depend on the computation point P, which is rather impractical, as a new set of g_n values has to be computed for every new computation point. This difficulty can be overcome, e.g., by performing the analytical continuation in two steps, where the surface gravity anomalies are first analytically continued to zero level (giving $\Delta g^{0'}$), then the Stokes operator is applied to compute the harmonically continued disturbing potential $T^{0'}$, and finally $T^{0'}$ is analytically continued back to the telluroid (Sideris 1987; Forsberg and Tscherning 1997), yielding

$$T = T^{0'} + \sum_{n=1}^{\infty} (H^N)^n L_n(T^{0'}) \,. \tag{5.136}$$

The two-step formulas are much better suited for numerical computations and allow the application of Fast Fourier Transform (FFT) techniques (Sideris 1987).

Furthermore, the first order solution of (5.135) is known as the gradient solution (Moritz 1980), which can be expressed as

$$T \approx \mathbf{S}(\Delta g + g_1) = \mathbf{S}\left(\Delta g - \frac{\partial \Delta g}{\partial h}(H^N - H_P^N)\right) . \tag{5.137}$$

Regarding the gradient solution, it is also worth mentioning that

$$\mathbf{S}(g_1) \approx \mathbf{S}(c) \tag{5.138}$$

holds in linear approximation for gravity anomalies Δg linearly dependent on the elevations, where c is the classical terrain correction (Moritz 1980). Besides this, the gravity anomalies $\Delta g_{\mathrm{Faye}} = \Delta g + c$ are denoted as Faye anomalies, giving

$$T \approx \mathbf{S}(\Delta g + c) = \mathbf{S}(\Delta g_{Faye}). \tag{5.139}$$

For the two-step procedure based on (5.135) and (5.136), the gradient solution leads to

$$T \approx \mathbf{S}\left(\Delta g - \frac{\partial \Delta g}{\partial h} H^N\right) + \frac{\partial T^{0'}}{\partial h} H^N. \tag{5.140}$$

Considering the relations (5.92), (5.91), and (5.95), yielding

$$\frac{\partial T}{\partial h} = -\delta g \,, \qquad \frac{\partial \zeta}{\partial h} = -\frac{1}{\gamma}\Delta g, \tag{5.141}$$

the gradient solution results in a very simple scheme for the computation of the height anomalies (Forsberg and Tscherning 1997):

1. Predict vertical gradient $\partial \Delta g/\partial h \approx -T_{zz}$ from Δg.
2. Continue Δg downward, giving $\Delta g^{0'} \approx \Delta g - (\partial \Delta g/\partial h)\, H^N$.

3. Apply Stokes operator, yielding $\zeta^{0'} = \mathbf{S}(\Delta g^{0'})$.
4. Continue the height anomalies upward to obtain $\zeta = \zeta^{0'} - (\Delta g^{0'}/\gamma)\, H^N$.

The above procedure is preferable over (5.139), as it contains fewer approximations, avoiding the assumption that the gravity anomalies are linearly dependent on the elevations.

Finally, it is noted that $\zeta^{0'}$, ζ, and the geoid undulation N (as well as the corresponding disturbing potentials $T^{0'}$, T, and T^0) are fundamentally different quantities. Initially, $T^{0'}$ is a disturbing potential obtained by harmonic continuation (to zero level), which is identical to T outside the Earth's surface; e.g., this is also the quantity obtained from a global geopotential model or least-squares collocation applied spatially. On the other hand, T^0 relates to the geoid (see next section), which is in general located inside the topography on the continents and hence not an equipotential surface of a harmonic function (because T^0 is not harmonic inside the topography). The differences between ζ and $\zeta^{0'}$ depend on the free-air gravity anomaly, see (5.141), while the differences between ζ and N depend on the Bouguer gravity anomaly, see (5.56). In typical mountainous areas with, e.g., 1 km changes in elevation, the differences between $\zeta^{0'}$, ζ, and N are at the level of 10 cm (Forsberg and Tscherning 1997).

3.5 Solutions to Stokes's Boundary Value Problem

Stokes's classical geodetic boundary value problem (GBVP) aims at the determination of the geoid from gravimetric data. The calculation of the geoid within the GBVP framework has two important consequences: first, the gravity values must refer to the geoid, which initially serves as the boundary surface, and second, there must be no masses outside the geoid. The latter requirement results from the boundary value problem approach of potential theory, which always involves harmonic functions satisfying the Laplace equation. Consequently, since no masses are allowed outside the geoid, the topography of the Earth must be eliminated mathematically by appropriate reductions.

Stokes's as well as Molodensky's GBVP may both be considered as scalar free problems (see above), the main difference being that Molodensky's problem is based on gravity data at the Earth's surface, while Stokes's problem involves gravity data at the geoid. Accordingly, Stokes problem is in principle easier to solve than Molodensky's problem, because the initial boundary surface, i.e., the geoid, is a level surface with the gravity vectors perpendicular to it; this corresponds to a BVP of the third kind (Robin problem) of potential theory, see boundary condition (5.95) or (5.99). However, the main drawback of Stokes's problem is that complicated topographic reductions are necessary.

For the solution of Stokes's GBVP, the geoid is approximated by an ellipsoid, and then after spherical and constant radius approximation the general solution is given by (5.126) with $r = R$, the right term being the Stokes integral (for further details see above). Now, in order to get the boundary values (gravity anomalies) at the geoid, the external masses outside the geoid must be either removed completely or moved inside the geoid. This so-called regularization obviously also changes the shape of the level surfaces and hence the geoid, leading to the cogeoid. In this context, it is advantageous to preserve the total mass of the Earth. Therefore, usually Helmert's second condensation reduction is applied, where the masses above the geoid are condensed onto a layer on the geoid (e.g., Sideris 1994, 2011a).

Accordingly, the entire procedure for the computation of the geoid may be described as follows (Heiskanen and Moritz 1967; Sideris 2011a):

1. The masses above the geoid are removed computationally, i.e., the attraction effect (A_P) is subtracted from the gravity value at the surface point P.
2. The gravity station is lowered from P to P_0 on the geoid using the free-air reduction (F) or harmonic downward continuation.
3. The topographic masses are condensed on the geoid, and the attraction effect at P_0 is restored ($A^c_{P_0}$).
4. The indirect effect (for details see Wichiencharoen 1982) on the potential (δT_{ind}) due to the shifting of the topographic masses is computed at P_0.
5. The indirect effect on gravity (δg_{ind}), which reduces gravity from the geoid to the cogeoid, is taken into account, finally yielding the gravity anomalies on the cogeoid as $\Delta g^c = g_P - A_P + F + A^c_{P_0} + \delta g_{\text{ind}} - \gamma_0$.
6. The disturbing potential for the cogeoid (T^{0c}) is computed by applying Stokes's operator to Δg^c.
7. The disturbing potential for the geoid (T^0) is computed by adding the indirect effect to the Stokes contribution, yielding

$$T^0 = \mathbf{S}(\Delta g^c) + \delta T_{\text{ind}} \ . \tag{5.142}$$

8. Finally, the geoid undulation (omitting the zero-degree term) is obtained from Bruns's formula as

$$N = \frac{T^0}{\gamma_0} = \frac{1}{\gamma_0}\mathbf{S}(\Delta g^c) + \frac{1}{\gamma_0}\delta T_{\text{ind}} \ . \tag{5.143}$$

In addition, with the quantity $\delta A = A^c_{P_0} - A_P \approx c$ and $\delta g_{\text{ind}} \approx 0$, the gravity anomaly at the cogeoid becomes $\Delta g^c \approx g_P + F - \gamma_0 + c \approx \Delta g_P + c = \Delta g_{\text{Faye}}$ (Forsberg and Tscherning 1997). Thus, the Faye anomalies play a role in the first order solutions of Stokes's as well as Molodensky's problem, see (5.139); further relationships and discussions on this matter can be found in Sideris (1994) and Forsberg and Tscherning (1997). Practical examples of large-scale geoid computations are the models derived for the United States (e.g., Smith and Roman 2001; Wang et al. 2011) and Canada (Véronneau and Huang 2007).

In conclusion, the computation of the geoid is in the details quite complicated and requires several approximations as well as assumptions about the density of the topographic masses above the geoid, which is the classical dilemma in determining the geoid as well as the orthometric heights. Furthermore, it is not an easy task to ensure that the orthometric heights and the corresponding geoid undulations are consistent, such that the equation $h = H + N$ is satisfied. This is also pointed out clearly by Forsberg and Tscherning (1997), mentioning that "if refined expansions are used for the downward continuation, consistency is lost with the conventional Helmert orthometric heights"; therefore, they suggest one works with height anomalies (or the disturbing potential) related to the Earth's surface as far as possible, and only at the end of the computation chain to shift back to the geoid if necessary, e.g., by using (5.56). In addition, this proposal is in accordance with the strategy applied for the modeling of the quasigeoid and geoid in Europe, performed at the IfE, LUH (e.g., Denker and Torge 1998; Denker et al. 2009).

3.6 The Spectral Combination Technique

The spectral combination technique encompasses all procedures to combine heterogeneous data by spectral weights (depending on spherical harmonic degree n); it was initially developed to combine terrestrial gravity data and a global geopotential model in an optimal way for the purpose of calculating the geoid or quasigeoid. The spectral combination approach is based on the Laplace surface harmonics derived from different data sets, which are then combined by employing spectral weights. The method was promoted mainly by Sjöberg (1980, 1981, 2003) and Wenzel (1981, 1982), with the basic idea of the procedure already being outlined in Moritz (1976).

The development of the basic formulas is based on the spherical harmonic expansion of the disturbing potential (5.110) as well as the spectral relations given in (5.116)–(5.120). The following derivations are consistently based on the spherical and constant radius approximation (see also previous sections), assuming that the observations are given on a sphere with radius $r = R$. The first data set to be considered is the global geopotential model, giving the disturbing potential Laplace surface harmonics referring to $r = R$ in the form

$$T_n^M(\theta, \lambda, R) = \left(\frac{a}{R}\right)^{n+1} T_n^M(\theta, \lambda, a) = \left(\frac{a}{R}\right)^{n+1} \sum_{m=-n}^{+n} \overline{T}_{nm}^M \, \overline{Y}_{nm}(\theta, \lambda) \,, \tag{5.144}$$

which can be evaluated by means of the given coefficients $\overline{T}_{nm}^M$.

On the other hand, corresponding surface harmonics can also be deduced from gravity anomalies in the form

$$T_n^G(\theta, \lambda, R) = \frac{R}{n-1} \Delta g_n^G(\theta, \lambda, R) = \frac{R}{4\pi} \frac{2n+1}{n-1} \iint_\sigma \Delta g(\theta', \lambda', R) \, P_n(\cos \psi) \, d\sigma \,, \tag{5.145}$$

where $\Delta g_n^G(\theta, \lambda, R)$ is derived analogous to (5.113).

The combined disturbing potential surface harmonic can now be computed as the weighted mean in the form

$$\widehat{T}_n(\theta,\lambda,R) = w_n^M T_n^M(\theta,\lambda,R) + w_n^G T_n^G(\theta,\lambda,R)\,, \tag{5.146}$$

where w_n^M and w_n^G are the spectral weights related to the geopotential model and the terrestrial gravity data, respectively. The spectral weights can either be determined empirically, e.g., as filter coefficients (Haagmans et al. 2003), or within the framework of a least-squares adjustment or a least-squares collocation solution (see also Kern 2004). The least-squares methods allow the taking into account of the error estimates of the spectral components T_n^M and T_n^G, which are represented by the corresponding error degree variances (referring to the radius $r = R$)

$$\sigma_n^2(\varepsilon_{T^M},R) = M\{\varepsilon_{T^M}^2(\theta,\lambda,R)\} = \left(\frac{a}{R}\right)^{2(n+1)} \sum_{m=-n}^{+n} \sigma^2_{\overline{T}^M_{nm}} \tag{5.147}$$

for the global geopotential model, and

$$\begin{aligned} \sigma_n^2(\varepsilon_{T^G},R) &= \left(\frac{R}{n-1}\right)^2 \sigma_n^2(\varepsilon_{\Delta g^G},R) \\ &= \left(\frac{R}{n-1}\right)^2 \frac{2n+1}{2} \int_0^{\pi} Cov(\varepsilon_{\Delta g^G},\varepsilon'_{\Delta g^G},\psi,R)\, P_n(\cos\psi)\sin\psi\, \mathrm{d}\psi \end{aligned} \tag{5.148}$$

for the terrestrial gravity data; $M\{\cdot\}$ is a homogeneous and isotropic averaging (mean value) operator for the sphere (for details see Moritz 1980), and $\sigma_n^2(\varepsilon_{\Delta g^G},R)$ are the gravity anomaly error degree variances. For the evaluation of the above two equations, the standard deviations $\sigma_{\overline{T}^M_{nm}}$ of the corresponding coefficients $\overline{T}^M_{nm}$, see (5.144), and an (isotropic) error covariance function of the terrestrial gravity data are required. In this context, the degree variance approach neglects possibly existing error correlations between individual geopotential model coefficients, and, in addition, error correlations between different data sets (here geopotential model and terrestrial gravity observations) are usually disregarded due to lacking information.

In principle, the above scheme can also be extended to employ further data sets (e.g., satellite altimetry; Wenzel 1982) for the derivation of corresponding disturbing potential surface harmonics and error degree variances. The general least-squares adjustment and collocation solutions with N components can be found, e.g., in Wenzel (1982) and Denker (1988), respectively. Furthermore, the truncation error due to a limited integration cap size may be considered (e.g., Sjöberg 2003), but is omitted here, as it can be made negligibly small (see Sect. 4.4). Returning to the case of only two different data sources, the least-squares adjustment solution gives the following spectral weights for the gravity components:

$$w_n^G = \frac{\sigma_n^2(\varepsilon_{T^M}, R)}{\sigma_n^2(\varepsilon_{T^M}, R) + \sigma_n^2(\varepsilon_{T^G}, R)}. \tag{5.149}$$

Correspondingly, the weights w_n^M can be derived, and the sum of both weights may be expressed as

$$s_n = w_n^M + w_n^G, \tag{5.150}$$

yielding $s_n = 1.0$ for the least squares adjustment solution (as well as the case of empirically determined weights; e.g., Haagmans et al. 2003), and $s_n \leq 1.0$ for the collocation solution due to the smoothing property inherent in this method.

Combining (5.146) and (5.150) in the form $w_n^M = s_n - w_n^G$ gives the following result for the combined disturbing potential surface harmonics:

$$\begin{aligned}\widehat{T}_n(\theta,\lambda,R) &= s_n T_n^M(\theta,\lambda,R) + w_n^G\left(T_n^G(\theta,\lambda,R) - T_n^M(\theta,\lambda,R)\right)\\ &= \widehat{T}_n^M(\theta,\lambda,R) + \widehat{T}_n^G(\theta,\lambda,R).\end{aligned} \tag{5.151}$$

The major advantage of rewriting the surface harmonics terms in the above equation is, that this basically results in a remove-compute-restore (RCR) procedure, i.e., the first part of (5.151) is the usual geopotential model component (for $s_n = 1.0$), and the second part is related to the difference between the terrestrial gravity anomalies and the corresponding global model values. This yields significant advantages in the numerical evaluation, because the difference quantities are small and average out at larger distances (see also below). Now, summing up all combined surface harmonics $\widehat{T}_n(\theta,\lambda,R)$ from degrees 2 to ∞, and considering (5.144) and (5.145), yields the final computation formulas for the disturbing potential:

$$\widehat{T}(\theta,\lambda,r) = \widehat{T}^M(\theta,\lambda,r) + \widehat{T}^G(\theta,\lambda,r), \tag{5.152}$$

with the contribution from the global geopotential model

$$\widehat{T}^M(\theta,\lambda,r) = \sum_{n=2}^{\infty} s_n \left(\frac{a}{r}\right)^{n+1} \sum_{m=-n}^{+n} \overline{T}_{nm}^M \overline{Y}_{nm}(\theta,\lambda), \tag{5.153}$$

and the terrestrial gravity data

$$\widehat{T}^G(\theta,\lambda,r) = \frac{R}{4\pi}\iint_\sigma \left(\Delta g\left(\theta',\lambda',R\right) - \Delta g^M\left(\theta',\lambda',R\right)\right) W(\psi,r)\,d\sigma, \tag{5.154}$$

associated with the (modified Stokes) kernel

$$W(\psi,r) = \sum_{n=2}^{\infty} \left(\frac{R}{r}\right)^{n+1} w_n^G \frac{2n+1}{n-1} P_n(\cos\psi). \tag{5.155}$$

In (5.154), Δg^M are the gravity anomalies related to the global geopotential model. The global model is used in a RCR fashion, i.e., the residual anomalies are used to compute the residual disturbing potential, and finally the disturbing potential contribution from the global model is added. However, this procedure can be applied to Stokes's formula as well. Hence, the only difference between the spectral combination approach and Stokes's formula relates to the spectral weights w_n^G in (5.155); thus Stokes's formula results as a special case of the above formulas by setting all weights $w_n^G = 1.0$ for degrees n equal 2 to ∞. Therefore, while the Stokes formula always extracts all degrees from 2 to ∞ from the terrestrial gravity data, the spectral weights allow control of which degrees are taken from the terrestrial gravity data; e.g., only the short wavelengths should be computed from the gravity data, while the long wavelength structures should be defined mainly by a global geopotential model (e.g., from GRACE, GOCE, etc.). Another important feature of the (modified Stokes) kernel in (5.155) is that the kernel function remains finite if the weights go to zero for very high degrees or the summation is limited to some maximum degree (e.g., because mean gravity anomalies are utilized); hence, in principle, no special consideration of the inner zone contribution is required, but due to the rapid change of the integration kernel near $\psi = 0°$ it is recommended to integrate numerically the kernel function within the innermost zone (see also Sect. 4.4). Lastly, the zero-degree and first-degree terms of T are omitted again in the above equations (see notes related to the Stokes integral in Sect. 3.3).

In addition, the spectral combination technique also permits the derivation of error estimates for the results based on the degree variance approach. Based on (5.151) and (5.152), the error estimates for the combined disturbing potential $\widehat{T}$ can be derived by straightforward error propagation. The error covariance function (related to points P and P') is given by

$$Cov(\varepsilon_{\widehat{T}}, \varepsilon'_{\widehat{T}}, \psi, r, r') = \sum_{n=2}^{\infty} \left(\frac{R^2}{r\,r'}\right)^{n+1} \sigma_n^2(\varepsilon_{\widehat{T}}, R)\, P_n(\cos\psi), \tag{5.156}$$

with the error degree variances of the combined disturbing potential

$$\sigma_n^2(\varepsilon_{\widehat{T}}, R) = \begin{cases} (s_n - w_n^G)^2 \sigma_n^2(\varepsilon_{T^M}, R) + (w_n^G)^2 \sigma_n^2(\varepsilon_{T^G}, R) & \text{for } n \le n_{\max} \\ \sigma_n^2(\varepsilon_{T^G}) & \text{for } n > n_{\max} \end{cases}, \tag{5.157}$$

where $n_{\max}$ is the maximum degree of the global geopotential model employed.

The spectral combination approach results in integral formulas, which initially have to be evaluated over the entire unit sphere σ. However, as residual anomalies $\Delta g - \Delta g^M$ are employed in (5.154), a limited integration to some maximum distance $\psi_{\max}$ (spherical cap σ_0) should lead to only a small truncation error. This truncation error (also denoted as omission error) may be estimated based on Molodensky's truncation coefficients (e.g., Sjöberg 2003) or by means of the frequency transfer function (Wenzel 1982). The latter approach is outlined in the following. Assuming that a limited integration of (5.154) over a spherical cap σ_0 results in $\tilde{T}^G$ (instead of $\widehat{T}^G$), which may be expressed by corresponding Laplace surface harmonics $\tilde{T}_n^G$ (related to $r = R$), the truncation error can be expressed as

$$\begin{aligned}\varepsilon_n^{tr} &= \widehat{T}_n^G - \widetilde{T}_n^G = w_n^G(T_n^G - T_n^M) - \widetilde{T}_n^G = (T_n^G - T_n^M)\left(w_n^G - \frac{\widetilde{T}_n^G}{T_n^G - T_n^M}\right) \\ &= (T_n^G - T_n^M)(w_n^G - FTF_n),\end{aligned} \tag{5.158}$$

where the relation (5.151) is considered with the independent variables being dropped for the sake of simplicity. After some mathematical manipulations, the frequency transfer function, known from signal processing, follows as

$$FTF_n = \frac{n-1}{2} \int_{\psi=0}^{\psi_{\max}} W(\psi, R)\, P_n(\cos\psi)\, \sin\psi\, \mathrm{d}\psi, \tag{5.159}$$

and, corresponding to (5.156), the covariance function of the truncation error is given by

$$Cov(\varepsilon^{tr}, \varepsilon^{tr'}, \psi, r, r') = \sum_{n=2}^{\infty} \left(\frac{R^2}{r r'}\right)^{n+1} \sigma_n^2(T^G - T^M, R)\, (w_n^G - FTF_n)^2 P_n(\cos\psi), \tag{5.160}$$

with

$$\sigma_n^2(T^G - T^M, R) = \begin{cases} \sigma_n^2(\varepsilon_{T^M}, R) + \sigma_n^2(\varepsilon_{T^G}, R) & \text{for } n \leq n_{\max} \\ \sigma_n^2(T, R) + \sigma_n^2(\varepsilon_{T^G}) & \text{for } n > n_{\max} \end{cases}, \tag{5.161}$$

where $\sigma_n^2(T, R)$ are the disturbing potential (signal) degree variances, which have to be derived from a given model, e.g., the degree variance model of Tscherning and Rapp (1974). Finally, the spectral combination approach can be extended to other input and output gravity field parameters; examples can be found in Wenzel (1982), Denker (2003), and Wolf (2007). Moreover, an extension of the spectral combination technique by a heterogeneous error model, with several accuracy classes for the gravity data, was investigated by Behrend (1999), while the use of a full error covariance matrix for the geopotential model has not yet been attempted. In addition, numerical results are provided in Sect. 4.

3.7 Least-Squares Collocation

Within the framework of physical geodesy, least-squares collocation (LSC) is a method for determining the anomalous gravity field by a combination of geodetic observations of different kinds (Moritz 1980). LSC allows the calculation of unknown deterministic parameters (e.g., station coordinates), and besides being able to propagate the input data noise into the results, it can utilize as input as well as predict (output) heterogeneous signals related to the anomalous gravity field;

hence, LSC can be considered as a method, combining least-squares adjustment, filtering, and prediction (Moritz 1980). The mathematical foundation of LSC is related to the fields of statistics and functional analysis, in particular the theory of reproducing kernel Hilbert spaces. The method was introduced in geodesy by Moritz (1962), while Krarup (1969) succeeded in unifying the functional analytic and statistical viewpoints. Further information on the mathematical foundation of LSC can be found, e.g., in the articles from Tscherning (1985, 1986, 1994), Sansò (1986), and the textbooks from Meschkowski (1962) and Moritz (1980).

The simultaneous determination of station coordinates and gravity field quantities is denoted as "integrated" or "operational geodesy" (e.g., Eeg and Krarup 1973; Grafarend 1978b; Hein 1986); however, this approach has not gained much acceptance in practice because it is extremely computation-intensive. Therefore, only the parameter-less case of LSC (least-squares prediction) is discussed briefly in the following. The basic formula for the prediction of signals in unsurveyed points is

$$\widehat{\mathbf{s}} = \mathbf{C_{st}}(\mathbf{C_{tt}} + \mathbf{C_{nn}})^{-1}\mathbf{l} = \mathbf{C_{st}}\,\bar{\mathbf{C}}^{-1}\,\mathbf{l}\,, \qquad \bar{\mathbf{C}} = \mathbf{C_{tt}} + \mathbf{C_{nn}}, \tag{5.162}$$

where $\mathbf{C_{tt}}$ and $\mathbf{C_{st}}$ are the auto and cross covariance matrices related to the signals $\mathbf{t}$ at the observation sites and $\mathbf{s}$ at the unsurveyed stations, $\mathbf{l}$ is the observation vector, consisting of a signal and noise component (i.e., $\mathbf{l} = \mathbf{s} + \mathbf{n}$), and $\mathbf{C_{nn}}$ is the noise covariance matrix. The LSC solution is based on a least-squares hybrid minimum condition on the weighted quadratic sum of the signal and noise parts, or equivalently, the prediction results satisfy the least error variance condition (Moritz 1980). The input and output signals may be heterogeneous (e.g., T, Δg, ξ, η, etc.), but as all these quantities depend on the (harmonic) disturbing potential T, corresponding relations must also be considered for the calculation of the signal covariances in the matrices $\mathbf{C_{tt}}$ and $\mathbf{C_{st}}$. In this context, the covariance function of the disturbing potential, $K(P,P')$, is typically chosen as the basic covariance function, from which all other required covariances are derived by covariance propagation, considering the harmonicity of T and the analytical relations between T and its functionals. Moreover, it is assumed that the signal and noise quantities have an expected (or mean) value equal to zero; regarding the disturbing potential T, this condition is fulfilled at the global scale if T does not contain a zero-degree harmonic.

The (spatial) homogeneous and isotropic covariance function of the disturbing potential is given by

$$\begin{aligned} K(P,P') &= Cov(T,T',\psi,r,r') = M\{T(\theta,\lambda,r)\cdot T(\theta',\lambda',r')\} \\ &= \sum_{n=2}^{\infty}\left(\frac{R^2}{r\,r'}\right)^{n+1} \sigma_n^2(T,R)\,P_n(\cos\psi), \end{aligned} \tag{5.163}$$

with the disturbing potential degree variances

$$\sigma_n^2(T,R) = M\{T_n^2(\theta,\lambda,R)\}, \tag{5.164}$$

where $M\{\cdot\}$ is the homogeneous and isotropic averaging (mean value) operator. In principle, the covariance function cannot be exactly determined empirically, as this would require a complete knowledge of the disturbing potential function. Therefore, an empirical covariance function has to be used, which is typically obtained by fitting an analytical expression to the given data within the area of interest (local covariance function; for details see Goad et al. 1984). In this context, the degree variance model of Tscherning and Rapp (1974), resulting in closed formulas for all covariance expressions, is widely used, and a corresponding planar model was derived by Forsberg (1987). Furthermore, for the residual disturbing potential with respect to a global geopotential model, the degree variances in (5.163) have to be replaced by corresponding error degree variances of the global model up to the maximum degree $n_{\max}$ of the global model. Finally, it is noted that the LSC estimates have the minimum variance property if the kernel function is identified with the empirical covariance function, while this property is lost if arbitrary kernel functions are employed according to the analytical aspect of collocation (Moritz 1980).

LSC also allows the computation of error estimates, e.g., the error covariance matrix for the signals estimated by (5.162) is given by

$$\mathbf{E_{ss}} = \mathbf{C_{ss}} - \mathbf{C_{st}}(\mathbf{C_{tt}} + \mathbf{C_{nn}})^{-1}\mathbf{C_{ts}}\,. \tag{5.165}$$

The main advantage of LSC is its flexibility, being able to handle all quantities related to the disturbing potential as input and output data, including the associated error estimates. The data may be located at arbitrary (discrete) points in space, and hence no additional gridding is necessary. Furthermore, the varying heights of the observation and prediction sites are taken into account if LSC is applied spatially; the method inherently includes the harmonic continuation, and therefore the non-level surface corrections (Molodensky terms) are irrelevant in this case (Forsberg and Tscherning 1997). Usually, the spatial covariance functions, e.g., (5.163), are based on spherical approximations, but this is not considered as a serious problem when residuals with respect to a global geopotential model are processed (for further discussions see Tscherning 2004).

The main drawback of LSC is, however, that a system of equations as large as the number of observations has to be solved, see (5.162). In addition, numerical problems may arise from identical points or points at a short distance, when, depending on the input data noise, the matrix $\bar{\mathbf{C}}$ may become nearly singular; therefore, such duplicate points should be excluded from the input data. Today, with modern computers, several thousands of observations can be handled without problems. Moreover, the computational effort can be significantly reduced for the simple case of interpolating a single gravity field quantity (typically gravity anomalies), as then the input data can be restricted to several points near the prediction site. Furthermore, the fast collocation method can be applied, requiring gridded data sets (Sansò and Tscherning 2003).

Finally, the LSC method may be considered as data-driven, starting from discrete data and information about their noise and signal covariances, while the GBVP solutions by integral formulas may be characterized as model-driven (Sideris 2011a).

For the limiting case of homogeneous and continuous data, least-squares collocation transforms into integral formulas such as Stokes's integral, etc. (Moritz 1976).

3.8 Astronomical Leveling

More recently, the method of astronomical leveling has again attracted some interest within the framework of special projects, including the high-precision geoid and quasigeoid determination along selected lines as well as the independent validation of corresponding regional gravimetric models and global geopotential models (e.g., Hirt and Flury 2008; Hirt et al. 2008; Ihde et al. 2010). The efficiency of astronomical latitude and longitude observations could be substantially improved by using transportable zenith cameras together with CCD technology, precise electronic tilt-meters, and GNSS for timing and positioning, associated with automated processing of the digital images. At present, two such systems exist, one at the IfE, LUH, and the other at the ETH (Eidgenössische Technische Hochschule), Zurich, Switzerland (Bürki et al. 2004; Hirt 2004). Besides the enhanced observation efficiency, allowing the occupation of 10–20 stations per night, the accuracy could also be substantially improved; while the new digital zenith camera systems reach an observation accuracy of $0.05''$ (1 h observation time) to $0.08''$ (20 min observation time; see Hirt and Seeber 2008), the standard classical analog zenith camera, astrolabe, and theodolite systems attained an accuracy of at most $0.5''$ (e.g., Wildermann 1988), often accompanied by systematic errors, in particular in the longitudes (Bäumker 1984).

The astronomical latitude and longitude provide the direction of the gravity vector and hence the inclination of the corresponding level surface. The deflection of the vertical describes the angle between the actual plumb line and a reference direction; the deflection of the vertical is a vectorial quantity and usually expressed by its components in north–south and east–west direction. Besides the deflection of the vertical according to Molodensky (see Sect. 3.2), the deflection of the vertical at the Earth's surface with respect to the ellipsoidal normal at P (Helmert's definition) and the corresponding quantity at the geoid (Pizzetti's definition) are to be distinguished (e.g., Torge 2001). Thus the components of the deflection of the vertical according to Helmert are given by

$$\xi_P^H = \Phi_P - \varphi_P \,, \qquad \eta_P^H = (\Lambda_P - \lambda_P)\cos\varphi_P \,. \tag{5.166}$$

Integrating the deflections of the vertical along a path on the geoid or the Earth's surface yields the geoid undulation difference

$$\Delta N_{12} = N_2 - N_1 = -\int_1^2 \varepsilon^0 \, ds = -\int_1^2 \varepsilon^H \, ds - \int_1^2 \delta\varepsilon \, ds = -\int_1^2 \varepsilon^H \, ds - E_{12} \,, \tag{5.167}$$

where ε^0 and ε^H are the azimuthal deflection components (in the azimuth α) according to Pizetti and Helmert at the geoid and Earth's surface, respectively, e.g.,

$$\varepsilon^H = \xi^H \cos\alpha + \eta^H \sin\alpha \,, \tag{5.168}$$

$\delta\varepsilon$ is the correction for the curvature of the plumb line, and E_{12} is the orthometric height reduction known from geometric leveling (for further details see Torge 2001).

Regarding the height anomalies, the relation

$$\mathrm{d}\zeta = \frac{\mathrm{d}\zeta}{\mathrm{d}s}\mathrm{d}s + \frac{\mathrm{d}\zeta}{\mathrm{d}h}\mathrm{d}h \tag{5.169}$$

has to be utilized (Molodenskii et al. 1962), which considers that the height anomalies are not related to a level surface, in contrast to the geoid undulations treated above (i.e., $\mathrm{d}N = (\mathrm{d}N/\mathrm{d}s)\,\mathrm{d}s = \varepsilon^0\,\mathrm{d}s$ can be considered as a special case of (5.169)). Considering that $\mathrm{d}\zeta/\mathrm{d}s$ is the azimuthal deflection component according to Molodensky, while taking $\mathrm{d}\zeta/\mathrm{d}h$ from (5.141), leads to

$$\Delta\zeta_{12} = \zeta_2 - \zeta_1 = -\int_1^2 \varepsilon^M\,\mathrm{d}s - \int_1^2 \frac{\Delta g}{\gamma}\,\mathrm{d}h \,. \tag{5.170}$$

Alternatively, utilizing the deflections of the vertical according to Helmert, the above equation can be expressed as

$$\Delta\zeta_{12} = \zeta_2 - \zeta_1 = -\int_1^2 \varepsilon^H\,\mathrm{d}s - E_{12}^N \,, \tag{5.171}$$

where E_{12}^N is the normal height reduction; for a detailed derivation of the equations for the height anomalies see Campbell (1971) and Torge (2001). The negative sign of the ε components in (5.167), (5.170), and (5.171) follows from the sign conventions for the geoid/quasigeoid heights and the deflections of the vertical.

The accuracy of astronomical leveling mainly depends on the accuracy of the vertical deflections as well as the quality of the interpolation between the observation sites. The latter component, i.e., the interpolation error, arises from the approximation of the line integrals by discrete observations; it can be substantially reduced by considering topographic information. The purely random observation errors propagate with the square root of the number of individual set-ups. The standard deviation of ΔN_i or $\Delta\zeta_i$ from a single set-up is given by $\sigma_{\Delta N_i[\mathrm{mm}]} = 4.8\ \sigma_{\varepsilon['']}\ \Delta S_{[\mathrm{km}]}$, where $\sigma_{\varepsilon['']}$ is the vertical deflection standard deviation in arc seconds, and ΔS is the station distance in kilometers. Hence, for a line of length S with n segments ($n = S/\Delta S$), the standard deviation is given by $\sigma_{\Delta N[\mathrm{mm}]} = 4.8\ \sigma_{\varepsilon['']}\ \Delta S_{[\mathrm{km}]}\sqrt{n}$, assuming uncorrelated errors; on the other hand, errors with a correlation of $r = 1.0$ and systematic errors propagate linearly with n or the distance S, resulting in $\sigma_{\Delta N\,[\mathrm{mm}]} = 4.8\ \sigma_{\varepsilon['']}\ S_{[\mathrm{km}]}$. Regarding the correlation due to anomalous refraction, Hirt and Seeber (2008) mention values of $r \approx 0.05$ between single observations.

Regarding the digital zenith camera results, an accuracy of about $0.1''$ can be safely assumed for a standard observation period of 20 min (Hirt and Seeber 2008), which leads to a formal accuracy of the ΔN or $\Delta\zeta$ quantities at the few millimeters level for shorter lines up to about 100 km, depending on the station distance. Accordingly, for longer lines of 500–1,000 km, the random errors accumulate to a level of about 1–3 cm, and systematic errors become more critical, e.g., a systematic error of only $0.01''$ transforms into a corresponding ΔN or $\Delta\zeta$ error of 2.4 cm ($S = 500$ km) and 4.8 cm ($S = 1{,}000$ km), respectively. Thus, on long traverses of several 100 km, the astrogeodetic results are getting into competition with the combined satellite (e.g., GOCE) and gravimetric solutions, as indicated by the practical results for two 500 km long traverses in Germany (e.g., Ihde et al. 2010). On the other hand, an independent verification of the astrogeodetic results at the millimeter level for spatial scales of a few 100 m to several 10 km is extremely difficult, as practically no other comparable data exists. While geometric leveling is accurate at the millimeter level, the GNSS techniques usually give worse results due to station dependent effects, etc.; however, based on 48 h GPS observations and sophisticated new approaches to reduce station dependent effects, Hirt et al. (2010) quote an RMS difference between GPS/leveling and astrogeodetic results of about 2–3 mm (max. 5 mm). In addition, considering the spectral relation between gravity anomalies and vertical deflection components with a conversion factor of 6.7 mGal/$''$ for higher degrees n, it should also be possible to verify at least the fine structures of the astrogeodetic calculations, provided that the local and regional gravity anomaly field is reasonably represented by observations, while the far-zone is modeled by a global geopotential model; in principle, an accuracy of $0.1''$ for the astronomical observations corresponds to an accuracy of about 0.7 mGal in terms of gravity, which is easily achieved by regional gravimetric surveys, being in most instances more accurate by at least one order of magnitude. To the best knowledge of the author, such an experiment is still lacking, and therefore would be an interesting topic for the future.

The main advantage of the astrogeodetic method is its independence from any data outside the area of calculation, in contrast to the gravimetric method where basically global data coverage is needed. Therefore, the main applications for the astrogeodetic method are seen in local geoid and quasigeoid calculations (especially for the case that gravity field observations in the surrounding areas are lacking or inaccessible for political reasons) as well as the independent validation of corresponding gravimetric results and global geopotential models up to distances of a few 100 km. A regional gravity field survey, particularly for larger areas, can be performed more effectively by land, sea, and airborne gravimetry than by astronomical observations, which require more time per station occupation and have to be done during the night outside normal working hours.

3.9 The Remove-Compute-Restore Technique and Topographic Effects

In practice, local and regional gravity field modeling is usually based on discrete data covering the area of interest and the immediate surroundings. This leads to two difficulties. First, the very short wavelength gravity field information is not properly represented by the discrete observations, leading to aliasing effects; this problem is counteracted by employing digital elevation models to obtain the high-frequency gravity field signals. Second, as the observation data cover only a certain region, the long wavelength gravity field information (i.e., longer than the extent of the region) must be computed in another way; this problem is remedied by using a state-of-the-art global geopotential model. Hence, the short and long wavelength gravity field structures are obtained from digital elevation models and a global geopotential model, respectively, while the medium wavelength field structures are derived from the regional discrete gravity field observations. This leads directly to the remove-compute-restore (RCR) technique, in which the short and long wavelength information is first removed from the observations, then the residual quantities are used for gravity field modeling (e.g., transformation from residual gravity to height anomalies), while finally the short and long wavelength contributions are restored again.

The general scheme of the RCR technique is based on a residual disturbing potential given by

$$T^{\mathrm{res}} = T - T^M - T^T \,, \tag{5.172}$$

where T^M and T^T are the contributions from the global geopotential model and the topographic information (or more generally the anomalous masses; see below). Correspondingly, all observations, described as linear functionals L of T, are consistently reduced by

$$L_{\mathrm{obs}}(T^{\mathrm{res}}) = L_{\mathrm{obs}}(T) - L_{\mathrm{obs}}(T^M) - L_{\mathrm{obs}}(T^T) \,, \tag{5.173}$$

where the global geopotential model should be utilized rigorously as a high-degree reference field (see Sect. 3.2). Then, after applying the gravity field modeling techniques, the effects of the global geopotential model and the topography are restored, leading to the final predictions

$$L_{\mathrm{pred}}(\widehat{T}) = L_{\mathrm{pred}}(\widehat{T}^{\mathrm{res}}) + L_{\mathrm{pred}}(T^M) + L_{\mathrm{pred}}(T^T) \,. \tag{5.174}$$

The removal of the short and long wavelength gravity field information corresponds to a spectral (low-pass and high-pass) filtering; this leads to residual quantities which are typically much smaller and smoother (as well as statistically more homogeneous and isotropic) than the original ones, facilitating, e.g., the tasks of interpolation, gridding, as well as field transformations by integral formulas or LSC, with the additional side effect that the collection of observational data can be restricted to the region of interest plus a narrow edge zone around it. The RCR technique has become a standard procedure for local and regional gravity field modeling as documented in textbooks (e.g., Torge 2001) and numerous journal articles (e.g., Sideris 2011b; Forsberg 2010; Denker et al. 2009; Tscherning 2004; Smith and Roman 2001).

Until recently, high-degree geopotential models were mainly developed up to degree and order $n_{\max} = 360$ from satellite and terrestrial data, corresponding to a resolution of 0.5° or about 55 km; examples are the EGM1996 (Lemoine et al. 1998) as well as the EIGEN (e.g., Förste et al. 2008a, b) and GGM (e.g., Tapley et al. 2007) models. However, the situation changed considerably with the advent of the EGM2008 model (Pavlis et al. 2008), which includes coefficients up to spherical harmonic degree $n_{\max} = 2{,}190$, corresponding to a resolution of 5′ or about 9 km. Hence, regarding areas with high-quality data included in EGM2008, it should only be necessary to add the very short wavelength gravity field structures (about 2–3 cm RMS for the geoid/quasigeoid) by means of local gravity and terrain data, but the situation may be quite different in areas where only poor data were available for EGM2008. At present, there is not very much practical experience on how to make optimal use of the ultra-high-degree EGM2008 model in regional gravity field modeling, and certainly more investigations on this topic are needed (e.g., according to Forsberg 2010). Further discussions in this direction, e.g., related to the impact on topographic reductions, follow below and in Sect. 4.4.

In mountainous regions, the gravity field is strongly correlated with the local topography; the gravitational attraction of the topographic masses causes a strong signal, which dominates at shorter wavelengths. Therefore, topographic information can be used to smooth the gravity field prior to any modeling process in order to avoid aliasing effects. For example, gravity stations are by tendency located in valleys along roads, and thus the observations are related to a level which is systematically below the average topography. Such aliasing errors can be very big and devastating for gravity field modeling (Forsberg and Tscherning 1997). Besides the direct gravitational effect, the topography implies that the observations are related to a non-level surface, which can be considered by LSC, but requires additional corrections (Molodensky terms) for the integral formula approaches (see Sect. 3.4).

In addition to the topography, other information about local density anomalies, e.g., due to salt domes, etc., may also be taken into account (Denker 1988). This leads to a so-called mass model, which may be considered as "source information" (the masses are the sources of the gravitational field), in contrast to the "effect information" from a global geopotential model (Sünkel 1983). The effect of the anomalous masses can be computed by Newton's law according to (5.19), giving

$$L_{\text{obs}}(T^T) = G \iiint_V \Delta\rho \, L_{\text{obs}}\left(\frac{1}{l}\right) \mathrm{d}v \, , \tag{5.175}$$

where $\Delta\rho$ are (appropriately defined) density anomalies (see Forsberg 1984; Forsberg and Tscherning 1997). In this context, it is important to note that the topographic (or mass) potential T^T has to be a harmonic function, because otherwise the above RCR scheme is not valid (Forsberg and Tscherning 1997).

This condition is fulfilled if either a fixed area is taken into account (e.g., specified by latitude and longitude limits), or if the reductions are—at least in principle—computed globally. The classical terrain reduction as well as reductions based on fixed spherical cap sizes do not fit into this scheme, as for every new station another area of the mass model is evaluated.

Different terrain reduction schemes and the associated advantages and disadvantages are discussed in detail in Forsberg and Tscherning (1981, 1997) as well as Forsberg (1984). The complete topographic (or Bouguer) reduction is not suited for gravity field modeling because it causes a very large change of the potential function due to the complete removal of all topographic masses (ideally shifted to infinity). The topographic-isostatic reduction is useful for gravity field modeling as it provides a smooth residual field; however, it has the disadvantage of generating long wavelength signals, and thus also the potential coefficients of a global model need to be reduced accordingly, which is a costly and time-consuming procedure. Furthermore, as a global high-degree spherical harmonic reference expansion is used within the RCR procedure, which obviously also includes the effect of the global topography, preferably only short wavelength topographic effects should be considered, leading to the so-called residual terrain model (RTM) reduction, introduced by Forsberg and Tscherning (1981).

The RTM procedure is based on a smooth reference topography surface, usually obtained by applying a moving average or other filter (e.g., a Gaussian filter; see Forsberg 2010) to the existing high-resolution digital elevation model (DEM), and then only the differences between the actual topography and the reference topography are taken into account within the reduction process; this leads to balanced positive and negative density anomalies, representing areas where the topography is either above or below the reference topography, and hence the RTM effects will cancel out for zones at larger distances (Forsberg and Tscherning 1997). Thus, the remote contributions become negligibly small, and the RTM reduction computations can be limited to a distance of about two to three times the resolution of the reference topography (Forsberg and Tscherning 1997). In summary, the RTM reduction has the advantage that only short wavelength gravity field structures are considered (without changing the long wavelengths), and in addition the reductions can be limited to some distance $\psi_{\max}$, while still satisfying the harmonicity condition for the topography potential function T^T. Furthermore, the RTM reduced gravity anomalies based on a mean elevation surface of about 100 km resolution will typically be quite similar in magnitude to the isostatic anomalies (Forsberg and Tscherning 1997).

The mass displacements associated with the RTM technique (also denoted as regularization) lead to the situation that stations above the mean elevation surface are left in the mass-free domain, whereas the gravity field quantities at stations in valleys with $h_P < h_{\text{ref}}$ are transformed into corresponding quantities inside the masses, where the associated potential function is not harmonic. The latter problem is usually remedied by a simple "harmonic correction", which is based on downward continuation through a Bouguer plate, being valid only if the reference topography is sufficiently long-wavelength (Forsberg and Tscherning 1997;

Forsberg 2010). The non-harmonicity of the reduced potential below the reference height surface is considered today as a major theoretical problem with the RTM method (Forsberg and Tscherning 1997), and some attention was given to it recently by Elhabiby et al. (2009), Omang et al. (2012), and Forsberg (2010). In principle, the quantities related to a harmonically continued topography potential function T^T are needed, which may be obtained by first removing the effect of the entire topography, followed by a harmonic continuation of the observations to the reference elevation surface, plus finally the restoration of the effect of the reference topography. Another option would be to condense the masses of the reference topography at a lower level. For further details on this topic, requiring additional investigations, see the references mentioned above.

The RTM concept is widely applied in practice, employing a mean elevation surface (reference topography) with a resolution ranging typically from 100 to 50 km, but also smaller values have been applied (e.g., Forsberg and Tscherning 1997; Denker 1988; Forsberg 2010). In this context, Forsberg (2010) describes quasigeoid computations based on local gravity and terrain data in connection with the full EGM2008 model ($n_{\max} = 2{,}190$) and RTM reductions associated with a 5′ and 30′ reference topography, as well as a computation based on EGM2008 with $n_{\max} = 360$ and RTM 30′ reductions. In comparison to GPS/leveling data, the two solutions based on the RTM 30′ reductions and EGM2008 to $n_{\max} = 360$ and 2,190 showed the best performance (with a slight advantage for the results using EGM2008 to $n_{\max} = 2{,}190$), while the solution with the RTM 5′ reductions and EGM2008 ($n_{\max} = 2{,}190$) fitted less well, which was attributed to the approximation errors in the harmonic correction (as discussed above). Interestingly enough, the solution based on EGM2008 to $n_{\max} = 2{,}190$ and the RTM 30′ reductions showed the best performance, and Forsberg (2010) mentions that this is "indicating that the 'double accounting' of the topography does not matter in practice (which it should not, since the remove–restore principle will account for this)."

The practical computation of terrain effects requires high-resolution DEMs and is mostly based on prism integration in the space domain, augmented by the much faster (but sometimes more approximate) FFT methods (e.g., Forsberg 1984; Schwarz et al. 1990; Forsberg and Tscherning 1997; Sideris 2011b). Regarding a rectangular prism of constant density $\Delta\rho$, closed formulas exist for the gravitational potential and its derivatives (MacMillan 1958; Forsberg 1984; Denker 1988), e.g., the potential at a point P, located at the origin of a local Cartesian coordinate system (x, y, z), is given by

$$\begin{aligned} T^{\text{prism}} &= G\Delta\rho \,\Big|\Big|\Big|\, xy\ln(z+d) + xz\ln(y+d) + yz\ln(x+d) \\ &\quad - \frac{x^2}{2}\arctan\frac{yz}{xd} - \frac{y^2}{2}\arctan\frac{xz}{yd} - \frac{z^2}{2}\arctan\frac{xy}{zd}\,\Big|_{x_1}^{x_2}\Big|_{y_1}^{y_2}\Big|_{z_1}^{z_2}\,, \\ &\quad \text{with} \quad d = \sqrt{x^2+y^2+z^2}\,. \end{aligned} \tag{5.176}$$

The computation of terrain effects related to other gravity field parameters is described in detail in Forsberg (1984), including a computer program (TC) for the efficient evaluation of large terrain grids with sophisticated features such as an inner zone densification, the consideration of the given station elevations, a simple Earth curvature correction, as well as the use of a detailed and coarse grid for the inner and outer zones, respectively, associated with an automatic switching between different computation formulas in order to save computing time and to obtain stable results. The use of the rectangular prisms is considered as sufficiently accurate in connection with the RTM method, as the differences between rectangular and spherical bodies will also cancel out to some extent within the RCR procedure. However, further studies on this topic are necessary, such as that presented by Heck and Seitz (2007), using the so-called tesseroids based on spherical coordinates.

Finally, it should be noted that the RCR technique can be applied in combination with all gravity field modeling procedures described in the previous sections, ranging from the computation of spherical harmonic models, over the integral formula approaches, least-squares collocation, astronomical leveling, to the application of Molodensky's theory. The whole Molodensky theory may in principle be applied to the original surface free-air data as well as to terrain-reduced data, yielding much smaller correction terms g_n associated with a more stable solution scheme (e.g., Forsberg and Tscherning 1997; Denker and Tziavos 1999).

4 Practical Results

4.1 Data Requirements

Today's demands for accuracy within the scope of regional gravity field modeling are at the level of about 0.01–0.001 m for the geoid/quasigeoid, 1 mGal for gravity anomalies, and $1''$ for vertical deflection components, e.g., regarding applications such as height determination by GNSS or dynamic ocean topography (DOT) modeling, interpolation of gravity data (for leveling stations, etc.), and local geodetic networks. In this context, terrestrial surface free-air gravity anomalies form one of the most important input data sets because they exist in many regions of the world with good quality and coverage. Regarding the input data requirements with respect to accuracy and resolution, theoretical as well as numerical investigations, including spectral analysis can be utilized. An easy way to gain an idea about the necessary data quality is provided by the degree variance approach, based on the spectral decomposition of the anomalous gravity field. The root mean square (RMS) omission error, describing the gravity field signals above a certain harmonic degree $n_{\max}$ (i.e., the terms neglected in a spherical harmonic expansion complete up to degree and order $n_{\max}$), can be computed simply as the square root of the sum of the corresponding degree variances for degrees $n_{\max} + 1$ to infinity. The degree variance model of Tscherning and Rapp (1974) with $n_{\max} = 10{,}000$ (corresponding to a spatial resolution of 2 km) yields an omission error of about 0.001 m for geoid/quasigeoid heights, 1.4 mGal for gravity anomalies, and $0.2''$ for vertical deflection components, while the corresponding values for $n_{\max} = 5{,}000$ (corresponding to a spatial resolution of 4 km) are about 0.005 m, 4.8 mGal, and 0.7″, respectively. Regarding high-degree geopotential models, the omission errors for $n_{\max} = 360$ (standard over the past decades) and $n_{\max} = 2{,}190$ (related to EGM2008; Pavlis et al. 2008) are of interest; for the latter case, the omission error is 0.023 m for geoid/quasigeoid heights, 11.1 mGal for gravity anomalies, and 1.7″ for vertical deflection components, while the corresponding values for $n_{\max} = 360$ are 0.227 m, 25.2 mGal and 3.8″, respectively. The latter figures also document that the combination of a global geopotential model with local terrestrial data is important for modeling the complete gravity field spectrum; otherwise, for example, geoid/quasigeoid signals with a magnitude of about 2–4 cm for $n_{\max} = 2{,}190$ (see also Jekeli et al. 2009) or a few decimeters for $n_{\max} = 360$ are lacking.

The omission error for different gravity field parameters is depicted in Fig. 5.7, which also gives some insight into the spectral sensitivity of the quantities, the geoid/quasigeoid (or disturbing potential T) signal being concentrated mainly at the longer wavelengths, while for the first and second derivatives of T (Δg, ξ, η, and T_{zz}, respectively; 1 Eötvös (E) = 10^{-9} s^{-2} = 0.1 mGal/km) the signals are progressively focused at the shorter wavelengths. Another useful formula for estimating the omission error for geoid/quasigeoid heights can be deduced from Kaula's rule of thumb (a simple degree variance model, based on the assumption that the standard deviation of a single fully normalized coefficient of the gravitational potential is $\sigma\{\bar{C}_{nm}, \bar{S}_{nm}\} \approx 10^{-5}/n^2$; Kaula 1966), resulting in (see also Forsberg 1993)

$$\sigma_{\text{omission}}^{n_{\max}}(\zeta) = \sqrt{\sum_{n=n_{\max}+1}^{\infty} \sigma_n^2(\zeta)} \approx \frac{64_{[\text{m}]}}{n_{\max}}. \tag{5.177}$$

The above formula gives an omission error of 0.006 m for $n_{\max} = 10{,}000$, 0.013 m for $n_{\max} = 5{,}000$, 0.029 m for $n_{\max} = 2{,}190$, and 0.18 m for $n_{\max} = 360$, which is in reasonable agreement with the results based on the Tscherning and Rapp (1974) model. Aiming at the computation of the geoid or quasigeoid with an accuracy of 0.01 m or better, the preceding simple and straightforward considerations suggest that corresponding harmonics up to degree and order 5,000–10,000 have to be modeled, requiring input data with a spatial resolution of roughly 2–4 km, while, e.g., the accuracy demands for gravity data are only at the level of about 1 mGal and hence not very severe. However, it has to be noted that these simple thoughts consider only random errors, and, furthermore, the omission and observation errors superimpose each other in the calculations.

On the other hand, small systematic gravity errors, affecting large regions, may also integrate up to significant geoid or quasigeoid errors. A rough estimation of such effects is possible on the basis of the formula for the inner zone geoid/quasigeoid contribution (5.130). For example, considering a systematic gravity error of 0.1 mGal over a circular cap with radius 100 km or 0.02 mGal over a 500 km cap, respectively, leads in each case to a systematic geoid/quasigeoid error of 0.01 m. With regard to the combination of terrestrial data with GOCE and GRACE geopotential models, the chosen radii approximately correspond to the

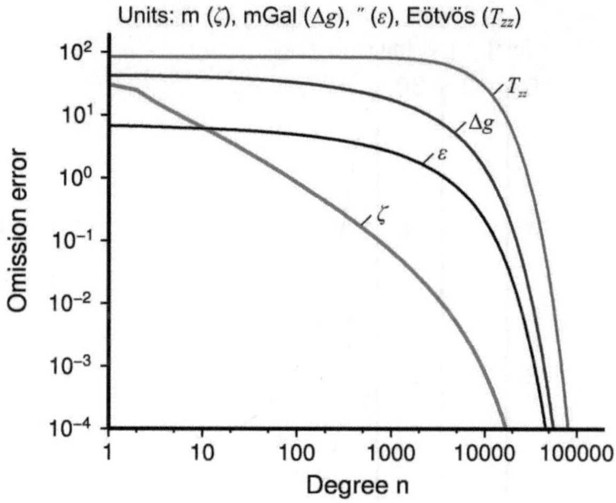

Fig. 5.7 Omission error for geoid/quasigeoid heights (N, ζ), gravity anomalies (Δg), single vertical deflection components (ε), and vertical gravity gradients (T_{zz})

resolutions where terrestrial data come into play. Therefore, it is desirable to connect at least the gravity base network to modern absolute gravity stations (with an accuracy of 0.01–0.02 mGal or better), while the requirements for regional detail surveys can be relaxed (with acceptable random errors up to about 1 mGal).

Another view on the data requirements is possible by looking at the gravitational effects of typical disturbing density anomalies. For example, in the north German lowlands, many saltdomes exist, which cause quite local gravity field structures. Figure 5.8 depicts the gravitational effect of a typical salt dome with an extension of $4 \times 4 \times 4$ km^3 and a density contrast of $\Delta\rho = 300$ kg/m^3, with the density of salt usually being lower than that of the surrounding rocks. Therefore, salt domes are frequently connected with (negative) gravity anomalies of about 20–30 mGal, associated with geoid/quasigeoid and vertical deflection effects of 5–6 cm and 2″, respectively. Hence, in order to capture such significant local gravity field variations, corresponding observations with a sufficient spatial resolution are needed, again leading to data spacings at the few kilometer level.

Besides the gravity observations themselves, the horizontal and vertical coordinates (either gravity field related heights from leveling or ellipsoidal heights from GNSS techniques) of the gravity sites must be known with sufficient accuracy. The station positions are required for the calculation of the normal gravity field parameters and the corresponding anomalous gravity field quantities (regarding the level ellipsoid, only the latitudes and heights are needed), as well as for the determination of the global geopotential model and terrain contributions within the framework of the RCR procedure. The horizontal and vertical position requirements should roughly conform with the gravity accuracy; considering, e.g., gravity observations with an accuracy of 0.01, 0.1, and 1.0 mGal, an actual gravity gradient of about 0.3 mGal/m for the vertical and 10 mGal/km (maximum) for the horizontal direction, respectively, results in position requirements of about 0.03, 0.3, and 3.0 m (vertically), and 1, 10, and 100 m (horizontally).

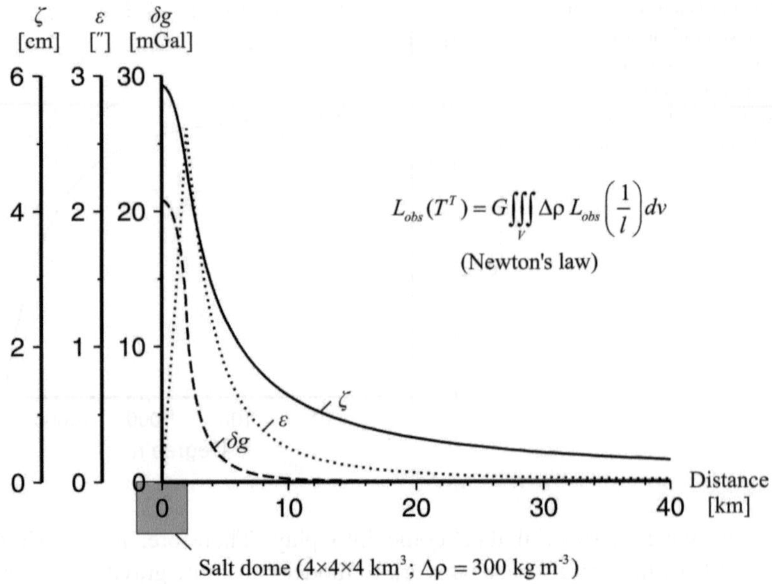

Fig. 5.8 Gravitational effect of a typical salt dome on height anomalies (ζ,), gravity disturbances (δg) and vertical deflections (ε)

In addition, within the framework of the remove-compute-restore (RCR) technique, the digital elevation models (DEMs) must also have a sufficient spatial resolution and accuracy. In this context, comprehensive numerical investigations are presented, e.g., in Li et al. (1995) and Grote (1996); in summary, regarding the targeted modeling accuracies (see above), the DEMs should have a spatial resolution of about 100–1,000 m for alpine to flat areas, while an accuracy of approximately 10 m is sufficient, provided that, especially for the gravity sites, the available station heights are employed in the computation of the terrain effects, as suggested by Forsberg (1984).

Furthermore, a comprehensive study on the effect of systematic gravity anomaly errors due to gravity, horizontal, and vertical datum inconsistencies is given in Heck (1990), with a similar investigation (for Europe) being presented by Denker (2001). The effect of such small but systematic gravity anomaly errors on the geoid/quasigeoid is predominantly of long wavelength nature, and hence it can be kept small (below the level of 0.01 m) by using a global high-degree satellite geopotential model (from the GRACE or GOCE mission) as a reference (e.g., in connection with the spectral combination approach), while the pure terrestrial solutions based on Stokes's integral may lead to errors at the decimeter level (for further details see Heck 1990 and Denker 2001).

In summary, gravity surveys should preferably be connected to a high-precision gravity base network (relying on absolute gravity observations) in order to avoid large-scale systematic errors; the corresponding accuracies should be about

0.01 mGal for gravity, 0.03 m for the heights, and 1.0 m for the horizontal coordinates, or better, respectively. On the other hand, the accuracy requirements for detail surveys may be relaxed to the level of about 1 mGal and accordingly for the vertical and horizontal positions (see above), provided that the errors are purely random. These figures are also supported by simulation studies based on least-squares collocation, etc. (examples can be found, e.g., in Denker 1988 and Forsberg 1993).

4.2 The European Gravity and Geoid Project

The historical development of geoid and quasigeoid modeling in Europe has been described, e.g., in Torge and Denker (1998). Since the beginning of the 1980s, the Institut für Erdmessung (IFE), Leibniz Universität Hannover (LUH), has been involved in such computations. The first important result was the "European Gravimetric Geoid Number One" (EGG1; Torge et al. 1982); it was based on mean gravity anomalies and had an accuracy at the level of several decimeters. Then, with the advent of the Global Positioning System (GPS), the accuracy demands for the geoid/quasigeoid increased to the centimeter level, which can be achieved only by combining high-resolution point gravity field data with corresponding topographic information and a global geopotential model. In this context, several investigations were carried out initially for very local areas, and regarding a small test network near Hannover, Germany, it was proven for the first time that an agreement between gravimetric and GPS/leveling results at the centimeter level is in fact possible (Denker and Wenzel 1987). These computations were subsequently extended to larger regions and lead to a new quasigeoid model for the whole of Germany (Denker 1989). Based on these experiences, IfE proposed to perform corresponding computations for the whole of Europe; finally, this task was supported by the International Association of Geodesy (IAG), Geoid Sub-Commission for Europe, and IfE served as the computing center in the period 1990–2003. A major result of this IAG enterprise was the high-resolution European geoid and quasigeoid model EGG1997 (Denker and Torge 1998), based on the global geopotential model EGM1996 (Lemoine et al. 1998) and high-resolution gravity and terrain data available at that time. The evaluation of EGG1997 by GPS and leveling data revealed the existence of long wavelength errors at the level of 0.1–1 ppm, while the agreement over shorter distances up to about 100 km was at the level of 0.01–0.02 m in many areas with good quality and coverage of the input data (Denker and Torge 1998; Denker 1998).

However, after a while, several advancements appeared to be possible, including strongly improved global satellite gravity fields (from the CHAMP, GRACE, GOCE missions), new or updated high-resolution data sets (gravity, terrain, satellite altimetry, GPS/leveling), as well as refined modeling techniques. Thus, a complete re-computation of the European geoid and quasigeoid was considered appropriate and promised significant accuracy improvements, especially at the longer wavelengths. As a result, after the IUGG General Assembly in

Sapporo, 2003, when the new structure of the IAG was implemented, it was decided to support the development of an improved European geoid/quasigeoid model in the form of an IAG Commission 2 Project, named "CP2.1—European Gravity and Geoid Project (EGGP)," and since 2011 this task has continued as IAG Sub-Commission 2.4a "Gravity and Geoid in Europe." The European geoid and gravity project has strong connections to the IAG International Gravity Field Service (IGFS) and its centers, as well as to several other IAG bodies, e.g., EUREF (IAG Reference Frame Sub-Commission for Europe). The project is chaired by H. Denker, IfE, and has about 50 national delegates (project members) from most of the countries in Europe. Due to the confidentiality of many data sets, only one data and computation center exists at IfE, Hannover. Further details on the project can be found, e.g., in Denker et al. (2009). Interim results and status reports of the project were presented roughly on an annual basis.

While the initial test computations within the framework of the EGGP were limited to central Europe, the first complete re-computation was finished in 2007 and is denoted as EGG2007 (European Gravimetric (Quasi)Geoid 2007). After that, the new global geopotential model EGM2008 (Pavlis et al. 2008) became available. As the comparisons of EGM2008 with the European gravity data sets revealed some systematic differences for a few gravity sources, these problem areas were corrected, and, besides a few other improvements, lead to a new computation, which was finished at the end of 2008 and is denoted as EGG2008. At present, investigations about the inclusion of a GOCE geopotential model and some further refinements are still going on. More details about the EGGP data and results are provided in the following sections.

4.3 The European Gravity and Terrain Data

Since the start of the European geoid project, significant improvements of the land gravity data base were made, including new or revised data sets for nearly all European countries. New gravity data sets became available for Austria, Belgium, Bulgaria, Croatia, Cyprus, Denmark, Estonia, Finland, France, Germany, Greece, Italy, Latvia, Luxemburg, the Netherlands, Norway, Portugal, Serbia, Slovenia, Spain, Sweden, Switzerland, and Turkey.

Significant progress was also made in the collection and reprocessing of marine gravity data. All marine gravity data collected until 2003 were edited and crossover adjusted (see Denker and Roland 2005), which led to significant data improvements. The comparisons with independent altimetric gravity anomalies from, e.g., the KMS2002 model (Andersen et al. 2005), showed an RMS difference of 18.0 mGal for the original data set, 10.2 mGal for the edited data set, and 7.8 mGal for the edited and crossover adjusted data set, respectively, which proves the effectiveness of the entire processing scheme (see Denker and Roland 2005).

In addition, after 2003, significant new marine gravity data sets became available, originating mainly from the authorities of the Scandinavian countries (coverage: Baltic Sea, North Sea, North Atlantic), France (coverage: western parts of the Mediterranean Sea, Atlantic), as well as the National Geospatial-Intelligence Agency (NGA), U.S.A. (coverage: central and eastern parts of the Mediterranean Sea). Moreover, some airborne data sets were also provided by the Scandinavian authorities, covering the Baltic Sea as well as parts of the North Atlantic and Greenland coastal waters. The aforementioned new gravity data sets were thus far not crossover adjusted together with the other marine gravity data sources, mainly because all of them are of high quality without the need for a crossover adjustment and also due to a lack of time.

In addition to this, the public domain data from the Arctic Gravity Project ArcGP (Forsberg and Kenyon 2004) were integrated in the project data base. Finally, data from the EGG1997 data base were utilized for some areas (e.g., Eastern Europe and Africa). Furthermore, in order to fill the remaining data voids in the marine gravity data, altimetric gravity anomalies were employed. Until 2007 (regarding the EGG2007 computation), the altimetric data set KMS2002 (Andersen et al. 2005) was used.

After that, when the global geopotential model EGM2008 (Pavlis et al. 2008) became available, some problem areas with systematic differences between the European gravity data sets and EGM2008 showed up. This led to the following updates of the gravity sources being carried out until September 2008 (related to the EGG2008 computation):

- The Greek and Turkish gravity values were corrected in absolute level (the errors were caused by imperfect gravity reference system information).
- A few minor new sources (nine) were added.
- The KMS2002 altimetric anomalies were replaced by a merger of the $1' \times 1'$ DNSC2008GRA (Andersen et al. 2010) and V18.1 (Sandwell and Smith 2009) data sets; the merging procedure was done in accordance with the EGM2008 approach (Pavlis et al. 2008), where the DNSC2008GRA data were used near the coast (out to 200 km distance) and the Sandwell and Smith data were employed over the open ocean; in addition, the altimetric data were completely edited out within a distance of 10 km from the shore line, as comparisons with some high-quality ship data within a 10 km wide coastal zone showed an RMS difference twice as high as on the open ocean (approximately 8 vs 4 mGal).
- The ship gravity data editing was improved and some sources with very poor quality were completely excluded (mainly older sources with RMS differences to the altimetric data exceeding about 10 mGal).
- Fill-in gravity anomalies were derived from the EGM2008 geopotential model for some $5' \times 5'$ cells in Africa, the Caucasus region, and parts of Asia, where no gravity observations were available within a distance of 15 km; this approach was selected to stay close to EGM2008, even in regions with large gravity data voids outside the main area of interest, being Europe and the surrounding waters.

The updates described above were taken into account for the EGG2008 computation, but not for EGG2007. Further information on the different data sets is included in Table 5.1. The final EGG2007 gravity data set consisted of 5,354,653

Table 5.1 Main characteristics of EGG1997/2007/2008

EGG1997	EGG2007	EGG2008
Gravity data		
Project data base		
2,684,133 (744 sources)	5,354,653 (709 sources)	5,355,206 (718 sources)
Other data sources		
–	195,840 (ArcGP)	195,840 (ArcGP)
335,124 (KMS1996)	951,251 (KMS2002)	13,222,260 (1′ × 1′ altimetry)
–	–	120,747 (EGM2008 fillins)
3,019,257 (Total)	6,501,744 (Total)	18,894,053 (Total)
Terrain data		
7.5″ ... 5′ grids	1″ ... 30″ grids	1″ ... 30″ grids
700 million elev.	8.3 billion elev.	8.3 billion elev.
15′ × 20′ RTM	30′ × 45′ RTM	15′ × 20′ RTM
Global geopotential model		
EGM1996 (n_{max} = 360)	EIGEN-GL04C (n_{max} = 360)	EGM2008 (n_{max} = 360/2190)
Computation procedure		
Remove-compute-restore technique, spectral combination (1D FFT)		
GRS80 normal potential, zero-tide system		
Computation grid		
25°–77°N, 35°W–67.4°E	25°–85°N, 50°W–70°E	25°–85°N, 50°W–70°E
1.0′ × 1.5′	1.0′ × 1.0′	1.0′ × 1.0′
3,120 × 4,096 pts.	3,600 × 7,200 pts.	3,600 × 7,200 pts.

observations from 709 sources, plus 195,840 gravity values from the ArcGP project and 951,251 altimetric anomalies from the KMS2002 data set. On the other hand, the corresponding EGG2008 data set consists of 5,355,206 gravity observations from 718 sources, plus 195,840 ArcGP, 13,222,260 (1′ × 1′) altimetric, as well as 120,747 EGM2008 fill-in values; thus, in comparison to the previous EGG1997 computation, the land and marine data from the project data base approximately doubled, while the total amount of data increased about sixfold.

The progress in the collection of gravity data is also documented for selected examples in Fig. 5.9. The left part of the figure shows the old status in 1997 (EGG1997; Denker and Torge 1998) and the right part shows the new status as of September 2008 (EGG2008) for the whole of Europe (top), Scandinavia (middle), and the Mediterranean Sea (bottom; ship data from Morelli et al. 1975 excluded). In this context, it should be noted that within the EGGP, the Morelli ship gravity data for the Mediterranean Sea was completely excluded, as comparisons with newer data sources revealed significant systematic discrepancies in several areas.

Finally, it is important to mention that all EGGP gravity sources were carefully checked regarding the underlying horizontal and vertical position as well as the gravity reference systems, and, if necessary, transformations were done to the target systems, being ETRS1989 (European Terrestrial Reference System 1989, coinciding with ITRS at epoch 1989.0, and co-moving with the stable part of the

Eurasian tectonic plate; for further details see http://www.euref.eu/), EVRS (European Vertical Reference System, based on the zero tide system; see Ihde et al. 2008) and a gravity datum based on absolute gravity observations. In the merging process of the various data sources, emphasis was placed on a thorough check with respect to systematic and gross errors, which was one of the most time-consuming steps.

Apart from the improved gravity data, comparable progress was also made in the collection of high-resolution digital elevation models (DEMs). For the EGG1997 computation, digital elevation models (DEMs) with a resolution of about 200 m were only available for Central and Western Europe, while coarser grids with a resolution of 0.5–10 km had to be used for the remaining parts of Europe. As of 1997, only Germany had released a very high-resolution DEM with a grid size of $1'' \times 1''$ (approx. 30 m), but meanwhile Switzerland and Austria also provided $1'' \times 1''$ DEMs for the EGGP. At present, high-resolution national DEMs do not exist or are confidential for large parts of Eastern Europe. Hence, in all areas not covered by high-resolution national DEMs, fill-ins from public domain data sets had to be utilized. However, compared to EGG1997, significantly improved fill-ins are available now, e.g., from the Shuttle Radar Topography Mission (SRTM) with a resolution of $3'' \times 3''$ (SRTM3; JPL 2007) or the global public domain model GTOPO30 with a resolution of $30'' \times 30''$ (USGS 2007). As the SRTM3 model covers only the latitudes between 60°N and 54°S, the GTOPO30 model had to be used for the regions in the far North.

All available DEMs were merged into a new European DEM with a common grid size of $3'' \times 3''$, covering the area 25°N–85°N and 50°W–70°E. Furthermore, for the area of Germany, Austria, and Switzerland, a corresponding $1'' \times 1''$ DEM was created. The $3'' \times 3''$ and $1'' \times 1''$ DEMs comprise about 6.6 and 1.7 billion elevations, respectively. In the merging process, the highest priority was given to the national DEMs, followed by the SRTM3 and GTOPO30 data. For testing purposes, a second $3'' \times 3''$ European DEM was created using only the public domain data sets SRTM3 and GTOPO30. Depth models were not considered so far, i.e., elevations for ocean cells were set equal to zero.

Within the merging process, the SRTM3 and GTOPO30 DEMs were also evaluated by comparisons with the high-resolution national DEMs. In Germany, the differences between the national and the SRTM3 DEMs showed a standard deviation of 7.9 m with maximum differences up to about 300 m. The largest differences were located in opencast mining areas and resulted from the different epochs of the data. Histograms of the differences showed a clear deviation from the normal distribution with a long tail towards too high SRTM3 elevations, which is expected due to the fact that SRTM is a "first return system," providing elevations of whatever the radar has bounced off from, and in many instances this is above the actual ground level (Denker 2005).

The evaluation of the GTOPO30 model by national and SRTM3 DEMs demonstrated that in large parts of Europe the longitudes of GTOPO30 should be increased by $30''$ (one block). In Central Europe, the longitude shift reduced the standard deviation of the differences to the national and SRTM3 models by roughly 75 % to about 10 m. Altogether, the national DEMs augmented by the

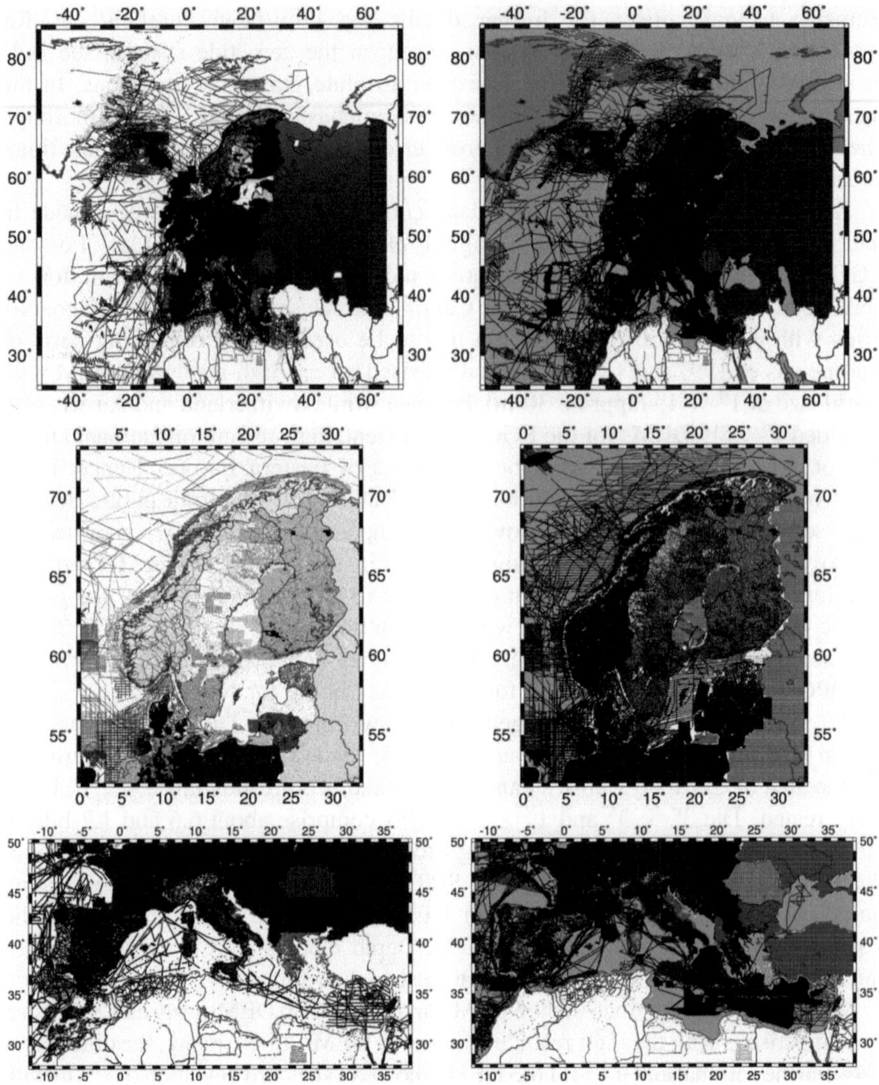

Fig. 5.9 Locations of terrestrial gravity data for entire Europe (*top*), Scandinavia (*middle*) and the Mediterranean Sea (*bottom*), excluding the Morelli data; the *left* part shows the status in 1997 (EGG1997) and the *right* part shows the status of September 2008 (EGG2008) with the ArcGP and altimetric data shown in *grey* and *blue*, respectively

SRTM3 and GTOPO30 data provide a significantly improved European DEM, as compared to EGG1997. The DEMs were not updated between 2007 and 2008, and hence both the EGG2007 and EGG2008 computations rely on the same DEM for Europe, which is depicted in Fig. 5.10.

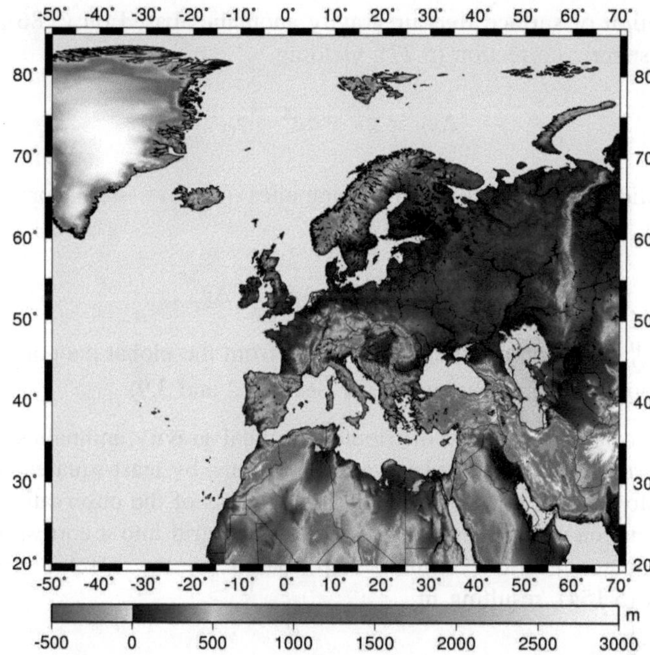

Fig. 5.10 Final digital elevation model (DEM) for Europe with a resolution of $3'' \times 3''$

4.4 Development of the European Quasigeoid Model EGG2008

The general computation strategy is based on the spatial gravity field modeling approach, aiming at the determination of the disturbing potential at the Earth's surface and the associated height anomalies or quasigeoid undulations (e.g., Denker et al. 2005). This concept has the advantage that only gravity field observations at the Earth's surface and in its exterior enter into the calculations, avoiding assumptions about the Earth's interior gravity field (as needed in connection with the orthometric heights and the geoid). A conversion of the height anomalies to geoid undulations can then be performed afterwards by introducing a density hypothesis, which should be consistent with that used for deriving the corresponding orthometric heights (e.g., the so-called Helmert heights based on (5.52); for further discussions see Sects. 2.4 and 3.5).

The remove-compute-restore (RCR) technique is utilized for combining the high-resolution terrestrial gravity and terrain data with a state-of-the-art global geopotential model. Terrain reductions are performed according to the residual terrain model (RTM) technique to smooth the data and to avoid aliasing effects (see Sect. 3.9). The transformation of the gravity anomalies into corresponding disturbing potential and height anomaly values is done by the spectral combination technique with integral formulas (see Sect. 3.6). In principle, the following steps are carried out:

1. Computation of surface free-air gravity anomalies based on (5.86), including the atmospheric correction (5.77), yielding

$$\Delta g_P = g_P + \delta g^A - \gamma_Q \; . \tag{5.178}$$

2. Computation of residual gravity anomalies (remove step) for all stations according to equation (5.173), giving

$$\Delta g_P^{\text{res}} = \Delta g_P - \Delta g_{Q*}^M - \Delta g_P^T \; , \tag{5.179}$$

   where Δg_{Q*}^M and Δg_P^T are the contributions from the global geopotential model and the topography, respectively (see Sects. 3.2 and 3.9).

3. Gridding of the irregularly distributed residual gravity anomalies, which are still referring to the actual observation positions, by least-squares collocation, taking into account the given standard deviations of the observations.
4. Transformation of the residual gravity anomaly grid into a corresponding disturbing potential grid (compute step) based on the spectral combination approach (5.154), resulting in

$$\widehat{T}_P^{\text{res}}(\theta, \lambda, r) = \frac{R}{4\pi} \iint_\sigma \Delta g_{P'}^{\text{res}} \, W(\psi, r) \, d\sigma \; . \tag{5.180}$$

5. Computation of the final disturbing potential by restoring the contributions from the geopotential model and the topography (restore step), yielding

$$\widehat{T}_P = \widehat{T}_P^{\text{res}} + T_P^M + T_P^T \; . \tag{5.181}$$

6. Conversion of the disturbing potential to height anomalies by means of (5.90).
7. Derivation of geoid undulations by introducing a density hypothesis, e.g., based on (5.56) and (5.52).

Regarding the details of the practical implementation of the above steps, the following more general notes are given:

- It is assumed that all station coordinates are based on the ETRS1989 reference system, the physical heights are normal heights based on the EVRS (zero tide system), and the gravity values are referring to an absolute gravity datum; regarding the accuracy requirements for the station coordinates, gravity values, etc.; see Sect. 4.1.
- The geodetic reference system GRS80 (e.g., Moritz 2000) is used as the normal gravity field in all computations, being the latest system recommended by IUGG and IAG; in addition, the GRS80 ellipsoid is recommended by IERS (2010) for

use with the ITRF solutions, and it is also mostly utilized for GNSS ellipsoidal coordinates.

- Atmospheric corrections for the observed gravity values are computed according to (5.77); the restore part δV^A (see (5.75), Sect. 2.6) is neglected due to its insignificant magnitude (of a few millimeters).
- The degree two zonal coefficient of the global geopotential model is always converted to the zero tide system, and hence the resulting height anomalies and geoid undulations also refer to the zero tide system.
- The different values of the constants GM (geocentric gravitational constant) and a (semimajor axis) in the global geopotential models and the GRS80 normal gravity field are handled rigorously throughout all computation steps; the resulting degree zero terms in the spherical harmonic expansions are taken into account in all calculations, see (5.110) and (5.111).
- The geopotential model parameters are so far not computed rigorously based on coordinate transformations of the relevant vectors and matrices (see Sects. 2.3 and 2.5), but instead ellipsoidal approximations based on Wenzel (1985) are employed, which also provide sufficiently accurate results. Furthermore, the geopotential model values were until now only computed utilizing the normal heights H^N, and not $H^N + \zeta_{Q*}$, as required within a rigorous linearization process with respect to a high-degree geopotential model (see Sect. 3.2). However, for future calculations the rigorous approach will be implemented, and it remains to be seen whether this also leads to improved results.
- Topographic reductions are computed based on the RTM technique (see Sect. 3.9), which results in a significant smoothing of the relevant gravity field quantities and reduces aliasing effects. The required reference topography was always computed by a moving average filter from the available DEMs; this ensures the consistency between the high-resolution DEMs and the reference topography, without creating undesirable long wavelength signals. The reference topography had a resolution (size of the moving average filter) of $30' \times 45'$ for EGG2007, and $15' \times 20'$ for EGG2008 (see also Table 5.1). The resolution of the reference topography was reduced for EGG2008, because even the RTM technique creates some small long wavelength signals, which lead to an inconsistency within the RCR procedure, as these signals are suppressed in the compute step (by the spectral combination approach), but are fully considered in the restore step on the other hand. All computations were done with the program TC based on prism integration; regarding the gravity stations, the DEM was forced to reproduce the given station elevations (see Sect. 3.9; Forsberg 1984). Figure 5.11 shows an example for the surface free-air gravity anomalies and the corresponding RTM reduced values; the figure clearly documents that the RTM reductions lead to a significant smoothing of the anomaly field, while preserving the long wavelength features.
- The spectral combination technique was employed so far for the compute step, because it allows an optimal combination of the terrestrial data with a global geopotential model based on the error characteristics of both data sets, and furthermore the resulting integral formulas can be evaluated rigorously and very effectively by 1D FFT techniques (Haagmans et al. 1993). In this context, the

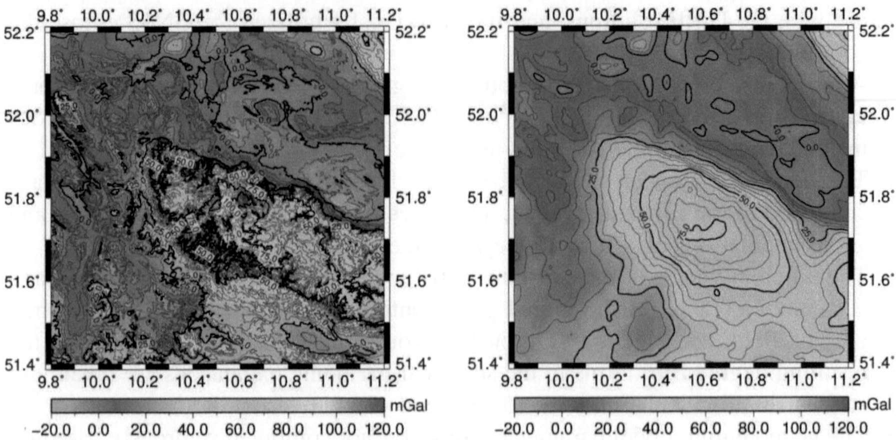

Fig. 5.11 Surface free-air gravity anomalies (*left*) and corresponding RTM 15′ × 20′ reduced values (*right*) for the area of the Harz mountains, Germany

regional terrestrial data, with possibly existing (small) long wavelength systematic errors, and a global geopotential model, being highly accurate at the very long wavelengths, complement each other in an optimal way; hence, within the combination process, the very accurate long wavelength gravity field structures of the present global geopotential models (based on the GRACE and GOCE missions) should be retained, while the terrestrial data should mainly contribute the shorter wavelength components. In addition, previous investigations clearly showed that the application of the original Stokes formula, implying that the complete disturbing potential spectrum (from degree 2 to ∞) is computed from the terrestrial gravity data plus the geopotential model in the outer zone, leads to unreasonable long wavelength distortions of the results, and therefore it is not well suited in this context. Further specific details on the implementation of the spectral combination procedure are given below.

- The computation area for the EGG2007 and EGG2008 models is 25°N–85°N and 50°W–70°E. The grid spacing is $1' \times 1'$, yielding 3,600 × 7,200 = 25,920,000 grid points (see also Table 5.1).

The key ingredients within the spectral combination approach are the spectral weights according to (5.149), which depend on the error degree variances associated with the terrestrial gravity data and the geopotential model. Regarding the terrestrial gravity data, the starting point was the following error covariance function:

$$Cov(\varepsilon_{\Delta g^G}, \varepsilon'_{\Delta g^G}, \psi, R) = 1_{\,[\mathrm{mGal}^2]}\, e^{-4\psi_{[^\circ]}} . \qquad (5.182)$$

This covariance model considers correlated noise and was originally suggested and applied by Weber (1984). Then, based on (5.148), corresponding error degree variances can be computed, and, together with the error degree variances for the

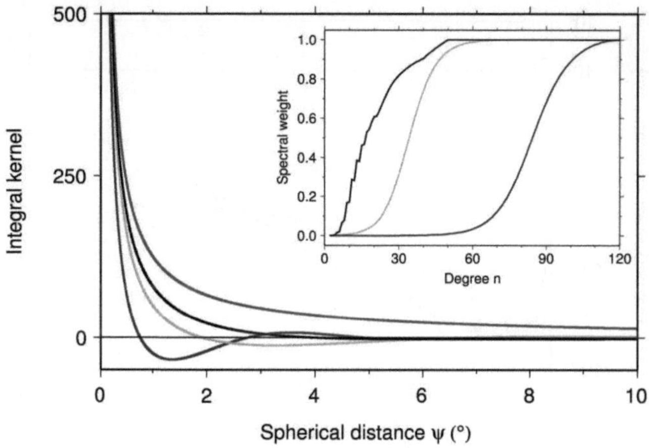

Fig. 5.12 Spectral weights and corresponding integral kernels related to a GRACE based geopotential model (*red*; used for EGG2007 and EGG2008), a recent CHAMP model (*green*), and EGM1996 (*black*; used for EGG1997), along with the classical Stokes kernel (*blue*)

geopotential model according to (5.147), the spectral weights can be estimated from (5.149).

For the computation of EGG2007, the geopotential model EIGEN-GL04C (Förste et al. 2008a) was employed, because at that time it was the latest available high-degree model based on GRACE and terrestrial data. The spectral weights, computed as described above, are shown in Fig. 5.12 together with corresponding values related to a recent CHAMP model and the EGM1996 model (used for the EGG1997 computation). In addition, Fig. 5.12 also depicts the modified integral kernels associated with the corresponding spectral weights as well as the original Stokes kernel. With respect to EGG2007, it was decided to do the combination only up to degree 120, while between degrees 120 and 10,000 (corresponding to the used grid size) full weight ($w_n^G = 1.0$) was given to the terrestrial gravity data in order to exploit fully the collected European gravity sources. However, this does not imply that the global model is completely disregarded above degree and order 120, as, e.g., in areas with larger data gaps the high-degree gravity information of the model is considered in the gridding process and thus practically taken over in the final combined solution. In this context, it is also worth mentioning that previous studies revealed that it is advantageous to use a high-degree model up to degree $n_{\max} = 360$, as this leads to smaller residual quantities accompanied with reduced effects of (e.g., linear) approximation errors in the mathematical modeling. In addition, a cosine tapering window was applied between degrees 10,000 and 30,000 in order to prevent oscillations of the integral kernel. As already mentioned in Sect. 3.6, the resulting modified integral kernels $W(\psi)$ remain finite if the weights go to zero for very high degrees or the summation is limited to some maximum degree; therefore, in principle, no special consideration of the inner zone contribution is required, but due to the rapid change of the integration kernel

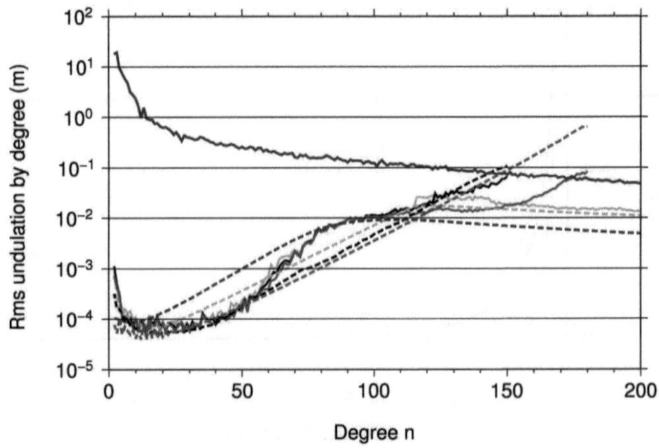

Fig. 5.13 Signal and error spectra for EGM2008 (*red thick* and *dashed lines*, respectively), along with the error spectra (*dashed lines*) for EIGEN-GL04C (*green*), EIGEN-5S (*black*), and ITG-GRACE2010 (*blue*; formal standard deviations scaled by factor 8), as well as the corresponding difference spectra with respect to EGM2008 (*thick lines*)

near $\psi = 0°$ (see Fig. 5.12) it is recommended to integrate numerically the kernel function within the innermost zone (see also Sect. 3.6). Consequently, a numerical kernel integration was implemented using 21×21 points for the innermost (central) grid cell and 11×11 points out to a distance of $\psi = 0.5°$; outside this distance, the kernel value is simply calculated based on the distance to the cell center. Moreover, to speed up the computations, internally a coarse grid is computed and employed for the remote zones, and in addition, kernel tabulation and interpolation are implemented.

In spring 2008, the new global geopotential model EGM2008 became available, given as spherical harmonic coefficients complete to degree and order 2,159, with additional coefficients going up to degree 2,190 and order 2,159 (Pavlis et al. 2008). The EGM2008 model is based on a corresponding GRACE-only model (ITG-GRACE03, $n_{\max} = 180$, including GRACE data from September 2002 to April 2007, computation method described in Mayer-Gürr 2006), along with its complete covariance matrix, and a new and comprehensive worldwide $5' \times 5'$ terrestrial gravity anomaly data set, combined by a least-squares adjustment, using internally ellipsoidal harmonic coefficients (Pavlis et al. 2008). At first the EGM2008 model was compared with corresponding GRACE static gravity field solutions. Figure 5.13 depicts the signal and error spectra related to EGM2008, along with the error spectra of the recent GRACE based models EIGEN-GL04C (Förste et al. 2008a; utilized for the EGG2007 calculation), EIGEN-5S (Förste et al. 2008b), and ITG-GRACE2010 (Mayer-Gürr et al. 2010; formal standard deviations scaled by factor 8), as well as the corresponding spectra of the differences to EGM2008, respectively. Now, regarding roughly the degree range 20–90, the EGM2008 error estimates are about five to six times higher than those from the more recent EIGEN-5S and ITG-GRACE2010 models, and about three times higher than those from the EIGEN-GL04C model. On the other hand, the depicted difference spectra are all very similar and show a reasonable agreement with the error spectra of the EIGEN-5S, ITG-GRACE2010 and EIGEN-GL04C models until about degrees 60–70, but the difference spectra significantly exceed the latter error curves in the degree range 60–100, while still being compatible with the EGM2008 error estimates; it is also worth mentioning that the same features show up with reference to the GGM03S/C models (Tapley et al. 2007), not shown in Fig. 5.13. Based on these findings, it appears that the EGM2008 error estimates are perhaps too pessimistic for the low degrees (degrees less than 60–70), while the bump in the difference signals (degrees 60–100) is probably related to the EGM2008 weighting procedure.

As a result, the spectral weights computed from (5.149) on the basis of the (perhaps too pessimistic) EGM2008 error estimates and the error covariance function for the terrestrial gravity data (1 mGal correlated noise; see (5.182)) turned out to be somewhat unrealistic, with too much weight given to the terrestrial data, dominating the combination solution (i.e., $w_n^G > 0.5$) already below degree 70. In principle, this could be counteracted by increasing the error estimates for the terrestrial data or decreasing the EGM2008 error estimates, as only the relative weighting matters. However, this was not attempted, but instead the spectral weights from the EGG2007 calculation were also adopted for EGG2008 in connection with the EGM2008 model, as these weights appear to be quite reasonable (with $w_n^G > 0.5$ at about degree 85; see Fig. 5.12); in addition, this weighting scheme is in very good agreement with the studies from Forsberg (2010), who found empirically from GPS/leveling comparisons that linearly increasing weights between degrees 80 ($w_n^G = 0.0$) and 90 ($w_n^G = 1.0$) are optimal. Furthermore, it is also noted that the combination solutions based on the original EGM2008 spectral weights, in contrast to those shown in Fig. 5.12, performed slightly worse in some of the GPS/leveling comparisons.

Another issue relates to the maximum degree of the employed geopotential model in conjunction with the reference topography used for the RTM reductions. This item, mainly concerning the ultra-high-degree model EGM2008, has been discussed already in Sect. 3.9 with reference to the studies in Forsberg (2010), showing "some inherent problems in implementing the RTM method for a highly varying reference topography" with "best results obtained for a relatively low-resolution (30′) reference height, irrespective of whether EGM2008 is used at a corresponding resolution ($n_{\max} = 360$) or to full resolution ($n_{\max} = 2{,}190$)." With respect to the European calculations, RTM reductions based on a 30' × 45' (EGG2007) and 15′ × 20′ (EGG2008) reference topography were tested and used in combination with the full resolution EGM2008 model ($n_{\max} = 2{,}190$); however, the GPS/leveling comparisons indicated a slight deterioration in a few but not all cases as compared to the solutions based on EGM2008 with $n_{\max} = 360$, and further studies are needed to understand fully this matter. Possible reasons may be related to the "double consideration" of the short wavelength topographic signals in the RTM reductions and the EGM2008 model, the non-rigorous linearization with respect to the geopotential model (i.e., the use of heights H^N instead of $H^N + \zeta_{Q*}$; see above and Sect. 3.2), as well as due to the RTM reductions themselves. Therefore, it was ultimately decided to employ the EGM2008 model only with $n_{\max} = 360$ in the EGG2008 calculation, accompanied by some fill-ins over large data voids (see above) to ensure that the full resolution EGM2008 model and the EGG2008 quasigeoid stay reasonably close together even there.

In principle, the Molodensky corrections terms, considering data on a non-level surface, also have to be taken into account in connection with the spectral combination approach (because the employed residual gravity anomalies are still referring to the actual observation positions). In this context, Molodensky's theory (see Sect. 3.4) and terrain reductions (see Sect. 3.9) are complementary, because the application of terrain reductions results in a significant smoothing of the gravity field observations, associated with a corresponding reduction of the Molodensky correction terms and a more stable computing scheme (Forsberg and Tscherning 1997). The magnitude of the Molodensky terms was studied in Denker and Tziavos (1999) in conjunction with different terrain reduction techniques, indicating that the maximum values associated with the RTM technique may reach about 5–10 cm in extremely rugged areas of the European Alps ($\approx$1 cm RMS) and about 1 cm in low mountain ranges ($\approx$1 mm RMS). However, utilizing the so-called gradient solution (see Sect. 3.4) for Switzerland and Austria did not lead to any improvements in the GPS/leveling comparisons, with similar findings reported by Forsberg (2010). Therefore, the Molodensky terms have been neglected so far in the EGG2007 and EGG2008 calculations.

One final item concerns the European Vertical Reference System (EVRS), where the vertical datum (zero level surface) is defined as the (zero tide) equipotential surface of the Earth's gravity field, which passes through the "Normaal Amsterdams Peil" (NAP; fundamental tide gauge in Amsterdam, the Netherlands) and which has the (constant) gravity potential W_0^{EVRS} (see Ihde et al. 2008). The latest EVRS realization is the EVRF2007 (European Vertical Reference Frame 2007), consisting of a set of points with precisely determined geopotential numbers and normal heights relative to the aforementioned zero level surface through the NAP at epoch 2000.0 (Ihde et al. 2008); EVRF2007 is the recommended reference frame for all pan-European applications. However, to remain general, the notation W_0^i is used for the potential of the zero level surface of a local vertical datum (i). Now W_0^i, being initially unknown, will in general differ from U_0, which has to be considered accordingly in (5.89) as well as (5.95) or (5.99) for the height and gravity anomalies, respectively. While the (small) constant term in the gravity anomaly equations has practically no effect on the computed disturbing potential T (because it is of long wavelength nature, and such signals are almost entirely defined by the global geopotential model; see above), and the zero-degree term of T (i.e., T_0 due to the different GM values; see (5.128)) is already taken into account within the calculation process (see above), Bruns's formula (5.89) together with (5.87) results in

$$\zeta_P^i = h_P - H^{N(i)} = \frac{T_P}{\gamma_Q} - \frac{W_0^i - U_0}{\gamma_Q} = \frac{T_P}{\gamma_Q} + \zeta_0^i , \tag{5.183}$$

where T_P is the disturbing potential (including the zero-degree term, as obtained directly from the above described computations) relative to the (GRS80) normal potential (U_P, U_0, respectively), $H^{N(i)}$ is the normal height based on the vertical datum (i), and ζ_0^i is a virtually constant term to account for the potential difference $W_0^i - U_0$. Regarding the EVRF2007, the latter constant was determined by comparisons with GPS and leveling data (i.e., $h_P - H^{N\,(\mathrm{EVRF2007})}$) from the European EUVN_DA data set (Kenyeres et al. 2010; see Sect. 4.5) as +0.302 m. As a result, a slightly rounded value for ζ_0^i (based also on an earlier EUVN_DA release) was employed for the computation of the final EGG2008 height anomalies (quasigeoid heights), yielding

$$\zeta_P^{EGG2008} = \frac{T_P}{\gamma_Q} + \zeta_0^{\mathrm{EVRF2007}} , \qquad \text{with} \quad \zeta_0^{\mathrm{EVRF2007}} = +0.300 \text{ m} . \tag{5.184}$$

The above correction ensures the compatibility between the European GPS (ETRS1989) and leveling data (EVRS, normal heights, zero tide system, etc.) on the one hand, and the EGG2008 height anomalies on the other hand.

Table 5.2 shows the statistics of the 18,154,254 irregularly distributed gravity anomalies without an error flag, which were used as input for the computation of the EGG2008 residual gravity anomaly grid; the number of points is smaller than that given in Table 5.1, because Table 5.2 excludes the stations which received an error flag (mainly edited altimetry data near the coast as well as some very bad and duplicate marine gravity sources; see Sect. 4.3). The standard deviation (std. dev.) of the original gravity anomalies is 34.11 mGal, which reduces to 14.95 mGal after subtracting EGM2008 ($n_{\max} = 360$) and 13.90 mGal after also subtracting the RTM contributions. The minimum and maximum values also reduce accordingly, and the mean value of the final residual anomalies is –0.57 mGal and thus reasonably close to zero (as it should be).

The statistics of the corresponding quasigeoid height or height anomaly terms for EGG2008 are given in Table 5.3. The major contribution to the final quasigeoid comes from the global geopotential model EGM2008 ($n_{\max} = 360$) with values ranging from –48.655 to +67.551 m and a standard deviation of 24.257 m. The standard deviations of the contributions from the topography and the terrestrial gravity data are 0.036 and 0.161 m, respectively. However, the maximum RTM effects are about 0.9 m, while the maximum effects of terrestrial gravity data are about 2.4 m (all large values are located in Asia or Africa). In addition, the final EGG2008 quasigeoid is depicted in Fig. 5.14.

The spectral combination technique also permits the derivation of error estimates for the computed quasigeoid heights on the basis of corresponding error degree variances. According to (5.156) and (5.157) the error degree variances related to the terrestrial gravity data and the global geopotential model are required. Table 5.4 shows the results based on 1 mGal correlated noise for the

Table 5.2 Statistics of 18,154,254 irregularly distributed gravity anomalies (without error flag) that were used as input for EGG2008; units are mGal

Parameter	Mean	Std. dev.	Minimum	Maximum
Δg	+6.59	34.11	−333.36	+498.88
Δg^M (EGM2008, $n_{\max} = 360$)	+7.49	31.27	−226.32	+236.25
Δg^T (RTM $15' \times 20'$)	−0.34	4.98	−215.21	+182.72
$\Delta g - \Delta g^M$	−0.90	14.95	−268.74	+290.52
$\Delta g - \Delta g^M - \Delta g^T = \Delta g^{\text{res}}$	−0.57	13.90	−163.43	+263.23

Std. dev.: standard deviation

Table 5.3 Statistics of 3,600 × 7,200 = 25,920,000 quasigeoid heights of the EGG2008 grid; units are m

Parameter	Mean	Std. dev.	Minimum	Maximum
$\zeta^{\text{res}} = \mathbf{S}(\Delta g^{\text{res}})$	0.000	0.161	−1.657	+2.361
ζ^T (RTM $15' \times 20'$)	0.000	0.036	−0.493	+0.934
ζ^M (EGM2008, $n_{\max} = 360$)	+26.498	24.257	−48.665	+67.551
$\zeta_0^{\text{EVRF2007}}$	+0.300	0.000	+0.300	+0.300
ζ (EGG2008)	+26.798	24.258	−48.858	+68.104

Std. dev.: standard deviation

terrestrial gravity data, see (5.182), and the error degree variances from different geopotential models; as the EGM2008 error estimates may be a little too pessimistic (see the discussion above), corresponding estimates from the EIGEN-GL04S1/GL04C and EIGEN-5S/5C models were also taken into consideration. In this context, the results based on the error estimates of the "S" (satellite-only) and "C" (combined) models do not differ significantly, because due to the selected weighting scheme (see Fig. 5.12), the error degree variances of the global model come into play only up to degree $n_{\max} = 120$, see (5.157), where the "S" and "C" values do not differ significantly. Table 5.4 gives a standard deviation for the quasigeoid heights of 3.1 cm related to the EGM2008 model and about 2.5 cm related to the EIGEN models, which can be viewed as the pessimistic and optimistic case, respectively. Table 5.4 also shows that the major error contributions are coming from the spectral band with $n = 50$–360, while today the very long wavelengths ($n < 50$) are accurately known from the GRACE mission and the short wavelengths ($n > 360$) can be obtained from high-quality terrestrial data. However, once GOCE can deliver the quasigeoid up to a resolution of 100 km ($n \approx 200$) with an accuracy of about 1 cm, the total error (complete spectrum) will reduce to 1.7 cm (for a corresponding study related to EGG1997, see Denker 1998). In addition to this, an error covariance function was derived for the quasigeoid heights based on (5.156), utilizing the more optimistic EIGEN-5S/5C error estimates; the result is depicted in Fig. 5.15, showing significant error correlations up to distances of about 300 km. Finally, it should be noted that the aforementioned error estimates apply only to those regions in Europe, where high-quality terrestrial gravity data exists, while in other areas (mainly Eastern Europe) less accurate results have to be expected. In the end, the future perspectives for

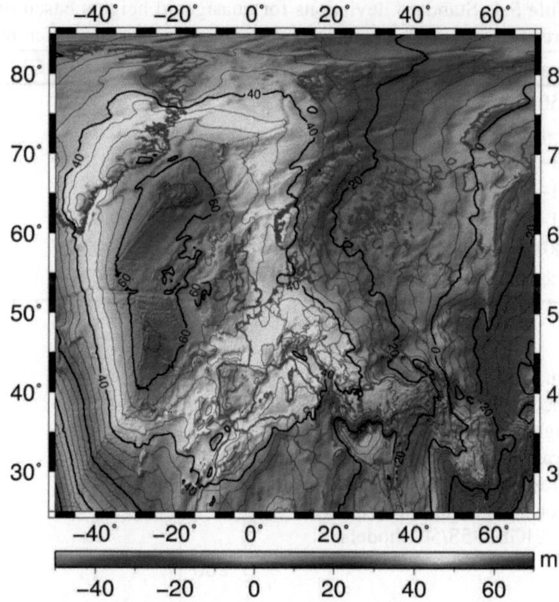

Fig. 5.14 Final EGG2008 quasigeoid referring to GRS80 (see text; units are m)

calculating gravimetric geoid/quasigeoid models with an accuracy of 1 cm are quite good with respect to well-surveyed regions, where such models can then replace geometric leveling and serve as a vertical datum, e.g., as planned in Canada and the U.S.A. (see Sect. 2.4).

Besides the commission error, the truncation error, resulting from the truncation of the kernel function (or integration) at some distance $\psi_{\max}$, is also of interest. The truncation error can be derived from (5.160) and (5.161) by means of the frequency transfer function, FTF_n, see (5.159). For $\psi_{\max} = 10$, 7.5, and 5.0°, the truncation error is estimated as 0.7, 1.6, and 3.5 mm, respectively. Moreover, the truncation error remains below 1 cm for $\psi_{\max}$ larger than about 3°, while truncation errors of about 18 mm and 25 mm are obtained for $\psi_{\max} = 2$ and 1° (all figures given in terms of standard deviation). In addition, the frequency transfer function (FTF_n) and the spectral weights differ by no more than about 1.5 % for $\psi_{\max} = 10°$. Finally, it is noted that in the practical computations, kernel truncation is not used at all, i.e., for every computation point the complete input grid is employed; this was done because kernel truncation offers no computational advantages and previous studies showed that kernel truncation may lead to unfavorable results, as for every computation point another input data field is utilized (which does not conform with the harmonicity condition, see the corresponding discussion on the computation of terrain reductions in Sect. 3.9 as well as Wolf 2008).

Table 5.4 Standard deviations for quasigeoid heights based on 1 mGal correlated noise for the terrestrial gravity data and error degree variances from different geopotential models; units are m

Degree range	EGM2008	EIGEN-GL04S1/C	EIGEN-5S/5C
2–50	0.0028	0.0012	0.0007
51–100	0.0213	0.0114	0.0097
101–200	0.0184	0.0184	0.0184
201–360	0.0115	0.0115	0.0115
361–2,000	0.0071	0.0071	0.0071
2,001–∞	0.0006	0.0006	0.0006
2–∞	0.0314	0.0256	0.0248

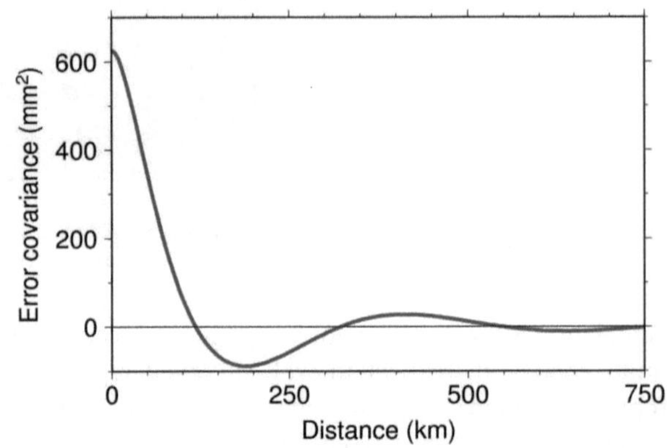

Fig. 5.15 Error covariance function for quasigeoid heights based on 1 mGal correlated noise for the terrestrial gravity data and error degree variances from the EIGEN-5S/5C models

4.5 Evaluation of the European Quasigeoid Model EGG2008

The EGG2008 quasigeoid model and all other interim solutions as well as the previous EGG1997 release were evaluated by independent national and European GPS and leveling data sets. The ellipsoidal GPS heights as well as the leveled heights (all given as normal heights) were converted to the zero tide system based on the transformation formulas published by Ihde et al. (2008). Regarding the GPS heights, it was generally assumed that they refer to the (conventional) tide-free system, as this is common practice and standard for the ITRF products (see Sect. 2.1); the transformation to the zero-tide system decreases the ellipsoidal heights by approximately 4.5 cm on average over the European continent. The leveling heights were usually treated as mean tide quantities, as this is a reasonable approximation for the typical case of not applying any tidal reductions to the leveling; the conversion from the mean tide system to the zero tide system was carried out relative to the central latitude of the GPS/leveling data set, which does not change the average height level of all stations, but in principle the corrections should be calculated relative to the fundamental datum point.

Table 5.5 shows the statistics of the differences between a German GPS/leveling data set, consisting of 907 stations (data from Bundesamt für Kartographie und

Table 5.5 Comparison of 907 GPS and leveling stations in Germany with quasigeoid models based on different terrestrial data sets and geopotential models; the GPS/leveling data were converted to the zero tide system; the differences are defined in the sense GPS/leveling minus gravimetric quasigeoid; the mean values refer to the raw differences without applying the constant $\zeta_0^{EVRF2007}$; the other statistical parameters are calculated after subtracting the mean value; units are m

Terrestrial data	Geopotential model	Mean	RMS	Minimum	Maximum
1997 (EGG1997)	EGM1996	+0.431	0.096	−0.188	+0.331
1997	EGM2008	+0.302	0.029	−0.095	+0.089
2007 (EGG2007)	EIGEN-GL04C	+0.298	0.036	−0.159	+0.075
2008	EGM1996	+0.416	0.074	−0.132	+0.300
2008	EIGEN-CHAMP03S	+0.288	0.050	−0.116	+0.262
2008	EIGEN-GRACE01S	+0.290	0.038	−0.111	+0.152
2008	EIGEN-GRACE02S	+0.295	0.037	−0.080	+0.123
2008	EIGEN-GL04S1	+0.299	0.029	−0.097	+0.086
2008	EIGEN-GL04C	+0.300	0.028	−0.093	+0.082
2008	EIGEN-5S	+0.300	0.027	−0.097	+0.073
2008	EIGEN-5C	+0.298	0.028	−0.098	+0.073
2008 (EGG2008)	EGM2008	+0.297	0.027	−0.091	+0.078
–	EGM2008 ($n_{max} = 2{,}190$)	+0.302	0.031	−0.110	+0.148

Geodäsie, BKG, Frankfurt; e.g., Liebsch et al. 2006), and various quasigeoid calculations based on different terrestrial data sets (1997, 2007, 2008) and geopotential models; the differences were always computed in the sense GPS/leveling minus gravimetric quasigeoid. All quasigeoid models were computed by the spectral combination technique based on 1 mGal correlated noise for the gravity data, see (5.182), and the error estimates for the geopotential model, where the spectral weights related to the EGM2008 geopotential model are a special case, as described in Sect. 4.4. As a result, the long wavelength components of the regional quasigeoid models and the underlying global geopotential model match to a great extent, and hence the comparisons of such regional quasigeoids with GPS/leveling data can also be considered as a validation tool for the respective global geopotential model. Table 5.5 provides the mean values of the raw differences without applying the constant ζ_0^i, while the other statistical parameters (RMS, Minimum, Maximum) are related to the centered differences (i.e., after subtracting the corresponding mean value). The results in Table 5.5 clearly demonstrate the enormous progress resulting from improved gravity and terrain data on the one hand and the global geopotential models based on the satellite missions CHAMP and GRACE on the other, with the RMS differences reducing by about a factor of 3.5 from 9.6 cm for EGG1997 to 2.7 cm for EGG2008. In detail, the combination of the older EGM1996 model with the terrestrial data sets from 1997 and 2008 leads to RMS differences of 9.6 and 7.4 cm, respectively, corresponding to an improvement of about 23 % (related to the updated terrestrial data), but, on the other hand, the combination of the 1997 terrestrial data with the EGM2008 model results in an RMS difference of only 2.9 cm (improvement 70 %); thus, most of the total improvement is due to the better satellite data in this case. Furthermore, considering the different EIGEN models from GFZ, the RMS differences improve in the course of time, where the newer models, associated with longer observation series, yield the best results; for example, the combination of the 2008 terrestrial data with EGM1996 gives an RMS difference of 7.4 cm, which reduces to 5.0 cm related to the CHAMP model (EIGEN-CHAMP03S), about 3.8 cm for the early GRACE models (EIGEN-GRACE01S/02S), and finally 2.7 cm for the recent GRACE based models (EIGEN-5S/C, EGM2008). Consequently, the new satellite mission data (CHAMP, GRACE) have a significant impact on the accuracy of regional quasigeoid models, and further improvements down to the level of about 1 cm are anticipated from the GOCE mission. In addition, the mean values associated with the CHAMP and GRACE based geopotential models are remarkably stable, differing by no more than about 1 cm. Besides the results given in Table 5.5, the differences between the German GPS/leveling data and EGG1997 as well as EGG2008 are illustrated in Fig. 5.16, showing again the progress from the 1997 to the 2008 quasigeoid models, especially with regard to the long wavelengths.

The statistics from further comparisons of national GPS and leveling data sets as well as the European EUVN_DA enterprise (Kenyeres et al. 2010) with selected quasigeoid solutions are presented in Tables 5.6 and 5.7. The comparisons were carried out in conformity with the German GPS/leveling data set (i.e., zero tide system, normal heights, no ζ_0^i constant considered, etc.), with the normal heights referring to the respective national vertical datums. Again, several combinations of the terrestrial data and geopotential models were utilized to assess the progress associated with the improved input data sets; besides the quasigeoid model EGG1997 (1997 terrestrial data, geopotential model EGM1996, denoted as EGG1997/EGM1996), the solutions EGG1997/EGM2008 (1997 gravity and terrain data, EGM2008 geopotential model), EGG2008 (2008 terrestrial data, EGM2008 geopotential model, denoted as EGG2008/EGM2008), as well as the pure EGM2008 model with $n_{\max} = 2{,}190$ are addressed in Tables 5.6 and 5.7. Besides a summary of the comparisons with the German GPS/leveling data, Table 5.6 includes corresponding results for the Netherlands, Belgium, a French traverse from Marseille to Dunkerque (1,100 km long) with new leveling data (NIREF; Duquenne et al., 2007), a French national data set (based on the older leveling network IGN69), Switzerland, Austria, and Russia (Demianov and Majorov 2004), while the results relating to the EUVN_DA project (Kenyeres et al. 2010) are listed in Table 5.7. The EUVN_DA project aimed at a densification of the previous EUVN campaign (Ihde et al. 2000) by collecting high-quality GPS and leveling data from participating European countries. At present, about 1,400 points are available with interstation distances ranging from about 50–100 km. The EUVN_DA data set is based on the reference systems ETRS1989 (GPS) and EVRS (leveling data, present realization EVRF2007; see Ihde et al. 2008), with the zero tide system implemented for both GPS and leveling data; the normal heights were derived in part directly from geopotential numbers as well as by simple transformations with up to three parameters (Kenyeres et al. 2010).

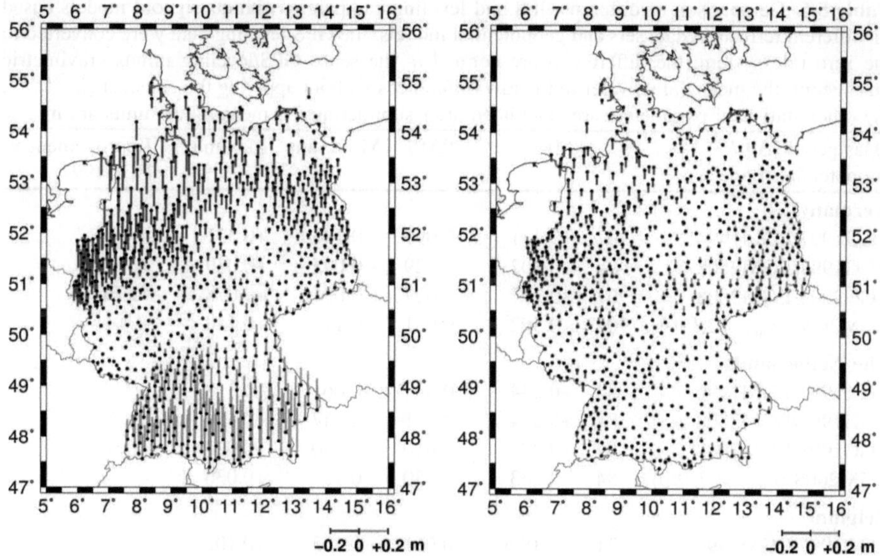

Fig. 5.16 Comparison of 907 GPS and leveling stations in Germany with the quasigeoid solutions EGG1997 (*left* part) and EGG2008 (*right* part); a constant bias is subtracted; stations (*dots*) and differences (positive: up in *red*; negative: down in *blue*) are depicted

In addition to the numerical results given in Tables 5.6 and 5.7, the differences of the EUVN_DA as well as the Russian GPS/leveling campaign with respect to EGG1997 and EGG2008 are illustrated in Fig. 5.17. On the whole, the EGG2008 model performs significantly better than EGG1997. The improvements result from the updated gravity and terrain data as well as from the utilization of better geopotential models (based on the GRACE mission). Tables 5.6 and 5.7 show that solely through the introduction of the (GRACE based) EGM2008 geopotential model (i.e., EGG1997/EGM1996 vs EGG1997/EGM2008), the RMS differences reduce by between 28 % (Switzerland) and 70 % (Germany). However, the update and re-processing of the gravity and terrain data also leads to substantial improvements in the GPS and leveling comparisons in all cases. The additional improvements from the upgraded terrestrial data (EGG1997/EGM2008 vs EGG2008/EGM2008) range from 2 % (Germany), 18 % (Russia; see Fig. 5.17, bottom), 20 % (EUVN_DA, excluding Great Britain and Italy, see below), 25 % (Austria), to about 33 % (the Netherlands); the improvements are particularly high in those areas where the data basis was significantly extended, e.g., in the Netherlands, Austria, Russia, as well as other regions of Europe. The overall improvement of EGG2008 over EGG1997 ranges from about 35–72 %, and also the long wavelength discrepancies are significantly reduced from 0.1 to 1.0 ppm for the EGG1997 model to typically below 0.1 ppm for all GRACE based solutions (see also Denker et al. 2009). Therefore, the consideration of additional tilt parameters in north–south and west–east directions between the respective quasigeoid heights from GPS/leveling and EGG2008 (see Sect. 2.1, formula (5.5))

Table 5.6 Comparison of different GPS and leveling campaigns with quasigeoid models based on different terrestrial data sets and geopotential models; the GPS/leveling data were converted to the zero tide system; the differences are defined in the sense GPS/leveling minus gravimetric quasigeoid; the mean values refer to the raw differences without applying the constant $\zeta_0^{EVRF2007}$; the other statistical parameters are calculated after subtracting the mean value; units are m

Quasigeoid (Δg / geopotential model)	#	Mean	RMS	Minimum	Maximum	Improvement vs EGG1997
Germany						
EGG1997/EGM1996	907	+0.431	0.096	−0.188	+0.331	–
EGG1997/EGM2008	907	+0.302	0.029	−0.095	+0.089	70 %
EGG2008/EGM2008	907	+0.297	0.027	−0.091	+0.078	72 %
EGM2008 (n_{max} =2,190)	907	+0.302	0.031	−0.110	+0.148	–
The Netherlands						
EGG1997/EGM1996	84	+0.244	0.034	−0.061	+0.118	–
EGG1997/EGM2008	84	+0.234	0.021	−0.047	+0.050	38 %
EGG2008/EGM2008	84	+0.255	0.010	−0.040	+0.027	71 %
EGM2008 (n_{max} = 2,190)	84	+0.263	0.030	−0.135	+0.036	–
Belgium						
EGG1997/EGM1996	31	−2.005	0.061	−0.103	+0.102	–
EGG1997/EGM2008	31	−2.054	0.031	−0.055	+0.046	49 %
EGG2008/EGM2008	31	−2.065	0.028	−0.053	+0.048	54 %
EGM2008 (n_{max} = 2,190)	31	−2.060	0.019	−0.047	+0.037	–
France(North-South traverse with new leveling)						
EGG1997/EGM1996	16	−0.027	0.086	−0.188	+0.124	–
EGG1997/EGM2008	16	−0.088	0.032	−0.051	+0.068	63 %
EGG2008/EGM2008	16	−0.097	0.026	−0.024	+0.059	70 %
EGM2008 (n_{max} = 2,190)	16	−0.100	0.038	−0.066	+0.082	–
France (Nationwide campaign)						
EGG1997/EGM1996	965	−0.132	0.125	−0.295	+0.351	–
EGG1997/EGM2008	965	−0.180	0.080	−0.227	+0.258	36 %
EGG2008/EGM2008	965	−0.188	0.076	−0.221	+0.191	39 %
EGM2008 (n_{max} = 2,190)	965	−0.181	0.084	−0.271	+0.346	–
Switzerland						
EGG1997/EGM1996	188	+0.535	0.080	−0.129	+0.258	–
EGG1997/EGM2008	188	+0.117	0.058	−0.201	+0.282	28 %
EGG2008/EGM2008	188	+0.174	0.052	−0.157	+0.230	35 %
EGM2008 (n_{max} = 2,190)	188	+0.141	0.056	−0.170	+0.170	
Austria						
EGG1997/EGM1996	170	+0.660	0.108	−0.202	+0.248	–
EGG1997/EGM2008	170	+0.356	0.064	−0.129	+0.197	41 %
EGG2008/EGM2008	170	+0.361	0.037	−0.098	+0.100	66 %
EGM2008 (n_{max} = 2,190)	170	+0.338	0.071	−0.212	+0.171	
Russia						
EGG1997/EGM1996	48	+0.574	0.256	−0.776	+0.707	–
EGG1997/EGM2008	48	+0.560	0.124	−0.261	+0.300	52 %
EGG2008/EGM2008	48	+0.555	0.076	−0.134	+0.163	70 %
EGM2008 (n_{max} = 2,190)	48	+0.555	0.072	−0.120	+0.141	–

Table 5.7 Comparison of the EUVN_DA GPS and leveling data set with quasigeoid models based on different terrestrial data sets and geopotential models; the GPS/leveling data were converted to the zero tide system; the differences are defined in the sense GPS/leveling minus gravimetric quasigeoid; the mean values refer to the raw differences without applying the constant $\zeta_0^{\mathrm{EVRF2007}}$; the other statistical parameters are calculated after subtracting the mean value; units are m

Quasigeoid (Δg / geopotential model)	#	Mean	RMS	Minimum	Maximum	Improvement vs EGG1997
EUVN_DA (all)						
EGG1997/EGM1996	1395	+0.287	0.243	−0.899	+0.708	–
EGG1997/EGM2008	1395	+0.253	0.188	−0.693	+0.527	23 %
EGG2008/EGM2008	1395	+0.250	0.173	−0.688	+0.443	29 %
EGM2008 ($n_{\max} = 2{,}190$)	1395	+0.254	0.171	−0.643	+0.481	–
EUVN_DA (excluding Great Britain and Italy)						
EGG1997/EGM1996	1139	+0.359	0.161	−0.599	+0.636	–
EGG1997/EGM2008	1139	+0.300	0.108	−0.607	+0.428	33 %
EGG2008/EGM2008	1139	+0.302	0.076	−0.302	+0.391	53 %
EGM2008 ($n_{\max} = 2{,}190$)	1139	+0.305	0.077	−0.250	+0.430	–

leads to only marginal improvements of the RMS differences in most cases, the only two exceptions being the French nationwide data set of 965 stations (as well as the corresponding French data set within EUVN_DA) and the British data set within EUVN_DA (see discussion below). Regarding the French nationwide data set, the RMS difference reduces from 7.6 cm (see Table 5.6) to 4.1 cm when considering additional tilt parameters; the tilt acts mainly in the north–south direction (about 0.25 m per 1000 km distance) and is related to the older leveling network (IGN69), as almost no tilt exists in the comparisons with the new leveling data (NIREF) available for the French traverse. For the latter data set, the RMS is 2.6 cm for the centered differences and 2.5 cm after considering additional tilt parameters, while the corresponding RMS value based on the older IGN69 heights is 8.0 cm for the centered differences, which clearly proves that the new French leveling is better than the old one (see also Sect. 2.4; Rebischung et al. 2008).

Of special interest are the results from the comparisons with the European EUVN_DA GPS/leveling data set, because it is based on common reference systems for GPS and leveling (see above). Table 5.7 and Fig. 5.17 (top) show that the EGG2008 model performs quite well over most parts of Europe, the main exception being Great Britain, but also over Italy and France (see previous paragraph) some systematic differences appear. Regarding Great Britain, the so-called second and third geodetic leveling differ by about 0.2 m in the north–south direction over 1,000 km distance (e.g., Kelsey 1972), and therefore the results from the third leveling were never used alone in practice; instead the results from the second leveling were held fixed and the third leveling was adjusted to it, leading to the "official ODN heights" from Ordnance Survey (Christie 1994). In addition, more recent studies suggest that the systematic error

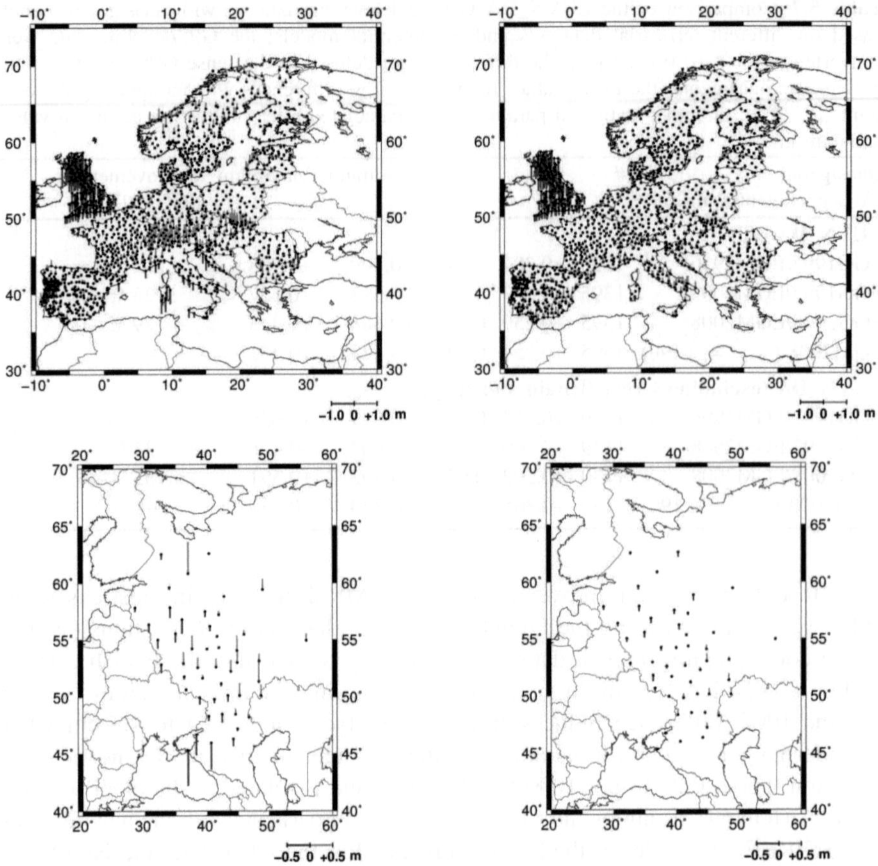

Fig. 5.17 Comparison of the EUVN_DA (*top*) and Russian (*bottom*) GPS and leveling data with the quasigeoid solutions EGG1997 (*left* part) and EGG2008 (*right* part); a constant bias is subtracted; stations (*dots*) and differences (positive: up in *red*; negative: down in *blue*) are depicted

is mainly related to the third leveling (Hipkin et al. 2004; Ziebart et al. 2008), and as the EUVN_DA data set is most probably based exclusively on the third leveling, significant systematic differences show up (see Fig. 5.17); thus for Great Britain, the RMS of the centered differences with respect to EGG2008 is 11.9 cm, which reduces to 3.8 cm after considering additional tilt parameters. The situation over Italy has improved with respect to earlier results (e.g., Denker et al. 2009) due to a recently performed update (replacement) of the entire Italian data set (Kenyeres et al. 2010), but some systematic differences remain (see Fig. 5.17), requiring further investigations. For the entire EUVN_DA data set (excluding Great Britain and Italy), the RMS difference reduces from 16.1 cm for EGG1997 to 7.6 cm for EGG2008; this means an improvement of about 53 % (Table 5.7). Furthermore, the comparisons on a country by country basis of the EUVN_DA data with EGG2008 give RMS values for the centered differences of less than 3 cm for Belgium, Denmark, Finland, Germany, Hungary, the Netherlands, Poland, Slovakia and Sweden, 3–6 cm for Austria, Croatia, the Czech Republic, Estonia, Lithuania, Norway and Switzerland, while the largest value is found for Romania (12.1 cm), which is certainly due to the low quality of the terrestrial data available for the EGG2008 development. Overall, the EUVN_DA comparison results are considered as quite satisfactory, in particular with regard to the very large area size (from the Iberian Peninsula to Northern Scandinavia, the Baltic States, Poland and Bulgaria) and the fairly small remaining systematic differences at the level of only about 1 dm (see also below).

Of significant importance is also the mean value between the EUVN_DA data (excluding Great Britain and Italy) and EGG2008 of +0.302 m, as it can be used to derive the potential value for the EVRF2007 zero level surface; formula (5.183) leads to

$$W_0^{\mathrm{EVRF2007}} = U_0 - \gamma_0\, \zeta_0^{\mathrm{EVRF2007}} , \tag{5.185}$$

where U_0 is the normal potential of the GRS80 level ellipsoid. Another option is to work with potential quantities only (a more strict approach), resulting in

$$W_0^{\mathrm{EVRF2007}} = W_P + C = U_P + T_P + C , \tag{5.186}$$

followed by an averaging over all stations. Both procedures lead to the same result (within the specified significant digits) of

$$W_0^{\mathrm{EVRF2007}} = 62,636,857.89 \pm 0.02\ \mathrm{m^2\,s^{-2}} , \tag{5.187}$$

where the latter figure is the empirical standard deviation of the mean value. The above potential value differs from that published by Denker et al. (2005) based on the earlier EUVN GPS/leveling data set (Ihde et al. 2000), and upon closer examination it turned out that the tide correction for the GPS heights had a sign error in the 2005 computations; after correcting this error, the data sets used in Denker et al. (2005), i.e., EUVN GPS/leveling data, EGG2004 quasigeoid, lead to a zero potential of 62,636,857.94 ± 0.16 $\mathrm{m^2\,s^{-2}}$, while the EUVN data together with EGG2008 yield a value of 62,636,858.03 ± 0.12 $\mathrm{m^2\,s^{-2}}$, both being in good agreement with the result given in (5.187). Furthermore, a good agreement exists with the values given in Bursa et al. (2001) for Germany (62,636,857.51 ± 0.54 $\mathrm{m^2\,s^{-2}}$) and the Netherlands (62,636,857.35 ± 0.70 $\mathrm{m^2\,s^{-2}}$), both based on the NAP (however, no information exists about the tidal systems for GPS and leveling). Hence, the NAP zero level surface is about 2 dm below the level surface defined by the global W_0 value of IERS (2010); see (5.48). Accordingly, the potential of the zero level surfaces can be derived for all other (national) vertical datums involved in Table 5.6.

Furthermore, the mean values of the differences between the EUVN_DA and EGG2008 data are quite consistent on a country by country basis, ranging from +0.241 to +0.377 m (excluding Great Britain and Italy). This suggests that systematic leveling errors over Europe are not very pronounced, besides the known problems in Great Britain, France and perhaps Italy (see above). This is also supported by Fig. 5.17 (right), showing only small long wavelength structures. In addition to this, the mean values listed in Tables 5.6 and 5.7 can also be employed to transform heights from one national or European height system to another one. For example, the largest mean value is found for Belgium (–2.065 m for EGG2008), which is due to the fact that the Belgian heights are referred to mean low water, while most other countries use MSL. Hence, in combination with the mean value for EVRF2007 of +0.302 m (EGG2008) it follows that the zero level surface of the Belgian heights is 2.363 m below the EVRF2007 zero level surface. This figure is in reasonable agreement with the results based exclusively on leveling, where the national heights are compared with the EVRF2007 (adjusted) leveling network, yielding a difference of 2.317 m (see http://www.crs-geo.eu); however, regarding the (small) difference between both figures, it should also be noticed that this is related to error contributions and time-variable effects from different epochs of all data sets involved, i.e., GPS, leveling, and gravimetric quasigeoid. The good agreement of the mean values related to EGG2008 for Germany (+0.297 m) and EUVN_DA (+0.302 m) is also remarkable, but both data sets rely on the Amsterdam tide gauge (NAP). On the other hand, the mean value for the Dutch data set is somewhat lower (+0.255 m), which is perhaps partly related to a subsidence of the area as well as different epochs of the involved data sets (see Ihde et al. 2008).

Finally, Tables 5.6 and 5.7 also include the comparison results with the complete EGM2008 model ($n_{\max} = 2{,}190$). The results show that EGM2008 is performing very well in all the comparisons, in the case of Belgium and Russia even better than EGG2008. On the other hand, EGG2008 is performing a little better than EGM2008 in the other cases, with the largest improvements seen for the Netherlands and Austria, the latter being probably related to the higher resolution of EGG2008.

4.6 Summary and Outlook

Significant progress was made within the framework of the European gravity and geoid project regarding the collection and homogenization of high-resolution gravity and terrain data, which was then utilized in combination with the global geopotential model EGM2008 to develop the completely updated European Gravimetric (Quasi)Geoid EGG2008, covering the whole of Europe and the surrounding marine areas. The evaluation of this model by independent GPS and leveling data showed that the new GRACE based geopotential models as well as the upgraded terrestrial gravity and terrain data both lead to substantial improvements compared to the previous model EGG1997 (in total by 35–72 %), and long wavelength errors, being the basic weakness of EGG1997, are virtually non-existent in the EGG2008 model. The RMS of the centered differences between national GPS/leveling data sets and EGG2008 range from about 1 cm to 5 cm for areas with a good data quality and coverage; the higher values are associated with Switzerland and Austria (high mountain regions), which is likely a consequence of both an insufficient gravity coverage and leveling quality in some local areas as well as theoretical shortcomings. On the other hand, the corresponding RMS differences exceed 10 cm for countries with a less favorable data quality (e.g., South-Eastern Europe). However, in this context it is also important to note that the differences from the GPS/leveling evaluation include error contributions and time-variable effects from different epochs of all data sets involved, i.e., GPS, leveling, and gravimetric quasigeoid. Taking this into account, the evaluation results indicate an accuracy potential of the gravimetric quasigeoid model EGG2008 in the order of 1–3 cm on a national basis, and 2–5 cm on continental scales, provided that high quality and resolution input data are available within the area of interest. These figures also conform to the internal error estimates of about 2–3 cm for the GRACE based calculations. In the end, the results obtained for large parts of Europe are about the optimum one can expect at present with up-to-date gravity, terrain, and GRACE data; further improvements are mainly anticipated from the GOCE and other future gravity field missions, as the terrestrial data can hardly be improved for the greater part of Europe, apart from a few exceptions (e.g., Eastern Europe).

Furthermore, a potential value for the EVRF2007 zero level surface was derived from the EGG2008 model as 62,636,857.89 $\pm$ 0.02 m^2s^{-2}, and the connection between national vertical datums was investigated by using GPS/leveling data and the EGG2008 model. In the future, the control and replacement of the costly geometric leveling, a differential technique susceptible to systematic errors, as well as the so-called "geoid based vertical datum" (implemented soon in Canada and the U.S.A.) will be interesting study topics. Finally, regional gravity field modeling on the basis of terrestrial and satellite data will certainly retain its importance in the future, as short wavelength gravity field structures (e.g., of the geoid and quasigeoid) with a resolution of a few kilometers can never be recovered from satellite data alone due to the necessary orbit heights of a few hundred kilometers.

References

Ågren J, Svensson R (2007) Postglacial land uplift model and system definition for the new Swedish height system RH 2000. Reports in geodesy and geographical information systems, Lantmateriet, Gävle

Andersen EG (1976) The effect of the topography on solution of Stokes' problem. Unisurv S-14, Sydney

Andersen EG, Rizos C, Mather RS (1975) Atmospheric effects in physical geodesy. Unisurv G-23, Sydney, pp 23–41

Andersen OB, Knudsen P, Berry PAM (2010) The DNSC08GRA global marine gravity field from double retracked satellite altimetry. J Geod 84:191–199

Andersen OB, Knudsen P, Trimmer R (2005) Improved high resolution altimetric gravity field mapping (KMS2002 global marine gravity field). In: Sansò F (ed) A window on the future of geodesy, IAG symposia, vol 128. Springer-Verlag, Berlin, Heidelberg, pp 326–331

Angermann D, Seitz M, Drewes H (2012) Global terrestrial reference systems and their realizations. Sciences of geodesy—II, this volume. Springer-Verlag, Berlin, Heidelberg

Bäumker M (1984) Zur dreidimensionalen Ausgleichung von terrestrischen und Satellitenbeobachtungen. Wiss. Arb. d. Fachr. Verm.wesen d. Univ. Hannover, Nr. 130, Hannover

Behrend D (1999) Untersuchungen zur Schwerefeldbestimmung in den europäischen Randmeeren. Wiss. Arb. d. Fachr. Verm.wesen d. Univ. Hannover, Nr. 229, Hannover

Bian S (1997) Some cubature formulas for singular integrals in physical geodesy. J Geod 71:443–453

BIPM (2006) Le Système international d'unités—The International System of Units (SI), 8th edn. Bureau international des poids et measures, Sèvres, France

Bosch W, Savcenko R (2010) On estimating the dynamic ocean topography—a profile based approach. In: Mertikas SP (ed) Gravity, geoid and earth observation, IAG symposia, vol 135. Springer-Verlag, Berlin, Heidelberg, pp 263–269

Bürki B, Müller A, Kahle H-G (2004) DIADEM: the new digital astronomical deflection measuring system for high-precision measurements of deflections of the vertical at ETH Zurich. In: Proceedings of gravity, geoid and space missions, GGSM 2004, IAG symposia, Porto, Portugal, 30 Aug–3 Sept (CD)

Bursa M, Groten E, Kenyon S, Kouba J, Radej K, Vatrt V, Voytiskova M (2002) Earth's dimension specified by geoidal potential. Stud Geophys Geod 46:1–8

Bursa M, Kenyon S, Kouba J, Radej K, Vatrt V, Voytiskova M (2001) In: Drewes H, Dodson A, Fortes LPS, Sanchez L, Sandoval P (eds) Vertical reference system, IAG symposia, vol 124. Springer-Verlag, Berlin, Heidelberg, pp 291–296

Campbell J (1971) Eine Erweiterung der Theorie des astronomischen Nivellements bei Einbeziehung von Schweremessungen. Wiss. Arb. d. Lehrstühle f. Geodäsie, Photogrammetrie und Kartographie a. d. Techn. Univ. Hannover, Nr. 49, Hannover

Cazenave A, Chen J (2010) Time-variable gravity from space and present-day mass redistribution in the earth system. Earth Planet Sci Lett 298:263–274

Christie RR (1994) A new geodetic heighting strategy for Great Britain. Surv Rev 32:328–343

Christodoulidis DC (1979) Influence of the atmospheric masses on the gravitational field of the Earth. Bull Géod 53:61–77

Condi F, Wunsch C (2004) Gravity field variability, the geoid, and ocean dynamics. In: Sansò F (ed) V Hotine-Marussi symposium on mathematical geodesy, IAG symposia, vol 127. Springer-Verlag, Berlin, Heidelberg, pp 285–292

Demianov GV, Majorov AN (2004) On the definition of a common world normal height system. In: Physical geodesy, scientific and technical reports on geodesy, photogrammetry and cartography, Federal Office of Geodesy and Cartography (Roskartographia), ZNIIGAiK, Moscow, pp 168–182 (in Russian)

Denker H (1988) Hochauflösende regionale Schwerefeldbestimmung mit gravimetrischen und topographischen Daten. Wiss. Arb. d. Fachr. Verm.wesen d. Univ. Hannover, Nr. 156, Hannover

Denker H (1989) A new gravimetric quasigeoid for the Federal Republic of Germany. Deutsche Geodät. Komm., Reihe B, Nr. 291, München

Denker H (1998) Evaluation and improvement of the EGG97 quasigeoid model for Europe by GPS and leveling data. In: Vermeer M, Ádám J (eds) Second continental workshop on the geoid in Europe, Proceedings Rep. Finnish Geod. Inst., vol 98, issue 4, Masala, pp 53–61

Denker H (2001) On the effect of datum inconsistencies in gravity and position on Europen geoid computations. Poster, IAG 2001 scientific assembly, 2–7 Sept 2001, Budapest

Denker H (2003) Computation of gravity gradients over Europe for calibration/validation of GOCE data. In: Tziavos IN (ed) Gravity and geoid 2002, 3rd Meeting of the Internat. Gravity and Geoid Commission, Ziti Editions, Thessaloniki, pp 287–292

Denker H (2005) Evaluation of SRTM3 and GTOPO30 terrain data in Germany. In: Jekeli C, Bastos L, Fernandes J (eds) Gravity, geoid and space missions, IAG symposia, vol 129. Springer-Verlag, Berlin, Heidelberg, pp 218–223

Denker H, Barriot J-P, Barzaghi R, Fairhead D, Forsberg R, Ihde J, Kenyeres A, Marti U, Sarrailh M, Tziavos IN (2009) The development of the European gravimetric geoid model EGG07. In: Sideris MG (ed) Observing our changing earth, IAG symposia, vol 133. Springer-Verlag, Berlin, Heidelberg, pp 177–186

Denker H, Barriot J-P, Barzaghi R, Forsberg R, Ihde J, Kenyeres A, Marti U, Tziavos IN (2005) Status of the European gravity and geoid project EGGP. In: Jekeli C, Bastos L, Fernandes J (eds) Gravity, geoid and space missions, IAG symposia, vol 129. Springer-Verlag, Berlin, Heidelberg, pp 125–130

Denker H, Roland M (2005) Compilation and evaluation of a consistent marine gravity data set surrounding Europe. In: Sansò F (ed) A window on the future of geodesy, IAG symposia, vol 128. Springer-Verlag, Berlin, Heidelberg, pp 248–253

Denker H, Torge W (1998) The European gravimetric quasigeoid EGG97—An IAG supported continental enterprise. In: Forsberg R, Feissel M, Dietrich R (eds) Geodesy on the move—Gravity, geoid, geodynamics and Antarctica, IAG symposia, vol 119. Springer-Verlag, Berlin, Heidelberg, pp 249–254

Denker H, Tziavos IN (1999) Investigation of the Molodensky series terms for terrain reduced gravity field data. Boll Geof Teor Appl 40:195–203

Denker H, Wenzel H-G (1987) Local geoid determination and comparison with GPS results. Bull Géod 61:349–366

Duquenne H, Duquenne F, Rebischung P (2007) New scientific levelling network NIREF. Data set and internal report, prepared for European gravity and geoid project (EGGP)

Ecker E, Mittermayer E (1969) Gravity corrections for the influence of the atmosphere. Boll Geof Teor Appl 11:70–80

Eeg J, Krarup T (1973) Integrated geodesy. The Danish Geodetic Institute, Internal Rep. No. 7, Copenhagen

Ekman M (1989a) The impact of geodynamic phenomena on systems for height and gravity. In: Andersen OB (ed) Modern techniques in geodesy and surveying, National Survey and Cadastre, KMS, Denmark, Publ. 4. Series vol. 1. Copenhagen, Denmark, pp 109–167

Ekman M (1989b) Impacts of geodynamic phenomena on systems for height and gravity. Bull Géod 63:281–296

Ekman M (1996) The permanent problem of the permanent tide: what to do with it in geodetic reference systems? Marées Terrestres. Bull d'Inf 125:9508–9513 (Bruxelles)

Elhabiby M, Sampietro D, Sansò F, Sideris MG (2009) BVP, global models and residual terrain correction. In: Sideris MG (ed) Observing our changing earth, IAG symposia, vol 133. Springer-Verlag, Berlin, Heidelberg, pp 211–217

ESA (1999) Gravity field and steady-state ocean circulation mission. Reports for mission selection, the four candidate earth explorer core missions, ESA SP-1233(1)

Featherstone WE, Kirby JF, Hirt C, Filmer MS, Claessens SJ, Brown NJ, Hu G, Johnston GM (2011) The AUSGeoid09 model of the Australian height datum. J Geod 85:133–150

Flechtner F, Thomas M, Dobslaw H (2010) Improved non-tidal atmospheric and oceanic de-aliasing for GRACE and SLR satellites. In: Flechtner F, Gruber Th, Güntner A, Mandea M, Rothacher M, Schöne T, Wickert J (eds) System earth via geodetic-geophysical space techniques, advanced technologies in earth sciences. Springer-Verlag, Berlin, pp 131–142

Flury J, Rummel R (2009) On the geoid-quasigeoid separation in mountain areas. J Geod 83:829–847

Forsberg R (1984) A study of terrain reductions, density anomalies and geophysical inversion methods in gravity field modeling. Report Department of Geodetic Science, no 355, The Ohio State University, Columbus, Ohio, USA

Forsberg R (1987) A new covariance model for inertial gravimetry and gradiometry. J Geophys Res B 92:1305–1310

Forsberg R (1993) Modelling of the fine-structure of the geoid: methods, data requirements and some results. Surv Geophys 14:403–418

Forsberg R (2010) Geoid determination in the mountains using ultra-high resolution spherical harmonic models—the Auvergne case. In: Contadakis ME, Kaltsikis C, Spatalas S, Tokmakidis K, Tziavos IN (eds) The apple of the knowledge, In Honor of Professor Emeritus Demetrius N. Arabelos, pp 101–111. Ziti Editions (ISBN: 978-960-243-674-5), Thessaloniki

Forsberg R, Kenyon S (2004) Gravity and geoid in the Arctic region—the northern GOCE polar gap filled. In: Proceedings of 2nd international GOCE workshop, Esrin, 8–10 March 2004, CD-ROM

Forsberg R, Tscherning CC (1981) The use of height data in gravity field approximation. J Geophys Res B 86:7843–7854

Forsberg R, Tscherning CC (1997) Topographic effects in gravity field modelling for BVP. In: Sansò F, Rummel R (eds) Geodetic boundary value problems in view of the one centimeter geoid, Lecture notes in earth sciences, vol 65. Springer-Verlag, Berlin, Heidelberg, pp 241–272

Förste C, and 12 others (2008a) The GeoForschungsZentrum Potsdam/Groupe de Recherche de Gèodésie Spatiale satellite-only and combined gravity field models: EIGEN-GL04S1 and EIGEN-GL04C. J Geod 82:331–346

Förste C, and 12 others (2008b) EIGEN-GL05C—a new global combined high-resolution GRACE-based gravity field model of the GFZ-GRGS cooperation. In: General assembly European Geosciences Union (Vienna, Austria 2008), Geophysical research abstracts, vol 10, Abstract No. EGU2008-A-06944

Gitlein O, Timmen L (2006) Atmospheric mass flow reduction for terrestrial absolute gravimetry in the Fennoscandian land uplift network. In: Tregoning P, Rizos C (eds) Dynamic planet, IAG symposia, vol 130. Springer-Verlag, Berlin, Heidelberg, pp 461–466

Goad CC, Tscherning CC, Chin MM (1984) Gravity empirical covariance values for the continental United States. J Geophys Res B 89:7962–7968

Grafarend EW (1978a) The definition of the telluroid. Bull Géod 52:25–37

Grafarend EW (1978b) Operational geodesy. In: Moritz H, Sünkel H (eds) Approximation methods in geodesy, Sammlung Wichmann, Neue Folge, Bd. 10, H. Wichmann Verlag, Karlsruhe, pp 235–284

Grafarend EW (1988) The geometry of the Earth's surface and the corresponding function space of the terrestrial gravitational field. In: Festschrift Rudolf Sigl zum 60. Geburtstag, Deutsche Geod. Komm., Reihe B, Heft Nr. 287, pp 76–94 (München)

Grote T (1996) Regionale Quasigeoidmodellierung aus heterogenen Daten mit "cm"-Genauigkeit. Wiss. Arb. d. Fachr. Verm.wesen d. Univ. Hannover, Nr. 212, Hannover

Haagmans R, de Min E, von Gelderen M (1993) Fast evaluation of convolution integrals on the sphere using 1D FFT, and a comparison with existing methods for Stokes' integral. Manuscripta geodaetica 18:227–241

Haagmans R, Prijatna K, Omang O (2003) An alternative concept for validation of GOCE gradiometry results based on regional gravity. In: Tziavos IN (ed) Gravity and geoid 2002, 3rd meeting of the Internat. Gravity and Geoid commission, pp 281–286, Ziti Editions, Thessaloniki

Heck B (1986) A numerical comparison of some telluroid mappings. In: Proceedings of I Hotine-Marussi symposium on mathematical geodesy (Roma, 3–6 June 1985), vol. 1. Milano, pp 19–38

Heck B (1990) An evaluation of some systematic error sources affecting terrestrial gravity anomalies. Bull Géod 64:88–108

Heck B (1991) On the linearized boundary value problems of physical geodesy. Report Department of Geodetic Science and Surveying, no 407, The Ohio State University, Columbus, Ohio, USA

Heck B (1997) Formulation and linearization of boundary value problems: from observables to a mathematical model. In: Sansò F, Rummel R (eds) Geodetic boundary value problems in view of the one centimeter geoid, Lecture notes in earth sciences, vol 65. Springer-Verlag, Berlin, Heidelberg, pp 121–160

Heck B (2003) Rechenverfahren und Auswertemodelle der Landesvermessung—Klassische und moderne Methoden, 3rd edn. Wichmann Verlag, Heidelberg

Heck B (2004) Problems in the definition of vertical reference frames. In: Sansò F (ed) V Hotine-Marussi symposium on mathematical geodesy, IAG symposia, vol 127. Springer-Verlag, Berlin, Heidelberg, pp 164–173

Heck B, Rummel R (1990) Strategies for solving the vertical datum problem using terrestrial and satellite geodetic data. In: Sünkel H, Baker T (eds) Sea surface topography and the geoid, IAG symposia, vol 104. Springer-Verlag, Berlin, Heidelberg, pp 116–128

Heck B, Seitz K (2007) A comparison of the tesseroid, prism and point-mass approaches for mass reductions in gravity field modeling. J Geod 81:121–136

Hein GW (1986) Integrated geodesy—state-of-the-art 1986 reference text. In: Sünkel H (ed) Mathematical and numerical techniques in physical geodesy, lecture notes in earth sciences, vol 7. Springer-Verlag, Berlin, Heidelberg, pp 505–548

Heiskanen WA, Moritz H (1967) Physical geodesy. W.H. Freeman and Company, San Francisco

Hipkin RG (2004) Ellipsoidal geoid computation. J Geod 78:167–179

Hipkin R, Haines K, Beggan C, Bingley R, Hernandez F, Holt J, Baker T (2004) The geoid EDIN2000 and mean sea surface topography around the British Isles. Geophys J Int 157: 565–577

Hirt C (2004) Entwicklung und Erprobung eines digitalen Zenitkamerasystems für die hochpräzise Lotabweichungsbestimmung. Wiss. Arb. d. Fachr. Verm.wesen d. Univ. Hannover, Nr. 253, Hannover

Hirt C, Feldmann-Westendorff U, Denker H, Flury J, Jahn C-H, Lindau A, Seeber G, Voigt C (2008) Hochpräzise Bestimmung eines astrogeodätischen Quasigeoidprofils im Harz für die Validierung des Quasigeoidmodells GCG05. Zeitschrift f. Verm.wesen (zfv) 133:108–119

Hirt C, Flury J (2008) Astronomical-topographic levelling using high-precision astrogeodetic vertical deflections and digital terrain model data. J Geod 82:231–248

Hirt C, Seeber G (2008) Accuracy analysis of vertical deflection data observed with the Hannover digital zenith camera system TZK2-D. J Geod 82:347–356

Hirt C, Schmitz M, Feldmann-Westendorff U, Wübbena G, Jahn C-H, Seeber G (2010) Mutual validation of GNSS height measurements from high-precision geometric-astronomical levelling. GPS Solutions 15:149–159

Hotine M (1969) Mathematical geodesy. ESSA Monograph 2, U.S. Department of Commerce, Washington

IAG (1970) Geodetic reference system 1967. Publ. Spéciale du Bull Géod, Paris

IAG (1984) Resolutions of the XVIII general assembly of the International Association of Geodesy, Hamburg, Germany, 15–27 Aug 1983. J Geod 58:309–323

IERS (2010) IERS conventions (2010). In: Petit G, Luzum B (eds) IERS technical note no. 36, Verlag des Bundesamtes für Kartographie und Geodäsie, Frankfurt am Main

Ihde J (2009) Inter-commission project 1.2: vertical reference frames. In: Drewes H, Hornik H (eds) International Association of Geodesy, Travaux, vol 36

Ihde J, Adam J, Gurtner W, Harsson BG, Sacher M, Schlüter W, Wöppelmann G (2000) The height solution of the European vertical reference network (EUVN). Veröff. Bayer. Komm. für die Internat. Erdmessung, Astronom. Geod. Arb., Nr. 61, München pp 132–145

Ihde J, Mäkinen J, Sacher M (2008) Conventions for the definition and realization of a European vertical reference system (EVRS)—EVRS conventions 2007. EVRS conventions V5.1, Bundesamt für Kartographie und Geodäsie, Finnish Geodetic Institute, publication date 17 Dec 2008

Ihde J, Sanchez L (2005) A unified global height reference system as a basis for IGGOS. J Geodyn 40:400–413

Ihde J, Wilmes H, Müller J, Denker H, Voigt C, Hosse M (2010) Validation of satellite gravity field models by regional terrestrial data sets. In: Flechtner F, Gruber Th, Güntner A, Mandea M, Rothacher M, Schöne T, Wickert J (eds) System earth via geodetic-geophysical space techniques, advanced technologies in earth sciences. Springer-Verlag, Berlin, pp 277–296

Jekeli C (1983) A numerical study of the divergence of the spherical harmonic series of the gravity and height anomalies at the Earth's surface. Bull Géod 57:10–28

Jekeli C (2009) Potential theory and static gravity field of the Earth. In: Herring T (vol ed) Treatise on geophysics, vol 3, Geodesy. Elsevier, Amsterdam, pp 11–42

Jekeli C, Yang HJ, Kwon JH (2009) Evaluation of EGM08—globally and locally in South Korea. In: External quality evaluation reports of EGM08, Newton's Bulletin, vol 4, pp 38–49

JPL (2007) SRTM—The mission to map the world. Jet Propulsion Laboratory, California Institute of Technology. http://www2.jpl.nasa.gov/srtm/index.html

Kaula WM (1966) Theory of satellite geodesy. Blaisdell Publ. Comp., Waltham (Mass.)

Kellog OD (1953) Foundations of potential theory. Dover Publ. Inc., New York

Kelsey J (1972) Geodetic aspects concerning possible subsidence in Southeastern England. Phil Trans R Soc Lond 272:141–149. doi:10.1098/rsta.1972.0040

Kenyeres A, Sacher M, Ihde J, Denker H, Marti U (2010) EUVN Densification action—final report. http://www.bkg.bund.de/nn_166762/geodIS/EVRS/EN/Projects/02EUVN-DA/01 Introduction/introduction__node.html__nnn=true

Kern M (2004) A comparison of data weighting methods for the combination of satellite and local gravity data. In: Sansò F (ed) V Hotine-Marussi Symposium on mathematical geodesy, IAG symposia, vol 127. Springer-Verlag, Berlin, Heidelberg, pp 137–144

Kovalevsky J, Seidelmann PK (2004) Fundamentals of astrometry. Cambridge University Press, Cambridge

Krarup T (1969) A contribution to the mathematical foundation of physical geodesy. Geodætisk Institut, Meddelelse No. 44, Copenhagen

Kumar M (2005) When ellipsoidal heights will do the job, why look elsewhere? Surv Land Inf Sci 65:91–94

Kurtenbach E, Mayer-Gürr T, Eicker A (2009) Deriving daily snapshots of the Earth's gravity field from GRACE L1B data using Kalman filtering. Geophys Res Lett 36:L17102. doi:10.1029/2009GL039564

Lemoine FG, and 14 others (1998) The development of the joint NASA GSFC and the National Imagery and Mapping Agency (NIMA) geopotential model EGM96. NASA/TP-1998-206861, Greenbelt, MD, USA

Li YC, Sideris MG, Schwarz K-P (1995) A numerical investigation on height anomaly prediction in mountainous areas. Bull Géod 69:143–156

Liebsch G, Schirmer U, Ihde J, Denker H, Müller J (2006) Quasigeoidbestimmung für Deutschland. DVW-Schriftenreihe 49:127–146

MacMillan WD (1958) Theoretical mechanics, vol 2., The theory of the potential. Dover Publ. Inc., New York

Mäkinen J, Ihde J (2009) The permanent tide in height systems. In: Sideris MG (ed) Observing our changing earth, IAG symposia, vol 133. Springer-Verlag, Berlin, Heidelberg, pp 81–87

Marti U, Schlatter A (2001) The new height system in Switzerland. In: Drewes H, Dodson A, Fortes LPS, Sanchez L, Sandoval P (eds) Vertical reference system, IAG symposia, vol 124. Springer-Verlag, Berlin, Heidelberg, pp 50–55

Marussi A (1985) Intrinsic geodesy. Springer-Verlag, Berlin

Mayer-Gürr T (2006) Gravitationsfeldbestimmung aus der Analyse kurzer Bahnbögen am Beispiel der Satellitenmissionen CHAMP und GRACE. Dissertation, University of Bonn

Mayer-Gürr T, Kurtenbach E, Eicker A (2010) ITG-Grace2010 gravity field model. URL: www.igg.uni-bonn.de/apmg/index.php

Meschkowski H (1962) Hilbertsche Räume mit Kernfunktionen. Springer-Verlag, Berlin

Mohr PJ, Taylor BN, Newell DB (2008) CODATA recommended values of the fundamental physical constants: 2006. Rev Mod Phys 80:633–730

Molodenskii MS, Eremeev VF, Yurkina MI (1962) Methods for study of the external gravitational field and figure of the Earth. Translation of the Russian Book, Israel Program for Scientific Translations, Jerusalem

Morelli C, Pisani M, Gantar C (1975) Geophysical anomalies and tectonics in the Western Mediterranean. Boll Geof Teor Appl 18(67):211–249 (Trieste)

Moritz H (1962) Interpolation and prediction of gravity and their accuracy. Report Department of Geodetic Science, no. 24, The Ohio State University, Columbus, Ohio, USA

Moritz H (1971) Kinematical geodesy II. Report Department of Geodetic Science, no. 165, The Ohio State University, Columbus, Ohio, USA

Moritz H (1976) Integral formulas and collocation. Manuscripta Geod 1:1–40

Moritz H (1980) Advanced physical geodesy. Wichmann, Karlsruhe

Moritz H (2000) Geodetic reference system 1980. J Geod 74:128–133

Morse PM, Feshbach H (1953) Methods of theoretical physics. McGraw-Hill, New York

Müller J, Soffel M, Klioner SA (2008) Geodesy and relativity. J Geod 82:133–145

NIMA (1997) Department of Defense World Geodetic System 1984—its definition and relationships with local geodetic datums. NIMA National Imagery and Mapping Agency, Technical Report, NIMA TR8350.2, 3rd Edition, 4 July 1997 (Amendment 1, 3 Jan. 2000; Amendment 2, 23 June 2004)

NIMA (2002) Addendum to NIMA TR 8350.2: implementation of the World Geodetic System 1984 (WGS 84) reference frame G1150. http://earth-info.nga.mil/GandG/publications/tr8350.2/tr8350_2.html

Omang OCD, Tscherning CC, Forsberg R (2012) Generalizing the harmonic reduction procedure in residual topographic modeling. In: Sneeuw N, Novák P, Crespi M, Sansò F (eds) VII Hotine-Marussi symposium on mathematical geodesy, IAG symposia, vol 137. Springer-Verlag, Berlin, Heidelberg, pp 233–238

Pail R, Bruinsma S, Migliaccio F, Foerste C, Goiginger H, Schuh W-D, Hoeck E, Reguzzoni M, Brockmann JM, Abrikosov O, Veicherts M, Fecher T, Mayrhofer R, Krasbutter I, Sansò F, Tscherning CC (2011) First GOCE gravity field models derived by three different approaches. J Geod 85:819–843

Pavlis NK, Holmes SA, Kenyon SC, Factor JK (2008) An earth gravitational model to degree 2160: EGM2008. Presentation, 2008 general assembly of the European Geosciences Union, Vienna, Austria, 13–18 Apr 2008

Poutanen M, Vermeer M, Mäkinen J (1996) The permanent tide in GPS positioning. J Geod 70:499–504

Rapp RH (I983a) The need and prospects for a world vertical datum. In: Proceedings lAG symposia, IUGG XVIII General Assembly, Hamburg, FRG, 1983, vol 2. Report Department of Geodetic Science and Surveying. The Ohio State University, Columbus, Ohio, USA, pp 432-445

Rapp RH (1983b) Tidal gravity computations based on recommendations of the standard earth tide committee. Bull. d'Inf., Marées Terrestres 89: 5814–5819 (Bruxelles)

Rapp RH (1995) A world vertical datum proposal. Allg Verm Nachr 102:297–304

Rapp RH (1997) Use of potential coefficient models for geoid undulation determinations using a spherical harmonic representation of the height anomaly/geoid undulation difference. J Geod 71:282–289

Rapp RH, Balasubramania N (1992) A conceptual formulation of a world height system. Report Department of Geodetic Science and Surveying, no. 421. The Ohio State University, Columbus, Ohio, USA

Rapp RH, Nerem RS, Shum CK, Klosko SM, Williamson RG (1991) Consideration of permanent tidal deformation in the orbit determination and data analysis for the Topex/Poseidon mission. NASA Technical Memorandum 100775, Goddard Space Flight Center, Greenbelt, MD, USA

Rebischung P, Duquenne H, Duquenne F (2008) The new French zero-order levelling network—first global results and possible consequences for UELN. EUREF 2008 symposium, Brussels, Belgium, 18–21 June 2008

Rummel R (1988) Zur iterativen Lösung der geodätischen Randwertaufgabe. In: Festschrift Rudolf Sigl zum 60. Geburtstag, Deutsche Geod. Komm., Reihe B, Heft Nr. 287, pp 175–181 (München)

Rummel R (1995) The first degree harmonics of the Stokes problem—what are the practical implications? In: Festschrift E. Groten on the occasion of his 60th anniversary, pp 98–106 (Munich)

Rummel R (1997) Spherical spectral properties of the Earth's gravitational potential and its first and second derivatives. In: Sansò F, Rummel R (eds) Geodetic boundary value problems in view of the one centimeter geoid, lecture notes in earth sciences, vol 65. Springer-Verlag, Berlin, Heidelberg, pp 359–404

Rummel R, Rapp RH (1976) The influence of the atmosphere on geoid and potential coefficient determinations from gravity data. J Geophys Res 81:5639–5642

Rummel R, Teunissen P (1988) Height datum definition, height datum connection and the role of the geodetic boundary value problem. Bull Géod 62:477–498

Rummel R, van Gelderen M (1995) Meissl scheme—spectral characteristics of physical geodesy. Manuscripta Geod 20:379–385

Rummel R, Weijong Y, Stummer C (2011) GOCE gravitational gradiometry. J Geod 85:777–790

Sanchez L (2008) Approach for the establishment of a global vertical reference level. In: Xu P, Liu J, Dermanis A (eds) VI Hotine-Marussi symposium on theoretical and computational geodesy, IAG symposia, vol 132. Springer-Verlag, Berlin, Heidelberg, pp 119–125

Sandwell DT, Smith WHF (2009) Global marine gravity from retracked Geosat and ERS-1 altimetry: ridge segmentation versus spreading rate. J Geophys Res B 114:B01411. doi:10.1029/2008JB006008

Sansò F (1986) Statistical methods in physical geodesy. In: Sünkel H (ed) Mathematical and numerical techniques in physical geodesy, lecture notes in earth sciences, vol 7. Springer-Verlag, Berlin, Heidelberg, pp 49–155

Sansò F (1995) The long road from measurements to boundary value problems in physical geodesy. Manuscripta Geod 20:326–344

Sansò F, Tscherning CC (2003) Fast spherical collocation: theory and examples. J Geod 77:101–112

Schwarz KP, Sideris MG, Forsberg R (1990) The use of FFT techniques in physical geodesy. Geophys J Int 100:485–514

Seitz K (1997) Ellipsoidische und topographische Effekte im geodätischen Randwertproblem. Deutsche Geod. Komm., Reihe C, Nr. 483, München

Sideris MG (1987) Spectral methods for the numerical solution of Molodensky's problem. UCSE Reports, No. 20024, Department of Surveying Engineering, The University of Calgary

Sideris MG (1994) Regional geoid determination. In: Vanicek P, Christou NT (eds) Geoid and its geophysical interpretations. CRC Press, Inc, Boca Raton, FL, USA, pp 77–94

Sideris MG (2011a) Geoid determination, theory and principles. In: Gupta HK (ed) Encyclopedia of Solid Earth Geophysics. Springer-Verlag, Berlin, Heidelberg, pp 356–362

Sideris MG (2011b) Geoid, computational method. In: Gupta HK (ed) Encyclopedia of Solid Earth Geophysics. Springer-Verlag, Berlin, Heidelberg, pp 366–371

Sigl R (1985) Introduction to potential theory. Abacus Press in association with H. Wichmann Verlag

Sjöberg L (1980) Least squares combination of satellite harmonics and integral formulas in physical geodesy. Gerlands Beitr. d. Geophysik 89:371–377

Sjöberg L (1981) Least squares spectral combination of satellite and terrestrial data in physical geodesy. Annales de Géophysique 37:25–30

Sjöberg LE (2003) A general model for modifying Stokes' formula and its least-squares solution. J Geod 77:459–464

Sjöberg LE (2010) A strict formula for geoid-to-quasigeoid separation. J Geod 84:699–702

Sjöberg LE, Nahavandchi H (2000) The atmospheric geoid effects in Stokes' formula. Geophys J Int 140:95–100

Smith DA, Roman DR (2001) GEOID99 and G99SSS: 1-arc-minute geoid models for the United States. J Geod 75:469–490

Smith D, Véronneau M, Roman D, Huang JL, Wang YM, Sideris M (2010) Towards the unification of the vertical datums over the North American continent. IAG Commission 1 symposium 2010, reference frames for applications in geosciences (REFAG2010), Marne-La-Vallée, France, 4–8 Oct 2010

Steinberg G, Papo H (1998) Ellipsoidal heights: the future of vertical geodetic control. GPS World 9(2):41–43

Strange WE (1982) An evaluation of orthometric height accuracy using bore hole gravimetry. Bull Géod 56:300–311

Sünkel H (1983) The geoid in Austria. In: Proceedings IAG symposia, IUGG XVIII General Assembly, Hamburg, FRG, 1983, vol 1. Report Department of Geodetic Science and Surveying, The Ohio State University, Columbus, Ohio, USA, pp 348–364

Tapley BD, Bettadpur S, Ries JC, Thompson PF, Watkins M (2004a) GRACE measurements of mass variability in the earth system. Science, vol 305, issue 5683, pp 503–505, 23 July 2004

Tapley BD, Bettadpur S, Watkins M, Reigber C (2004b) The gravity recovery and climate experiment: mission overview and early results. Geophys Res Lett 31(9):L09607. doi:10.1029/2004GL019920

Tapley B, Ries J, Bettadpur S, Chambers D, Cheng M, Condi F, Poole S (2007) The GGM03 mean earth gravity model from GRACE. Eos Trans. AGU, vol 88(52), Fall Meeting Supplement, Abstract G42A-03

Tenzer R, Vaníček P, Santos M, Featherstone WE, Kuhn M (2005) The rigorous determination of orthometric heights. J Geod 79:82–92

Timmen L (2010) Absolute and relative gravimetry. In: Xu G (ed) Sciences of geodesy—I, advances and future directions. Springer-Verlag, Berlin, pp 1–48

Torge W (1977) Untersuchungen zur Höhen- und Geoidbestimmung im dreidimensionalen Testnetz Westharz. Zeitschrift f. Verm.wesen (zfv) 102:173–186

Torge W (1982) Zur Geoidbestimmung im Meeresbereich. In: Geodaesia Universalis—Festschrift Karl Rinner zum 70. Geburtstag, Mitt. Geod. Inst. T.U. Graz, Folge 40, pp 346–355

Torge W (1989) Gravimetry. W. de Gruyter, Berlin

Torge W (2001) Geodesy, 3rd edn. W. de Gruyter, Berlin

Torge W, Denker H (1998) The European geoid—development over more than 100 years and present status. In: Vermeer M, Ádám J (eds) Second continental workshop on the geoid in Europe, Proceedings Rep. Finnish Geod. Inst., vol 98, issue 4. Masala, pp 47–52

Torge W, Röder RH, Schnüll M, Wenzel H-G, Faller JE (1987) First results with the transportable absolute gravity meter JILAG-3. Bull Géod 61:161–176

Torge W, Weber G, Wenzel H-G (1982) Computation of a high resolution European geoid (EGG1). In: Proceedings of 2nd international symposium on the geoid in Europe and Mediterranian area, Rome. Istituto Geografico Militare, Italy, Florence, pp 437–460

Tscherning CC (1976a) Computation of second-order derivatives of the normal potential based on the representation by a Legendre series. Manuscripta Geod 1:71–92

Tscherning CC (1976b) On the chain-rule method for computing potential derivatives. Manuscripta Geod 1:125–141
Tscherning CC (1985) Local approximation of the gravity potential by least squares collocation. In: Schwarz KP (ed) Proceedings of international summer school on local gravity field approximation, Beijing, 21 Aug–4 Sept 1984, The University of Calgary, Publ. 60003, pp 277−361
Tscherning CC (1986) Functional methods for gravity field approximation. In: Sünkel H (ed) Mathematical and numerical techniques in physical geodesy. Lecture notes in earth sciences, vol 7. Springer-Verlag, Berlin, Heidelberg, pp 3–47
Tscherning CC (1994) Geoid determination by least-squares collocation using GRAVSOFT. In: International school for the determination and use of the geoid. Lecture notes, Milan, 10–15 Oct 1994, pp 135–164
Tscherning CC (2004) A discussion of the use of spherical approximation or no approximation in gravity field modeling with emphasis on unsolved problems in least-squares collocation. In: Sansò F (ed) V Hotine-Marussi symposium on mathematical geodesy, IAG symposia, vol 127. Springer-Verlag, Berlin, Heidelberg, pp 184–188
Tscherning CC, Rapp RH (1974) Closed covariance expressions for gravity anomalies, geoid undulations, and deflections of the vertical implied by anomaly degree variance models. Report Department of Geodetic Science, no. 208, The Ohio State University, Columbus, Ohio, USA
USGS (2007) Global 30 arc-second elevation data set GTOPO30. U.S. Geological Survey, Center for Earth Resources Observation and Science (EROS). http://edc.usgs.gov/products/elevation/gtopo30/gtopo30.html
Vaníček P (1998) The height of reason (a letter to the editor). GPS World 9(4):14
Véronneau M, Duvai R, Huang J (2006) A gravimetric geoid model as a vertical datum in Canada. Geomatica 60:165–172
Véronneau M, Huang J (2007) The Canadian gravimetric geoid model 2005 (CGG2005). Report, Geodetic Survey Division, Earth Sciences Sector, Natural Resources Canada, Ottawa, Canada
Wahr JM (2009) Time variable gravity from satellites. In: Herring T (vol ed) Treatise on geophysics, vol 3, geodesy. Elsevier, Amsterdam, pp 213–237
Wang YM, Saleh J, Li X, Roman DR (2011) The US Gravimetric geoid of 2009 (USGG2009): model development and evaluation. J Geod. doi:10.1007/s00190-011-0506-7
Weber G (1984) Hochauflösende Freiluftanomalien und gravimetrische Lotabweichungen für Europa. Wiss. Arb. d. Fachr. Verm.wesen d. Univ. Hannover, Nr. 135, Hannover
Wenzel HG (1981) Zur Geoidbestimmung durch Kombination von Schwereanomalien und einem Kugelfunktionsmodell mit Hilfe von Integralformeln. Zeitschrift f. Verm.wesen (zfv) 106:102–111
Wenzel HG (1982) Geoid computation by least squares spectral combination using integral formulas. In: Proceedings of general meeting of the IAG, Tokyo, 7–15 May 1982, pp 438–453
Wenzel H-G (1985) Hochauflösende Kugelfunktionsmodelle für das Gravitationspotential der Erde. Wiss. Arb. d. Fachr. Verm.wesen d. Univ. Hannover, Nr. 137, Hannover
Wenzel H-G (1989) On the definition and numerical computation of free air gravity anomalies. Bull. d'Information 64. Bureau Gravimetrique Internationale, Toulouse, France, pp 23−40
Wichiencharoen C (1982) The indirect effects on the computation of geoid undulations. Report Department of Geodetic Science, no. 336, The Ohio State University, Columbus, Ohio, USA
Wildermann E (1988) Untersuchungen zur lokalen Schwerefeldbestimmung aus heterogenen Daten dargestellt am Beispiel der Geotraverse venezolanische Anden. Wiss. Arb. d. Fachr. Verm.wesen d. Univ. Hannover, Nr. 155, Hannover
Wolf H (1974) Über die Einführung von Normalhöhen. Zeitschrift f. Verm.wesen (zfv) 99:1–5
Wolf KI (2007) Kombination globaler Potentialmodelle mit terrestrischen Schweredaten für die Berechnung der zweiten Ableitungen des Gravitationspotentials in Satellitenbahnhöhe. Wiss. Arb. d. Fachr. Geodäsie u. Geoinformatik d. Leibniz Univ. Hannover, Nr. 264, Hannover
Wolf KI (2008) Evaluation regionaler Quasigeoidlösungen in synthetischer Umgebung. Zeitschrift f. Verm.wesen (zfv) 133:52–63

Wunsch C, Gaposchkin EM (1980) On using satellite altimetry to determine the general circulation of the oceans with application to geoid improvement. Rev Geophys 18:725–745

Zeman A (1987) Definition of the normal gravity field including the constant part of tides. Stud Geophys Geod 31:113–120

Ziebart MK, Iliffe JC, Forsberg R, Strykowski G (2008) Convergence of the UK OSGM05 GRACE-based geoid and the UK fundamental benchmark network. J Geophys Res B 113:B12401. doi:10.1029/2007JB004959

Zilkoski DB, Richards JH, Young GM (1995) A summary of the results of the general adjustment of the North American vertical datum of 1988. Allg Verm Nachr 102:310–321

Chapter 6
Regularization and Adjustment

Yunzhong Shen and Guochang Xu

Part I: Regularized Solution to Ill-Posed Problems

1 Introduction

The linear observation equation is usually expressed as

$$l = Ax + e \tag{6.1}$$

where the non-random design matrix $A \in R^{m\times n}$, the vector of unknown parameters $x \in R^{n\times 1}$, the vector of measurements $l \in R^{m\times 1}$and the random error vector $e \in R^{m\times 1}$ with zero mean and variance–covariance matrix $\sigma_0^2 P^{-1}$, where P is the weight matrix and σ_0^2 is the variance of unit weight. If the design matrix A possesses very big condition number, the observation equation (6.1) is called ill-conditioned, which is defined as ill-posed problems by Hadamard (1932). In geodesy ill-posed problems are frequently encountered in satellite gravimetry due to downward continuation, or in geodetic date procession due to the colinearity among parameters that are to be estimated. The so-called ill-posed problem violates at least one of the following conditions:

Y. Shen
Department of Surveying and Geo-Informatics Engineering,
TongJi University, 1239 Siping Road, 200092, Shanghai, PR. China

G. Xu (✉)
Deptartment 1 Geodesy and Remote Sensing, GFZ German Research Centre for Geosciences, Telegrafenberg , 14473, Potsdam, Brandenburg, Germany
e-mail: xu@gfz-potsdam.de

Y. Shen · G. Xu
Chinese Academy of Space Technology, Beijing, 10081, P.R. China

G. Xu (ed.), *Sciences of Geodesy - II*, DOI: 10.1007/978-3-642-28000-9_6,

- Solution is existent
- Solution is unique
- Solution is stable.

Geodetic ill-posed problems have unstable solutions, i.e., the least squares adjustment cannot be used to solve such problems. In order to stabilize the solution, the method called regularization initially proposed by Tikhonov (1963a, b) must be applied, which is based on the criterion of minimizing the following Tikhonov smoothing function:

$$\Phi^{\alpha}(\boldsymbol{x}) = \|\boldsymbol{A}\boldsymbol{x} - \boldsymbol{l}\|_{\boldsymbol{P}}^2 + \alpha\|\boldsymbol{x}\|_{\boldsymbol{K}}^2 = \min \tag{6.2}$$

where $\|\bullet\|_{\boldsymbol{P}}^2$ and $\|\bullet\|_{\boldsymbol{K}}^2$ denote the second norm with respect to kernel $\boldsymbol{P}$ and $\boldsymbol{K}$, respectively, α is the regularization parameter, and the other symbols are as in (6.1). This smoothing function (6.2) has been successfully applied for solving ill-posed problems and a lot of papers have been published on this topic. For details on ill-posed problems one can refer to the books Tikhonov and Arsenin (1977); Tikhonov et al. (1995); and Morozov (1984). It is worth mentioning that Hoerl and Kennard (1970) first presented the formula for computing the biases and proposed the mean squared error (MSE) for estimating the accuracy for regularized estimates. In geodesy, especially in satellite gravimetry, there also exist many inverse ill-posed problems (see e.g., Schaffrin 1980, 2008; Xu 1992, 1998; Xu and Rummel 1994; Reigber et al. 2005; Xu et al. 2006). Since the ambiguity and coordinates parameters are highly correlated in fast GNSS positioning, regularization is also applied for deriving stable real ambiguity solutions (Shen and Li 2007; Li et al. 2010).

In the following sections we first analyze the unstable characteristics of least squares solution, when the observation equations are ill-posed. Then the regularized solution is introduced and the algorithms of regularized parameter are outlined together with the formula for efficiently calculating the regularized parameter based on minimizing the MSE. Then the biasness of residuals for regularized solution is also analyzed and the unbiased estimate of the variance of unit weight based on the bias-removed residuals is discussed. Finally the numerical examples for demonstrating the performance of regularized are presented.

2 Unstable Analysis of Least Squares Solution to Ill-Posed Observation Equation

The least squares solution to (6.1) can be expressed as

$$\boldsymbol{x}_L = (\boldsymbol{A}^{\mathrm{T}}\boldsymbol{P}\boldsymbol{A})^{-1}\boldsymbol{A}^{\mathrm{T}}\boldsymbol{P}\boldsymbol{l} \tag{6.3}$$

where $\boldsymbol{x}_L$ indicates the least squares solution. The covariance matrix $\boldsymbol{D}_{\boldsymbol{x}_L}$ of $\boldsymbol{x}_L$ is

$$\boldsymbol{D}_{\boldsymbol{x}_L} = \sigma_0^2(\boldsymbol{A}^{\mathrm{T}}\boldsymbol{P}\boldsymbol{A})^{-1} \tag{6.4}$$

and σ_0^2 is variance of unit weight and can be estimated from residuals as:

$$\hat{\sigma}_0^2 = \frac{\hat{\boldsymbol{e}}_L^{\mathrm{T}}\boldsymbol{P}\hat{\boldsymbol{e}}_L}{m-n} \tag{6.5a}$$

and

$$\hat{\boldsymbol{e}}_L = \boldsymbol{A}\boldsymbol{x}_L - \boldsymbol{l} = \left(\boldsymbol{A}\left(\boldsymbol{A}^{\mathrm{T}}\boldsymbol{P}\boldsymbol{A}\right)^{-1}\boldsymbol{A}^{\mathrm{T}}\boldsymbol{P} - \boldsymbol{I}_m\right)\boldsymbol{l}. \tag{6.5b}$$

Here, $\boldsymbol{I}_m$ is an $m \times m$ identity matrix. If (6.1) is ill-posed, the singular values of the design matrix $\boldsymbol{A}$ will monotonously decrease to zero, and, as a result, small errors in the observational vector cause a big disturbance in the estimated parameters. Therefore, the least squares solution cannot produce a satisfactory result. In order to analyze the cause, the solution (6.3) is rewritten as

$$\boldsymbol{x}_L = \left(\tilde{\boldsymbol{A}}^{\mathrm{T}}\tilde{\boldsymbol{A}}\right)^{-1}\tilde{\boldsymbol{A}}^{\mathrm{T}}\tilde{\boldsymbol{l}} \tag{6.6}$$

where $\tilde{\boldsymbol{A}} = \boldsymbol{P}^{1/2}\boldsymbol{A}$, $\tilde{\boldsymbol{l}} = \boldsymbol{P}^{1/2}\boldsymbol{l}$. Then the design matrix $\tilde{\boldsymbol{A}}$ is decomposed by using singular value decomposition (SVD) as follows:

$$\tilde{\boldsymbol{A}} = \boldsymbol{U}\boldsymbol{W}\boldsymbol{V}^{\mathrm{T}} \tag{6.7}$$

where $\boldsymbol{U} \in R^{m\times n}$ and $\boldsymbol{U}^{\mathrm{T}}\boldsymbol{U} = \boldsymbol{I}_n$, $\boldsymbol{V} \in R^{n\times n}$ and $\boldsymbol{V}^{\mathrm{T}}\boldsymbol{V} = \boldsymbol{V}\boldsymbol{V}^{\mathrm{T}} = \boldsymbol{I}_n$, and $\boldsymbol{W} \in R^{n\times n}$ is a diagonal matrix with elements $\mathrm{diag}\,(\boldsymbol{W}) = (\lambda_1 \quad \lambda_2 \quad \cdots \quad \lambda_n)$, λ_i is the singular value of matrix $\tilde{\boldsymbol{A}}$ and satisfies the condition $\lambda_1 \geq \lambda_2 \geq \cdots \geq \lambda_n$. If (6.1) is rank-deficient, there exists a number $k<n$, when $i > k$, such that $\lambda_i = 0$. If (6.1) is ill-posed, then the smallest eigenvalue λ_n tends toward 0. If the matrices $\boldsymbol{U}$, $\boldsymbol{V}$ are expressed with column vector $\boldsymbol{u}_i \in R^{m\times 1}$ and $\boldsymbol{v}_i \in R^{n\times 1}$ as $\boldsymbol{U} = (\boldsymbol{u}_1 \quad \boldsymbol{u}_2 \quad \cdots \quad \boldsymbol{u}_n)$ and $\boldsymbol{V} = (\boldsymbol{v}_1 \quad \boldsymbol{v}_2 \quad \cdots \quad \boldsymbol{v}_n)$, then (6.7) can be represented as:

$$\tilde{\boldsymbol{A}} = \sum_{i=1}^{n} \boldsymbol{u}_i\lambda_i\boldsymbol{v}_i^{\mathrm{T}} \tag{6.8}$$

and $\boldsymbol{u}_i, \boldsymbol{v}_i$ are called left- and right- singular vectors. If the measurement vector $\tilde{\boldsymbol{l}}$ is decomposed into two parts,

$$\tilde{\boldsymbol{l}} = \bar{\boldsymbol{l}} + \tilde{\boldsymbol{e}} \tag{6.9}$$

where $\bar{\boldsymbol{l}}$ is the exact value of $\tilde{\boldsymbol{l}}$ and $\tilde{\boldsymbol{e}} = \boldsymbol{P}^{1/2}\boldsymbol{e}$. Inserting (6.9) and (6.8) into (6.6), we obtain the spectral representation of least squares solution as follows:

$$\boldsymbol{x}_L = \sum_{i=1}^{n} \frac{\boldsymbol{u}_i^{\mathrm{T}} \tilde{\boldsymbol{l}}}{\lambda_i} \boldsymbol{v}_i = \sum_{i=1}^{n} \frac{\boldsymbol{u}_i^{\mathrm{T}} \bar{\boldsymbol{l}}}{\lambda_i} \boldsymbol{v}_i + \sum_{i=1}^{n} \frac{\boldsymbol{u}_i^{\mathrm{T}} \tilde{\boldsymbol{e}}}{\lambda_i} \boldsymbol{v}_i. \tag{6.10}$$

The covariance matrix of estimated parameters can also be expressed as spectral:

$$\boldsymbol{D}_{\boldsymbol{x}_L} = \sigma_0^2 \left(\boldsymbol{A}^{\mathrm{T}} \boldsymbol{P} \boldsymbol{A}\right)^{-1} = \sigma_0^2 \left(\tilde{\boldsymbol{A}}^{\mathrm{T}} \tilde{\boldsymbol{A}}\right)^{-1} = \sigma_0^2 \sum_{i=1}^{n} \frac{\boldsymbol{v}_i \boldsymbol{v}_i^{\mathrm{T}}}{\lambda_i^2} \tag{6.11}$$

and the estimate of the variance of unit weight is

$$\hat{\sigma}_0^2 = \frac{\hat{\boldsymbol{e}}_L^{\mathrm{T}} \boldsymbol{P} \hat{\boldsymbol{e}}_L}{m-n} = \frac{\hat{\tilde{\boldsymbol{e}}}_L^{\mathrm{T}} \hat{\tilde{\boldsymbol{e}}}_L}{m-n} \tag{6.12}$$

where

$$\hat{\tilde{\boldsymbol{e}}}_L = \boldsymbol{P}^{1/2} \hat{\boldsymbol{e}}_L = \left(\tilde{\boldsymbol{A}} \left(\tilde{\boldsymbol{A}}^{\mathrm{T}} \tilde{\boldsymbol{A}}\right)^{-1} \tilde{\boldsymbol{A}}^{\mathrm{T}} - \boldsymbol{I}_m\right) \tilde{\boldsymbol{l}}. \tag{6.13}$$

By substituting (6.8) into (6.13), we have

$$\hat{\tilde{\boldsymbol{e}}}_L = \left(\sum_{i=1}^{n} \boldsymbol{u}_i \boldsymbol{u}_i^{\mathrm{T}} - \boldsymbol{I}_m\right) \tilde{\boldsymbol{l}}. \tag{6.14}$$

In (6.10), the signal spectra $\boldsymbol{u}_i^{\mathrm{T}} \bar{\boldsymbol{l}}$ decays faster than singular value λ_i, but the noise spectra $\boldsymbol{u}_i^{\mathrm{T}} \tilde{\boldsymbol{e}}$ has the same amplitude in all spectral domains due to its being white noise. Therefore, the solution is mainly contaminated by high frequency noise because the singular value in high frequency domain is very small. However, the residual vector as shown (6.14) is not dominated by high frequency noises.

The so-called regularized solution is to find a filter factor τ_i, which approaches 1 in low frequency domain but decays faster than a singular value in order to filter out the high frequency noises. After introducing such a filter factor into (6.10), we obtain the so-called regularized solution in spectral representation as follows:

$$\boldsymbol{x}_\tau = \sum_{i=1}^{n} \tau_i \frac{\boldsymbol{u}_i^{\mathrm{T}} \tilde{\boldsymbol{l}}}{\lambda_i} \boldsymbol{v}_i \tag{6.15}$$

where, $\boldsymbol{x}_\tau$ denotes the regularized solution using filter factor τ_i. Considering the covariance $\boldsymbol{D}_{\tilde{l}} = \boldsymbol{P}^{1/2} \boldsymbol{D}_l \boldsymbol{P}^{1/2} = \sigma_0^2 \boldsymbol{P}^{1/2} \boldsymbol{P}^{-1} \boldsymbol{P}^{1/2} = \sigma_0^2 \boldsymbol{I}_n$, we can derive the covariance matrix of $\boldsymbol{x}_\tau$ according to the law of error propagation:

$$\boldsymbol{D}_{\boldsymbol{x}_\tau} = \sigma_0^2 \sum_{i=1}^{n} \tau_i^2 \frac{\boldsymbol{v}_i \boldsymbol{v}_i^{\mathrm{T}}}{\lambda_i^2} \tag{6.16}$$

and the residual vector is

$$\hat{\tilde{e}}_\tau = \sum_{i=1}^{n} \tau_i \boldsymbol{u}_i^{\mathrm{T}} \tilde{\boldsymbol{l}} \boldsymbol{u}_i - \tilde{\boldsymbol{l}}. \tag{6.17}$$

3 Regularized Solution to Ill-Posed Observation Equations

The regularized solution $\boldsymbol{x}_\tau$ is biased and its bias vector $\boldsymbol{b}_\tau$ can be estimated from (6.15) as

$$\boldsymbol{b}_\tau = E(\boldsymbol{x}_\tau) - \bar{\boldsymbol{x}} = \sum_{i=1}^{n} (\tau_i - 1) \frac{\boldsymbol{u}_i^{\mathrm{T}} \bar{\boldsymbol{l}}}{\lambda_i} \boldsymbol{v}_i = \sum_{i=1}^{n} (\tau_i - 1) \boldsymbol{v}_i^{\mathrm{T}} \bar{\boldsymbol{x}} \boldsymbol{v}_i \tag{6.18}$$

where $\bar{\boldsymbol{x}}$ and $\bar{\boldsymbol{l}}$ are the true values of parameter vector $\boldsymbol{x}$ and measurement vector $\tilde{\boldsymbol{l}}$, respectively. The equation $\frac{\boldsymbol{u}_i^{\mathrm{T}} \bar{\boldsymbol{l}}}{\lambda_i} = \boldsymbol{v}_i^{\mathrm{T}} \bar{\boldsymbol{x}}$ used in deriving (6.18) can be derived by left-applying the orthogonal vector $\boldsymbol{v}_i^{\mathrm{T}}$ to $\bar{\boldsymbol{x}} = \sum_{i=1}^{n} \frac{\boldsymbol{u}_i^{\mathrm{T}} \bar{\boldsymbol{l}}}{\lambda_i} \boldsymbol{v}_i$, which can be derived from (6.10). Because the estimated vector $\boldsymbol{x}_\tau$ is biased, the bias term should be considered in estimating the accuracy of $\boldsymbol{x}_\tau$. The mean square error (MSE), which consists of covariance matrix and the bias part, is used to estimate the accuracy of biased estimates. Considering (6.16) and (6.18), the spectral representation of the MSE is expressed as follows:

$$\boldsymbol{M}_{\boldsymbol{x}_\tau} = \boldsymbol{D}_{\boldsymbol{x}_\tau} + \boldsymbol{b}_\tau \boldsymbol{b}_\tau^{\mathrm{T}} = \sigma_0^2 \sum_{i=1}^{n} \frac{\tau_i^2 \boldsymbol{v}_i \boldsymbol{v}_i^{\mathrm{T}}}{\lambda_i^2} + \sum_{i=1}^{n} (\tau_i - 1)^2 \left(\boldsymbol{v}_i^{\mathrm{T}} \bar{\boldsymbol{x}}\right)^2 \boldsymbol{v}_i \boldsymbol{v}_i^{\mathrm{T}} \tag{6.19}$$

Equations (6.15)–(6.19) are the general formulae of regularized solution by using the filter factor. The filter factor must have different forms with different criteria of regularization.

3.1 Solution to Rank-Deficient Observation Equations

If the observational equation is rank-deficient with rank $k<n$, the filter factor τ_i can be defined as

$$\tau_i = \begin{cases} 1, & 1 \le i \le k \\ 0, & k < i \le n \end{cases} \tag{6.20}$$

Substituting (6.20) into (6.15)–(6.19), we obtain the spectral representation of the solution of rank-deficient observational equation as

$$x_k = \sum_{i=1}^{k} \frac{u_i^{\mathrm{T}}\tilde{l}}{\lambda_i} v_i \tag{6.21a}$$

$$D_k = \sigma_0^2 \sum_{i=1}^{k} \frac{v_i v_i^{\mathrm{T}}}{\lambda_i^2} \tag{6.21b}$$

$$\hat{\hat{e}}_\tau = \sum_{i=1}^{k} u_i^{\mathrm{T}}\tilde{l} u_i - \tilde{l} \tag{6.21c}$$

$$b_k = \sum_{i=k+1}^{n} \frac{u_i^{\mathrm{T}}\tilde{l}}{\lambda_i} v_i = \sum_{i=k+1}^{n} v_i^{\mathrm{T}}\bar{x} v_i \tag{6.21d}$$

$$M_k = \sigma_0^2 \sum_{i=1}^{k} \frac{v_i v_i^{\mathrm{T}}}{\lambda_i^2} + \sum_{i=k+1}^{n} \left(v_i^{\mathrm{T}}\bar{x}\right)^2 v_i v_i^{\mathrm{T}} \tag{6.21e}$$

Here, the subscript k denotes the rank-deficient solution.

3.2 Regularized Solution to Ill-Posed Observation Equations

If the observational equation is ill-posed, regularization approach must be applied to stabilize the solution. The criterion of Tikhonov regularization is based on the smoothing function (6.2); its solution can be easily derived from the smoothing function (6.2) as

$$x_\alpha = \left(A^{\mathrm{T}}PA + \alpha K\right)^{-1} A^{\mathrm{T}}Pl \tag{6.22a}$$

where α is a small positive real number called the regularization parameter, and the subscript α denotes the regularized solution. Once the regularization parameter is determined, one can compute the estimate vector x_α with (6.22a), and its covariance matrix D_α, bias vector b_α, MSE matrix M_α and the residual vector $\hat{e}_\alpha$, with the following expressions:

$$D_\alpha = \sigma_0^2 \left(A^{\mathrm{T}}PA + \alpha K\right)^{-1} A^{\mathrm{T}}PA \left(A^{\mathrm{T}}PA + \alpha K\right)^{-1} \tag{6.22b}$$

$$b_\alpha = -\alpha \left(A^{\mathrm{T}}PA + \alpha K\right)^{-1} Kx \tag{6.22c}$$

$$\begin{aligned} M_\alpha = D_\alpha + b_\alpha b_\alpha^{\mathrm{T}} &= \sigma_0^2 \left(A^{\mathrm{T}}PA + \alpha K\right)^{-1} A^{\mathrm{T}}PA \left(A^{\mathrm{T}}PA + \alpha K\right)^{-1} \\ &+ \alpha^2 \left(A^{\mathrm{T}}PA + \alpha K\right)^{-1} Kx\bar{x}^{\mathrm{T}}K \left(A^{\mathrm{T}}PA + \alpha K\right)^{-1} \end{aligned} \tag{6.22d}$$

$$\hat{\boldsymbol{e}}_\alpha = \left(\boldsymbol{A}\left(\boldsymbol{A}^{\mathrm{T}}\boldsymbol{P}\boldsymbol{A} + \alpha\boldsymbol{K}\right)^{-1}\boldsymbol{A}^{\mathrm{T}}\boldsymbol{P} - \boldsymbol{I}_m\right)\boldsymbol{l} \tag{6.22e}$$

If matrix $\boldsymbol{K}$ is an identity matrix, (6.22a)–(6.22e) can be simplified as

$$\boldsymbol{x}_\alpha = \left(\boldsymbol{A}^{\mathrm{T}}\boldsymbol{P}\boldsymbol{A} + \alpha\boldsymbol{I}_n\right)^{-1}\boldsymbol{A}^{\mathrm{T}}\boldsymbol{P}\boldsymbol{l} = \left(\tilde{\boldsymbol{A}}^{\mathrm{T}}\tilde{\boldsymbol{A}} + \alpha\boldsymbol{I}_n\right)^{-1}\tilde{\boldsymbol{A}}^{\mathrm{T}}\tilde{\boldsymbol{l}} \tag{6.23a}$$

$$\begin{aligned}\boldsymbol{D}_\alpha &= \sigma_0^2\left(\boldsymbol{A}^{\mathrm{T}}\boldsymbol{P}\boldsymbol{A} + \alpha\boldsymbol{I}_n\right)^{-1}\boldsymbol{A}^{\mathrm{T}}\boldsymbol{P}\boldsymbol{A}\left(\boldsymbol{A}^{\mathrm{T}}\boldsymbol{P}\boldsymbol{A} + \alpha\boldsymbol{I}_n\right)^{-1}\\ &= \sigma_0^2\left(\tilde{\boldsymbol{A}}^{\mathrm{T}}\tilde{\boldsymbol{A}} + \alpha\boldsymbol{I}_n\right)^{-1}\tilde{\boldsymbol{A}}^{\mathrm{T}}\tilde{\boldsymbol{A}}\left(\tilde{\boldsymbol{A}}^{\mathrm{T}}\tilde{\boldsymbol{A}} + \alpha\boldsymbol{I}_n\right)^{-1}\end{aligned} \tag{6.23b}$$

$$\boldsymbol{b}_\alpha = -\alpha\left(\boldsymbol{A}^{\mathrm{T}}\boldsymbol{P}\boldsymbol{A} + \alpha\boldsymbol{I}_n\right)^{-1}\bar{\boldsymbol{x}} = -\alpha\left(\tilde{\boldsymbol{A}}^{\mathrm{T}}\tilde{\boldsymbol{A}} + \alpha\boldsymbol{I}_n\right)^{-1}\bar{\boldsymbol{x}} \tag{6.23c}$$

$$\begin{aligned}\boldsymbol{M}_\alpha &= \sigma_0^2\left(\tilde{\boldsymbol{A}}^{\mathrm{T}}\tilde{\boldsymbol{A}} + \alpha\boldsymbol{I}_n\right)^{-1}\tilde{\boldsymbol{A}}^{\mathrm{T}}\tilde{\boldsymbol{A}}\left(\tilde{\boldsymbol{A}}^{\mathrm{T}}\tilde{\boldsymbol{A}} + \alpha\boldsymbol{I}_n\right)^{-1}\\ &\quad+\alpha^2\left(\tilde{\boldsymbol{A}}^{\mathrm{T}}\tilde{\boldsymbol{A}} + \alpha\boldsymbol{I}_n\right)^{-1}\bar{\boldsymbol{x}}\bar{\boldsymbol{x}}^{\mathrm{T}}\left(\tilde{\boldsymbol{A}}^{\mathrm{T}}\tilde{\boldsymbol{A}} + \alpha\boldsymbol{I}_n\right)^{-1}\end{aligned} \tag{6.23d}$$

$$\hat{\tilde{\boldsymbol{e}}}_\alpha = \left(\tilde{\boldsymbol{A}}\left(\tilde{\boldsymbol{A}}^{\mathrm{T}}\tilde{\boldsymbol{A}} + \alpha\boldsymbol{I}_n\right)^{-1}\tilde{\boldsymbol{A}}^{\mathrm{T}} - \boldsymbol{I}_m\right)\tilde{\boldsymbol{l}}. \tag{6.23e}$$

Inserting the SVD decomposition formula (6.8) into (6.23a), we obtain the spectral representation of Tikhonov regularization solution as

$$\boldsymbol{x}_\tau = \sum_{i=1}^{n}\frac{\lambda_i\boldsymbol{u}_i^{\mathrm{T}}\tilde{\boldsymbol{l}}}{\lambda_i^2 + \alpha}\boldsymbol{v}_i. \tag{6.24}$$

Comparing (6.24) with (6.15), we get the filter factor for Tikhonov regularization as

$$\tau_i = \frac{\lambda_i^2}{\lambda_i^2 + \alpha} \tag{6.25}$$

It is obvious that the filter factor of Tikhonov regularization in (6.25) decays faster than singular value λ_i due to $\alpha > 0$. According to the characteristic of the filter factor which approaches 1 in a low frequency domain and 0 in a high frequency domain, it can be derived from (6.25) that the regularization parameter α should fit the condition

$$0 < \alpha/\lambda_1^2 << 1 \quad 0 < \lambda_n^2/\alpha << 1. \tag{6.26}$$

The value of the regularization parameter depends on the convergence rate of singular value λ_i and the sample value of white noise. Substituting (6.25) into (6.16)–(6.19), we get the corresponding spectral forms as follows:

$$\boldsymbol{D}_\alpha = \sigma_0^2\sum_{i=1}^{n}\frac{\lambda_i^2\boldsymbol{v}_i\boldsymbol{v}_i^{\mathrm{T}}}{\left(\lambda_i^2 + \alpha\right)^2} \tag{6.27}$$

$$\hat{\tilde{e}}_\alpha = \sum_{i=1}^{n} \frac{\lambda_i^2}{\lambda_i^2 + \alpha} \boldsymbol{u}_i^{\mathrm{T}} \tilde{\boldsymbol{l}} \boldsymbol{u}_i - \tilde{\boldsymbol{l}} \tag{6.28}$$

$$\boldsymbol{b}_\alpha = -\alpha \sum_{i=1}^{n} \frac{\boldsymbol{v}_i^{\mathrm{T}} \bar{\boldsymbol{x}}}{\lambda_i^2 + \alpha} \boldsymbol{v}_i \tag{6.29}$$

$$\boldsymbol{M}_\alpha = \sigma_0^2 \sum_{i=1}^{n} \frac{\lambda_i^2 \boldsymbol{v}_i \boldsymbol{v}_i^{\mathrm{T}}}{\left(\lambda_i^2 + \alpha\right)^2} + \alpha^2 \sum_{i=1}^{n} \frac{\left(\boldsymbol{v}_i^{\mathrm{T}} \bar{\boldsymbol{x}}\right)^2 \boldsymbol{v}_i \boldsymbol{v}_i^{\mathrm{T}}}{\left(\lambda_i^2 + \alpha\right)^2} \tag{6.30}$$

Inserting the SVD decomposition formula (6.8) into (6.23a), we can get the same formulae as (6.24) and those from (6.27) to (6.30). Because of $\boldsymbol{b}_\alpha < 0$, the regularization solution has less power than the 'true' one. Substituting $\tilde{\boldsymbol{l}} = \bar{\boldsymbol{l}} + \tilde{\boldsymbol{e}}$ into (6.23a), we get the error vector of the Tikhonov regularized solution as

$$\boldsymbol{x}_\alpha - \bar{\boldsymbol{x}} = -\alpha \left(\tilde{\boldsymbol{A}}^{\mathrm{T}} \tilde{\boldsymbol{A}} + \alpha \boldsymbol{I}_n\right)^{-1} \bar{\boldsymbol{x}} + \left(\tilde{\boldsymbol{A}}^{\mathrm{T}} \tilde{\boldsymbol{A}} + \alpha \boldsymbol{I}_n\right)^{-1} \tilde{\boldsymbol{A}}^{\mathrm{T}} \tilde{\boldsymbol{e}} \tag{6.31}$$

where the expression $\bar{\boldsymbol{l}} = \tilde{\boldsymbol{A}} \bar{x}$ is used. The spectral representation of (6.31) can be expressed as

$$\boldsymbol{x}_\alpha - \bar{\boldsymbol{x}} = -\alpha \sum_{i=1}^{n} \frac{\boldsymbol{v}_i^{\mathrm{T}} \bar{\boldsymbol{x}}}{\lambda_i^2 + \alpha} \boldsymbol{v}_i + \sum_{i=1}^{n} \frac{\lambda_i \boldsymbol{u}_i^{\mathrm{T}} \tilde{\boldsymbol{e}}}{\lambda_i^2 + \alpha} \boldsymbol{v}_i. \tag{6.32}$$

The first term at the right-hand side of (6.32) denotes the error caused by regularization and the second term denotes the error caused by measurement error. The scale value of the error vector is measured by its norm and because the spaces of the two kinds of errors are orthogonal, this norm can be expressed as

$$\|\boldsymbol{x}_\alpha - \bar{\boldsymbol{x}}\|^2 = \alpha^2 \sum_{i=1}^{n} \frac{\left(\boldsymbol{v}_i^{\mathrm{T}} \bar{\boldsymbol{x}}\right)^2}{\left(\lambda_i^2 + \alpha\right)^2} + \sum_{i=1}^{n} \frac{\lambda_i^2 \left(\boldsymbol{u}_i^{\mathrm{T}} \tilde{\boldsymbol{e}}\right)^2}{\left(\lambda_i^2 + \alpha\right)^2} \tag{6.33}$$

Taking the expectation to (6.33), we can obtain the following expression:

$$E\left(\|\boldsymbol{x}_\alpha - \bar{\boldsymbol{x}}\|^2\right) = \alpha^2 \sum_{i=1}^{n} \frac{\left(\boldsymbol{v}_i^{\mathrm{T}} \bar{\boldsymbol{x}}\right)^2}{\left(\lambda_i^2 + \alpha\right)^2} + \sigma_0^2 \sum_{i=1}^{n} \frac{\lambda_i^2}{\left(\lambda_i^2 + \alpha\right)^2} \tag{6.34}$$

From (6.34), it can be seen that the error caused by regularization, the first term on the right, is a monotonously increasing function with respect to regularization parameter α; when $\alpha \to 0$, this error approaches minimal value 0 and when $\alpha \to \infty$, it approaches maximal value $\bar{\boldsymbol{x}}^{\mathrm{T}} \bar{\boldsymbol{x}}$. In contrast, observational error is a monotonously decreasing function with respect to α; when $\alpha \to \infty$, this error approaches minimal value 0 and when $\alpha \to 0$, it approaches maximal value

$\sigma_0^2 \sum_{i=1}^{n} \frac{1}{\lambda_i^2}$. The regularization parameter α is used to balance the two kinds of errors and makes the total error $E\left(\|\boldsymbol{x}_\alpha - \bar{\boldsymbol{x}}\|^2\right)$ approach a minimum.

It is worth mentioning here that we can derive the same formula as (6.34) by taking the trace of MSE in (6.30). This indicates that minimizing the trace of MSE is equivalent to minimizing the total error of the regularized solution.

In many real applications, the matrix $\boldsymbol{K}$ is not an identity matrix, but a symmetric matrix. For example, in recovering the gravitational potential model from satellite to satellite tracking data, a regularization approach must be applied for stabilizing the solution due to the downward continuation. In this case, the matrix $\boldsymbol{K}$ is a diagonal matrix; its diagonal elements k_i are determined by Kaula's rule as

$$k_i = 2 \times 10^{10} l^2, \quad \left(l^2 - 3 \leq i \leq l^2 + 2l - 3\right) \tag{6.35}$$

where l is the degree of gravitational potential model. In order to derive the spectral expression for the diagonal matrix $\boldsymbol{K}$, we first decompose $\boldsymbol{K}$ as

$$\boldsymbol{K} = \boldsymbol{P}_K^{\mathrm{T}} \boldsymbol{P}_K. \tag{6.36}$$

It is obvious that $\boldsymbol{P}_K$ is also a diagonal matrix; its diagonal elements are the square roots of the correspondent elements of$\boldsymbol{K}$. Inserting (6.36) into (6.22d), we obtain

$$\begin{aligned} \boldsymbol{M}_\alpha = {} & \sigma_0^2 \boldsymbol{P}_K^{-1} \left(\boldsymbol{A}'^{\mathrm{T}} \boldsymbol{P} \boldsymbol{A}' + \alpha \boldsymbol{I}_n\right)^{-1} \boldsymbol{A}'^{\mathrm{T}} \boldsymbol{P} \boldsymbol{A}' \left(\boldsymbol{A}'^{\mathrm{T}} \boldsymbol{P} \boldsymbol{A}' + \alpha \boldsymbol{I}_n\right)^{-1} \left(\boldsymbol{P}_K^{-1}\right)^{\mathrm{T}} \\ & + \alpha^2 \boldsymbol{P}_K^{-1} \left(\boldsymbol{A}'^{\mathrm{T}} \boldsymbol{P} \boldsymbol{A}' + \alpha \boldsymbol{I}_n\right)^{-1} \boldsymbol{P}_K \bar{\boldsymbol{x}} \bar{\boldsymbol{x}}^{\mathrm{T}} \boldsymbol{P}_K^{\mathrm{T}} \left(\boldsymbol{A}'^{\mathrm{T}} \boldsymbol{P} \boldsymbol{A}' + \alpha \boldsymbol{I}_n\right)^{-1} \left(\boldsymbol{P}_K^{-1}\right)^{\mathrm{T}} \end{aligned} \tag{6.37}$$

where $\boldsymbol{A}' = \boldsymbol{A} \boldsymbol{P}_K^{-1}$. $\boldsymbol{A}'^{\mathrm{T}} \boldsymbol{P} \boldsymbol{A}'$is a symmetric and positive definite matrix which can be decomposed as

$$\boldsymbol{A}'^{\mathrm{T}} \boldsymbol{P} \boldsymbol{A}' = \boldsymbol{G}^{\mathrm{T}} \boldsymbol{W} \boldsymbol{G} \tag{6.38}$$

where $\boldsymbol{G} \in R^{n \times n}$, $\boldsymbol{G}^{\mathrm{T}} \boldsymbol{G} = \boldsymbol{G} \boldsymbol{G}^{\mathrm{T}} = \boldsymbol{I}_n$, and $\boldsymbol{W}$ is a diagonal matrix, its diagonal elements λ_i^2 being the eigenvalues of the matrix $\boldsymbol{A}'^{\mathrm{T}} \boldsymbol{P} \boldsymbol{A}'$. Substituting (6.38) into (6.37) and introducing symbols $\boldsymbol{G}_K = \boldsymbol{P}_K^{-1} \boldsymbol{G}$, $\boldsymbol{y} = \boldsymbol{G}^{\mathrm{T}} \boldsymbol{P}_K \bar{\boldsymbol{x}}$, we obtain the following expression:

$$\begin{aligned} \boldsymbol{M}_\alpha = {} & \sigma_0^2 \boldsymbol{G}_K (\boldsymbol{W} + \alpha \boldsymbol{I}_n)^{-1} \boldsymbol{G}_K^{\mathrm{T}} - \alpha \sigma_0^2 \boldsymbol{G}_K (\boldsymbol{W} + \alpha \boldsymbol{I}_n)^{-2} \boldsymbol{G}_K^{\mathrm{T}} \\ & + \alpha^2 \boldsymbol{G}_K (\boldsymbol{W} + \alpha \boldsymbol{I}_n)^{-1} \boldsymbol{y} \boldsymbol{y}^{\mathrm{T}} (\boldsymbol{W} + \alpha \boldsymbol{I}_n)^{-1} \boldsymbol{G}_K^{\mathrm{T}} \end{aligned}. \tag{6.39}$$

Taking the trace to$\boldsymbol{M}_\alpha$, we can derive the following expression:

$$Tr(\boldsymbol{M}_\alpha) = \sum_{i=1}^{n} \frac{\lambda_i^2 \sigma_0^2 + \alpha^2 y_i^2}{\left(\lambda_i^2 + \alpha\right)^2 k_i} \tag{6.40}$$

where Tr denotes the trace operator, and y_i is the ith element of vector $\boldsymbol{y}$.

4 Determination of the Regularization Parameter

Hoerl and Kennard (1970) proposed a criterion to estimate the regularization parameter by minimizing the MSE of a regularized solution. If the regularization parameter α is known, the regularized solution can be obtained with (6.22a) or (6.23a). Therefore, the key problem now is how to determine the regularization parameter.

The regularization parameter depends not only on the ill-conditioned observational equation, but also on the measurement error. Therefore, ill-conditioned observational equation and the variance of the unit weight measurement error σ_0 are the necessary and sufficient quantities in determining the regularization parameter. In most cases σ_0 needs to be estimated from the measurements, and σ_0^2 should be substituted with the estimated value $\hat{\sigma}_0^2$ (6.14) shows that the residual vector $\hat{\tilde{\boldsymbol{e}}}_L$ is not amplified by the small singular value; therefore even least squares adjustment can provide a satisfactory estimation of the standard deviation. This estimate $\hat{\sigma}_0^2$ can be used to calculate the regularization parameter.

There are many criteria for the determination of the regularization parameter, such as the L-curve method, generalized discrepancy principle, iterative ridge method, and generalized cross validation (GCV) method. Different methods can produce a different regularization parameter, and the best regularization parameter α^* should satisfy the following condition:

$$E\left(\|\boldsymbol{x}_{\alpha*} - \bar{\boldsymbol{x}}\|^2\right) \leq E\left(\|\boldsymbol{x}_\alpha - \bar{\boldsymbol{x}}\|^2\right), \quad \forall\alpha \in R^+ \tag{6.41}$$

where R^+ denotes the positive real set. In other words, the best regularization parameter can be expressed as the solution of the following criterion:

$$E\left(\|\boldsymbol{x}_\alpha - \bar{\boldsymbol{x}}\|^2\right) = \min \quad \alpha \in R^+ \tag{6.42}$$

The iterative ridge method originally proposed by Hoerl and Kennard (1970) in estimation of regression parameters and successfully used by Xu (1992), Xu and Rummel (1994), Xu (1998) in the simulation study of recovery of gravity anomaly from gradiometer measurements is based on the this criterion. Unfortunately, this criterion needs the exact parameters, i.e., the vector $\bar{\boldsymbol{x}}$. However it is actually impossible; otherwise the parameters need not be estimated. If the exact parameters in this criterion are substituted with their prior-values or with those from the least squares solution, if this solution exists, the estimated regularization parameter based on this criterion is not an optimal one. In fact, each criterion has its advantage and disadvantage and has a special occasion to be used. No criterion is superior to others all the time. For this reason, we introduce here four criteria to compute the regularization. The first is called the generalized discrepancy principle, which needs σ_0 to be known, the second is the L-curve method, which does not need information about σ_0, the third is the formal discussed iterative ridge method, which needs not only σ_0 but also the exact parameter vector $\bar{\boldsymbol{x}}$, and the fourth is the GCV method, which also needs neither σ_0, nor $\bar{\boldsymbol{x}}$.

Generalized Discrepancy Principle

The criterion of generalized discrete discrepancy principle is represented by the following formula (Hansen 1996):

$$\rho(\alpha) = \|\boldsymbol{A}\boldsymbol{x}_\alpha - \boldsymbol{l}\|_{\boldsymbol{P}}^2 - m\sigma_0^2 = 0 \tag{6.43}$$

where $\boldsymbol{x}_\alpha$ can be substituted by (6.22a) and m is the number of measurements. The only unknown in (6.43) is the regularization parameter α. Because the function $\rho(\alpha)$ is a monotonous increase function with $\rho(0)<0$, and $\exists\alpha_1,\ \rho(\alpha_1) > 0$; the symbol $\exists$ denotes existence, so that (6.43) exists as a unique solution. In practice, the regularization parameter α derived from (6.43) is somewhat bigger, because $m\sigma_0^2$ is not the unbiased estimator of $\|\boldsymbol{A}\boldsymbol{x}_\alpha - \boldsymbol{l}\|_{\boldsymbol{P}}^2$. Therefore, (6.43) can be modified as follows:

$$\rho(\alpha) = \|\boldsymbol{A}\boldsymbol{x}_\alpha - \boldsymbol{l}\|_{\boldsymbol{P}}^2 - \sigma_0^2 T(\alpha) = 0 \tag{6.44}$$

with

$$\begin{aligned} T(\alpha) &= Tr\left(\boldsymbol{I}_m - \boldsymbol{A}\left(\boldsymbol{A}^{\mathrm{T}}\boldsymbol{P}\boldsymbol{A} + \alpha\boldsymbol{I}_n\right)^{-1}\boldsymbol{A}^{\mathrm{T}}\boldsymbol{P}\right) \\ &= Tr\left(\boldsymbol{I}_m - \tilde{\boldsymbol{A}}\left(\tilde{\boldsymbol{A}}^{\mathrm{T}}\tilde{\boldsymbol{A}} + \alpha\boldsymbol{I}_n\right)^{-1}\tilde{\boldsymbol{A}}^{\mathrm{T}}\right) = m - \sum_{i=1}^{n}\frac{\lambda_i^2}{\lambda_i^2+\alpha} \end{aligned} \tag{6.45}$$

The modified formula, which is called the compensated discrepancy principle, can get better results due to its being almost unbiased. In other words, the variance of unit weight of the regularized solution should be estimated by the following formula:

$$\hat{\sigma}_0^2 = \frac{\|\boldsymbol{A}\boldsymbol{x}_\alpha - \boldsymbol{l}\|_{\boldsymbol{P}}^2}{T(\alpha)} \tag{6.46}$$

It can be verified that $T(\alpha)$ reduces to $m - n$ on condition that $\alpha = 0$. Considering the expression $\|\boldsymbol{A}\boldsymbol{x}_\alpha - \boldsymbol{l}\|_{\boldsymbol{P}}^2 = \hat{\boldsymbol{e}}_\alpha^{\mathrm{T}}\boldsymbol{P}\hat{\boldsymbol{e}}_\alpha = \hat{\tilde{\boldsymbol{e}}}_\alpha^{\mathrm{T}}\hat{\tilde{\boldsymbol{e}}}_\alpha$, and inserting (6.45) into (6.44), we obtain the spectral representation of (6.44) as follows:

$$\rho(\alpha) = \hat{\tilde{\boldsymbol{e}}}_\alpha^{\mathrm{T}}\hat{\tilde{\boldsymbol{e}}}_\alpha - \sigma_0^2\left(m - \sum_{i=1}^{n}\frac{\lambda_i^2}{\lambda_i^2+\alpha}\right) = 0 \tag{6.47}$$

Because (6.47) is a nonlinear equation, it is difficult to derive its explicit solution, the numerical method such as Newton iteration or bisection method should be used. Therefore, the computation of regularization parameter with the spectral form (6.47) is much faster than that in matrix form due to the iteration. The procedure of bisection method is summarized as the following steps:

1. The basic ridge parameter or regularization parameter $\alpha_0 = n\sigma_0^2/\boldsymbol{x}_L^{\mathrm{T}}\boldsymbol{x}_L$ is taken as the initial value in order to speed up the iteration. If the least squares solutions $\boldsymbol{x}_L$ does not exist, any value $\alpha_0 \in R^+$can be used as initial value.
2. If $\rho(\alpha_0) > 0$, taking $\alpha_2 = \alpha_0,\ \alpha_1 = c\alpha_0,\quad (0<c<1)$, where, c is selected to make $\rho(\alpha_1)<0$. Otherwise, if $\rho(\alpha_0)<0$, taking $\alpha_1 = \alpha_0,\ \alpha_2 = c\alpha_0,\quad (c > 1)$ and c is selected to make $\rho(\alpha_2) > 0$. Because $\rho(\alpha)$ is a monotonous increasing function, a unique solution must exist within interval$[\alpha_1,\ \alpha_2]$.
3. Taking $\alpha = (\alpha_1 + \alpha_2)/2$.
4. If $\rho(\alpha) > 0$, taking $\alpha_2 = \alpha$, otherwise taking $\alpha_1 = \alpha$.
5. Repeating steps 3 and 4 until the length of interval $[\alpha_1,\ \alpha_2]$ is less than a given value of limitation.

L-Curve Method

In the L-curve method $\log\|\boldsymbol{x}_\alpha\|$ is taken as the vertical axis η and $\log\|\boldsymbol{A}\boldsymbol{x}_\alpha - \boldsymbol{l}\|_P$ is taken as horizontal axis ξ. The curve drawn with respect to α is shown in Fig. 6.1 and the optimal α lies on the place of the biggest curve. As shown in Fig. 6.1, the form of the curve is just like L, so it is called the L-curve method. The L-curve method can be expressed as the solution of the following criterion:

$$\kappa(\alpha) = \frac{\dot{\xi}\ddot{\eta} - \ddot{\eta}\dot{\xi}}{\left(\dot{\xi}^2 + \dot{\eta}^2\right)^{3/2}} = \max \tag{6.48}$$

where the dot is a differential operator. After $\xi = \log\|\boldsymbol{A}\boldsymbol{x}_\alpha - \boldsymbol{l}\|_P$ and $\eta = \log\|\boldsymbol{x}_\alpha\|$ are inserted into (6.48), the only unknown in (6.48) is the regularization parameter α, which can be solved numerically from

$$\dot{\kappa}(\alpha) = 0. \tag{6.49}$$

Unfortunately, this solution is not unique. The optimal ones should be searched for in the area around the basic ridge parameter $\alpha_0 = n\sigma_0^2/\boldsymbol{x}_L^T\boldsymbol{x}_L$. Miller (1970) first

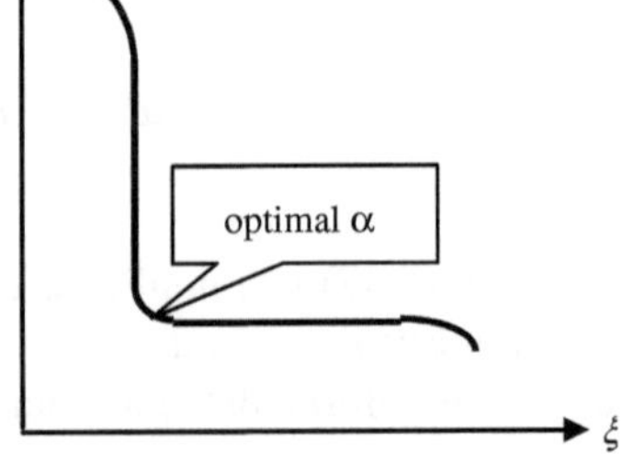

Fig. 6.1 L-curve method

used this method to solve ill-posed equations. For the details about this method, one can refer to Hansen (1996).

Iterative Ridge Method

The iterative ridge method is based on the criterion of minimizing the traced mean square error, which is expressed as

$$Tr(\boldsymbol{M}_\alpha) = \sigma_0^2 \sum_{i=1}^{n} \frac{\lambda_i^2}{\left(\lambda_i^2 + \alpha\right)^2} + \alpha^2 \sum_{i=1}^{n} \frac{\left(\boldsymbol{v}_i^{\mathrm{T}} \bar{\boldsymbol{x}}\right)^2}{\left(\lambda_i^2 + \alpha\right)^2} = \min \tag{6.50}$$

As mentioned above, $Tr(\boldsymbol{M}_\alpha)$ exactly equals $E\left(\|\boldsymbol{x}_\alpha - \bar{\boldsymbol{x}}\|^2\right)$, which is expressed with (6.34). Its first and second order derivatives are derived as follows:

$$\frac{dTr(\boldsymbol{M}_\alpha)}{d\alpha} = \sum_{i=1}^{n} \frac{\lambda_i^2 \left(\left(\boldsymbol{v}_i^{\mathrm{T}} \bar{\boldsymbol{x}}\right)^2 \alpha - \sigma_0^2\right)}{\left(\lambda_i^2 + \alpha\right)^3} \tag{6.51}$$

$$\frac{d^2 Tr(\boldsymbol{M}_\alpha)}{d\alpha^2} = \sum_{i=1}^{n} \frac{\lambda_i^2 \left(\lambda_i^2 \left(\boldsymbol{v}_i^{\mathrm{T}} \bar{\boldsymbol{x}}\right)^2 \alpha + 3\sigma_0^2\right)}{\left(\lambda_i^2 + \alpha\right)^4} \tag{6.52}$$

It is obvious that $\forall \alpha > 0$, $\frac{d^2 Tr(\boldsymbol{M}_\alpha)}{d\alpha^2} > 0$. This indicates that the function $Tr(\boldsymbol{M}_\alpha)$ is strongly convex. Therefore the criterion (6.50) has a unique minimum and the corresponding regularization parameter is numerically solved with the following equation:

$$f(\alpha) = \frac{dTr(\boldsymbol{M}_\alpha)}{d\alpha} = \sum_{i=1}^{n} \frac{\lambda_i^2 \left(\left(\boldsymbol{v}_i^{\mathrm{T}} \bar{\boldsymbol{x}}\right)^2 \alpha - \sigma_0^2\right)}{\left(\lambda_i^2 + \alpha\right)^3} = 0. \tag{6.53}$$

Equation (6.53) can also be solved with the bisection method. As already mentioned, it needs the exact parameters $\bar{\boldsymbol{x}}$ and variance of unit weight σ_0^2. σ_0^2 can be substituted with the estimated value $\hat{\sigma}_0^2$, which can be computed by least squares adjustment even when the equation is ill-posed, because it is not amplified by small eigenvalues.

If the matrix $\boldsymbol{K}$ is a diagonal matrix, the trace of MSE is shown in (6.40). To minimize $Tr(\boldsymbol{M}_\alpha)$ in (6.40), it can be converted to the following algebraic equation:

$$\frac{dTr(\boldsymbol{M}_\alpha)}{d\alpha} = 2 \sum_{i=1}^{n} \frac{\lambda_i^2 \left(\alpha y_i^2 - \sigma_0^2\right)}{\left(\lambda_i^2 + \alpha\right)^3 k_i} = 0. \tag{6.54}$$

Equation (6.54) can also be solved by the bisection method.

GCV Method

The GCV method minimizes the following GCV function to solve the regularized parameter α :

$$\mathrm{GCV}(\alpha) = \frac{\frac{1}{n}\|(\boldsymbol{I}_m - \boldsymbol{H}(\alpha))\boldsymbol{l}\|_{\boldsymbol{P}}^2}{\left(\frac{1}{n}Tr(\boldsymbol{I}_m - \boldsymbol{H}(\alpha))\right)^2} = \min \tag{6.55}$$

where $\boldsymbol{H}(\alpha) = \boldsymbol{A}\left(\boldsymbol{A}^{\mathrm{T}}\boldsymbol{P}\boldsymbol{A} + \lambda\boldsymbol{I}_n\right)^{-1}\boldsymbol{A}^{\mathrm{T}}\boldsymbol{P}$.

Unbiased Variance of Unit Weight

Now the only open problem in the regularization solution is how to estimate the variance $\hat{\sigma}_0^2$ of the unit weight. Although this value can be substituted by that estimated by the least squares adjustment, it is still worth getting the unbiased estimation in the regularized solution. As already mentioned in discussing the generalized discrepancy principle, the variance estimated from (6.56) is biased:

$$\hat{\sigma}_0^2 = \frac{\|\boldsymbol{A}\boldsymbol{x}_\alpha - \boldsymbol{l}\|_{\boldsymbol{P}}^2}{m-n} = \frac{\hat{\boldsymbol{e}}_\alpha^{\mathrm{T}}\boldsymbol{P}\hat{\boldsymbol{e}}_\alpha}{m-n} = \frac{\hat{\bar{\boldsymbol{e}}}_\alpha^{\mathrm{T}}\hat{\bar{\boldsymbol{e}}}_\alpha}{m-n}. \tag{6.56}$$

Since the residual vector, $\hat{\boldsymbol{e}}_\alpha$ or $\hat{\bar{\boldsymbol{e}}}_\alpha$, is biased. The simple way to get the unbiased estimate of the variance of unit weight is to remove the biases from the residuals, which is proposed by Xu et al. (2006). From (6.22e), we derive the bias of the residual vector as follows:

$$\begin{aligned} \delta\hat{\boldsymbol{e}}_\alpha &= E(\hat{\boldsymbol{e}}_\alpha) = \left(\boldsymbol{A}\left(\boldsymbol{A}^{\mathrm{T}}\boldsymbol{P}\boldsymbol{A} + \alpha\boldsymbol{I}_n\right)^{-1}\boldsymbol{A}^{\mathrm{T}}\boldsymbol{P} - \boldsymbol{I}_m\right)E(l) \\ &= \boldsymbol{A}\left(\boldsymbol{A}^{\mathrm{T}}\boldsymbol{P}\boldsymbol{A} + \alpha\boldsymbol{I}_n\right)^{-1}\boldsymbol{A}^{\mathrm{T}}\boldsymbol{P}\boldsymbol{A}\bar{\boldsymbol{x}} - \boldsymbol{A}\bar{\boldsymbol{x}} = -\alpha\boldsymbol{A}\left(\boldsymbol{A}^{\mathrm{T}}\boldsymbol{P}\boldsymbol{A} + \alpha\boldsymbol{I}_n\right)^{-1}\bar{\boldsymbol{x}} \end{aligned} \tag{6.57}$$

After the bias vector $\delta\hat{\boldsymbol{e}}_\alpha$ in (6.57) is removed from $\hat{\boldsymbol{e}}_\alpha$, one can use the bias-removed residual vector to compute the variance of unit weight with (6.56). Unfortunately, computing the bias vector $\delta\hat{\boldsymbol{e}}_\alpha$, one needs the exact parameter vector $\bar{\boldsymbol{x}}$, which is impossible in practice. What we can do is only substitute the exact parameter vector with its estimate. In this case, there still exist higher order biases.

5 Numerical Cases

In order to demonstrate the performance of regularized solution in fast GNSS positioning, we use ten epochs of GPS L1 carrier phase measurements to compute the float ambiguities. The design matrix $\boldsymbol{A}$ and measurement vector $\boldsymbol{l}$ are as below:

$$
A = \begin{bmatrix}
-0.0953 & 0.4594 & 0.6947 & -0.1903 & 0 & 0 & 0 & 0 \\
-0.5089 & 0.2043 & 1.2862 & 0 & -0.1903 & 0 & 0 & 0 \\
-0.6157 & 0.1098 & 0.4187 & 0 & 0 & -0.1903 & 0 & 0 \\
0.5549 & -0.1547 & 0.0865 & 0 & 0 & 0 & -0.1903 & 0 \\
-0.6869 & -0.6729 & 0.1710 & 0 & 0 & 0 & 0 & -0.1903 \\
-0.0952 & 0.4594 & 0.6949 & -0.1903 & 0 & 0 & 0 & 0 \\
-0.5088 & 0.2042 & 1.2863 & 0 & -0.1903 & 0 & 0 & 0 \\
-0.6156 & 0.1097 & 0.4185 & 0 & 0 & -0.1903 & 0 & 0 \\
0.5549 & -0.1547 & 0.0864 & 0 & 0 & 0 & -0.1903 & 0 \\
-0.6868 & -0.6730 & 0.1710 & 0 & 0 & 0 & 0 & -0.1903 \\
-0.0951 & 0.4594 & 0.6950 & -0.1903 & 0 & 0 & 0 & 0 \\
-0.5087 & 0.2041 & 1.2864 & 0 & -0.1903 & 0 & 0 & 0 \\
-0.6154 & 0.1095 & 0.4184 & 0 & 0 & -0.1903 & 0 & 0 \\
0.5549 & -0.1546 & 0.0863 & 0 & 0 & 0 & -0.1903 & 0 \\
-0.6867 & -0.6731 & 0.1711 & 0 & 0 & 0 & 0 & -0.1903 \\
-0.0950 & 0.4594 & 0.6951 & -0.1903 & 0 & 0 & 0 & 0 \\
-0.5085 & 0.2040 & 1.2865 & 0 & -0.1903 & 0 & 0 & 0 \\
-0.6153 & 0.1094 & 0.4182 & 0 & 0 & -0.1903 & 0 & 0 \\
0.5550 & -0.1546 & 0.0863 & 0 & 0 & 0 & -0.1903 & 0 \\
-0.6867 & -0.6732 & 0.1711 & 0 & 0 & 0 & 0 & -0.1903 \\
-0.0949 & 0.4594 & 0.6953 & -0.1903 & 0 & 0 & 0 & 0 \\
-0.5084 & 0.2040 & 1.2866 & 0 & -0.1903 & 0 & 0 & 0 \\
-0.6152 & 0.1093 & 0.4181 & 0 & 0 & -0.1903 & 0 & 0 \\
0.5550 & -0.1546 & 0.0862 & 0 & 0 & 0 & -0.1903 & 0 \\
-0.6866 & -0.6733 & 0.1712 & 0 & 0 & 0 & 0 & -0.1903 \\
-0.0948 & 0.4594 & 0.6954 & -0.1903 & 0 & 0 & 0 & 0 \\
-0.5082 & 0.2039 & 1.2867 & 0 & -0.1903 & 0 & 0 & 0 \\
-0.6150 & 0.1092 & 0.4179 & 0 & 0 & -0.1903 & 0 & 0 \\
0.5550 & -0.1546 & 0.0861 & 0 & 0 & 0 & -0.1903 & 0 \\
-0.6865 & -0.6734 & 0.1712 & 0 & 0 & 0 & 0 & -0.1903 \\
-0.0947 & 0.4594 & 0.6955 & -0.1903 & 0 & 0 & 0 & 0 \\
-0.5081 & 0.2038 & 1.2868 & 0 & -0.1903 & 0 & 0 & 0 \\
-0.6149 & 0.1091 & 0.4178 & 0 & 0 & -0.1903 & 0 & 0 \\
0.5550 & -0.1546 & 0.0860 & 0 & 0 & 0 & -0.1903 & 0 \\
-0.6864 & -0.6735 & 0.1713 & 0 & 0 & 0 & 0 & -0.1903 \\
-0.0946 & 0.4594 & 0.6956 & -0.1903 & 0 & 0 & 0 & 0 \\
-0.5079 & 0.2037 & 1.2869 & 0 & -0.1903 & 0 & 0 & 0 \\
-0.6148 & 0.1090 & 0.4176 & 0 & 0 & -0.1903 & 0 & 0 \\
0.5551 & -0.1546 & 0.0859 & 0 & 0 & 0 & -0.1903 & 0 \\
-0.6863 & -0.6736 & 0.1713 & 0 & 0 & 0 & 0 & -0.1903 \\
-0.0945 & 0.4594 & 0.6958 & -0.1903 & 0 & 0 & 0 & 0 \\
-0.5078 & 0.2036 & 1.2870 & 0 & -0.1903 & 0 & 0 & 0 \\
-0.6146 & 0.1089 & 0.4175 & 0 & 0 & -0.1903 & 0 & 0 \\
0.5551 & -0.1546 & 0.0859 & 0 & 0 & 0 & -0.1903 & 0 \\
-0.6863 & -0.6737 & 0.1714 & 0 & 0 & 0 & 0 & -0.1903 \\
-0.0944 & 0.4594 & 0.6959 & -0.1903 & 0 & 0 & 0 & 0 \\
-0.5077 & 0.2036 & 1.2871 & 0 & -0.1903 & 0 & 0 & 0 \\
-0.6145 & 0.1088 & 0.4174 & 0 & 0 & -0.1903 & 0 & 0 \\
0.5551 & -0.1546 & 0.0858 & 0 & 0 & 0 & -0.1903 & 0 \\
-0.6862 & -0.6738 & 0.1714 & 0 & 0 & 0 & 0 & -0.1903
\end{bmatrix}, \quad
l = \begin{bmatrix}
-0.1797 \\ 0.2722 \\ -0.5077 \\ -0.4296 \\ -0.5321 \\
-0.1818 \\ 0.2678 \\ -0.5088 \\ -0.4364 \\ -0.5305 \\
-0.1783 \\ 0.2690 \\ -0.5081 \\ -0.4344 \\ -0.5265 \\
-0.1820 \\ 0.2718 \\ -0.5068 \\ -0.4356 \\ -0.5249 \\
-0.1835 \\ 0.2702 \\ -0.5121 \\ -0.4396 \\ -0.5304 \\
-0.1819 \\ 0.2686 \\ -0.5101 \\ -0.4372 \\ -0.5390 \\
-0.1814 \\ 0.2705 \\ -0.5111 \\ -0.4390 \\ -0.5378 \\
-0.1780 \\ 0.2709 \\ -0.5087 \\ -0.4355 \\ -0.5282 \\
-0.1827 \\ 0.2715 \\ -0.5115 \\ -0.4410 \\ -0.5380 \\
-0.1837 \\ 0.2749 \\ -0.5085 \\ -0.4369 \\ -0.5337
\end{bmatrix}
$$

In the design matrix, the first three columns are coefficients of position parameters, and the rest columns are the coefficients of five ambiguities. The formal normal matrix and the constant vector are as follows:

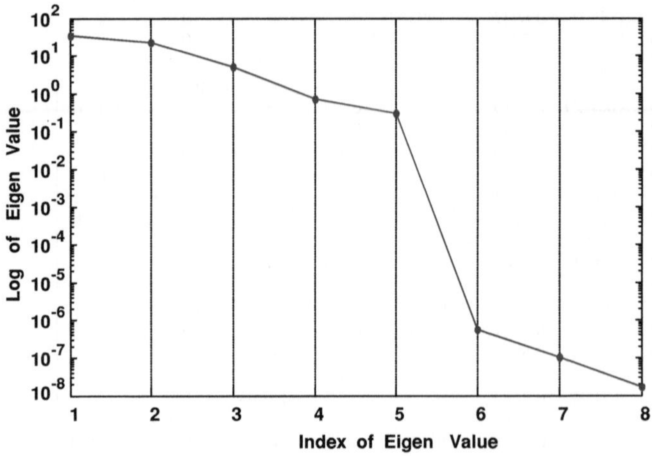

Fig. 6.2 Eigenvalues of normal matrix for position and ambiguity parameters

$$N = A^{\mathrm{T}}A = \begin{bmatrix} 22.4288 & 2.9916 & -8.9799 & -0.4953 & 1.0784 & 1.4848 & -2.9684 & 1.7567 \\ 2.9916 & 14.8283 & 10.4683 & -1.7836 & -0.8112 & -0.4511 & 0.5534 & 2.5276 \\ -8.9799 & 10.4683 & 23.4697 & -0.9607 & -3.2112 & 0.0946 & 1.3577 & 1.0340 \\ -0.4953 & -1.7836 & -0.9607 & 0.6035 & -0.1207 & -0.1207 & -0.1207 & -0.1207 \\ 1.0784 & -0.8112 & -3.2112 & -0.1207 & 0.6035 & -0.1207 & -0.1207 & -0.1207 \\ 1.4848 & -0.4511 & 0.0946 & -0.1207 & -0.1207 & 0.6035 & -0.1207 & -0.1207 \\ -2.9684 & 0.5534 & 1.3577 & -0.1207 & -0.1207 & -0.1207 & 0.6035 & -0.1207 \\ 1.7567 & 2.5276 & 1.0340 & -0.1207 & -0.1207 & -0.1207 & -0.1207 & 0.6035 \end{bmatrix}$$

$$C = A^{\mathrm{T}}l = [1.7961 \quad 6.1980 \quad 8.7025 \quad -0.2409 \quad -1.7418 \quad 1.2268 \quad 0.5175 \quad 1.1400]^{\mathrm{T}}$$

The normal matrix is seriously ill-posed with conditional numbers as big as 2×10^9, and their eigenvalues are plotted in Fig. 6.2. The ambiguities solved by least squares adjustment are $[8.369, -5.506, -8.188, 12.576, -23.400]^{\mathrm{T}}$ and their accuracies are $[10.40, 16.55, 14.32, 10.31, 32.33]^{\mathrm{T}}$, while the true values of these ambiguities are $[1, -1, 3, 2, 3]^{\mathrm{T}}$. Therefore, the least squares solution cannot produce satisfactory results in this case.

After eliminating three position parameters, we can derive the normal equation only for ambiguities:

$$\begin{bmatrix} 0.3831 & -0.1664 & -0.1919 & -0.0601 & 0.1936 \\ -0.1664 & 0.1026 & -0.0017 & -0.0302 & -0.0758 \\ -0.1919 & -0.0017 & 0.3349 & 0.1881 & -0.1203 \\ -0.0601 & -0.0302 & 0.1881 & 0.1139 & -0.0458 \\ 0.1936 & -0.0758 & -0.1203 & -0.0458 & 0.1001 \end{bmatrix} \begin{bmatrix} n_1 \\ n_2 \\ n_3 \\ n_4 \\ n_5 \end{bmatrix} = \begin{bmatrix} 0.4068 \\ -0.5488 \\ 0.8407 \\ 0.6270 \\ 0.1038 \end{bmatrix}$$

where n_i ($i = 1, 2, \ldots, 5$) denotes the ambiguity parameters. The condition number of the above normal equation is 4×10^7 and five eigenvalues are shown in Fig. 6.3.

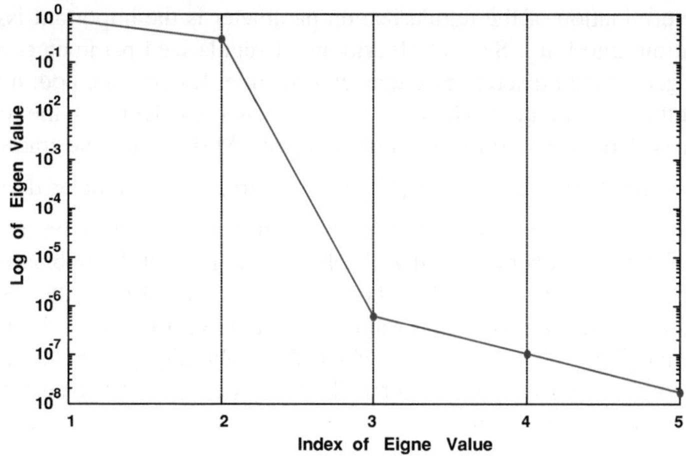

Fig. 6.3 Eigenvalues of normal matrix for ambiguity parameters

In order to improve the accuracy of solved ambiguities, we use Tikhonov regularization to reduce the condition number and stabilize the solution. The regularization parameter is 1.91×10^{-5}, the ambiguities of regularized solution are [1.588, −1.762, 2.419, 2.075, 0.485], and their accuracies are [0.68, 0.97, 0.66, 0.95, 1.05], respectively. Therefore, the quality of regularized solution is greatly improved.

6 Summary

In the first part of this chapter, we discuss the regularized solution and its spectral decomposition formulae to the ill-posed linear observation equations. We first briefly introduce the ill-posed problems and its applications in solving ill-conditioned geodetic observation model, and present the related literature.

The unstable characteristics of least squares solution are analyzed using singular value decomposition, when the observation equations are ill-posed. It shows that the least squares solution is mainly contaminated by high frequency noises, since the singular value in high frequency domain is very small. Then a filter factor, which is close to 0 and 1 for low and high frequency terms, respectively, is introduced to preserve the low frequency signals and filter out the high frequency noises. As a special case, we present the filter factor for a rank deficient model.

Then we discuss the regularized solution to ill-posed observation equations and the spectral decomposition representations, and establish the relationship between the filter factor and the regularization parameter. Since the regularized solution is biased, the MSE is proposed to evaluate the quality of the solution.

The determination of the regularization parameter is the important issue of the regularization algorithm. Several algorithms of regularized parameters are given, including generalized discrete discrepancy principle, L-curve method, minimizing MSE, and the GCV method. The formula for efficiently calculating the regularized parameter is derived based on minimizing traced MSE. Since the traced MSE is equal to the total error $E\left(\|\boldsymbol{x}_\alpha - \bar{\boldsymbol{x}}\|^2\right)$, the regularization parameter derived from minimizing traced MSE can probably give better results. Then we analyze the biasness of the residuals of regularized solution and present the unbiased estimate of the variance of unit weight after the biases are removed from the residuals.

Finally we present a numerical example of fast GPS positioning with ten epochs of L1 carrier phase observations, in which the ambiguity parameters are highly collinear with the position parameters. The results show that the performance of regularized solution is excellent.

Part II: Adjustment

7 Introduction

Most useful and necessary adjustment algorithms for data processing are outlined in the second part of this chapter. The adjustment algorithms discussed here include least squares adjustment, sequential application of least squares adjustment via accumulation, sequential least squares adjustment, conditional least squares adjustment, a sequential application of conditional least squares adjustment, block-wise least squares adjustment, a sequential application of block-wise least squares adjustment, an equivalent algorithm to form the eliminated observation equation system and the algorithm to diagonalize the normal equation and equivalent observation equation.

A priori constrained adjustments are discussed for solving the rank deficient problems. After a general discussion on the a priori parameter constraints, a special case of the so-called a priori datum method is given. A quasi-stable datum method is also discussed.

A summary is given at the end of this part of this chapter.

8 Least Squares Adjustment

The principle of least squares adjustment can be summarized as below (Gotthardt 1978; Cui et al. 1982).

1. The linearized observation equation system can be represented by

$$V = L - AX,\ P \tag{6.1}$$

where

L is the observation vector of dimension m,
A is the coefficient matrix of dimension $m \times n$,
X is an unknown parameter vector of dimension n,
V is the residual vector of dimension m,
n is the number of unknowns,
m is the number of observations, and
P is the symmetric and definite weight matrix of dimension $m \times m$

2. The least squares criterion for solving the observation equations is well-known as

$$V^{\mathrm{T}}PV = \min, \tag{6.2}$$

where V^{T} is the transpose of the related vector V.

3. To solve X and compute V, a function F is set as

$$F = V^{\mathrm{T}}PV. \tag{6.3}$$

The function F reaches minimum value if the partial differentiation of F with respect to X equals zero, i.e.,

$$\frac{\partial F}{\partial X} = 2V^{\mathrm{T}}P(-A) = 0$$

or

$$A^{\mathrm{T}}PV = 0, \tag{6.4}$$

where A^{T} is the transpose matrix of A.

4. Multiplying $A^{\mathrm{T}}P$ with (6.1), one has

$$A^{\mathrm{T}}PAX - A^{\mathrm{T}}PL = -A^{\mathrm{T}}PV \tag{6.5}$$

Setting (6.4) into (6.5), one has

$$A^{\mathrm{T}}PAX - A^{\mathrm{T}}PL = 0. \tag{6.6}$$

5. For simplification, let $M = A^{\mathrm{T}}PA$, $Q = M^{-1}$, where superscript $^{-1}$ is an inverse operator, and M is usually called a normal matrix. The least squares solution of (6.1) is then

$$X = Q(A^{\mathrm{T}}PL). \tag{6.7}$$

6. The precision of the ith element of the estimated parameter is

$$p[i] = m_0\sqrt{Q[i][i]}, \tag{6.8}$$

where i is the element index of a vector or a matrix, m_0 is the so-called standard deviation (or sigma), $p[i]$ is the ith element of the precision vector, $Q[i][i]$ is the i-th diagonal element of the cofactor matrix Q, and

$$m_0 = \sqrt{\frac{V^{\mathrm{T}}PV}{m-n}}, \quad \text{if } (m > n). \tag{6.9}$$

7. For convenience of sequential computation, $V^{\mathrm{T}}PV$ can be calculated by using

$$V^{\mathrm{T}}PV = L^{\mathrm{T}}PL - (A^{\mathrm{T}}PL)^{\mathrm{T}}X. \tag{6.10}$$

This can be obtained by substituting (6.1) into $V^{\mathrm{T}}PV$ and considering (6.4).

The complete formulas of least squares adjustment have now been derived.

8.1 Least Squares Adjustment with Sequential Observation Groups

Suppose one has two sequential observation equation systems:

$$V_1 = L_1 - A_1X \qquad \text{and} \tag{6.11}$$

$$V_2 = L_2 - A_2X, \tag{6.12}$$

with weight matrices P_1 and P_2. These two equation systems are uncorrelated or independent and have the common unknown vector X. The combined problem can be represented as

$$\begin{pmatrix} V_1 \\ V_2 \end{pmatrix} = \begin{pmatrix} L_1 \\ L_2 \end{pmatrix} - \begin{pmatrix} A_1 \\ A_2 \end{pmatrix} X \quad \text{and} \quad P = \begin{pmatrix} P_1 & 0 \\ 0 & P_2 \end{pmatrix}. \tag{6.13}$$

The least squares normal equation can be formed then as

$$\begin{pmatrix} A_1^{\mathrm{T}} & A_2^{\mathrm{T}} \end{pmatrix} \begin{pmatrix} P_1 & 0 \\ 0 & P_2 \end{pmatrix} \begin{pmatrix} A_1 \\ A_2 \end{pmatrix} X = \begin{pmatrix} A_1^{\mathrm{T}} & A_2^{\mathrm{T}} \end{pmatrix} \begin{pmatrix} P_1 & 0 \\ 0 & P_2 \end{pmatrix} \begin{pmatrix} L_1 \\ L_2 \end{pmatrix}$$

or

$$(A_1^{\mathrm{T}}P_1A_1 + A_2^{\mathrm{T}}P_2A_2)X = (A_1^{\mathrm{T}}P_1L_1 + A_2^{\mathrm{T}}P_2L_2). \tag{6.14}$$

This is indeed the accumulation of the two least squares normal equations formed from (6.11) and (6.12), respectively:

$$(A_1^T P_1 A_1)X = A_1^T P_1 L_1 \tag{6.15}$$

and

$$(A_2^T P_2 A_2)X = A_2^T P_2 L_2. \tag{6.16}$$

The solution is then

$$X = (A_1^T P_1 A_1 + A_2^T P_2 A_2)^{-1}(A_1^T P_1 L_1 + A_2^T P_2 L_2). \tag{6.17}$$

The precision of the ith element of the estimated parameter is

$$p[i] = m_0 \sqrt{Q[i][i]}, \tag{6.18}$$

where

$$m_0 = \sqrt{\frac{V^T P V}{m-n}}, \quad \text{if } (m > n), \tag{6.19}$$

and

$$Q = (A_1^T P_1 A_1 + A_2^T P_2 A_2)^{-1}, \tag{6.20}$$

where m is the number of total observations and n is the number of unknowns. $V^T PV$ can be calculated by using

$$\begin{aligned} V^T PV &= V_1^T P_1 V_1 + V_2^T P_2 V_2 \\ &= L_1^T P_1 L_1 + L_2^T P_2 L_2 - (A_1^T P_1 L_1)^T X - (A_2^T P_2 L_2)^T X \\ &= (L_1^T P_1 L_1 + L_2^T P_2 L_2) - (A_1^T P_1 L_1 + A_2^T P_2 L_2)^T X. \end{aligned} \tag{6.21}$$

Equation (6.17) indicates that the sequential least squares problem can be solved by simply accumulating the normal equations of the observation equations. The weighted squares residuals can also be computed by accumulating the individual quadratic forms of the residuals using (6.21).

For further sequential and independent observation equation systems,

$$V_1 = L_1 - A_1 X, \quad P_1, \tag{6.22}$$

$$V_2 = L_2 - A_2 X, \quad P_2, \tag{6.23}$$
$$\ldots$$

$$V_i = L_i - A_i X, \quad P_i, \tag{6.24}$$

the solution can be similarly derived as

$$X = (A_1^T P_1 A_1 + A_2^T P_2 A_2 + \cdots + A_i^T P_i A_i)^{-1}(A_1^T P_1 L_1 + A_2^T P_2 L_2 + \cdots + A_i^T P_i L_i) \tag{6.25}$$

and

$$V^{\mathrm{T}}PV = (L_1^{\mathrm{T}}P_1L_1 + L_2^{\mathrm{T}}P_2L_2 + \cdots + L_i^{\mathrm{T}}P_iL_i) - (A_1^{\mathrm{T}}P_1L_1 + A_2^{\mathrm{T}}P_2L_2 + \cdots + A_i^{\mathrm{T}}P_iL_i)^{\mathrm{T}}X. \tag{6.26}$$

It is obvious that if the solution is needed for every epoch, then the accumulated equation system has to be solved at each epoch. The accumulations always have to be made with the sequential normal equations. Of course, the solutions can be computed after a defined epoch or at the last epoch. This could be very useful if the solution of the problem is unstable at the beginning.

9 Sequential Least Squares Adjustment

Recalling the discussions in Sect. 8, one has sequential observation equation systems

$$V_1 = L_1 - A_1X, \quad P_1 \tag{6.27}$$

and

$$V_2 = L_2 - A_2X, \quad P_2. \tag{6.28}$$

These two equation systems are uncorrelated. The sequential problem can be then solved by accumulating the individual normal equations as discussed in Sect. 8:

$$\left(A_1^{\mathrm{T}}P_1A_1 + A_2^{\mathrm{T}}P_2A_2\right)X = \left(A_1^{\mathrm{T}}P_1L_1 + A_2^{\mathrm{T}}P_2L_2\right) \tag{6.29}$$

or

$$X = \left(A_1^{\mathrm{T}}P_1A_1 + A_2^{\mathrm{T}}P_2A_2\right)^{-1}\left(A_1^{\mathrm{T}}P_1L_1 + A_2^{\mathrm{T}}P_2L_2\right). \tag{6.30}$$

$V^{\mathrm{T}}PV$ can be calculated by using

$$V^{\mathrm{T}}PV = \left(L_1^{\mathrm{T}}P_1L_1 + L_2^{\mathrm{T}}P_2L_2\right) - \left(A_1^{\mathrm{T}}P_1L_1 + A_2^{\mathrm{T}}P_2L_2\right)^{\mathrm{T}}X. \tag{6.31}$$

If (6.27) is solvable, then the least squares solution can be represented as

$$X = \left(A_1^{\mathrm{T}}P_1A_1\right)^{-1}\left(A_1^{\mathrm{T}}P_1L_1\right). \tag{6.32}$$

and

$$V^{\mathrm{T}}PV = L_1^{\mathrm{T}}P_1L_1 - \left(A_1^{\mathrm{T}}P_1L_1\right)^{\mathrm{T}}X. \tag{6.33}$$

For convenience, the estimated vector of X by using the first group of observations is denoted by X_1 and the quadratic form of the residuals by $(V^{\mathrm{T}}PV)_1$ as well as $Q_1 = \left(A_1^{\mathrm{T}}P_1A_1\right)^{-1}$.

Using the formula (Cui et al. 1982; Gotthardt 1978)

$$(D+ACB)^{-1} = D^{-1} - D^{-1}AKB\,D^{-1}, \tag{6.34}$$

where A and B are any matrices, C and D are matrices that can be inversed, and

$$K = (C^{-1} + B\,D^{-1}A)^{-1}, \tag{6.35}$$

the inversion of the accumulated normal matrix can be represented as Q:

$$\begin{aligned} Q &= \left(A_1^{\mathrm{T}}P_1A_1 + A_2^{\mathrm{T}}P_2A_2\right)^{-1} \\ &= \left(A_1^{\mathrm{T}}P_1A_1\right)^{-1} - \left(A_1^{\mathrm{T}}P_1A_1\right)^{-1}A_2^{\mathrm{T}}KA_2\left(A_1^{\mathrm{T}}P_1A_1\right)^{-1} \\ &= Q_1 - Q_1A_2^{\mathrm{T}}KA_2Q_1 \\ &= \left(E - Q_1A_2^{\mathrm{T}}KA_2\right)Q_1. \end{aligned} \tag{6.36}$$

and

$$K = \left(P_2^{-1} + A_2Q_1A_2^{\mathrm{T}}\right)^{-1}, \tag{6.37}$$

where E is an identity matrix. The total term in the parenthesis on the right-hand side of (6.36) can be interpreted as a modifying factor for the Q_1 matrix; in other words, due to the sequential (6.28), the Q matrix can be computed by multiplying a factor to the Q_1 matrix. So sequential least squares solution of (6.27) and (6.28) can be obtained:

$$\begin{aligned} X &= \left(Q_1 - Q_1A_2^{\mathrm{T}}KA_2Q_1\right)\left(A_1^{\mathrm{T}}P_1L_1 + A_2^{\mathrm{T}}P_2L_2\right) \\ &= \left(E - Q_1A_2^{\mathrm{T}}KA_2\right)X_1 + Q\left(A_2^{\mathrm{T}}P_2L_2\right). \end{aligned} \tag{6.38}$$

Mathematically, the solutions of the sequential problem of (6.27) and (6.28) that are solved by using accumulation of the least squares method as discussed in Sect. 8.1 or using sequential adjustment as discussed above shall be the same. However, in practice, accuracy of the computation is always limited by the effective digits of the computer being used. Such a limit on the effective digits causes an inaccuracy of numerical computation. This inaccuracy will be accumulated and propagated in further computing processes. By comparing the results obtained with the above-mentioned methods, it is noticed that the sequential method will give a drift in the results. The drift increases with time and is generally not negligible after a long time interval.

10 Conditional Least Squares Adjustment

The principle of least squares adjustment with condition equations can be summarized as below (Gotthardt 1978; Cui et al. 1982):

1. The linearized observation equation system can be represented by (6.1) (see Sect. 8).
2. The corresponding condition equation system can be written as

$$CX - W = 0, \tag{6.39}$$

where
C is the coefficient matrix of dimension $r \times n$,
W is the constant vector of dimension r, and
R is thenumber of conditions

3. The least squares criterion for solving the observation equations with condition equations is well-known as

$$V^{\mathrm{T}}PV = \min, \tag{6.40}$$

where V^{T} is the transpose of the related vector V.

4. To solve X and compute V, a function F can be formed as

$$F = V^{\mathrm{T}}PV + 2K^{\mathrm{T}}(CX - W), \tag{6.41}$$

where K is a gain vector (of dimension r) to be determined. The function F reaches minimum value if the partial differentiation of F with respect to X equals zero, i.e.,

$$\frac{\partial F}{\partial X} = 2V^{\mathrm{T}}P(-A) + 2K^{\mathrm{T}}C = 0;$$

then one has

$$-A^{\mathrm{T}}PV + C^{\mathrm{T}}K = 0 \tag{6.42}$$

or

$$A^{\mathrm{T}}PAX + C^{\mathrm{T}}K - A^{\mathrm{T}}PL = 0, \tag{6.43}$$

where A^{T}, C^{T} are transpose matrices of A and C, respectively.

5. Combining (6.43) and (6.39) together, one has

$$A^{\mathrm{T}}PAX + C^{\mathrm{T}}K - A^{\mathrm{T}}PL = 0 \tag{6.44}$$

and

$$CX - W = 0. \tag{6.45}$$

6. For simplification, let $M = A^{\mathrm{T}}PA$, $W_1 = A^{\mathrm{T}}PL$, $Q = M^{-1}$, where superscript $^{-1}$ is an inverse operator. The solutions of (6.44) and (6.45) are then

$$K = \left(CQC^{\mathrm{T}}\right)^{-1}(CQW_1 - W), \tag{6.46}$$

$$X = -Q(C^T K - W_1)$$

or

$$\begin{aligned} X &= (A^{\mathrm{T}}PA)^{-1}(A^{\mathrm{T}}PL) - (A^{\mathrm{T}}PA)^{-1}C^{\mathrm{T}}K \\ &= (A^{\mathrm{T}}PA)^{-1}(A^{\mathrm{T}}PL - C^{\mathrm{T}}K) \end{aligned}. \tag{6.47}$$

7. The precisions of the solutions are then

$$p[i] = m_0\sqrt{Q_{\mathrm{c}}[i][i]}, \tag{6.48}$$

where i is the element index of a vector or a matrix, $\sqrt{\ }$ is the square root operator, m_0 is the so-called standard deviation (or sigma), $p[i]$ is the ith element of the precision vector, $Q_{\mathrm{c}}[i][i]$ is the ith diagonal element of the quadratic matrix Q_{c}, and

$$Q_{\mathrm{c}} = Q - QC^{\mathrm{T}}Q_2CQ, \tag{6.49}$$

and

$$Q_2 = \left(CQC^{\mathrm{T}}\right)^{-1} \tag{6.50}$$

$$m_0 = \sqrt{\frac{V^{\mathrm{T}}PV}{m - n + r}}, \quad \text{if } (m > n - r). \tag{6.51}$$

8. For convenience of sequential computation, $V^{\mathrm{T}}PV$ can be calculated by using

$$V^{\mathrm{T}}PV = L^{\mathrm{T}}PL - \left(A^{\mathrm{T}}PL\right)^{\mathrm{T}}X - W^{\mathrm{T}}K. \tag{6.52}$$

This can be obtained by substituting (6.1) into $V^{\mathrm{T}}PV$ and using the relations of (6.39) and (6.42).

The complete formulas of conditional least squares adjustment have now been derived.

10.1 Sequential Application of Conditional Least Squares Adjustment

Recalling the least squares adjustment discussed in Sect. 8, the linearized observation equation system

$$V = L - AX, \ P \tag{6.53}$$

has the solution

$$X = \left(A^{\mathrm{T}}PA\right)^{-1}\left(A^{\mathrm{T}}PL\right). \tag{6.54}$$

The precisions of the solutions can be obtained by

$$p[i] = m_0\sqrt{Q[i][i]}, \tag{6.55}$$

where

$$m_0 = \sqrt{\frac{V^{\mathrm{T}}PV}{m-n}}, \quad \text{if } (m > n), \tag{6.56}$$

and $V^{\mathrm{T}}PV$ can be calculated by using

$$V^{\mathrm{T}}PV = L^{\mathrm{T}}PL - \left(A^{T}PL\right)^{\mathrm{T}}X. \tag{6.57}$$

For convenience, the least squares solution vector is denoted by X_0 and weighted residuals square by $(V^{\mathrm{T}}PV)_0$.

Similarly, in the conditional least squares adjustment discussed in Sect. 10, the linearized observation equation system and conditional equations read

$$V = L - AX \tag{6.58}$$

and

$$CX - W = 0; \tag{6.59}$$

the solution follows

$$X = \left(A^{\mathrm{T}}PA\right)^{-1}\left(A^{\mathrm{T}}PL - C^{\mathrm{T}}K\right), \tag{6.60}$$

where K is the gain, and

$$K = \left(CQC^{\mathrm{T}}\right)^{-1}(CQW_1 - W). \tag{6.61}$$

The precision vector of the solution vector can be obtained by using (6.48)–(6.52). Using the notations obtained in least squares solution, one has

$$X = X_0 - QC^{\mathrm{T}}K \tag{6.62}$$

and

$$V^{\mathrm{T}}PV = (V^{\mathrm{T}}PV)_0 + (A^{\mathrm{T}}PL)^{\mathrm{T}}QC^{\mathrm{T}}K - W^{\mathrm{T}}K. \tag{6.63}$$

Equation (6.62) indicates that the conditional least squares problem can be solved first without the conditions, and then through the gain K to compute a modification term. The change of the solution is caused by the conditions. For computing the weighted squares of the residuals, (6.63) can be used (by adding two modification terms to the weighted squares of residuals of the least squares solution). This property is very important for many practical applications such as ambiguity fixing or coordinates fixing. For example, after the least squares solution and fixing the ambiguity values, one needs to compute the ambiguity fixed solution. Of course, one can put the fixed ambiguities as known parameters and go back to solve the problem once again. However, by using the above formulas, one can use the fixed ambiguities as conditions to compute the gain and the modification terms to get the ambiguity fixed solution directly. Similarly, this property can also be used for solutions with some fixed station coordinates.

11 Block-Wise Least Squares Adjustment

The principle of block-wise least squares adjustment can be summarized as below (Gotthardt 1978; Cui et al. 1982):

1. The linearized observation equation system can be represented by (6.1) (see Sect. 8).
2. The unknown vector X and observable vector L is rewritten as two sub-vectors:

$$\begin{pmatrix} V_1 \\ V_2 \end{pmatrix} = \begin{pmatrix} L_1 \\ L_2 \end{pmatrix} - \begin{pmatrix} A_{11} & A_{12} \\ A_{21} & A_{22} \end{pmatrix} \begin{pmatrix} X_1 \\ X_2 \end{pmatrix} \quad \text{and} \quad P = \begin{pmatrix} P_1 & 0 \\ 0 & P_2 \end{pmatrix}. \tag{6.64}$$

The least squares normal equation can then be formed as

$$\begin{pmatrix} A_{11} & A_{12} \\ A_{21} & A_{22} \end{pmatrix}^{\mathrm{T}} \begin{pmatrix} P_1 & 0 \\ 0 & P_2 \end{pmatrix} \begin{pmatrix} A_{11} & A_{12} \\ A_{21} & A_{22} \end{pmatrix} \begin{pmatrix} X_1 \\ X_2 \end{pmatrix} = \begin{pmatrix} A_{11} & A_{12} \\ A_{21} & A_{22} \end{pmatrix}^{\mathrm{T}} \begin{pmatrix} P_1 & 0 \\ 0 & P_2 \end{pmatrix} \begin{pmatrix} L_1 \\ L_2 \end{pmatrix}. \tag{6.65}$$

The normal equation can be denoted by

$$\begin{pmatrix} M_{11} & M_{12} \\ M_{21} & M_{22} \end{pmatrix} \begin{pmatrix} X_1 \\ X_2 \end{pmatrix} = \begin{pmatrix} B_1 \\ B_2 \end{pmatrix} \tag{6.66}$$

or

$$M_{11}X_1 + M_{12}X_2 = B_1 \tag{6.67}$$

and

$$M_{21}X_1 + M_{22}X_2 = B_2, \tag{6.68}$$

where

$$M_{11} = A_{11}^{\mathrm{T}} P_1 A_{11} + A_{21}^{\mathrm{T}} P_2 A_{21}, \tag{6.69}$$

$$M_{12} = M_{21}^{\mathrm{T}} = A_{11}^{\mathrm{T}} P_1 A_{12} + A_{21}^{\mathrm{T}} P_2 A_{22}, \tag{6.70}$$

$$M_{22} = A_{12}^{\mathrm{T}} P_1 A_{12} + A_{22}^{\mathrm{T}} P_2 A_{22} \tag{6.71}$$

$$B_1 = A_{11}^{\mathrm{T}} P_1 L_1 + A_{21}^{\mathrm{T}} P_2 L_2 \tag{6.72}$$

and

$$B_2 = A_{12}^{T} P_1 L_1 + A_{22}^{T} P_2 L_2. \tag{6.73}$$

3. Normal equations (6.67) and (6.68) can be solved as follows: from (6.67), one has

$$X_1 = M_{11}^{-1}(B_1 - M_{12}X_2). \tag{6.74}$$

Substituting X_1 into (6.68), one gets a normal equation related to the second block of unknowns:

$$M_2 X_2 = R_2, \tag{6.75}$$

where

$$M_2 = M_{22} - M_{21} M_{11}^{-1} M_{12} \tag{6.76}$$

and

$$R_2 = B_2 - M_{21} M_{11}^{-1} B_1. \tag{6.77}$$

The solution of (6.75) is then

$$X_2 = M_2^{-1} R_2. \tag{6.78}$$

From (6.78) and (6.74), the block-wise least squares solution of (6.1) and (6.64) can be computed. For estimating the precision of the solved vector, one has (see discussion in Sect. 8):

$$p[i] = m_0 \sqrt{Q[i][i]} \tag{6.79}$$

where

$$m_0 = \sqrt{\frac{V^T P V}{m - n}}, \qquad \text{if} \qquad (m > n). \tag{6.80}$$

Q is the inversion of the total normal matrix M, m is the number of total observations, and n is the number of unknowns.

Furthermore,

$$Q = \begin{pmatrix} M_{11} & M_{12} \\ M_{21} & M_{22} \end{pmatrix}^{-1} = \begin{pmatrix} Q_{11} & Q_{12} \\ Q_{21} & Q_{22} \end{pmatrix} \text{ is denoted,} \tag{6.81}$$

where (Gotthardt 1978; Cui et al. 1982)

$$Q_{11} = \left(M_{11} - M_{12}M_{22}^{-1}M_{21}\right)^{-1}, \tag{6.82}$$

$$Q_{22} = \left(M_{22} - M_{21}M_{11}^{-1}M_{12}\right)^{-1}, \tag{6.83}$$

$$Q_{12} = M_{11}^{-1}(-M_{12}Q_{22}), \tag{6.84}$$

and

$$Q_{21} = M_{22}^{-1}(-M_{21}Q_{11}). \tag{6.85}$$

In addition, $V^{\mathrm{T}}PV$ can be calculated by using

$$V^TPV = L^TPL - (A^TPL)^TX. \tag{6.86}$$

One finds very important applications in GPS data processing by separating the unknowns into two groups, which will be discussed in the next section.

11.1 Sequential Solution of Block-Wise Least Squares Adjustment

Suppose one has two sequential observation equation systems

$$V_{t1} = L_{t1} - A_{t1}Y_{t1} \tag{6.87}$$

and

$$V_{t2} = L_{t2} - A_{t2}Y_{t2}, \tag{6.88}$$

with weight matrices P_{t1} and P_{t2}. The unknown vector Y can be separated into two sub-vectors; one is sequential dependent and the other is time independent. Let us assume

$$Y_{t1} = \begin{pmatrix} X_{t1} \\ X_2 \end{pmatrix} \quad \text{and} \quad Y_{t2} = \begin{pmatrix} X_{t2} \\ X_2 \end{pmatrix}, \tag{6.89}$$

where X_2 is the common unknown vector, and X_{t1} and X_{t2} are sequential (time) independent unknowns (i.e., they are different from each other).

Equations (6.87) and (6.88) can be solved separately by using the block-wise least squares method as follows (see Sect. 11):

$$X_{t1} = (M_{11})_{t1}^{-1}(B_1 - M_{12}X_2)_{t1}, \tag{6.90}$$

$$(M_2)_{t1}X_2 = (R_2)_{t1} \quad \text{and} \tag{6.91}$$

$$X_2 = (M_2)_{t1}^{-1}(R_2)_{t1}, \tag{6.92}$$

and

$$X_{t2} = (M_{11})_{t2}^{-1}(B_1 - M_{12}X_2)_{t2}, \tag{6.93}$$

$$(M_2)_{t2}X_2 = (R_2)_{t2} \quad \text{and} \tag{6.94}$$

$$X_2 = (M_2)_{t2}^{-1}(R_2)_{t2}, \tag{6.95}$$

where indices $t1$ and $t2$ outside of the parenthesis indicate that the matrices and vectors are related to (6.87) and (6.88), respectively.

The combined solution of (6.87) and (6.88) can then be derived as

$$X_{t1} = (M_{11})_{t1}^{-1}((B_1)_{t1} - (M_{12})_{t1}(X_2)_{ta}), \tag{6.96}$$

$$X_{t2} = (M_{11})_{t2}^{-1}((B_1)_{t2} - (M_{12})_{t2}(X_2)_{ta}), \tag{6.97}$$

$$((M_2)_{t1} + (M_2)_{t2})(X_2)_{ta} = (R_2)_{t1} + (R_2)_{t2} \quad \text{and} \tag{6.98}$$

$$(X_2)_{ta} = ((M_2)_{t1} + (M_2)_{t2})^{-1}((R_2)_{t1} + (R_2)_{t2}), \tag{6.99}$$

where index ta means that the solution is related to all equations. The normal equations related to the common unknowns are accumulated and solved for. The solved common unknowns are used for computing sequentially different unknowns.

In the case of many sequential observations, a combined solution could be difficult or even impossible because of the large number of unknowns and the requirement of the computing capacities. Therefore, a sequential solution could be a good alternative. For the sequential observation equations

$$V_{t1} = L_{t1} - A_{t1}Y_{t1}, \quad P_{t1}, \tag{6.100}$$

$$\ldots$$

$$V_{ti} = L_{ti} - A_{ti}Y_{ti}, \quad P_{ti}, \tag{6.101}$$

the sequential solutions are

$$X_{t1} = (M_{11})_{t1}^{-1}(B_1 - M_{12}X_2)_{t1}, \tag{6.102}$$

$$(M_2)_{t1}X_2 = (R_2)_{t1}, \tag{6.103}$$

$$X_2 = (M_2)_{t1}^{-1}(R_2)_{t1}, \tag{6.104}$$

$$\ldots$$

$$X_{ti} = (M_{11})_{ti}^{-1}((B_1)_{ti} - (M_{12})_{ti}X_2), \tag{6.105}$$

$$((M_2)_{t1} + \cdots + (M_2)_{ti})X_2 = (R_2)_{t1} + \cdots + (R_2)_{ti}, \tag{6.106}$$

and

$$X_2 = ((M_2)_{t1} + \cdots + (M_2)_{ti})^{-1}((R_2)_{t1} + \cdots + (R_2)_{ti}). \tag{6.107}$$

It is notable that the sequential solution of the second unknown sub-vector X_2 is exactly the same as the combined solution at the last step. The only difference between the combined solution and the sequential solution is that the X_2 used are different. In the sequential solution, only the up-to-date X_2 is used. Therefore at the end of the sequential solution (6.107), the last X_2 obtained has to be substituted into all X_{tj} computing formulas, where $j < i$. This can be done in two ways. The first way is to remember all formulas for computing X_{tj}, after X_2 is obtained from (6.107), using X_2 to compute X_{tj}. The second way is to go back to the beginning after the X_2 is obtained, and use X_2 as the known vector to solve X_{tj} once again. In these ways, the combined sequential observation equations can be solved exactly in a sequential way.

12 Equivalently Eliminated Observation Equation System

In least squares adjustment, the unknowns can be divided into two groups and then solved in a block-wise manner as discussed in Sect. 11. In practice, sometimes only one group of unknowns is of interest, and it is better to eliminate the other group of unknowns (called nuisance parameters) because of its size, for example. In this case, using the so-called equivalently eliminated observation equation system could be very beneficial (Wang et al. 1988; Xu and Qian 1986; Zhou 1985). The nuisance parameters can be eliminated directly from the observation equations instead of from the normal equations.

The linearized observation equation system can be represented by

$$V = L - \begin{pmatrix} A & B \end{pmatrix} \begin{pmatrix} X_1 \\ X_2 \end{pmatrix}, \quad P. \tag{6.108}$$

where

L is the observational vector of dimension m,
A, B are the coefficient matrices of dimension $m \times (n\text{–}r)$ and $m \times r$,
X_1, X_2 are unknown vectors of dimension n–r and r,
V is the residual vector of dimension m,
n is the number of total unknowns,

m	is the number of observations, and
P	is the symmetric and definite weight matrix, of dimension $m \times m$

The least squares normal equation can then be formed by

$$\begin{pmatrix} M_{11} & M_{12} \\ M_{21} & M_{22} \end{pmatrix} \begin{pmatrix} X_1 \\ X_2 \end{pmatrix} = \begin{pmatrix} B_1 \\ B_2 \end{pmatrix}, \tag{6.109}$$

where

$$\begin{pmatrix} M_{11} & M_{12} \\ M_{21} & M_{22} \end{pmatrix} = \begin{pmatrix} A^{\mathrm{T}}PA & A^{\mathrm{T}}PB \\ B^{\mathrm{T}}PA & B^{\mathrm{T}}PB \end{pmatrix}, \tag{6.110}$$

$$B_1 = A^{\mathrm{T}}PL, \quad B_2 = B^{\mathrm{T}}PL.$$

The elimination matrix is formed,

$$\begin{pmatrix} E & 0 \\ -Z & E \end{pmatrix} \tag{6.111}$$

where E is the identity matrix, 0 is a zero matrix, and $Z = M_{21}M_{11}^{-1}$. M_{11}^{-1} is the inversion of M_{11}. Multiplying the elimination matrix (6.111) to the normal equation (6.109) one has

$$\begin{pmatrix} E & 0 \\ -Z & E \end{pmatrix} \begin{pmatrix} M_{11} & M_{12} \\ M_{21} & M_{22} \end{pmatrix} \begin{pmatrix} X_1 \\ X_2 \end{pmatrix} = \begin{pmatrix} E & 0 \\ -Z & E \end{pmatrix} \begin{pmatrix} B_1 \\ B_2 \end{pmatrix},$$

or

$$\begin{pmatrix} M_{11} & M_{12} \\ 0 & M_2 \end{pmatrix} \begin{pmatrix} X_1 \\ X_2 \end{pmatrix} = \begin{pmatrix} B_1 \\ R_2 \end{pmatrix} \tag{6.112}$$

where

$$\begin{aligned} M_2 &= -M_{21}M_{11}^{-1}M_{12} + M_{22} \\ &= B^{\mathrm{T}}PB - B^{\mathrm{T}}PAM_{11}^{-1}A^{\mathrm{T}}PB = B^{\mathrm{T}}P(E - AM_{11}^{-1}A^{\mathrm{T}}P)B. \end{aligned} \tag{6.113}$$

$$R_2 = B_2 - M_{21}M_{11}^{-1}B_1 = B^{\mathrm{T}}P(E - AM_{11}^{-1}A^{\mathrm{T}}P)L. \tag{6.114}$$

If one is only interested in the unknown vector X_2, one just needs to solve the second equation of (6.112). The solution is identical to that of solving the whole (6.112). The above eliminating process is similar with the Gauss-Jordan algorithm, which has often been used for the inversion of the normal matrix (or for solving linear equation system). Indeed, the second equation of (6.112) is identical to (6.75) derived in the block-wise least squares adjustment (see Sect. 11). Letting

$$J = AM_{11}^{-1}A^{T}P, \tag{6.115}$$

one has properties of

$$J^2 = (AM_{11}^{-1}A^{T}P)(AM_{11}^{-1}A^{T}P) = AM_{11}^{-1}A^{T}PAM_{11}^{-1}A^{T}P = AM_{11}^{-1}A^{T}P = J,$$

$$(E - J)(E - J) = E^2 - 2EJ + J^2 = E - 2J + J = E - J$$

and

$$[P(E - J)]^{\mathrm{T}} = (E - J^{\mathrm{T}})P = P - (AM_{11}^{-1}A^{\mathrm{T}}P)^{\mathrm{T}}P = P - PAM_{11}^{-1}A^{\mathrm{T}}P = P(E - J),$$

i.e., matrices J and $(E - J)$ are idempotent and $(E - J)^{\mathrm{T}}P$ is symmetric, or

$$J^2 = J, \quad (E - J)^2 = E - J \quad \text{and} \quad (E - J)^{\mathrm{T}}P = P(E - J). \tag{6.116}$$

Using the above derived properties, M_2 in (6.113) and R_2 in (6.114) can be rewritten as

$$M_2 = B^{\mathrm{T}}P(E - J)B = B^{\mathrm{T}}P(E - J)(E - J)B = B^{\mathrm{T}}(E - J)^{\mathrm{T}}P(E - J)B \tag{6.117}$$

and

$$R_2 = B^{\mathrm{T}}P(E - J)L = B^{\mathrm{T}}(E - J)^{\mathrm{T}}PL. \tag{6.118}$$

Denoting

$$D_2 = (E - J)B, \tag{6.119}$$

then the eliminated normal equation (the second equation of (6.112)) can be rewritten as

$$B^{\mathrm{T}}(E - J)^{\mathrm{T}}P(E - J)BX_2 = B^{\mathrm{T}}(E - J)^{\mathrm{T}}PL \tag{6.120}$$

or

$$D_2^{\mathrm{T}}PD_2X_2 = D_2^{\mathrm{T}}PL. \tag{6.121}$$

This is the least squares normal equation of the following linear observation equation:

$$U_2 = L - D_2X_2, \ P \tag{6.122}$$

or

$$U_2 = L - (E - J)BX_2, \ P, \tag{6.123}$$

where L and P are the original observational vector and weight matrix, and U_2 is the residual vector, which has the same property as V in (6.108).

The advantage of using (6.123) is that the unknown vector X_1 has been eliminated; however, L vector and P matrix remain the same as the originals.

12.1 Diagonalized Normal Equation and the Equivalent Observation Equation

In least squares adjustment, the unknowns can be divided into two groups. One group of unknowns can be eliminated by matrix partitioning to obtain an equivalently eliminated normal equation system of the other group of unknowns. Using the elimination process twice for the two groups of unknowns respectively, the normal equation can be diagonalized. The algorithm can be outlined as follows.

A linearized observation equation and the normal equations can be represented by (6.108) and (6.109). From the first equation of (6.109), one has

$$X_1 = M_{11}^{-1}(B_1 - M_{12}X_2). \tag{6.124}$$

Setting X_1 into the second equation of (6.109), one gets an equivalently eliminated normal equation of X_2:

$$M_2X_2 = R_2, \tag{6.125}$$

where

$$\begin{aligned} M_2 &= M_{22} - M_{21}M_{11}^{-1}M_{12} \\ R_2 &= B_2 - M_{21}M_{11}^{-1}B_1. \end{aligned} \tag{6.126}$$

Similarly, from the second equation of (6.109), one has

$$X_2 = M_{22}^{-1}(B_2 - M_{21}X_1). \tag{6.127}$$

Setting X_2 into the first equation of (6.109), one gets an equivalently eliminated normal equation of X_1:

$$M_1X_1 = R_1, \tag{6.128}$$

where

$$\begin{aligned} M_1 &= M_{11} - M_{12}M_{22}^{-1}M_{21} \\ R_1 &= B_1 - M_{12}M_{22}^{-1}B_2. \end{aligned} \tag{6.129}$$

Combining (6.128) and (6.125) together, one has

$$\begin{pmatrix} M_1 & 0 \\ 0 & M_2 \end{pmatrix}\begin{pmatrix} X_1 \\ X_2 \end{pmatrix} = \begin{pmatrix} R_1 \\ R_2 \end{pmatrix}, \tag{6.130}$$

where (see, e.g., Cui et al. 1982; Gotthardt 1978)

$$\begin{aligned} &Q_{11} = M_1^{-1} && Q_{22} = M_2^{-1} \\ &Q_{12} = -M_{11}^{-1}(M_{12}Q_{22}), && Q_{21} = -M_{22}^{-1}(M_{21}Q_{11}). \end{aligned} \tag{6.131}$$

It is obvious that (6.109) and (6.130) are two equivalent normal equations. The solutions of the both equations are identical. Equation (6.130) is a diagonalized normal equation related to X_1 and X_2. The process of forming (6.130) from (6.109) is called the diagonalization process of a normal equation.

As discussed in Sect. 12, the equivalently eliminated observation equation of the second equation of (6.130) is (6.123). Similarly, if one denotes

$$I = BM_{22}^{-1}B^{\mathrm{T}}P$$

and

$$D_1 = (E - I)A,$$

then the equivalently eliminated observation equation of the first normal equation of (6.130) has a form of

$$U_1 = L - (E - I)AX_1, P. \tag{6.132}$$

where U_1 is a residual vector which has the same property as V in (6.108). L and P are the original observational vector and weight matrix.

The above equation and (6.123) can be written together as

$$\begin{pmatrix} U_1 \\ U_2 \end{pmatrix} = \begin{pmatrix} L \\ L \end{pmatrix} - \begin{pmatrix} D_1 & 0 \\ 0 & D_2 \end{pmatrix}\begin{pmatrix} X_1 \\ X_2 \end{pmatrix}, \quad \begin{pmatrix} P & 0 \\ 0 & P \end{pmatrix}. \tag{6.133}$$

Equation (6.133) is derived from the normal equation (6.130); therefore, it is true inversely, i.e., (6.130) is the least squares normal equation of the observation equation (6.133). Equations (6.109) and (6.130) are normal equations of the observation equations (6.108) and (6.133). So (6.133) is an equivalent observation equation of (6.108). Equations (6.130) and (6.133) are called diagonalized equations of (6.109) and (6.108), respectively.

13 A Priori Constrained Least Squares Adjustment

Up to now in this chapter, several adjustment methods have been discussed. All of them are methods suitable for full rank linear equation problems. A full rank quadratic matrix means such a matrix can be inversed to obtain its inversion. A rank deficient linear equation system is sometimes called an over-parameterized problem. Except for the conditional least squares adjustment method, all other methods discussed above cannot be directly used for solving a rank deficient problem. The conditional least squares adjustment method with extra conditions can make the problem solvable. The conditions, of course, should be well-formulated mathematically and well-reasoned physically. In other words, the conditions are considered as exactly known. In practice, quite often, the conditions are known with certain a priori precision. Adjustment, which uses such a priori information as constraints, is called a priori constrained adjustment, which will be discussed in this section.

13.1 A Priori Parameter Constraints

1. A linearized observation equation system can be represented by

$$V = L - AX, \quad P_L, \tag{6.134}$$

where P_L is the symmetric and definite weight matrix of dimension $m \times m$.

2. The corresponding a priori condition equation system can be written as

$$U = W - BX, \quad P_W, \tag{6.135}$$

where

B is the coefficient matrix of dimension $r \times n$,
W is the constant vector of dimension r,
U is the residual vector of dimension r,
P_W is the a priori (symmetric and definite) weight matrix of dimension $r \times r$, and
r is the number of condition equations; $r < n$

3. One may interpret the constraints of (6.135) as additional pseudo-observations or as fictitious observations. This leads to the total observation equations:

$$\begin{pmatrix} V \\ U \end{pmatrix} = \begin{pmatrix} L \\ W \end{pmatrix} - \begin{pmatrix} A \\ B \end{pmatrix} X, \quad P = \begin{pmatrix} P_L & 0 \\ 0 & P_W \end{pmatrix}. \tag{6.136}$$

Then the least squares normal equations are well-known as (see, e.g., Sect. 8.1):

$$\begin{pmatrix} A^T & B^T \end{pmatrix} \begin{pmatrix} P_L & 0 \\ 0 & P_W \end{pmatrix} \begin{pmatrix} A \\ B \end{pmatrix} X = \begin{pmatrix} A^T & B^T \end{pmatrix} \begin{pmatrix} P_L & 0 \\ 0 & P_W \end{pmatrix} \begin{pmatrix} L \\ W \end{pmatrix}$$

or

$$(A^T P_L A + B^T P_W B)X = (A^T P_L L + B^T P_W W). \tag{6.137}$$

For convenience, a factor k (here $k = 1$) is introduced in (6.137):

$$(A^T P_L A + kB^T P_W B)X = (A^T P_L L + kB^T P_W W). \tag{6.138}$$

Equation (6.138) shows that the a priori information constraints can be added to the original least squares normal equations. In other words, the a priori information can be used for solving the rank deficient problem and making it possible for the normal matrix to be inversed. Of course, these a priori information constraints should be reasonable and realistic ones; otherwise the solutions could be disturbed by worse a priori constraints. In case of $k = 0$, the normal equation (6.138) turns out to be the original one and will give the free solution (without any a priori constraints).

The solution of the a priori constrained least squares solution is then

$$X = (A^T P_L A + kB^T P_W B)^{-1}(A^T P_L L + kB^T P_W W), \tag{6.139}$$

where $k = 1$. Generally, the a priori weight matrix is given by covariance matrix Q_W and

$$P_W = Q_W^{-1}. \tag{6.140}$$

The a priori constraints only cause two additional terms in both sides of the normal equations; therefore, all the above discussed adjustment methods can be directly used for solving the a priori constrained problem.

13.2 A Priori Datum

Suppose the B matrix in the a priori constraints of (6.135) is an identity matrix, and the parameter vector W is just a coordinate sub-vector of the total parameter vector. Then it turns out to be a special case called a priori datum. The observation equations and a priori constraints may be rewritten as

$$V = L - \begin{pmatrix} A_1 & A_2 \end{pmatrix} \begin{pmatrix} X_1 \\ X_2 \end{pmatrix}, \quad P_L \quad \text{and} \tag{6.141}$$

$$U = \bar{X}_2 - X_2, \quad P_2, \tag{6.142}$$

where $\bar{X}_2$ is the "observed" parameter sub-vector, P_2 is the weight matrix with respect to the parameter sub-vector X_2 and is generally a diagonal matrix, and U is a residual vector that has the same property as V. Usually, $\bar{X}_2$ is "observed" independently, so P_2 is a diagonal matrix. If X_2 is a sub-vector of station coordinates, then the constraint of (6.142) is called the datum constraint. (This is also the reason why the name a priori datum is used).

The least squares normal equation of problems (6.141) and (6.142) can then be formed (similar to what was discussed in Sect. 13.1) as

$$\begin{pmatrix} M_{11} & M_{12} \\ M_{21} & M_{22} \end{pmatrix} \begin{pmatrix} X_1 \\ X_2 \end{pmatrix} = \begin{pmatrix} B_1 \\ B_2 \end{pmatrix} \tag{6.143}$$

or

$$M_{11}X_1 + M_{12}X_2 = B_1 \tag{6.144}$$

and

$$M_{21}X_1 + M_{22}X_2 = B_2, \tag{6.145}$$

where

$$M_{11} = A_1^{\mathrm{T}} P_L A_1, \tag{6.146}$$

$$M_{12} = M_{21}^{\mathrm{T}} = A_1^{\mathrm{T}} P_L A_2, \tag{6.147}$$

$$M_{22} = A_2^{\mathrm{T}} P_L A_2 + P_2, \tag{6.148}$$

$$B_1 = A_1^{\mathrm{T}} P_L L \tag{6.149}$$

and

$$B_2 = A_2^{\mathrm{T}} P_L L + P_2 \bar{X}_2. \tag{6.150}$$

The least squares principle used here is

$$V^{\mathrm{T}} P_L V + U^{\mathrm{T}} P_2 U = \min . \tag{6.151}$$

The normal equation (6.143) can also be derived by differentiating (6.151) with respect to X, and then letting it equal zero and taking (6.142) into account. In practice, the sub-vector $\bar{X}_2$ is usually a zero vector; this can be achieved through careful initialization by forming the observation equation (6.141). Comparing the normal equation system of the a priori datum problem of (6.141) and (6.142) with the normal equation of (6.141), the only difference is that the a priori weight matrix P_2 has been added to M_{22}. This indicates that the a priori datum problem can be dealt with simply by adding P_2 to the normal equation of the observation equation (6.141).

If some diagonal components of the weight matrix P_2 are set to zero, then the related parameters (X_2) are free parameters (or free datum) of the adjustment problem (without a priori constraints). Otherwise, parameters with a priori constraints are called a priori datum. Large weight indicates strong constraint and small weight indicates soft constraint. The strongest constraint is to keep the datum fixed.

13.3 Quasi-Stable Datum

The quasi-stable datum method was proposed by Zhou et al. (1997). The basic idea is that the network is a dynamic one, i.e., most parameters are changing all the time. However, a few points are relatively stable, or their geometric center is relatively stable. All the assumptions and observation equations are the same as in Sect. 13.2:

$$V = L - \begin{pmatrix} A_1 & A_2 \end{pmatrix} \begin{pmatrix} X_1 \\ X_2 \end{pmatrix}, \quad P_L \tag{6.152}$$

and

$$U = \bar{X}_2 - X_2, \quad P_2. \tag{6.153}$$

The least squares principles for the quasi-stable datum are

$$V^{\mathrm{T}} P_L V = \min \tag{6.154}$$

and

$$U^{\mathrm{T}} P_2 U = \min\,. \tag{6.155}$$

Equation (6.154) is the same as the original least squares principle. From (6.154), one has the normal equation

$$\begin{pmatrix} M_{11} & M_{12} \\ M_{21} & M_{22} \end{pmatrix} \begin{pmatrix} X_1 \\ X_2 \end{pmatrix} = \begin{pmatrix} B_1 \\ B_2 \end{pmatrix}, \tag{6.156}$$

where

$$M_{11} = A_1^{\mathrm{T}} P_L A_1,$$

$$M_{12} = M_{21}^{\mathrm{T}} = A_1^{\mathrm{T}} P_L A_2,$$

$$M_{22} = A_2^{\mathrm{T}} P_L A_2,$$

$$B_1 = A_1^{\mathrm{T}} P_L L$$

and

$$B_2 = A_2^{\mathrm{T}} P_L L. \tag{6.157}$$

Even if (6.156) is a rank deficient equation, one may first solve (6.156) to get an explicit expression for X_2. Recalling the discussion in Sect. 11, one gets a normal equation related to X_2:

$$M_2 X_2 = R_2, \tag{6.158}$$

where

$$M_2 = M_{22} - M_{21}M_{11}^{-1}M_{12}$$

and

$$R_2 = B_2 - M_{21}M_{11}^{-1}B_1^{\mathrm{T}}. \tag{6.159}$$

The new condition can be considered by forming

$$F = U^{\mathrm{T}}P_2U + 2K^{\mathrm{T}}(M_2X_2 - R_2)$$

and

$$\frac{\partial F}{\partial X} = 2U^{\mathrm{T}}P_2 + 2K^{\mathrm{T}}M_2 = 0.$$

Considering the symmetry of M_2, we have

$$U = -P_2^{-1}M_2K. \tag{6.160}$$

Substituting (6.160) into (6.153), one gets

$$X_2 = \bar{X}_2 + P_2^{-1}M_2K \tag{6.161}$$

or

$$M_2X_2 = M_2\bar{X}_2 + M_2P_2^{-1}M_2K. \tag{6.162}$$

Substituting (6.158) into (6.162), one has

$$K = (M_2P_2^{-1}M_2)^{-1}(M_2\bar{X}_2 - R_2). \tag{6.163}$$

Thus,

$$X_2 = \bar{X}_2 + P_2^{-1}M_2K, \tag{6.164}$$

and

$$X_1 = M_{11}^{-1}(A_1^{\mathrm{T}}P_LL - M_{12}X_2) \tag{6.165}$$

$$m_0 = \sqrt{\frac{V^{\mathrm{T}}P_LV}{n-r}}, \tag{6.166}$$

where m_0 is the standard deviation, n is the number of observations, and r is the summation of the both ranks of the matrices A_1 and A_2.

14 Summary

In the second part of this chapter, the most applicable and necessary algorithms for data processing are outlined.

Least squares adjustment is the most basic adjustment method. It starts by establishing observation equations and forming normal equations; then it solves the unknowns. The sequential application of least squares adjustment by accumulating the sequential normal equations makes applications of least squares adjustment more effective. Normal equations can be formed epoch-wise and then accumulated. This method can be used not only for solving the problem at the end, but also for obtaining epoch-wise solutions. The equivalent sequential least squares adjustment, which can be read from different publications, is also derived. Xu (author) and Morujao (Coimbra, Portugal) have independently pointed out that, by applying such an algorithm, the results obtained compared with those obtained by the accumulating method will have differences. The differences increase with time and are generally not negligible. Therefore by using this method, the numerical process has to be carefully examined to avoid the accumulation of numerical errors.

Conditional least squares adjustment is needed if there are some constraints that have to be taken into account. The sequential application of conditional least squares adjustment is discussed because of practical needs. The problem may be solved first without conditions, and then the conditions may be applied afterward.

Block-wise least squares adjustment is discussed for separating the unknowns into two groups. The sequential application of block-wise least squares adjustment makes it possible to give up some unknowns (say, out of date unknowns, such as past coordinates) and keep the information related to the common unknowns during the processing process. This method avoids the problem that may be caused by a rapid increase of the number of unknowns. There are two ways to keep the solution equivalent with a solution that is not sequential. One is to use the time independent unknowns at the end of data processing as known, and then go back to process the data once again. The other is to remember all sequential normal equations until the best solution of the time independent unknowns are obtained, and then the coordinates can be recomputed.

The equivalently eliminated observation equation system is discussed for eliminating some nuisance parameters. This method is nearly the same as block-wise least squares adjustment if one carefully compares the normal equations of the second group of unknowns (see Sect. 11) and the eliminated normal equations (see Sect. 12). However, the most important point is that the equivalently eliminated observation equations have been derived here. Instead of solving the original problem, one may directly solve the equivalently eliminated observation equations, where the unknowns are greatly reduced, whereas the observation vector and weight matrix remain the originals (i.e., the problem remains uncorrelated). The precision estimation can also be made more easily by using the formulas derived in least squares adjustment. The derivation of such an equivalent observation equation was first made by Zhou (1985) and applied in GPS theory by Xu (2002a). The unified GPS data processing method is derived by using this principle (see Xu 2007). Based on the derivation of the equivalent equation, a diagonalization algorithm of the normal equation and the observation equation are discussed. The diagonalization algorithm can be used for separating one adjustment problem into two sub-problems.

A priori constrained least squares adjustment is discussed in Sect. 13 for solving the rank deficient problems. A general discussion on the a priori parameter constraints is given. This method makes it possible to form the observation equations in a general way, and then a priori information can be added to keep some references fixed, such as the clock error of the reference satellite and the coordinates of the reference station. As a special case of a priori parameter constraints, a so-called a priori datum method is discussed. The advantage of this method is that the a priori constraints just change the normal equation by adding a term (the a priori weight matrix) so that all discussed least squares adjustment methods can be directly used for solving the rank deficient problems. Linear conditions related to the coordinate parameters can be introduced by using this method. A quasi-stable datum method is also discussed. From the point of view of the dynamic Earth, all stations are not fixed stations. The quasi-stable datum method takes such dynamic behavior of the stations into account.

Bibliography

Abramowitz M, Stegun IA (1965) Handbook of mathematical functions. Dover Publications Inc., New York

Albertella A, Sacerdote F (1995) Spectral analysis of block averaged data in geopotential global model determination. J Geodesy 70(3):166–175

Arikan F, Erol CB, Arikan O (2003) Regularized estimation of vertical total electron content from global positioning system data. J Geophys Res 108(A12): SIA20/1–12

Axelsson O (1994) Iterative solution methods. Cambridge University Press, Cambridge

Ayres F (1975) Differential- und Integralrechnung, schaum's outline. McGraw-Hill Book, New York

Blewitt G (1998) GPS data processing methodology. In: Teunissen PJG, Kleusberg A (eds) GPS for geodesy. Springer, Berlin, pp 231–270

Bronstein IN, Semendjajew KA (1987) Taschenbuch der Mathematik.B. G. Teubner Verlagsgesellschaft, Leipzig, ISBN 3-322-00259-4

Cross PA, Ramjattan AN (1995) A Kalman filter model for an integrated land vehicle navigation sys-tem. In: Proceedings of the 3rd international workshop on high precision navigation: High precision navigation 95. University of Stuttgart, April 1995, Bonn, pp 423–434

Cui X, Yu Z, Tao B, Liu D (1982) Adjustment in surveying. Surveying Press, Peking, (in Chinese)

Davis P, Rabinowitz P (1984) Methods of numerical integration, 2nd edn. Academic, New York

Davis PJ (1963) Interpolation and approximation. Dover Publications Inc., New York

Ding X, Coleman R (1996) Multiple outlier detection by evaluating redundancy contributions of observations. J Geodesy 708:489–498

Faruqi FA, Turner KJ (2000) Extended Kalman filter synthesis for integrated global positioning/iner-tial navigation systems. Appl Math Comput 115(2–3):213–227

Gleason DM (1996) Avoiding numerical stability problems of long duration DGPS/INS Kalman filters. J Geodesy 70(5):263–275

Gotthardt E (1978) Einführung in die Ausgleichungsrechnung. Herbert Wichmann Verlag, Karlsruhe

Hadamard J (1932) Lecture on Cauchy's problem in linear partial differential equations, Yale University Press, reprinted by Dover, New York, 1952

Hansen P (1996) Rank-deficient and ill-posed problems, PHD thesis of the technical university of Denmark

Hoerl AE, Kennard RW (1970) Ridge regression: biased estimation for nonorthogonal problems. Technometrics 12:55–67

Hostetter GH (1987) Handbook of digital signal processing, engineering applications. Academic, New York

Hotine M (1991) Differential geodesy. Springer, Berlin

Huber PJ (1964) Robust estimation of a location parameter. Ann Math Stat 35:73–101

Knickmeyer ET, Knickmeyer EH, Nitschke M (1996) Zur Auswertung kinematischer Messungen mit dem Kalman-Filter. Schriftenreihe des Deutschen Vereins für Vermessungswesen, Bd. 22, Stuttgart, pp 141–166

Koch KR (1980) Parameterschätzung und Hypothesentests in linearen Modellen. Dümmler-Verlag, Bonn

Koch KR (1988) Parameter estimation and hypothesis testing in linear models. Springer, Berlin

Koch KR (1996) Robuste Parameterschätzung. Allgemeine Vermessungsnachrichten 103(1): 1–18

Koch KR, Yang Y (1998a) Konfidenzbereiche und Hypothesentests für robuste Parameterschätzungen. ZfV 123(1):20–26

Koch KR, Yang Y (1998b) Robust Kalman filter for rank deficient observation model. J Geodesy 72:436–441

Lemmens R (2004) Book review: GPS—theory, algorithms and applications, Xu G 2003. Int J Appl Earth Obs Geoinf 5: 165–166

Li B, Shen Y, Feng Y (2010) Fast GNSS ambiguity resolution as an ill-posed problem. J Geodesy 84:683–698

Ludwig R (1969) Methoden der Fehler- und Ausgleichsrechnung. Vieweg and Sohn, Braunschweig

Masreliez CJ, Martin RD (1977) Robust Bayesian estimation for the linear model and robustifying the Kalman filter. IEEE T Automat Contr AC-22:361–371

Miller K (1970) Least squares methods for ill-posed problems with a prescribed bound, SIAM J. Math Anal 1:52–74

Mohamed AH, Schwarz KP (1999) Adaptive Kalman filtering for INS/GPS. J Geodesy 73:193–203

Morozov VA (1984) Methods for solving incorrectly posed problems. Springer, Berlin

Ou JK, Wang ZJ (2004) An improved regularization method to resolve integer ambiguity in rapid positioning using single frequency GPS receivers. Chinese Sci Bull 49(2):196–200

Reigber C, Schmidt R, Flechtner F, König R, Meyer U, Neumayer KH, Schwintzer P, Zhu SY (2005) An earth gravity field model complete to degree and order 150 from GRACE: EIGEN-GRACE02S. J Geodynamics 39:1–10

Rothacher M, Schaer S (1995) GPS-Auswertetechniken. Schriftenreihe des Deutschen Vereins für Vermessungswesen, Bd. 18, pp 107–121

Schaffrin B (1980) Tikhonov regularization in geodesy, an example. Boll Geod Sci Aff 39:207–216

Schaffrin B (1991) Generating robustified Kalman filters for the integration of GPS and INS. Techni-cal Report, No. 15, Institute of Geodesy, University of Stuttgart

Schaffrin B (1995) On some alternative to Kalman filtering. In: Sanso F (ed) Geodetic theory today. Springer, Berlin, pp 235–245

Schaffrin B (2008) Minimum mean squared error (MSE) adjustment and the optimal Tikhonov-Phillips regularization parameter via reproducing best invariant quadratic uniformly unbiased estimates (repro-BIQUE). J. Geod 82:113–121

Schaffrin B, Grafarend E (1986) Generating classes of equivalent linear models by nuisance parameter elimination. Manuscr Geodaet 11:262–271

Shen Y, Li B (2007) Regularized solution to fast GPS ambiguity resolution. J Surveying Eng 133(4):168–172

Strang G, Borre K (1997) Linear algebra, geodesy, and GPS. Cambridge Press, Wellesley

Tarantola A (2005) Inverse problem theory. SIAM, Phildelphia

Teunissen PJG (1995) The least-squares ambiguity decorrelation adjustment: a method for fast GPS integer ambiguity estimation. J Geodesy 70(1–2):65–82

Tikhonov AN (1963a) Regularization of ill-posed problems, English translation of Dokl. Akad Nauk SSSR 151(1):49–52

Tikhonov AN (1963b) Solution of incorrectly formulated problems and the regularization method, English translation of Dokl. Akad Nauk SSSR 151(3):501–504

Tikhonov AN, Arsenin VY (1977) Solutions of ill-posed problems. Wiley, New York

Tikhonov AN, Goncharsky AV, Steppanov VV, Yagola AG (1995) Numerical methods for the solution of ill-posed problems. Kluwer Academic Publishers, Netherlands

Tsai C, Kurz L (1983) An adaptive robustifing approach to Kalman filtering. Automatica 19: 279–288

Wahba G (1983) Bayesian "confidence intervals" for the cross-validated smoothing spline. J R Stat Soc B45:133–150

Wang LX, Fang ZD, Zhang MY, Lin GB, Gu LK, Zhong TD, Yang XA, She DP, Luo ZH, Xiao BQ, Chai H, Lin DX (1979) Mathematic handbook. Educational Press, Peking. ISBN 13012-0165

Wang G, Chen Z, Chen W, Xu G (1988) The principle of GPS precise positioning system. Surveying Press, Peking, p 345. ISBN 7-5030-0141-0/P.58 (in Chinese)

Xu G (2002a) GPS data processing with equivalent observation equations. GPS Solutions, vol 6, No. 1–2, 6:28–33

Xu G (2002b) A general criterion of integer ambiguity search. J GPS 1(2):122–131

Xu G (2003) A diagonalization algorithm and its application in ambiguity search. J. GPS 2(1):35–41

Xu G (2007) GPS–theory, algorithms and applications. Springer, Berlin, pp xix + 340

Xu G, Qian Z (1986) The application of block elimination adjustment method for processing of the VLBI Data. Crustal Deformation and Earthquake, Vol. 6, No. 4, (in Chinese)

Xu P (1992) Determination of surface gravity anomalies using gradiometric observables. Geophys J Int 110:321–332

Xu P (1998) Truncated SVD methods for discrete linear ill-posed problems. Geophys J Int 135:505–514

Xu P, Rummel R (1994) Generalized ridge regression with applications in determination of potential fields. Manuscr Geod 20:8–20

Xu P, Shen Y, Fukuda Y, Liu Y (2006) Variance components estimation in linear inverse ill-posed models. J Geod 80:69–81

Yang Y (1991) Robust Bayesian estimation. B Geod 65:145–150

Yang Y (1993) Robust estimation and its applications. Bayi Publishing House, Peking

Yang Y (1994) Robust estimation for dependent observations. Manuscr Geodaet 19:10–17

Yang Y (1997a) Estimators of covariance matrix at robust estimation based on influence functions. ZfV 122(4):166–174

Yang Y (1997b) Robust Kalman filter for dynamic systems. J Zhengzhou Inst Surveying Mapping 14:79–84

Yang Y (1999) Robust estimation of geodetic datum transformation. J Geodesy 73:268–274

Yang Y, Cui X (2008) Adaptively robust filter with multi adaptive factors. Surv Rev 40(309):260–270

Yang Y, Gao W (2005) Comparison of adaptive factors on navigation results. J Navigation 58:471–478

Yang Y, Gao W (2006) An optimal adaptive kalman filter with applications in navigation. J Geodesy 80:177–183

Yang Y, Gao W (2006) A new learning statistic for adaptive filter based on predicted residuals. Prog Nat Sci 16(8):833–837

Yang Y, He H, Xu G (2001) Adaptively robust filtering for kinematic geodetic positioning. J Geodesy 75:109–116

Yang Y, Tang Y, Li Q, Zou Y (2006) Experiments of adaptive filters for kinemetic GPS positioning applied in road information updating in GIS. J Surv Eng (in press)

Yang YX et al (2005a) Combined adjustment project of national astronomical geodetic networks and 2000' national GPS control network. Prog Nat Sci 15(5):435–441

Yang YX, Xu TH, Song LJ (2005) Robust estimation of variance components with application in global positioning system network adjustment. J Surv Eng ASCE 131(4): 107–112

Zhou J (1985) On the Jie factor. Acta Geodaetica et Geophysica 5 (in Chinese)

Zhou J (1989) Classical theory of errors and robust estimation. Acta Geod Cartogr Sinica 18: 115–120

Zhou J, Huang Y, Yang Y, Ou J (1997) Robust least squares method. Publishing House of Huazhong University of Science and Technology, Wuhan

Zhu J (1996) Robustness and the robust estimate. J Geodesy 70(9):586–590

Chapter 7
Very Long Baseline Interferometry for Geodesy and Astrometry

Harald Schuh and Johannes Böhm

Very Long Baseline Interferometry (VLBI) is a microwave-based space geodetic technique that measures the difference in arrival times of signals from a radio source by cross correlation. Most commonly the observed radio sources are extragalactic objects but beacons from satellites have also been used. VLBI plays a unique role in the practical realization and maintenance of the International Celestial Reference Frame (ICRF) and contributes significantly to the International Terrestrial Reference Frame (ITRF), in particular for its scale. It is the only technique that provides the full set of Earth orientation parameters, which are indispensable for positioning and navigation on Earth and in space. In addition, VLBI allows access to valuable information concerning interactions within the Earth system. In particular, direct measurements of nutation parameters and of the Earth rotation angle (UT1–UTC) are uniquely provided by VLBI. Furthermore, several other geodynamic, atmospheric, and astronomical parameters can be derived from the long history of VLBI measurements starting in the late 1970s. In 1999, the International Association of Geodesy (IAG) accepted the International VLBI Service for Geodesy and Astrometry (IVS) as an official IAG service and the IVS was also approved as a service of the International Astronomical Union (IAU). Since then, the coordination of world-wide VLBI observation and analysis has improved significantly, leading to valuable results for the wider scientific community. Since 2005, the IVS has been working on a new VLBI system in terms of hardware, software, and operational procedures, known as VLBI2010. The IVS recommended a review of all current VLBI systems and processes from antennas to analysis and outlined the path to the

H. Schuh (✉) · J. Böhm
Institute of Geodesy and Geophysics, University of Technology Vienna,
Gusshausstrasse 27-29 E128, A-1040 Wien, Austria
e-mail: harald.schuh@tuwien.ac.at

J. Böhm
e-mail: johannes.boehm@tuwien.ac.at

G. Xu (ed.), *Sciences of Geodesy - II*, DOI: 10.1007/978-3-642-28000-9_7,

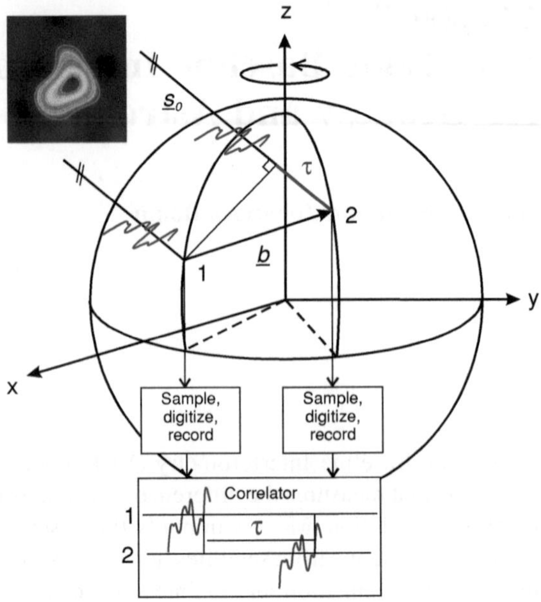

Fig. 7.1 Geometric VLBI model

next-generation system with unprecedented new capabilities envisaged: 1 mm position and 0.1 mm/year velocity accuracy on global scales, continuous measurements to obtain uninterrupted time series of station positions and Earth orientation parameters, and a turnaround time from the observations to initial geodetic results of less than 24 h. This new system will be realized in the coming years.

1 Introduction

1.1 Geometric Principle

The geometric principle of VLBI is simple and straightforward. The radiation from extragalactic radio sources arrives on Earth as plane wavefronts. This is different from nearby Earth satellites such as those of the Global Navigation Satellite Systems (GNSS) where the finite distance to the emitter produces parallactic angles. The basic triangle for the determination of the baseline vector reduces to a rectangular one providing a direct relation between the baseline vector $\underline{b}$ and the direction to the radio source $\underline{s}_0$ (Campbell 2000). The scalar product τ represents the observed delay between the reception times t_1 and t_2 at stations 1 and 2 (see Fig. 7.1; (7.1)) with the sign convention $\tau = t_2 - t_1$ and the velocity of light c.

$$\tau = -\frac{\underline{b} \cdot \underline{s}_0}{c} = t_2 - t_1. \tag{7.1}$$

The delay τ is time-dependent, and the largest contribution to its variation is due to the fact that the interferometer is fixed to the Earth's surface and thus follows its diurnal rotation with respect to the celestial reference system that is realized by positions of radio sources. The geodetic VLBI concept uses two or more radio telescopes to observe numerous extragalactic radio sources distributed across the skies, mostly quasars or radio galaxies. In geodetic VLBI since the end of the 1970s the observations are done within S-band (2.3 GHz) and X-band (8.4 GHz),[1] and the data are recorded and time-tagged using very stable and precise time signals obtained from hydrogen masers. These data are then sent to particular correlation centers for cross-correlation to generate so-called fringes and to obtain the group delay observable τ which is relevant for geodetic and astrometric applications. From these delays, the baseline lengths b and other geodetic parameters can be derived nowadays with sub-centimeter accuracy. The VLBI technique measures very accurately the angle between the Earth-fixed baseline vector $\underline{b}$ and the space-fixed radio sources $\underline{s}_0$ which have to be transformed into a common system for the evaluation of (7.1) by parameter estimation techniques. Thus, even the most subtle changes in the baseline lengths and in the angles between the reference systems can be detected, and the main geodynamic phenomena such as Earth orientation parameters (EOP) can be monitored with unprecedented accuracy (Schuh 2000). However, '... if we leave the Euclidean geometry in empty space and return to the real world with curved space, flickering quasars, billowing atmospheres, wobbling axes, and drifting continents, we have to delve into layers of complexity, fortunately not only as a chore but also as an opportunity to gain a wealth of new knowledge about our system Earth' (Campbell 2000). More details about the complexity of VLBI are provided in the next sections.

1.2 History and Technological Developments

In this section we summarize the early history of geodetic VLBI and of the VLBI technique. The interested reader may find further details on the history in Sovers et al. (1998), Campbell (2000), and Kellermann and Moran (2001), and references therein. For details on technology, we refer to the textbooks by Thompson et al. (1986) and Takahashi et al. (2000), and references therein.

VLBI is an outgrowth of radio interferometry with cable-connected elements designed to overcome the limited resolution of single dish radio telescopes (Cohen

---

[1] A change of the frequency setup, e.g., observing on a frequency band between 2 and 14 GHz, is envisaged for the next VLBI generation, VLBI2010 (Petrachenko et al. 2009).

et al. 1968). However, to reveal the structure of extremely compact radio sources, the resolving power of Connected Element Radio Interferometers (CERI) was insufficient, even at higher frequencies (Campbell 2000). The advent in the late 1960s of high-speed tape recorders and high-stability atomic frequency/time standards made possible the construction of phase-coherent, Michelson-type interferometers whose elements required no physical connection between them and hence could be spaced arbitrarily far apart. In 1967, several groups working independently in Canada and the United States developed and successfully operated two-station interferometers (Bare et al. 1967; Broten et al. 1967; Moran et al. 1967; Brown et al. 1968). Signals received at each station were down-converted in frequency, time-tagged, and recorded on tape for subsequent playback at a correlator center, where the common signal received from a radio source at two (or more) antennas was detected by cross-correlation and integration although the signal itself is very much weaker than the background noise. This technique eliminated the need for a real-time phase-stable connection between radio telescopes.

Potential geophysical applications of geodetic VLBI were recognized early (Gold 1967; Shapiro and Knight 1970). The first experiments that were explicitly aimed at achieving geodetic accuracy on long baselines were conducted by the Haystack/MIT group on the 845 km baseline between the Haystack Observatory in Northern Massachusetts and the National Radio Astronomy Observatory of Green Bank, West Virginia, U.S.A. (Hinteregger et al. 1972). Since that time the station position precision improved dramatically from a few meters to the current level of better than 1 cm. A major factor in the improved precision was made possible by equipment improvements such as wider spanned and recorded bandwidths, dual-frequency observations, lower system temperatures, and phase calibration. As an example the first geodetic observations used the MkI system (Whitney et al. 1976) which could record only 0.72 Mbits/s, whereas modern systems allow one to record at 1024 Mbits/s or even faster. Other factors included improvements in observing strategies, analysis methods, and modeling of physical processes.

The key to the high group delay precision of 1 ns (30 cm) attained in these experiments was the invention of the so-called bandwidth synthesis technique (Rogers 1970), which helped to overcome the limitations of tape recording equipment in terms of recordable bandwidth (Campbell 2000). A milestone was reached when the first significant estimates of the length change on the transatlantic baseline Haystack–Onsala (Sweden) were announced. A baseline rate of 17 mm/year with a statistical standard deviation of ±2 mm/year derived from 31 experiments between September 1980 and August 1984 was published by Herring et al. (1986). However, they reported that the systematic error could be as large as 10 mm/year. In comparison, Fig. 7.2 indicates session-wise baseline length estimates from 1984 to 2011 between the stations Wettzell (Germany) and Westford (Massachusetts, U.S.A.) determined by the VLBI group at TU Wien, Vienna, Austria. Clearly visible is the continuously improving accuracy, in particular during the first decade of the time series, and the seasonal variation of the length estimates, which is due to modeling

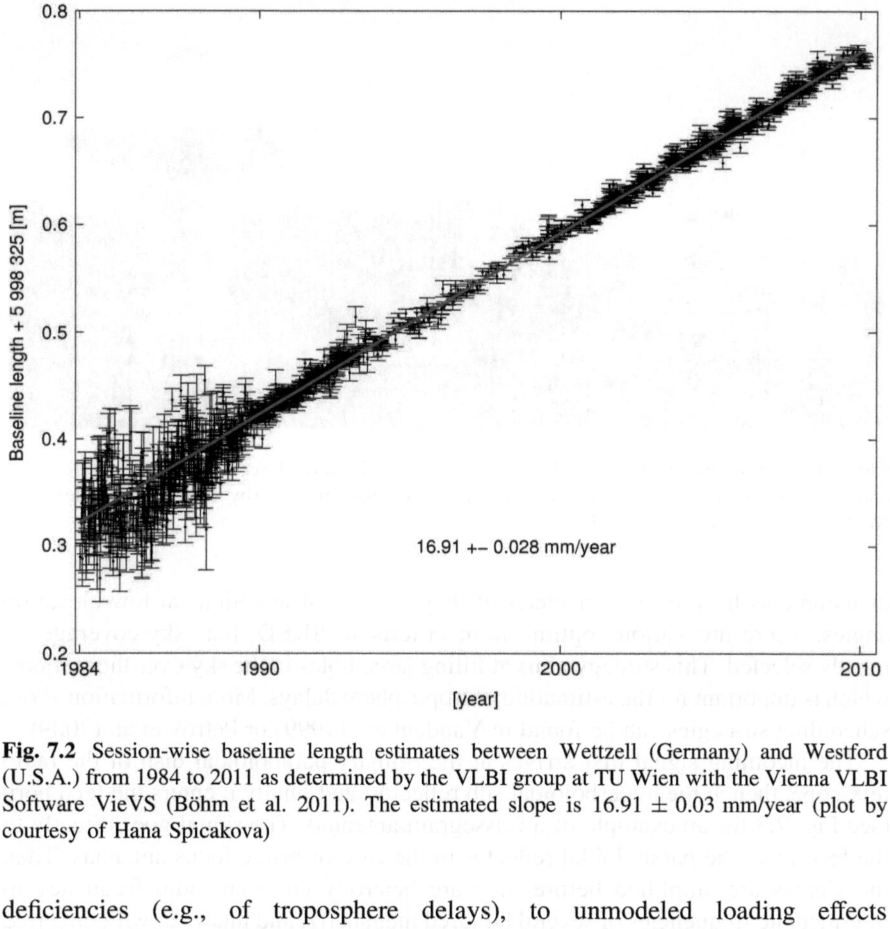

Fig. 7.2 Session-wise baseline length estimates between Wettzell (Germany) and Westford (U.S.A.) from 1984 to 2011 as determined by the VLBI group at TU Wien with the Vienna VLBI Software VieVS (Böhm et al. 2011). The estimated slope is 16.91 ± 0.03 mm/year (plot by courtesy of Hana Spicakova)

deficiencies (e.g., of troposphere delays), to unmodeled loading effects (e.g., atmosphere or hydrology loading), or a combination of both.

1.3 Data Acquisition

Geodetic VLBI is an active observing technique which needs to control the radio telescopes and steer them to various positions on the sky in a predefined observing schedule; thus, scheduling is a very important part of VLBI. The package SKED (Vandenberg 1999) developed at NASA Goddard Space Flight Center is widely used within the geodetic community to generate the observing plans for radio telescopes. At any instant, different subsets of antennas will be observing different sources. (All observations to one source at a time form a so-called 'scan'.) The integration time varies from antenna to antenna to reach the required signal-to-noise-ratio (SNR) (Petrov et al. 2009). The elevation mask is usually set to 5° but any obstacles

Fig. 7.3 New skyline at site Wettzell with the 'old' VLBI radio telescope (20 m) at the right side and the new twin radio telescopes (13.2 m) in the background (by courtesy of Alexander Neidhardt; picture taken in 2011)

or mountains have to be considered if they prevent observations at low elevation angles. There are various optimization criteria in SKED, but 'sky-coverage' is mostly selected. This strategy aims at filling large holes in the sky over the stations, which is important for the estimation of troposphere delays. More information about scheduling strategies can be found in Vandenberg (1999) or Petrov et al. (2009).

The incoming signal first arrives at the primary paraboloidal dish of the radio telescope, then at the hyperboloidal sub reflector, and finally it enters the feed horn (see Fig. 7.3 for an example of a Cassegrain antenna). The signal goes directly to the feed from the paraboloidal reflector in the case of prime focus antennas. Then the signals are amplified before they are heterodyned from radio frequency to intermediate frequencies of several hundred megahertz, and finally down-converted to baseband frequencies (simultaneously in multiple frequency sub-bands or channels), where the signal is band-limited to a width of a few megahertz, sampled and digitized (Sovers et al. 1998). The system temperatures typically range from 20 to 100 K for S- and X-band.

Finally, the signals are formatted and recorded on magnetic disks (or on tapes in the early years). Nowadays, data from shorter sessions can also be transferred to the correlators via high-speed broadband communication links. However, the majority of electronic transfer of the raw VLBI data is still asynchronous, i.e., the transfer is started during or after the observation but then needs more time than the actual observation (termed e-transfer). Only in a few experimental sessions, such as those described by Sekido et al. (2008), was the transmission carried out in real-time (termed e-VLBI).

VLBI radio telescopes need to have large collecting areas as well as high sampling and recording rates because the signal flux density is in the order of 1 Jansky (1 Jy $= 10^{-26}$ Wm^{-2} Hz^{-1}) or even lower. On the other hand, the structure of the antennas has to be sufficiently stable to allow slewing between widely separated sources within a few minutes or faster (Sovers et al. 1998). (See Sect. 2.5 for more information on antenna deformation.)

Phase shifts caused by the instrumentation have to be calibrated to take full advantage of the precision of current frequency standards (e.g., hydrogen masers stable to 10^{-14} at 50 min or better). Otherwise, those phase shifts can corrupt the estimated phase and group delay of the incoming signal. The technique of phase calibration (Rogers 1975) compensates for the instrumental phase errors by generating a signal of known phase, injecting this signal into the front end of the VLBI signal path, and examining the phase after the signal has traversed the instrumentation. This calibration signal is embedded in the broadband VLBI data stream as a set of low-level monochromatic tones along with the signal of the radio source (Sovers et al. 1998). These tones are used at a later stage by the post-correlation software. Furthermore, the length variations in the cables from the clocks to the antennas (called cable delays) have to be corrected properly.

In the next step the signals recorded at the antennas are combined pair-wise, producing an interference pattern. These installations are called correlators, and they are presently run world-wide, e.g., in the U.S.A. (Haystack Observatory, Westford; U.S. Naval Observatory, Washington D.C.), in Germany (Max Planck Institute for Radio Astronomy, Bonn), and in Japan (National Institute of Information and Communications Technology, Kashima). They are made up of special hardware that is used to determine the difference in arrival times at the two stations by comparing the recorded bit streams. If $V_1(t)$ and $V_2(t)$ are the antenna voltages as functions of time t, T is the averaging interval, and the asterisk denotes the complex conjugate, the group delay τ can be determined by maximizing the cross-correlation function R (Sovers et al. 1998):

$$R(\tau) = \frac{1}{T} \int_0^T V_1(t) \cdot V_2^*(t - \tau) \cdot dt. \tag{7.2}$$

Due to Doppler shifts caused by the Earth's rotation, VLBI observations at X-band (8.4 GHz) would be oscillating at several kilohertz if not 'counter-rotated' first (Sovers et al. 1998).

In recent years, software correlators have also been developed (e.g., Kondo et al. 2004; Tingay et al. 2009) because correlation algorithms for geodetic VLBI can be effectively implemented on parallel computers or on distributed systems. Software correlation is already well beyond the development stage. For example, the Bonn correlator has processed all IVS (International VLBI Service for Geodesy and Astrometry) sessions with the DiFX software correlator from November 2010 onwards.

During the correlation process, amplitudes and phases are determined every 1–2 s in parallel for typically 14 frequency channels ω_i. The post-correlation software applies corrections for the phase calibration and fits the phase ϕ_0, the group delay τ_{gd}, and the phase rate τ'_{pd}. to the phase samples $\phi(\omega_i, t_j)$ from the various frequency channels ω_i and times t_j. The phase-derived observables are determined (for phase ϕ and circular frequency ω) from a bilinear least-squares fit to the measured phases $\phi(\omega, t)$ with (Sovers et al. 1998)

$$\phi(\omega,t) = \phi_0(\omega_0,t_0) + \frac{d\phi}{d\omega}(\omega - \omega_0) + \frac{d\phi}{dt}(t - t_0), \tag{7.3}$$

where the phase delay τ_{pd}, group delay τ_{gd}, and phase delay rate τ'_{pd} are defined, respectively, as

$$\tau_{pd} = \frac{\phi_0}{\omega_0},\ \tau_{gd} = \frac{d\phi}{d\omega},\ \tau'_{pd} = \frac{1}{\omega_0}\frac{d\phi}{dt}. \tag{7.4}$$

The group delay rate τ'_{gd} is not accurate enough to be useful for geodetic or astrometric purposes; however, it is needed to resolve group delay ambiguities in a first solution step. The amplitudes are usually not used in geodetic/astrometric VLBI.

The natural ultra-wide band continuum radiation provides the means to use the essentially unambiguous wide band group delay as the prime geodetic VLBI observable. The group delay resolution is proportional to the inverse of the signal-to-noise-ratio (SNR) and the root mean square (rms) of the frequency about the mean (sometime called rms spanned bandwidth) B_{eff} (Rogers 1970):

$$\sigma_\tau = \frac{1}{2\pi} \cdot \frac{1}{\mathrm{SNR} \cdot B_{\mathrm{eff}}}. \tag{7.5}$$

If we increase the rms spanned bandwidth B_{eff} at a given SNR by a factor of ten, the group delay uncertainty will be reduced by the same factor, a relation with tremendous consequences (Campbell 2000). On the other hand, the SNR is dependent on the recorded bandwidth B with

$$\mathrm{SNR} = \eta \cdot \rho_0 \cdot \sqrt{2 \cdot B \cdot T}, \tag{7.6}$$

where T is the so-called coherent integration time, η is the digital loss factor, and ρ_0 is the correlation amplitude which depends on the system noise temperatures and on the equivalent noise temperature of the source signal (Takahashi et al. 2000). There are virtually no limitations to improving the statistical precision of the geodetic group delay further, except technological constraints and costs. For the upcoming VLBI2010 system, a four-band system is recommended that uses a broadband feed to span the entire frequency range from 2 to 14 GHz (see Sect. 6). This will also allow the use of phase delays, which are still an issue of research with the current system where they provide very high accuracy, but only on very short baselines (Herring 1992a; Petrov 1999), as on longer baselines the phase ambiguities are still an unsolved problem.

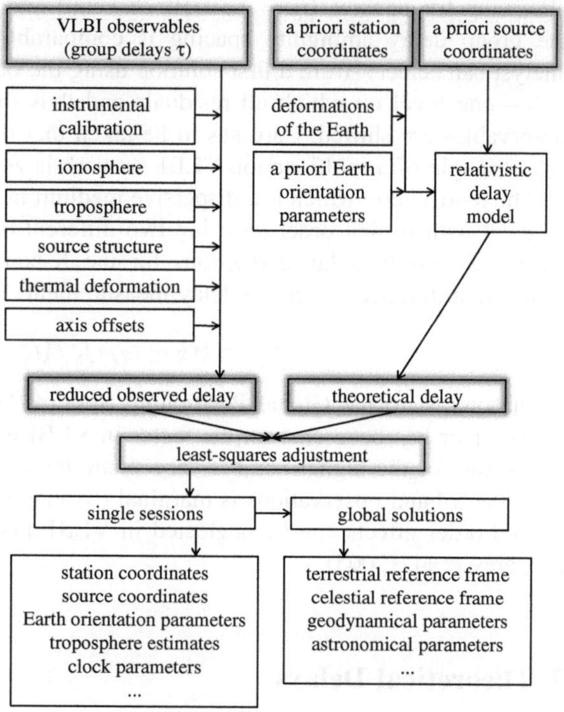

Fig. 7.4 Flow diagram of geodetic VLBI data analysis (according to Schuh 1987)

1.4 Data Analysis

The VLBI data analysis model is developed using the best presently available knowledge to recreate mathematically, as closely as possible, the situation at the time of observation (see Sect. 2). Then, a least-squares parameter estimation algorithm or other estimation methods can be used to determine the best values of the quantities to be solved (Sect. 3). Before this process starts, the raw observations have to be cleaned from several systematic effects, which in fact limit the final accuracy of the results (Schuh 2000).

The flow diagram of a geodetic VLBI data analysis (according to Schuh 1987) is shown in Fig. 7.4. The system can be seen to have two main streams, one containing the actual observations which undergo instrumental and environmental corrections to obtain the reduced delay observables, and the other to produce the theoretical delays, starting with the a priori parameter values, a set of initial values for the parameters of the VLBI model. Both streams converge at the entrance to the parameter estimation algorithm, e.g., the least-squares fit, where the 'observed minus computed' values are formed. The instrumental effects include systematic clock instabilities, electronic delays in cables and circuitry, and the group delay ambiguities. The latter are due to observation by the multichannel frequency setup described in Sect. 1.2 covering the total spanned bandwidth around each of the two observing frequencies ($f_S = 2.3$ GHz (S-band) and $f_X = 8.4$ GHz (X-band)). As the group delay ambiguity spacing is comparably large and well-known, the analyst can select—from a first solution using the observed group delay rates τ'_{gd} only—one level on which all residuals and thus the corresponding group delay observables are shifted. Care has to be taken that the group delay closure within each triangle of a multi-station VLBI network is zero (Schuh 2000).

The ionosphere, which is a dispersive medium in the radio frequency band, can be dealt with to first order by using two different observing frequencies, i.e., the ionosphere group delay corrections for the X-band observations are computed from the differences of group delay measurements at X- and S-band:

$$\Delta\tau_X^{\text{ion}} = (\tau_X - \tau_S) \cdot f_S^2 / (f_X^2 - f_S^2). \tag{7.7}$$

In contrast to the Global Positioning System (GPS), where a very close frequency pair has been chosen, the factor in VLBI to convert the difference into a correction for the higher band is very small, 0.081, so that the error contribution from the S-band observations is marginal (Schuh 2000). Unlike GPS, ionospheric second order effects can be neglected in VLBI analysis as was demonstrated by Hawarey et al. (2005).

2 Theoretical Delays

In order to calculate observed minus computed values for the least-squares adjustment (see Sect. 3), several models need to be applied. At first, the station coordinates at the observation epoch have to be determined (Sect. 2.1). Then the station coordinates are rotated from the terrestrial to the celestial system (Sect. 2.2) where the computed delays between two stations forming a baseline and a radio source are built (Sect. 2.3), taking into account relativistic corrections and applying troposphere delay (Sect. 2.4), and other corrections. For all the modeling details the reader is referred to the Conventions of the International Earth Rotation and Reference Systems Service (Petit and Luzum 2010) and its online updates at http://tai.bipm.org/iers/convupdt/convupdt.html, as well as to special IVS Conventions such as those for the treatment of the thermal expansion of radio telescopes (Nothnagel 2009).

2.1 Station Coordinates at the Time of Observation

At first, coordinates (valid at a reference epoch; e.g., J2000.0 = 1 January 2000 at 12 h Terrestrial Time TT) and velocities of a specific realization of the International Terrestrial Reference System (ITRS) are taken to determine the mean coordinates at the time of observation. It should be mentioned here that, typically, these realizations are TT frames and that the coordinates are provided in a conventional tide-free system. Examples are the International Terrestrial Reference Frame ITRF2008 (ITRF2008; Altamimi et al. 2011) or specific VLBI realizations like the VTRF2008 (Böckmann et al. 2010), the VLBI contribution to ITRF2008.

Then several corrections are added to get closer to the true station coordinates at the observation epoch. These corrections include periodic and aperiodic deformations of the Earth's crust. The largest periodic corrections are for the solid Earth tides, ocean tide loading, and pole tide loading. Solid Earth tides show mainly diurnal and semidiurnal oscillations which cause vertical deformations within a range of ± 20 cm and horizontal displacements of about 30% of the vertical effect (e.g., Mathews et al. 1997). More difficult to model is the loading by the water masses of ocean tides and currents (ocean loading), which amounts to as much as 1 dm on some coastal or island sites (e.g., Scherneck 1991; Schuh 2000) Additionally, there is also a periodic deformation at the S1 (24 h) and S2 (12 h) periods caused by pressure tides due to thermal heating of the atmosphere. All these effects, which should be corrected at the observation level, are described in detail in the IERS Conventions 2010 (Petit and Luzum 2010) and its electronic updates.

The analysis strategy is not so clear with aperiodic deformations, e.g., with non-tidal atmosphere (but also non-tidal ocean and hydrological) loading, although their significance in VLBI analysis has been shown repeatedly (e.g., Rabbel and Schuh 1986; vanDam and Herring 1994). So far there has been no general agreement within the international space geodesy community as to whether these corrections should be applied at the observation level: Arguments against the application of non-tidal atmosphere loading at the observation level are that there is no consensus model available, that the accuracy of existing models (e.g., Petrov and Boy 2004) is still not sufficient, and that geophysicists are interested in station coordinate time series which would show the loading signals. On the other hand—and this holds in particular for VLBI with a small number of stations (6–8) taking part in typical 24 h sessions—neglected a priori atmosphere loading is partly absorbed by no-net-rotation (NNR) and no-net-translation (NNT) conditions and does not show up in the estimated station coordinate time series (Böhm et al. 2009a). Furthermore, there is significant variation in non-tidal atmospheric loading corrections within 24 h which would be neglected if the correction was applied at a later stage of data analysis.

2.2 Earth Orientation

In the next step, we need to transform the station coordinates from the ITRS into the Geocentric Celestial Reference System (GCRS) at the epoch of the observation t. The transformation matrix can be written as

$$[\text{GCRS}] = Q(t) \cdot R(t) \cdot W(t) \cdot [\text{ITRS}], \tag{7.8}$$

where $Q(t)$, $R(t)$, and $W(t)$ are the transformation matrices arising from the motion of the celestial pole in the celestial reference system, from the rotation of the Earth around the axis associated with the pole, and from polar motion respectively (Petit and Luzum 2010). Matrix W ('wobble') includes as parameters the coordinates x_p and y_p (polar motion) of the Celestial Intermediate Pole (CIP) in the Earth-fixed frame and the correction angle s' which locates the position of the Terrestrial Intermediate Origin (TIO) on the equator of the CIP. Terrestrial (TIO) and Celestial Intermediate Origin (CIO) realize reference meridians in the respective systems. These terms are part of the 'CIO-based' transformation concept following the Non-Rotating Origin (NRO), which replaces the former 'equinox-based' transformation. Matrix $R(t)$ is the Earth rotation matrix with the angle θ between TIO and CIO. The conventional relationship defining UT1 from the Earth's rotation angle θ is given by Capitaine (2000) as

$$\theta(T_u) = 2\pi \cdot (0.7790572732640 + 1.00273781191135448 \cdot T_u), \tag{7.9}$$

where T_u = (Julian UT1 date—2451545.0), and UT1 = UTC + (UT1–UTC). The difference between Universal Time (UT1) and Universal Time Coordinated (UTC), which differs by a known integer number of SI-seconds from TAI, the International Atomic Time (realized as a weighted mean of signals provided by atomic clocks located all over the world) can be uniquely observed by VLBI. All satellite techniques like the GNSS or Satellite Laser Ranging (SLR)— due to the direct dependence between UT1 and the right ascension of the ascending node of the satellite orbit—can only observe length-of-day, which is the negative time derivative of (UT1–UTC), but need external information about UT1–UTC every few days.

The precession/nutation matrix, denoted $Q(t)$, includes the rotations around the angles X and Y (which are the coordinates of the CIP in the celestial frame) and the correction angle s which positions the CIO on the equator of the CIP. The CIP is the reference pole for space geodetic measurements, i.e., it defines the observed axis. This is a pure convention realized by an appropriate theory of precession and nutation as will be described below. The orientation of the CIP does not coincide with that of a physical axis like the rotation axis, the figure axis, or the angular momentum axis, but it can be related to all of them. By definition the CIP is an intermediate pole which divides the motion of the pole of the ITRS w.r.t. the GCRS into a celestial and a terrestrial part. The celestial part (precession and nutation, [X, Y]) includes all motions with periods >2 days observed in the celestial frame, and this corresponds to all frequencies between −0.5 (retrograde) and +0.5 (prograde) cycles per sidereal day in the GCRS. The terrestrial part (polar motion, [x_p, y_p]) includes all motions outside of the retrograde daily band in the ITRS, i.e., it includes frequencies below −1.5 and above −0.5 cycles per sidereal day in the ITRS.

The largest part of the celestial motion of the CIP can be calculated with a conventional precession/nutation model. Presently the model IAU 2006/2000A is recommended by the IERS Conventions 2010. However, remaining unmodeled parts of the celestial motion can be observed with VLBI and are provided by the IERS as so-called celestial pole offsets [δX, δY]. These offsets stem from residual errors of the a priori precession/nutation model and the phenomenon of the Free Core Nutation (FCN) which is a resonance mode due to the deviation of the rotation axis of the mantle from the rotation axis of the core (Dehant and Mathews 2009). This retrograde motion with a period of about 430 days in the GCRS and a varying amplitude of up to 0.3 mas ($\sim$10 mm on the Earth's surface) (e.g., Herring et al. 2002) is not predictable and cannot be neglected if someone wants to achieve the highest positioning accuracies. Thus, neither precession/nutation nor polar motion and UT1–UTC can be predicted accurately enough with models but have to be observed by space geodetic techniques. A combination of these estimates is provided by the IERS, e.g., in the IERS 05 C04 series (Bizouard and Gambis 2009), which can be used by space geodetic techniques as a priori information. If not estimated from the observations, the standard pole coordinates to be used are those published by the IERS $(x, y)_{\mathrm{IERS}}$ with additional terms to account for the diurnal and semi-diurnal variations caused by ocean tides $(\Delta x, \Delta y)_{\mathrm{ocean\_tides}}$ (Englich et al. 2008; see also Fig. 7.8) and for libration $(\Delta x, \Delta y)_{\mathrm{libration}}$:

$$(x_p, y_p) = (x, y)_{\mathrm{IERS}} + (\Delta x, \Delta y)_{\mathrm{ocean\_tides}} + (\Delta x, \Delta y)_{\mathrm{libration}}. \tag{7.10}$$

Here $(\Delta x, \Delta y)_{\mathrm{libration}}$ are the forced variations in pole coordinates corresponding to motions with periods less than 2 days in space that is not part of the IAU 2000A nutation model. The IERS EOP Product Center provides a subroutine to interpolate ('Lagrange' interpolation) in the $(x, y)_{\mathrm{IERS}}$ pole coordinates which are typically released at midnight. However, this kind of interpolation (unlike linear interpolation) does not allow the rigorous estimation of estimated polar motion rates to be used for Earth rotation excitation studies.

The situation is similar for the Earth rotation angle θ with models for the effects of ocean tides and libration that have to be added to the IERS 05 C04 values of UT1–UTC. In the case of the Earth rotation angle, tidal terms (with periods from 5 to 35 days) are usually removed before the Lagrange interpolation of the IERS values, and are restored afterwards.

2.3 General Relativistic Model for the VLBI Time Delay

The general relativistic model for the time delay is developed within the frame of the IAU Resolutions, i.e., general relativity using the Barycentric Celestial Reference System (BCRS) and the GCRS. The procedure to compute the VLBI time delay according to the so-called consensus model is taken from Eubanks (1991), and it is summarized in the IERS Conventions 2010 (Chap. 11; Petit and Luzum 2010). The model has been developed for VLBI observations of extragalactic radio sources taken from the Earth's surface, but not for observations of objects in our solar system like Earth- or Moon-orbiting satellites. In the model, which is accurate to the picosecond-level, it is assumed that the ionospheric delays have already been removed (see (7.7) in Sect. 1).

Theoretically, the VLBI time delays are measured in proper times of the station clocks, whereas the VLBI model (7.19) is expressed in terms of coordinate time in a given reference system. We consider the VLBI time delay from the correlator to be equal to the Terrestrial Time (TT, agrees with SI second on the geoid and is equal to TAI, apart from a constant offset: TT = TAI + 32.184 s) coordinate time interval d_{TT} between the arrival of a radio signal at the reference point of the first station and the arrival of the same signal at the reference point of the second station. From a TT coordinate interval, d_{TT}, the Geocentric Coordinate Time (TCG) coordinate interval, d_{TCG}, can be determined by scaling: $d_{TCG} = d_{TT}/(1 - L_G)$ with $L_G = 6.969290134 \times 10^{-10}$ (Petit and Luzum 2010).

The Terrestrial Reference System (TRS) space coordinates from the analysis of VLBI observations, x_{VLBI}, are termed 'consistent with TT' if derived from d_{TT} intervals, and the TRS coordinates recommended by the IAU and IUGG resolutions, x_{TCG}, can be derived with $x_{\mathrm{TCG}} = x_{\mathrm{VLBI}}/(1 - L_G)$ (see Petit 2000). Presently, all VLBI analysis centers provide their coordinate solutions as consistent with TT. Since SLR analysis centers also submit their solutions consistent with TT, it was decided that the coordinates should not be re-scaled to x_{TCG} for the computation of ITRF2008 so that the scale of ITRF2008 (and earlier realizations) in this respect does not fully comply with IAU and IUGG resolutions (Petit and Luzum 2010).

In the remaining part of this section, we provide the equations for the general relativistic VLBI time delay model and follow the IERS Conventions 2010 (Petit and Luzum 2010). Although the delay to be calculated is by convention the time of arrival at station 2 minus the time of arrival at station 1, it is the time of arrival at station 1, t_1, which serves as the time reference for the measurement. Thus, all quantities are assumed to be calculated at t_1, including the effects of the troposphere. Assuming that the reference time is the UTC time of arrival of the VLBI signal at receiver 1, and that it is transformed to the appropriate time scale (e.g., UT1 or TT) to be used to compute each element of the geometric model, the following steps are carried out to calculate the VLBI time delay. First, the barycentric station vectors $\underline{X}_i(t_1)$ for the receivers are determined with

$$\underline{X}_i(t_1) = \underline{X}_\oplus(t_1) + \underline{x}_i(t_1), \tag{7.11}$$

where t_1 is the TCG time of arrival of the radio signal at the first receiver, $\underline{X}_\oplus(t_1)$ is the barycentric radius vector of the geocenter, and $\underline{x}_i(t_1)$ the GCRS radius vector of the i-th receiver. Then we calculate the vectors $\underline{R}_{iJ}$ from the Sun, the Moon, and each planet (with the barycentric coordinates $\underline{X}_J$) except the Earth to receivers 1 and 2. The time t_{1J} of the closest approach of the signal to the gravitating body J can be determined with

$$t_{1J} = \min\left[t_1, t_1 - \frac{\underline{K} \cdot (\underline{X}_J(t_1) - \underline{X}_1(t_1))}{c}\right] \tag{7.12}$$

so that

$$\underline{R}_{1J}(t_1) = \underline{X}_1(t_1) - \underline{X}_J(t_{1J}) \tag{7.13}$$

and

$$\underline{R}_{2J}(t_1) = \underline{X}_2(t_1) - \frac{\underline{V}_\oplus}{c}(\underline{K} \cdot \underline{b}) - \underline{X}_J(t_{1J}) \tag{7.14}$$

with $\underline{V}_\oplus$ the barycentric velocity of the geocenter, $\underline{b} = \underline{x}_2(t_1) - \underline{x}_1(t_1)$ the GCRS baseline vector at the TCG time of arrival t_1, $\underline{K}$ the unit vector from the barycenter to the source in the absence of gravitational or aberrational effects, and c the velocity of light. The differential TCB gravitational delay for each of those bodies J can then be calculated by

$$\Delta T_{\mathrm{grav},J} = (1+\gamma)\frac{GM_J}{c^3}\ln\frac{|\underline{R}_{1J}| + \underline{K} \cdot \underline{R}_{1J}}{|\underline{R}_{2J}| + \underline{K} \cdot \underline{R}_{2J}}, \tag{7.15}$$

where M_J is the rest mass of the Jth gravitating body and G is the gravitational constant (Petit and Luzum 2010).

According to the General Theory of Relativity (GRT), the so-called light deflection parameter γ is usually set to unity, but γ can also be estimated as a solve-for parameter in a VLBI global solution comprising many or all VLBI sessions that have ever been carried out. See Lambert and Le Poncin-Lafitte (2009), Heinkelmann and Schuh (2010), or Lambert and Le Poncin-Lafitte (2011) for further details.

Analogously, we determine the TCB gravitational delay due to the Earth with

$$\Delta T_{\mathrm{grav}\oplus} = (1+\gamma)\frac{GM_\oplus}{c^3}\ln\frac{|\underline{x}_1| + \underline{K} \cdot \underline{x}_1}{|\underline{x}_2| + \underline{K} \cdot \underline{x}_2}, \tag{7.16}$$

where $M_\oplus$ is the rest mass of the Earth, and we sum up the effects of all gravitating bodies to find the total differential TCB gravitational delay with

$$\Delta T_{\mathrm{grav}} = \sum_J \Delta T_{\mathrm{grav},J}. \tag{7.17}$$

We need to consider the Sun, Earth, Jupiter, the Earth's moon, and the other planets. If observations pass close to them, the major satellites of Jupiter, Saturn, and Neptune should also be added. If the ray path passes very close to some massive bodies, extra terms need to be included for accuracies better than 1 ps (see Klioner 1991). For observations made very close to the Sun, higher order relativistic time delay effects become increasingly important. The largest correction is due to the change in delay caused by the bending of the ray path by the gravitating body J described in Richter and Matzner (1983) and Hellings (1986). The correction is

$$\delta T_{\mathrm{grav},J} = \frac{4G^2M_J^2}{c^5}\frac{\underline{b} \cdot (\underline{N}_{1J} + \underline{K})}{\left(|\underline{R}|_{1J} + \underline{R}_{1J} \cdot \underline{K}\right)^2} \tag{7.18}$$

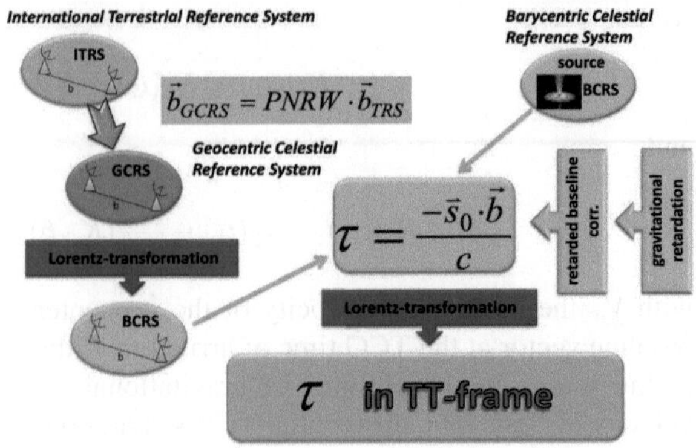

Fig. 7.5 Symbolic representation of the derivation of the time delay τ in the TT frame starting with station coordinates in the ITRS and source coordinates in the ICRS (by courtesy of Lucia Plank)

which should be added to ΔT_{grav} in (7.17). $\underline{N}_{1J}$ is the unit vector from the J-th gravitating body to the first receiver, and M_J is the rest mass of the J-th gravitating body. Next, we compute the vacuum delay between t_{vi}, which are the 'vacuum' TCG times of arrival of a radio signal at the i-th VLBI receiver including the gravitational delay but neglecting the troposphere propagation delay and the change in the geometric delay caused by the existence of the troposphere propagation delay:

$$t_{v2} - t_{v1} = \frac{\Delta T_{\text{grav}} - \frac{\underline{K}\cdot\underline{b}}{c}\left[1 - \frac{(1+\gamma)\cdot U}{c^2} - \frac{\left|\underline{V}_{\oplus}\right|^2}{2c^2} - \frac{\underline{V}_{\oplus}\underline{\omega}_2}{c^2}\right] - \frac{\underline{V}_{\oplus}\underline{b}}{c^2}\left(1 + \underline{K}\cdot\underline{V}_{\oplus}/2c\right)}{1 + \frac{\underline{K}\cdot\left(\underline{V}_{\oplus}+\underline{\omega}_2\right)}{c}}. \tag{7.19}$$

In (7.19), $\underline{\omega}_i$ is the geocentric velocity of the i-th receiver. The aberrated radio source vectors $\underline{k}_i$ (the unit vector from the i-th station to the radio source after aberration) for use in the determination of the troposphere propagation delays are calculated with

$$\underline{k}_i = \underline{K} + \frac{\underline{V}_{\oplus} + \underline{\omega}_i}{c} - \underline{K}\frac{\underline{K}\cdot\left(\underline{V}_{\oplus} + \underline{\omega}_i\right)}{c}. \tag{7.20}$$

Thus, we can add the geometric part of the troposphere propagation delay to the vacuum delay with

$$t_{g2} - t_{g1} = t_{v2} - t_{v1} + \Delta L_1 \cdot \frac{\underline{K}\cdot\left(\underline{\omega}_2 - \underline{\omega}_1\right)}{c}, \tag{7.21}$$

where ΔL_i is the troposphere propagation TCG delay for the i-th receiver ($=t_i - t_{gi}$). The total delay can be found by adding the best estimate of the troposphere propagation delay:

$$t_2 - t_1 = t_{g2} - t_{g1} + (\Delta L_2 - \Delta L_1). \tag{7.22}$$

The troposphere propagation delays in (7.21) and (7.22) need not be from the same model. The estimate in (7.22) should be as accurate as possible (see Sect. 2.4), while the ΔL_1 model in (7.21) need only be accurate to about 10 ns (Petit and Luzum 2010). Sections 2.2 (Earth orientation) and 2.3 (Relativistic delay model) are symbolically summarized in Fig. 7.5.

2.4 Troposphere Delay Modeling

The troposphere path delay $\Delta L(e)$ at the elevation angle e is usually represented as the product of the zenith delay ΔL^z and an elevation-dependent mapping function $mf(e)$:

$$\Delta L(e) = \Delta L^z \cdot mf(e). \tag{7.23}$$

This concept is not only used to determine a priori slant delays for the observations, but the mapping function is also the partial derivative to estimate residual zenith delays, typically every 20–60 min. In the analysis of space geodetic observations, not only zenith delays are estimated, but also other parameters like stations heights and clocks. Whereas the partials for the clocks (=1) and the station heights (=sin(e)) are exactly known, the partial derivative for the zenith delays (i.e., the mapping function) is only known with limited accuracy. Via the correlations between zenith delays, station heights, and clocks, any imperfection of the mapping function is also manifested as station height error (and clock error) (Nothnagel et al. 2002). To reduce these correlations, observations at low elevations need to be included in the analysis; however, care has to be taken because mapping function errors increase rapidly at very low elevations, i.e., at 5° or below. Presently, the best trade-off between reduced correlations and increasing mapping function errors is found for cutoff elevation angles of about 7° (MacMillan and Ma 1994; Teke et al. 2008) or by appropriate down-weighting (Gipson 2009). Simulations of VLBI2010 observing scenarios indicate that in the future, with faster antennas and significantly more observations, the cutoff elevation angle can be increased so that the mapping function is less critical for the accuracy of VLBI analysis (Petrachenko et al. 2009).

Considering (7.23) we find the following relationship: if the erroneous mapping function was too large, the estimated zenith delay ΔL^z becomes too small, because the observed troposphere delay $\Delta L(e)$ stays the same. Consequently, the estimated station height moves up to account for the reduced zenith delay. MacMillan and Ma (1994) set up a rule of thumb specifying that the error in the station height is approximately 0.22 of the delay error at the lowest elevation angle included in the analysis. Böhm (2004) confirmed this rule of thumb for VLBI analysis (and a cutoff elevation angle of 5°) specifying that the station height error is about one-fifth of the delay error at 5° elevation angle. The corresponding decrease of the zenith delay becomes about one half of the station height increase.

Assuming azimuthal symmetry of the neutral atmosphere around the station (i.e., at a constant elevation angle the delay is not dependent on the azimuth of the observation), the approach as described in (7.24) (e.g., Davis et al. 1985) is generally applied:

$$\Delta L(e) = \Delta L_h^z \cdot mf_h(e) + \Delta L_w^z \cdot mf_w(e), \tag{7.24}$$

where $\Delta L(e)$ is the total path delay of the microwaves in the neutral atmosphere and e is the elevation angle of the observation to the quasar (vacuum or geometric elevation angle). ΔL_h^z and ΔL_w^z are the a priori zenith hydrostatic and the estimated zenith wet delays, and $mf_h(e)$ and $mf_w(e)$ are the mapping functions which provide the ratio of the slant delay to the delay in the zenith direction. The input to both mapping functions is the vacuum elevation angle e, because the bending effect is accounted for by the hydrostatic mapping function. The underlying continued fraction form to all mapping functions is that proposed by Herring (1992b):

$$mf(e) = \frac{1 + \frac{a}{1+\frac{b}{1+c}}}{\sin(e) + \frac{a}{\sin(e)+\frac{b}{\sin(e)+c}}}. \tag{7.25}$$

At present the most accurate mapping functions globally available are the Vienna Mapping Functions 1 (VMF1; Böhm et al. 2006). Whereas the b and c coefficients of (7.25) are provided as analytical functions depending on day of the year and station latitude, the a coefficients (hydrostatic and wet) are provided as a time series with a 6 h time resolution on global (2.0° in latitude times 2.5° in longitude) grids as well as for all VLBI sites for the complete history of VLBI observations. The coefficients are available from the website http://ggosatm.hg.tuwien.ac.at/ of the Vienna University of Technology as derived from operational analysis data as well as from forecast data of the European Centre for Medium-range Weather Forecasts (ECMWF). VMF1 coefficients from forecast data can be used for VLBI real-time applications without significant loss of accuracy as shown by Böhm et al. (2009b). Böhm et al. (2006) tested the concept of a 'total' VMF1, i.e., using the same function for mapping the a priori zenith delay to the elevation of the observation and for the estimation of the residual zenith delays. This concept was not as successful (in terms of baseline length repeatability, which is the standard deviation after removing a linear trend from a time series of baseline lengths) as the separation into a hydrostatic and a wet part, because the variation of the zenith wet delay is faster than can be described by coefficients with a 6 h time resolution. A total mapping function (which is close to the hydrostatic mapping function) cannot account for this variation whereas the wet mapping function is able to account for it by estimating zenith wet delays every hour or even faster.

The a priori zenith hydrostatic delays in (7.24) can be determined very accurately from the atmosphere pressure at the site (see Saastamoinen 1972; Davis et al. 1985). If locally recorded pressure values are not available, it is recommended to take values retrieved from numerical weather models such as, e.g., those provided by the Vienna University of Technology at http://ggosatm.hg.tuwien.ac.at/ together with the coefficients of the VMF1. If those are also not accessible, it is recommended to use an analytical expression like the Global Pressure and Temperature Model (GPT; Böhm et al. 2007). Zenith wet delays are usually fully estimated in the VLBI analysis, although Gipson (2009) found a slight improvement if zenith wet delays were already added to the a priori delays and only residual zenith delays were estimated. Approximate values of zenith wet delays are provided with the VMF1 files.

Errors in the zenith hydrostatic delays or the mapping functions have an influence on station heights as can be described with the rule of thumb by Böhm (2004) mentioned above. Exemplarily, the zenith hydrostatic and wet delays shall be 2000 and 200 mm, respectively, the minimum elevation angle is 5°, and the corresponding values for the hydrostatic and wet mapping functions are 10.15 ($mf_h(5°)$) and 10.75 ($mf_w(5°)$).

1. We consider an error in the wet mapping function of 0.1 ($mf_w(5°) = 10.85$ instead of 10.75) or in the hydrostatic mapping function of 0.01 ($mf_h(5°) = 10.16$ instead of 10.15). The error at 5° elevation in both cases is +20 mm, i.e., the error in the station height is approximately +4 mm.

2. We assume an error in the total pressure measured at the station of −10 hPa. Let us assume that we take the 'mean' pressure from GPT (Böhm et al. 2007) although the real pressure at the site is larger by 10 hPa. −10 hPa correspond to about −20 mm zenith hydrostatic delay (Saastamoinen 1972), which is then mapped with the wrong mapping function (factor 0.6 = 10.75 − 10.15). At 5° elevation the mapping function error causes +12 mm delay error, and one-fifth of it, i.e., +2.4 mm, is the resulting station height error (Böhm et al. 2006).

On the other hand—due to atmosphere loading—the station height is decreased by about 4 mm (assuming a regression coefficient of −0.4 mm/hPa) if the actual pressure is larger than the mean pressure (from GPT) by 10 hPa. This implies that the application of GPT (or any other mean pressure model) for the determination of the a priori zenith hydrostatic delays of VLBI observables (or any other technique using microwave signals) partly corrects for atmosphere loading (see also Steigenberger et al. 2009). In our example 2.4 mm out of 4.0 mm are compensated, because the application of GPT causes an error that goes into the same direction as the atmosphere loading corrections.

In addition to the azimuthal symmetry of the troposphere delays according to (7.24), we need to take azimuthal asymmetry into account, which in a North–South direction is caused by the larger extension of the troposphere above the equator compared to the poles and which can also be due to local weather phenomena, e.g., if a VLBI site is located close to the coast. Typically, North and East gradients are estimated in VLBI analysis with a resolution of 2–24 h. MacMillan (1995) proposed to use the equation

$$\Delta L(a,e) = \Delta L_0(e) + mf_h(e) \cdot \cot(e) \cdot [G_n \cos(a) + G_e \sin(a)] \tag{7.26}$$

which goes back to Davis et al. (1993) with a denoting the azimuth and ΔL_0 the symmetric delay (see (7.24)). Chen and Herring (1997) published the formula

$$\Delta L(a,e) = \Delta L_0(e) + \frac{1}{\sin(e)\tan(e) + C} \cdot [G_n \cos(a) + \cdot G_e \sin(a)] \qquad (7.27)$$

and recommended the use of $C = 0.0032$ if the total (hydrostatic plus wet) gradients are estimated. Böhm and Schuh (2007) tested the application of a priori gradients derived from data of the ECMWF and found that estimating gradients from VLBI observations provides better baseline length repeatabilities than keeping those values fixed to non-zero a priori values derived from numerical weather models.

At present many investigations are carried out on 'direct ray-tracing', i.e., deriving the slant delay for every single observation from numerical weather models (Hobiger et al. 2008). Gipson and MacMillan (2009) found better baseline length repeatabilities for CONT08 (a 14-day VLBI campaign in August 2008) with slant delays derived from data of the Goddard Modelling and Assimilation Office (GMAO) compared to using VMF1. Böhm et al. (2010) obtained improvement for UT1 estimates from intensive sessions on the baseline Wettzell–Tsukuba if using ray-traced delays at station Tsukuba.

2.5 Antenna Deformation

It has already been mentioned that, in addition to the troposphere delays, there are other effects which depend on azimuth and elevation of the observations and need to be taken into account. For instance, deformations of the radio telescope structure which occur during the 24 h of an observing session or between the observing sessions have to be considered if we want to achieve the highest accuracies possible. They can be caused by snow and ice loading of the antenna (Haas et al. 1999) or thermal expansion of the radio telescopes (Nothnagel 2009). Wresnik et al. (2007) compared thermal deformation measurements with invar rods at Wettzell (Germany) and Onsala (Sweden) to thermal deformation models applying local air temperatures as well as structure temperatures measured at those sites. They found an improvement if the air temperatures are first converted to structure temperatures before using them for modeling the thermal deformation of the radio telescopes (see Fig. 7.6). Recent investigations also deal with gravitational deformations (Abbondanza and Sarti 2010) which can significantly influence the estimated station heights if neglected in VLBI analysis (Sarti et al. 2011).

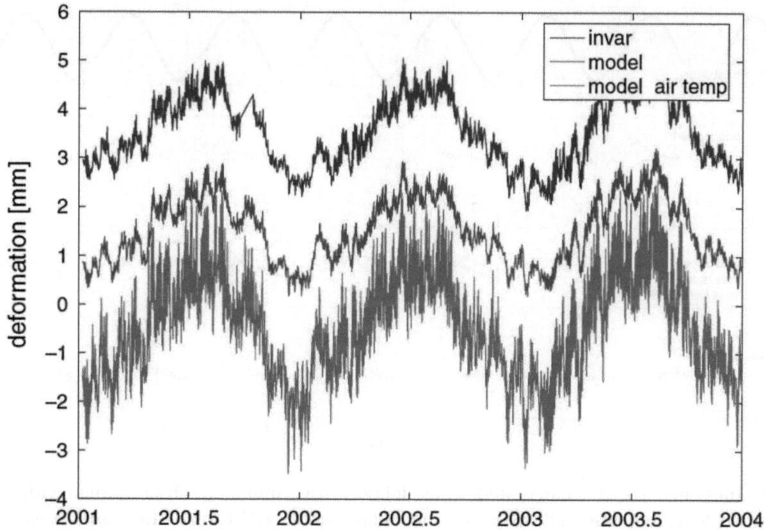

Fig. 7.6 Measured vertical deformation (invar rod, *upper curve*), calculated antenna deformation using the air temperature to model the structure temperature (*central curve*), and the thermal deformation using the measured air temperature directly at Wettzell (*lower curve*). The curves are offset by 2 mm for clarity (Wresnik et al. 2007)

2.6 Axis Offsets

At the radio telescope the distance between the feed horn and the axis intersection, which constitutes the baseline reference point, is assumed to be constant at the millimeter-level. In this case the corresponding time offset becomes part of the clock offset parameter. However, an axis offset model is applied to each antenna where the pointing axes do not intersect (Sovers et al. 1998). The axis offsets are provided by the IVS Analysis Coordinator together with the coefficients of the thermal expansion models (Nothnagel 2009). Large radio telescopes such as the Effelsberg 100-m antenna exhibit elevation-dependent changes in the focal distance which can however be modeled to a level of a few millimeters (Rius et al. 1987). In the case of radio telescopes of the type alt-azimuth, the correction $\Delta\tau$ due to an axis offset AO can be calculated with

$$\Delta\tau = -\text{AO} \cdot \sin(zd') \tag{7.28}$$

if zd' is the zenith distance corrected for refraction.

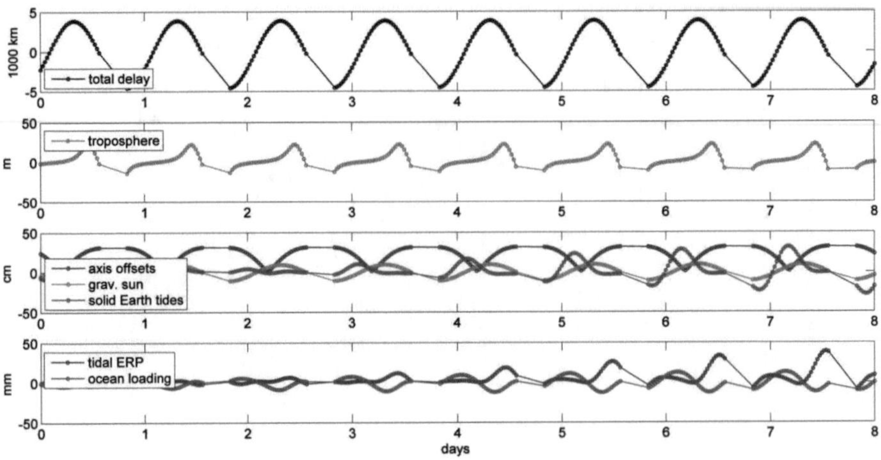

Fig. 7.7 Influence on the calculated time delay τ for simulated observations with stations Westford (U.S.A) and Wettzell (Germany) to radio source 0642+449. From *top* to *bottom*: total delay (mainly caused by Earth rotation), troposphere, axis offsets, gravitational effect of the Sun, solid Earth tides, high-frequency model for the Earth rotation caused by ocean tides, and ocean loading (by courtesy of Lucia Plank)

2.7 Source Structure

A major problem is that most of the observed radio sources tend to show structure at the level of a few milliarcseconds which often varies with time. These effects, in particular the changes in the source structure, pose a limit on the accuracy of the radio reference frame. For the highest accuracies, a regular monitoring of the structure, which is also accomplished by analyzing VLBI data, can be done in parallel to the geodetic analysis, thus providing a means to correct for the source structure effects (see Collioud and Charlot 2009 and references therein; Schuh 2000).

2.8 A Few Examples of Constituents of the Delay

The various effects that have been described in the previous sections on the calculated time delay τ for simulated observations with stations Westford (U.S.A) and Wettzell (Germany) to source 0642+449 are evident in Fig. 7.7. These effects are ordered by magnitude from top to bottom. The upper plot displays the total delay which is mainly caused by the Earth's rotation (Sect. 2.2) and can be as large as one Earth radius. The next plot depicts the contribution of the hydrostatic delays at both sites to the total delay, i.e., the delay at Wettzell minus the delay at Westford. At each site, the hydrostatic delay at low elevation angles can be larger than 20 m, and it increases rapidly as the elevation angle decreases (Sect. 2.4). The third subplot illustrates the influence of the axis offsets at Wettzell and Westford (Sect. 2.6), of the gravitational effect of the Sun (Sect. 2.3), and of the solid Earth tides (Sect. 2.1). The bottom plot contains the influence of the high-frequency Earth rotation model accounting for the ocean tides (Sect. 2.2), as well as the effect of ocean loading (Sect. 2.1) at the sites on the calculated delay τ.

3 Least-Squares Adjustment in VLBI

As can be seen in the flowchart of Fig. 7.4 the theoretical delays are then compared with the reduced observed delays by a parameter estimation process, e.g., a classical Gauß–Markov model as will be described below, or Kalman filtering (Herring et al. 1990), or collocation (Titov and Schuh 2000)—all three following the least-squares concept—or a square-root information filter (Bierman 1977).

The least-squares adjustment theory allows estimating unknown parameters in an over-determined system of equations. Since there are more equations than unknown parameters, the solution will not be exactly correct for each equation, but the adjustment provides a unique solution dx and minimizes the squared sum of the weighted residuals v. Functional and stochastic models are based on linearized observation equations:

$$A \cdot dx = l + v \quad \text{or} \quad \begin{bmatrix} A_{\mathrm{ro}} \\ A_{\mathrm{po}} \end{bmatrix} \cdot dx = \begin{bmatrix} l_{\mathrm{ro}} \\ l_{\mathrm{po}} \end{bmatrix} + \begin{bmatrix} v_{\mathrm{ro}} \\ v_{\mathrm{po}} \end{bmatrix}, \tag{7.29}$$

$$P = \begin{bmatrix} P_{\mathrm{ro}} & 0 \\ 0 & P_{\mathrm{po}} \end{bmatrix}. \tag{7.30}$$

The design matrix A_{ro} contains the first derivatives of the function of the real observations with respect to the estimated parameters. Short estimation intervals (e.g., 20 min for zenith wet delays) could lead to singularity problems if there was no observation within a time segment. Therefore, the A_{ro} matrix is extended by a pseudo-observation matrix A_{po}, which constrains the value of variability of the parameters, either by constraining the absolute values to zero (typically used for troposphere gradients in early VLBI sessions) or by constraining the relative variation of the continuous piecewise linear functions. The reduced observations, i.e., the observed minus calculated time delays, are listed in the l_{ro} vector (real observations) whereas the l_{po} vector (pseudo-observations) is typically filled with zeros. The weighting of the observations is done by the weight matrix P.

There are different groups of parameters. Auxiliary parameters, e.g., clock parameters and sometimes also the troposphere parameters like zenith wet delays or gradients, have to be computed but are usually not of interest for geodesists. As clock parameters, linear or quadratic polynomials (over the 24 h session accounting for clock offset, clock frequency offset, and clock frequency drift) are estimated. Additionally, continuous piecewise linear functions with, e.g., hourly

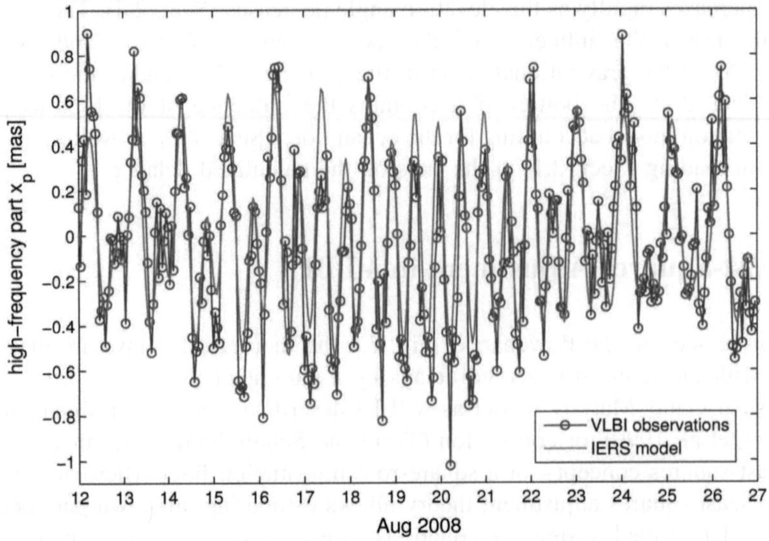

Fig. 7.8 Hourly x-pole values in mas estimated with VieVS for CONT08 in *red*. For comparison high-frequency (UT1–UTC) values as derived from the IERS conventions model (ocean tides and libration; Petit and Luzum 2010) are shown in *gray* (by courtesy of Sigrid Böhm)

segments, are estimated with respect to a reference clock that is set to zero in order to account for rapid clock instabilities. A typical constraint for the relative variation of the piecewise linear clock function is 0.5 ps^2/s; for time segments, e.g., of 60 min, this would correspond to a variance of 1800 ps^2 over that time span of 1 h, meaning that the difference between two adjacent clock offsets is 0 ± 13 mm. Special care has to be paid to the detection of clock breaks that can occur at some station clocks (hydrogen masers, described in Sect. 1.1) during a VLBI session. The epochs of those clock breaks have to be introduced in the analysis of VLBI sessions, because separate, quadratic polynomials are used to describe the behavior of the clock before and after the break.

3.1 The Concept of Piecewise Linear Offsets

In the Vienna VLBI Software (VieVS; Böhm et al. 2011) zenith wet delays are typically estimated every 20–60 min, and rather loose constraints are put on the variation of the zenith wet delays (0.7 ps^2/s). So-called 'piecewise linear offsets' are used in VieVS, i.e., the functional model is based on offsets only (no rates) (see (7.31)). These piecewise linear offsets are estimated at integer hours (e.g., at 18 UTC, 19 UTC,...), at integer fractions of integer hours (e.g., 18:20 UTC, 18:40 UTC,...), or at integer multiples of integer hours (e.g., 18:00 UTC, 0:00 UTC, 6:00 UTC,...). In VieVS, this representation is not only possible for troposphere zenith delays and gradients, station clocks, and Earth orientation parameters, but also for coordinates of selected stations and radio sources. Equation (7.31) denotes the functional model of the wet delay ΔL_w at one station represented by piecewise linear offsets x_1 and x_2 of the zenith wet delays at the integer hours t_1 and t_2. The wet mapping function at epoch t of an observation which is in between the integer hours is expressed by $mf_w(t)$. The partial derivatives which have to be entered in the design matrix are given in (7.32) and (7.33). This concept is similar for all parameters, and with this kind of parameterization all combinations (at the normal equation level) with other space geodetic techniques will be easy and straightforward.

$$\Delta L_w(t) = mf_w(t) \cdot x_1 + mf_w(t) \cdot \frac{t - t_1}{t_2 - t_1} \cdot (x_2 - x_1) \tag{7.31}$$

$$\frac{d\Delta L_w}{dx_1} = mf_w(t) - mf_w(t) \cdot \frac{t - t_1}{t_2 - t_1} \tag{7.32}$$

$$\frac{d\Delta L_w}{dx_2} = mf_w(t) \cdot \frac{t - t_1}{t_2 - t_1} \tag{7.33}$$

In addition to the clock and troposphere parameters (zenith wet delays and gradients) mentioned above, there are many other geodetic/astrometric parameters which can by estimated from VLBI sessions (see Sect. 4). EOP—although typically estimated once per session—can also be retrieved with a higher temporal resolution, applying the concept of the piecewise linear offsets. Figure 7.8 shows hourly estimates of (UT1–UTC) during CONT08.

3.2 Global VLBI Solutions

Other 'global' parameters such as station or source coordinates can in principle also be estimated from single VLBI sessions, but they are preferably determined in a global solution, i.e., from a large number of VLBI sessions connected to a common least-squares parameter estimation. Due to limited computer memory capacity it is essential to keep the equation system small. In VLBI analysis there are auxiliary parameters in the observation equations which cannot be fixed to a priori values, even if we are not interested in them, e.g., clock parameters. Therefore, a reduction algorithm is used which is based on a separation of the normal equation system into two parts. The first part contains parameters which we want to estimate and the second part contains parameters which will be reduced. Even if we 'reduce' parameters, they still belong to the functional model of unknown parameters and will be estimated implicitly.

$$\begin{bmatrix} N_{11} & N_{12} \\ N_{21} & N_{22} \end{bmatrix} \cdot \begin{bmatrix} dx_1 \\ dx_2 \end{bmatrix} = \begin{bmatrix} b_1 \\ b_2 \end{bmatrix}. \tag{7.34}$$

In (7.34) $N = A^T PA$ and $b = A^T Pl$, and the reduction of dx_2 is done by executing the matrix operation

$$\left(N_{11} - N_{12}N_{22}^{-1}N_{21}\right) \cdot dx_1 = b_1 - N_{12}N_{22}^{-1}b_2 \quad \text{or} \quad N_{\text{reduc}} \cdot dx_1 = b_{\text{reduc}}. \tag{7.35}$$

Stacking is used for combining normal equation systems if a parameter is contained in at least two normal equation systems and only one common value in the resulting combined system should be estimated. For a combined solution of the identical parameters (dx_1), the normal matrices (N_{reduc}) and the right hand side vectors (b_{reduc}) from n single sessions have to be summed up:

$$N_{\text{REDUC}} = N_{\text{reduc\_1}} + N_{\text{reduc\_2}} + \ldots + N_{\text{reduc\_}n}, \tag{7.36}$$

$$b_{\text{REDUC}} = b_{\text{reduc\_1}} + b_{\text{reduc\_2}} + \ldots + b_{\text{reduc\_}n}. \tag{7.37}$$

Conditions on the N_{REDUC} matrix are applied in order to prevent the matrix from being singular. From the analysis of VLBI sessions we get free station networks, which are the result of adjusting observations in a model where coordinates are unknowns without fixing the coordinate system (Sillard and Boucher 2001). With three-dimensional VLBI station networks the rank deficiency is six (the scale is determined from the observations), which means that at least six conditions have to be applied to remove the rank deficiency. In case of station coordinates three NNT and three NNR conditions are applied on selected datum stations, and in the case of source coordinates an NNR condition is usually applied on a selected set of datum sources. In the case of longer time spans NNR-rate and NNT-rate conditions are also applied on station coordinate velocities. It is very important to use stable stations and sources for the datum, because otherwise the quality of the terrestrial and celestial reference would be deteriorated. Moreover, it is absolutely necessary to take into account any episodic changes in the station coordinates, e.g., due to instrumental changes or earthquakes.

Unlike positions and velocities, no scale or scale rate parameters are estimated in VLBI, as the scale directly depends on the speed of light, c, one of the defining natural constants. The final solution is obtained by an inversion of the normal matrix:

$$dx_1 = N_{\text{REDUC}}^{-1} \cdot b_{\text{REDUC}}. \tag{7.38}$$

Since the least-squares adjustment minimizes the squared sum of weighted residuals, this value is used to scale the standard deviations of the estimates. It is determined with

$$v^T Pv = \left(l^T Pl\right)_{\text{REDUC}} - x_1^T b_{\text{REDUC}}, \tag{7.39}$$

where the first part $(l^T Pl)_{\text{REDUC}}$ depends only on observations; it has to be corrected for the influence of the reduced parameters which is known from the single normal equation systems:

$$(l^T Pl)_{\text{REDUC}} = \sum_{i=1}^{n} (l^T Pl - b_2^T N_{22}^{-1} b_2). \tag{7.40}$$

The second part in (7.39), $x_1^T b_{\text{REDUC}}$, depends on the combined solution. The a posteriori variance of unit weight σ_0^2 is a scaling factor for the inverse normal equation matrix, i.e., for the covariance matrix Q of the estimated parameters:

$$Q = \sigma_0^2 \cdot N^{-1}. \tag{7.41}$$

It is determined with

$$\sigma_0^2 = \frac{v^T P v}{k - u + d}, \tag{7.42}$$

where k is the number of observations, u the number of estimated and reduced parameters, and d the number of additional condition equations.

4 Results from Geodetic VLBI and the IVS

The VLBI technique has been employed for more than 30 years in geodesy, geophysics, and astronomy, and results of geodetic VLBI have been presented and interpreted in a multitude of publications by hundreds of authors. During the first two decades, most of the scientific and operational activities were organized through national or bi-lateral agreements only, which was not a basis sufficiently strong for carrying out VLBI sessions in global networks.

In 1999 the IVS was established to coordinate the global VLBI components and resources on an international basis. All international collaboration, in accordance with the IVS terms of reference, is based on a standing call for participation that was first issued in 1998. Any institution that is prepared to participate in IVS activities may join at any time after getting accepted by the IVS Directing Board (Schlüter and Behrend 2007). The inauguration of the IVS took place in March 1999, and the first meeting of the Directing Board was held at the Fundamental Station Wettzell, Germany. The IVS was approved as a service of the IAU, of the IAG, and of the Federation of Astronomical and Geophysical data analysis Services (FAGS); the latter was dissolved in 2010 and replaced in 2011 by the World Data System (WDS).

According to its terms of reference, the IVS is an international collaboration of organizations that operate or support VLBI components for geodetic and astrometric applications. Specific goals are to provide a service to support geodetic, geophysical, and astrometric research and operational activities, to promote research and development activities in all aspects of the geodetic and astrometric VLBI technique, and to interact with the community of users of VLBI products and to integrate VLBI into a global Earth observing system. Since 2003 the GGOS (GGOS;

Table 7.1 Summary of IVS main products (modified from Schlüter and Behrend 2007)

Products	Specification	Status 2010
Polar motion x_p, y_p	Accuracy	50–80 μas
	Product delivery	8–10 days
	Resolution	1 day
	Frequency of solution	~3 days/week
UT1–UTC	Accuracy	3–5 μs
	Product delivery	8–10 day
	Resolution	1 day
	Frequency of solution	~3 days/week
UT1–UTC (Intensives)	Accuracy	15–20 μas
	Product delivery	1 day
	Resolution	1 day
	Frequency of solution	7 days/week
Celestial pole dX, dY	Accuracy	50 μas
	Product delivery	8–10 days
	Resolution	1 day
	Frequency of solution	~3 days/week
TRF (x, y, z)	Accuracy	5 mm
CRF (α, δ)	Accuracy	40–250 μas
	Frequency of solution	1 year
	Product delivery	3 months

Plag and Pearlman 2009) has been developed as a main component of the IAG, and the IVS provides an essential contribution to it (Schlüter and Behrend 2007).

'Official IVS products' are the realization of the Celestial Reference Frame (CRF) through the positions of extragalactic radio sources, the maintenance of the Terrestrial Reference Frame (TRF), such as station positions and their changes with time, and the generation of series describing the Earth orientation (see Table 7.1).

Geodetic VLBI is the only space geodetic technique that allows the observation of the full set of EOP, and it is unique in providing UT1 (see Fig. 7.8) as well as celestial pole offsets over longer time spans. Figure 7.9 depicts polar motion estimates as determined from VLBI observations since 1984 with the Vienna VLBI software VieVS (Böhm et al. 2011). As mentioned above, the IVS plays a key role within GGOS and thus all IVS products are also considered GGOS products, today.

Moreover, VLBI is the only technique for the determination of the ICRF. The ICRF (ICRF1; Ma et al. 1998), defined by positions of 212 compact radio sources (out of a total of 608 radio sources), was the first realization at radio frequencies. Since its approval in 1997 by the IAU, the IVS has been in charge of the VLBI realization. At the XXVII General Assembly in 2009 the IAU adopted the ICRF2 including the positions of 3,414 compact radio astronomical sources. This is more than five times the number of sources in the ICRF1 (or its later extension, the ICRF1-Ext.2). The noise floor of the ICRF2 is at the level of 40 μas and the axis stability at the level of 10 μas (Fey et al. 2009). The ICRF2 has 295 defining

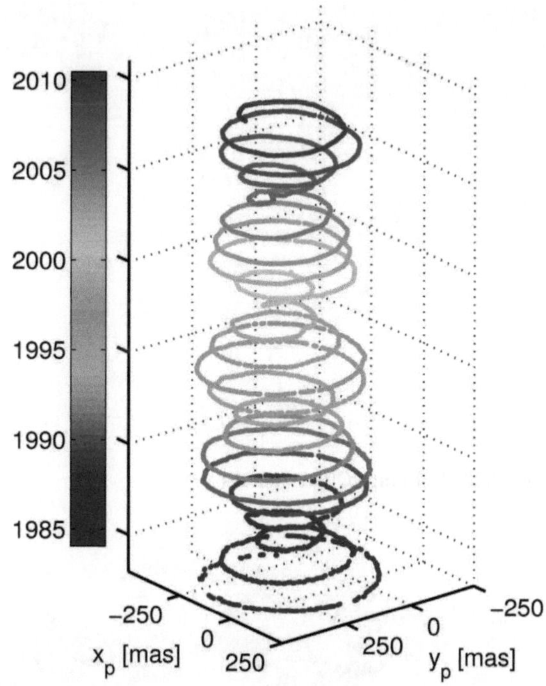

Fig. 7.9 Polar motion estimates in milliarcseconds as determined from VLBI observations since 1984. Clearly visible is the modulation between the Chandler wobble period of about 1.18 years and the annual variation (by courtesy of Sigrid Böhm)

sources with an equal distribution, in particular in the Southern celestial hemisphere, and smaller source structure effects, both weaknesses in the ICRF1 (Fey et al. 2009). (Fig. 7.10)

Geodetic VLBI also contributes to the realization of the ITRF by measuring long intercontinental baselines within global networks. Compared to those space geodetic techniques using satellites, VLBI has the principal advantage that its realization of the ITRF scale only depends on the speed of light c, which is used to transform the delay observables into metric units. There exists no evidence at all that during the last three decades a bias or rate of this conversion has occurred due to technical reasons. Figure 7.11 illustrates the horizontal velocities of the VLBI stations included in the VTRF2008 (Böckmann et al. 2010), the VLBI contribution to the ITRF2008 (Altamimi et al. 2011).

Table 7.1 provides a summary of current IVS main products (Schlüter and Behrend 2007). Observations of geodetic VLBI have been carried out for more than three decades providing a basis for the precise determination of geodynamic and astronomical parameters including their long-term variations. For example, VLBI can determine Love numbers h and l of the solid Earth tides model (Spicakova et al. 2010), ionosphere models (Hobiger 2006), or troposphere parameters. The long-term VLBI zenith wet delays are of interest for climatologists because they contain information about the precipitable water above the stations for their complete history (Heinkelmann 2008); they can also be used to

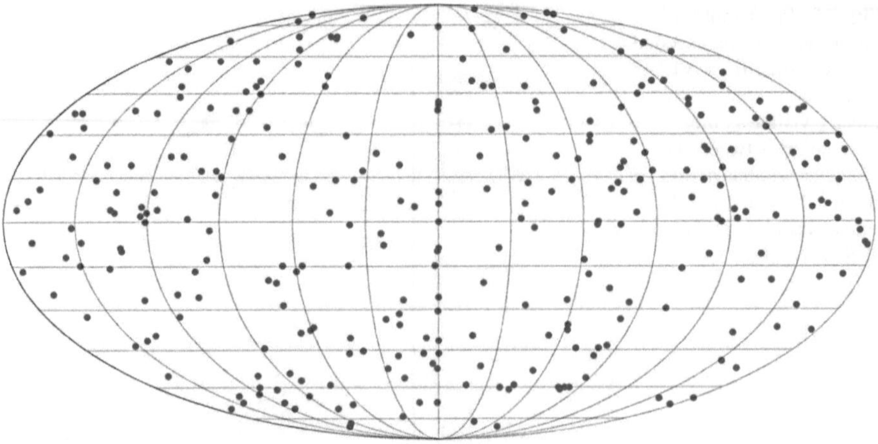

Fig. 7.10 Defining ICRF2 sources

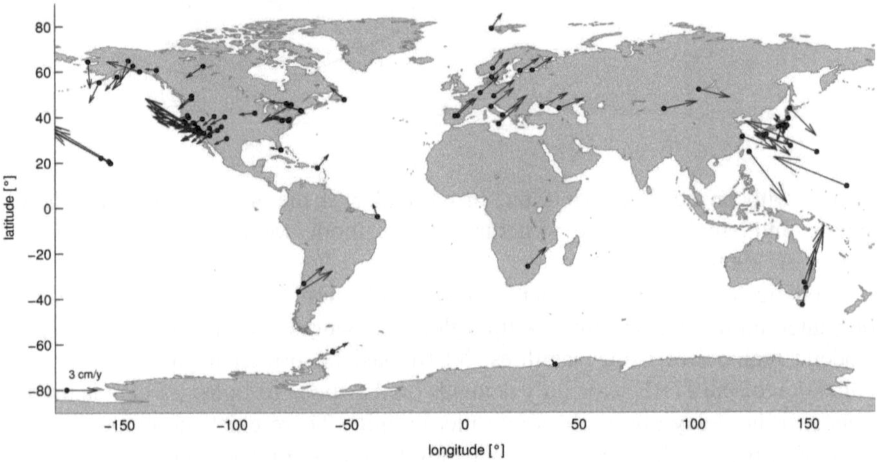

Fig. 7.11 VTRF2008 station distribution with horizontal velocities

validate troposphere parameters from other space geodetic techniques (Snajdrova et al. 2005; Teke et al. 2011).

Another interesting phenomenon, which can be observed by VLBI, is the gravitational deflection of radio waves by the solar gravity field according to general relativity. As described in Sect. 2.3, radio waves are subject to space–time curvature caused by any massive body (in our solar system mainly that by the Sun has to be considered but also that by Jupiter for close approaches). At the elongation angle of 2.5° to the Sun, which was the minimal angle of VLBI observations till 2002, the differential deflection reaches 150 mas (Robertson et al. 1991) causing a significant effect on the observed group delays. With respect to the noise floor of source coordinates, which is about 40 μas for the ICRF2 (Fey et al. 2009), analysis of source observations in the vicinity of the Sun allows the determination of the post-Newtonian parameter γ ('light deflection parameter') characterizing the space curvature due to gravity (see (7.15)). Although since 2002 the VLBI observations have been scheduled for a minimal angle of 15° to the Sun, the gravitational deflection still influences the measurements significantly and the most recent VLBI global solutions provided γ with a precision of 1×10^{-4} (Lambert and Le Poncin-Lafitte 2009, 2011). The series of VLBI data is also sensitive to a possible acceleration of the solar system barycenter which might cause a secular drift of aberration with a magnitude of 4 μas/year (Sovers et al. 1998; Titov 2010). Furthermore, the solar system motion relative to the cosmic microwave background might produce a dipole pattern that decreases with red shift (Titov 2010).

5 The Next Generation VLBI System, VLBI2010

In September 2005 the IVS Directing Board accepted the final report of its Working Group 3 (WG3) entitled "VLBI2010: Current and Future Requirements for Geodetic VLBI Systems" (Niell et al. 2006) which recommended a review of all current VLBI systems and processes from antennas to analysis and outlined a path to a next-generation system with unprecedented new capabilities: 1 mm position and 0.1 mm/year velocity accuracy on global scales, continuous measurements for time series of station positions and Earth orientation parameters, and a turnaround time to initial geodetic results of less than 24 h.

As a consequence, the IVS established the VLBI2010 Committee (V2C) to carry out a series of studies and to encourage the realization of the new vision for geodetic VLBI. Making rational design decisions for VLBI2010 requires an understanding of the impact of new strategies on the quality of VLBI products. For this purpose, Monte Carlo simulators were developed which have been used to study the effects of the dominant VLBI random error processes (related to the atmosphere, the reference clocks, and the delay measurement noise; Pany et al. 2010) and new approaches to reduce them, such as decreasing the source-switching interval and improving analysis and scheduling strategies. Shortening the source-switching interval results in a higher number of observables leading to a significant improvement in station position accuracy (Petrachenko et al. 2009). In any case, the simulators confirm that the dominant error source is the troposphere. It is recommended that research on analysis strategies for the atmosphere continues to be a priority for the IVS. Based on the findings from the Monte Carlo studies, the source-switching interval should be reduced. This includes decreasing both the on-source time needed to make a precise delay measurement and the time required to slew between sources. From these two somewhat competing goals, recommendations for the VLBI2010 antennas are emerging, e.g., either a single ~12 m diameter antenna or larger with very high slew rates, e.g., 12°/s in azimuth, or a pair of ~12 m diameter antennas (or larger), each with more moderate slew rates, e.g., 5°/s in azimuth (Petrachenko et al. 2009).

In order to reduce the on-source observing time, it is important to find a means for measuring the delay with the requisite precision even at a modest signal-to-noise ratio. To do this a new approach is being developed in which several widely spaced frequency bands are used to resolve unambiguously the interferometric phase. The new observable is being referred to as the broadband delay. A four-band system is recommended that uses a broadband feed to span the entire frequency range from 2 to 14 GHz (Petrachenko et al. 2009). A total instantaneous data rate as high as 32 Gbps and a sustained data storage or transmission rate as high as 8 Gbps are necessary to detect an adequate number of high-quality radio sources (Petrachenko et al. 2009). NASA is funding a proof-of-concept effort till 2012 to test the broadband delay technique, and first fringes have already been detected in all bands. In addition to random errors, systematic errors need to be reduced too. For example, updated calibration systems are being developed to account for electronic biases. Conventional surveying techniques have to be refined to observe antenna deformations, and the application of small reference antennas is considered for generating deformation models and establishing site ties. Furthermore, corrections based on images derived directly from the VLBI2010 observations are under study to mitigate errors due to source structure (Petrachenko et al. 2009).

The progress report of the IVS V2C (Petrachenko et al. 2009) recommends that a globally distributed network of at least 16 VLBI2010 antennas observes every day to determine the Earth orientation parameters, and that other antennas be added as needed for the maintenance of the celestial and terrestrial reference frames. Antennas with access to high-speed fiber networks are also required to enable daily delivery of initial IVS products in less than 24 h. A high priority is placed on increasing the number of stations in the Southern hemisphere. Since IVS products must be delivered without interruption, a transition period to VLBI2010 operations is required in which there will be a mix of antennas with current and next-generation receiving systems. For this period a compatibility mode of operation has been identified and tested to a limited extent with the NASA proof-of-concept system. In order to increase reliability and to reduce the cost of operations, enhanced automation will be introduced both at the stations and in the analysis process. Stations will be monitored centrally to ensure compatible operating modes, to update schedules as required, and to notify station staff when problems occur. Automation of the analysis process will benefit from the work of the current IVS Working Group 4, which is updating data structures and modernizing data delivery (Petrachenko et al. 2009).

For more details the authors refer to various reports, memos, and other documents describing the concept and realizations of VLBI2010. Many of those are accessible via the webpage of the IVS at http://ivscc.gsfc.nasa.gov/.

6 Concluding Remarks

VLBI plays a unique and fundamental role in the maintenance of the global (terrestrial and celestial) reference frames and in monitoring the Earth orientation parameters, which are required for precise positioning and navigation on Earth and in space. Furthermore, very valuable information on various time scales can be obtained about several other parameters needed for the investigation of phenomena such as meteorological and climatologic changes, and geodynamical or astronomical effects. Thus, geodetic VLBI is essential for the Global Geodetic Observing System (GGOS), the flagship component of the IAG. The IVS has served this task very successfully in the past, and with the upcoming VLBI2010 concept, it is advancing to a bright and challenging future.

Acknowledgments We are grateful to the IVS with all its components and contributing agencies for acquiring and providing the data used in the examples given in this chapter and for the very fruitful cooperation in the past decades. The authors would like to acknowledge all individuals mentioned in this chapter and many other VLBI experts who developed the VLBI technology, correlator, and data analysis since the mid-1960s. Their continuous efforts have made it such an exciting space geodetic technique, providing scientific results of the highest quality but also producing regular measurements with highest impact on society and our Earth in general. Also, we would like to thank all members of the VLBI group at the Vienna University of Technology who contributed with their efforts to this summary; their continuous enthusiasm is advancing VLBI. We are also grateful to Brian Corey (MIT Haystack Observatory), Axel Nothnagel (University Bonn), and Ludwig Combrinck (Hartebeesthoek Radio Astronomy Observatory) for their comments on the manuscript.

References

Abbondanza C, Sarti P (2010) Effects of illumination functions on the computation of gravity-dependent signal path variation models in primary focus and cassegrainian VLBI telescopes. J Geodesy 84(8):515–525. doi:10.1007/s00190-010-0389-z

Altamimi Z, Collilieux X, Métivier L (2011) ITRF2008: an improved solution of the international terrestrial reference frame. J Geodesy 85(8):457–473. doi:10.1007/s00190-011-0444-4

Bare CC, Clark BG, Kellerman KI, Cohen MH, Jauncey DL (1967) Interferometry experiment with independent local oscillators. Science 157:189

Bierman G (1977) Factorization methods for discrete sequential estimation. Academic, New York

Bizouard C, Gambis D (2009) The combined solution C04 for earth orientation parameters consistent with international terrestrial reference frame. In: Drewes H (ed) Geodetic Reference Frames, vol 134. IAG Symposium, Munich, Germany, pp 265–270, 9–14 Oct 2006

Böckmann S, Artz T, Nothnagel A (2010) VLBI terrestrial reference frame contributions to ITRF2008. J Geodesy 84(3):201–219

Böhm J (2004) Troposphärische Laufzeitverzögerungen in der VLBI. Geowissenschaftliche Mitteilungen, Heft Nr. 68, Schriftenreihe der Studienrichtung Vermessung und Geoinformation, Technische Universität Wien, ISSN 1811–8380 (in German)

Böhm J, Schuh H (2007) Troposphere gradients from the ECMWF in VLBI analysis. J Geodesy 81(6–8):403–408. doi:10.1007/s00190-007-0144-2

Böhm J, Werl B, Schuh H (2006) Troposphere mapping functions for GPS and very long baseline interferometry from European centre for medium-range weather forecasts operational analysis data. J Geophys Res 111:B02406. doi:10.1029/2005JB003629

Böhm J, Heinkelmann R, Schuh H (2007) Short note: a global model of pressure and temperature for geodetic applications. J Geodesy 81(10):679–683. doi:10.1007/s00190-007-0135-3

Böhm J, Heinkelmann R, Mendes Cerveira PJ, Pany A, Schuh H (2009a) Atmospheric loading corrections at the observation level in VLBI analysis. J Geodesy 83(11):1107–1113. doi:10.1007/s00190-009-0329-y

Böhm J, Kouba J, Schuh H (2009b) Forecast Vienna mapping functions 1 for real-time analysis of space geodetic observations. J Geodesy 86(5):397–401

Böhm J, Hobiger T, Ichikawa R, Kondo T, Koyama Y, Pany A, Schuh H, Teke K (2010) Asymmetric tropospheric delays from numerical weather models for UT1 determination from VLBI Intensive sessions on the baseline Wettzell-Tsukuba. J Geodesy 84(5):319–325. doi:10.1007/s00190-10-0370-x

Böhm J, Böhm S, Nilsson T, Pany A, Plank L, Spicakova H, Teke K, Schuh (2011) The new Vienna VLBI Software VieVS. In: Kenyon S, Pacino MC, Marti U (eds) Proceedings of the 2009 IAG Symposium, Buenos Aires, Argentina, 31 Aug–4 Sept 2009, International Association of Geodesy symposia, vol 136

Broten NW, Legg TH, Locke JL, McLeish CW, Richards RS, Chisholm RM, Gush RM, Yen JL, Galt JA (1967) Long baseline interferometry: a new technique. Science 156:1592

Brown GW, Carr TD, Block WF (1968) Long baseline interferometry of s-bursts from Jupiter. Astrophys Lett 1:89–94

Campbell J (2000) From quasars to benchmarks: VLBI links heaven and earth. In: Vandenberg NR, Baver KD (eds) IVS 2000 General Meeting proceedings, NASA/CP-2000-209893, pp 19–34

Capitaine N (2000) Definition of the celestial ephemeris pole and the celestial ephemeris origin. In: Johnston KJ, McCarthy DD, Luzum BJ, Kaplan GH (eds) Towards models and constants for sub-microarcsecond astrometry, U.S. Naval Observatory, pp 153–163

Chen G, Herring TA (1997) Effects of atmospheric azimuthal asymmetry on the analysis of space geodetic data. J Geophys Res 102(B9):20489–20502

Cohen MH, Jauncey DL, Kellerman KI, Clark BG (1968) Radio interferometry at one-thousandth second of arc. Science 162:88–94

Collioud A, Charlot P (2009) The Bordeaux VLBI image database. In: Bourda G, Charlot P, Collioud A (eds) Proceedings of the 19th European VLBI for Geodesy and Astrometry Working Meeting, Bordeaux, pp 19–22, 24–25 Mar 2009

Davis JL, Herring TA, Shapiro II, Rogers AEE, Elgered G (1985) Geodesy by radio interferometry: effects of atmospheric modeling errors on estimates of baseline length. Radio Sci 20(6):1593–1607

Davis JL, Elgered G, Niell AE, Kuehn CE (1993) Ground-based measurement of gradients in the "wet" radio refractivity of air. Radio Sci 28(6):1003–1018

Dehant V, Mathews PM (2009) Earth rotation variations. In: Herring TA (ed) Treatise on Geophysics, vol 3, Elsevier, Oxford

Englich S, Heinkelmann R, Schuh H (2008) Re-assessment of ocean tidal terms in high-frequency earth rotation variations observed by VLBI. In: Finkelstein A, Behrend D (eds) Measuring the future. Proceedings of the 5th IVS General Meeting, ISBN 978-5-02-025332-2, pp 314–318

Eubanks TM (ed) (1991) Proceedings of the U.S. naval observatory workshop on relativistic models for use in space geodesy. U.S. Naval Observatory, Washington D.C.

Fey A, Gordon D, Jacobs CS (eds) (2009) The second realization of the international celestial reference frame by very long baseline interferometry, Presented on behalf of the IERS/IVS Working Group. IERS Technical Note 35, Verlag des Bundesamts für Kartographie und Geodäsie, Frankfurt am Main

Gipson J, MacMillan DS (2009) Recent modeling improvements in solve analysis. In: Bourda G, Charlot P, Collioud A (eds) Proceedings of the 19th European VLBI for Geodesy and Astrometry Working Meeting, Bordeaux, pp 54–57, 24–25 Mar 2009

Gold T (1967) Radio method for the precise measurement of the rotation period of the earth. Science 157:302–304

Haas R et al (1999) Explanatory supplement to the section "antenna deformation'' of the IERS conventions (1996). In: Schuh H (ed) DGFI report no. 71, pp 26–29

Hawarey M, Hobiger T, Schuh H (2005) Effects on the 2nd order ionospheric terms on VLBI measurements. Geophys Res Lett 32:L11304. doi:10.1029/2005GL022729

Heinkelmann R (2008) Bestimmung des atmosphärischen Wasserdampfes mittels VLBI als Beitrag zur Klimaforschung. Heft Nr. 82, Schriftenreihe der Studienrichtung Vermessung und Geoinformation, Technische Universität Wien, ISSN 1811–8380 (in German)

Heinkelmann R, Schuh H (2010) Very long baseline interferometry (VLBI): accuracy limits and relativistic tests. In: Klioner S, Seidelmann PK, Soffel M (eds) Proceedings of the IAU Symposium, No. 261, pp 286–290

Hellings RW (1986) Relativistic effects in astronomical timing measurements. Astron J 91:650–659, Erratum, ibid. p 1446

Herring TA (1992a) Submillimeter horizontal position determination using very long baseline interferometry. J Geophys Res 97:1981–1990

Herring TA (1992b) Modeling atmospheric delays in the analysis of space geodetic data. In: DeMunck, Spoelstra (eds) Refraction of transatmospheric signals in geodesy, Netherlands Geodetic Commission, Publications on Geodesy, No. 36, pp 157–164

Herring TA, Davis JL, Shapiro II (1990) Geodesy by radio interferometry: the application of Kalman filtering to the analysis of very long baseline interferometry data. J Geophys Res 95(B8):12561–12581

Herring TA, Mathews PM, Buffett BA (2002) Modeling of nutation–precession: very long baseline interferometry results. J Geophys Res 107(B4):2069

Herring TA, Shapiro II, Clark TA, Ma C, Ryan JW, Schupler BR, Knight CA, Lundquist G, Shaffer DB, Vandenberg NR, Corey BE, Hinteregger HF, Rogers AEE, Webber JC, Whitney AR, Elgered G, Rönnäng BO, Davis JL (1986) Geodesy by radio interferometry: evidence for contemporary plate motion. J Geophys Res 91:8341–8347

Hinteregger HF, Shapiro II, Robertson DS, Knight CA, Ergas RA, Whitney AR, Rogers AEE, Moran JM, Clark TA, Burke BF (1972) Precision geodesy via radio interferometry. Science 178:396–398

Hobiger T (2006) VLBI as a tool to probe the ionosphere. Heft Nr. 75, Schriftenreihe der Studienrichtung Vermessung und Geoinformation, Technische Universität Wien, ISSN 1811–8380

Hobiger T, Ichikawa R, Koyama Y, Kondo T (2008) Fast and accurate ray-tracing algorithms for real-time space geodetic applications using numerical weather models. J Geophys Res 113, D20302. doi:10.1029/2008JD010503.

Kellermann KI, Moran JM (2001) The development of high-resolution imaging in radio astronomy. Ann Rev Astron Astrophys 39:457–509

Klioner SA (1991) General relativistic model of VLBI observables. In: Carter WE (ed) Proceedings of AGU Chapman Conference on geodetic VLBI: monitoring global change, NOAA technical report NOS 137 NGS 49, American Geophysical Union, Washington D.C., pp 188–202

Kondo T, Kimura M, Koyama Y, Osaki H (2004) Current status of software correlators developed at kashima space research center. In: Vandenberg NR, Baver KD (eds) International VLBI Service for Geodesy and Astrometry 2004 General Meeting proceedings, NASA/CP-2004-212255, pp 186–190

Lambert SB, Le Poncin-Lafitte C (2009) Determining the relativistic parameter γ using very long baseline interferometry. Astron Astrophys 499:331–335. doi:10.1051/0004-6361/200911714

Lambert SB, Le Poncin-Lafitte C (2011) Improved determination of γ by VLBI. Astron Astrophys 529:A70. doi:10.1051/0004-6361/201016370

Ma C, Arias EF, Eubanks TM, Fey AL, Gontier A-M, Jacobs CS, Sovers OJ, Archinal BA, Charlot P (1998) The international celestial reference frame as realized by very long baseline interferometry. Astron J 116:516–546

MacMillan DS (1995) Atmospheric gradients from very long baseline interferometry observations. Geophys Res Lett 22(9):1041–1044

MacMillan DS, Ma C (1994) Evaluation of very long baseline interferometry atmospheric modeling improvements. J Geophys Res 99(B1):637–651

Mathews PM, Dehant V, Gipson JM (1997) Tidal station displacement. J Geophys Res 102(B9):20469–20477

Moran JM, Crowther PP, Burke BF, Barrett AH, Rogers AEE, Ball JA, Carter JC, Bare CC (1967) Spectral line interferometry with independent time standards at stations separated by 845 kilometers. Science 157:676–677

Niell AE, Whitney A, Petrachenko B, Schlüter W, Vandenberg N, Hase H, Koyama Y, Ma C, Schuh H, Tuccari G (2006) VLBI2010: current and future requirements for geodetic VLBI systems. IVS memorandum 2006-008v01, ftp://ivscc.gsfc.nasa.gov/pub/memos/ivs-2006-008v01.pdf

Nothnagel A (2009) Conventions on thermal expansion modelling of radio telescopes for geodetic and astrometric VLBI. J Geodesy 83(8):787–792

Nothnagel A, Vennebusch M, Campbell J (2002) On correlations between parameters in geodetic VLBI data analysis. In: Vandenberg NR, Baver KD (eds) IVS 2002 general meeting proceedings, pp 260–264

Pany A, Böhm J, MacMillan DS, Schuh H, Nilsson T, Wresnik J (2010) Monte Carlo simulations of the impact of troposphere, clock and measurement errors on the repeatability of VLBI positions. J Geodesy 85(1):39–50. doi:10.1007/s00190-010-0415-1

Petit G (2000) Importance of a common framework for the realization of space-time reference systems. In: Rummel R, Drewes H, Bosch W, Hornik H (eds) IAG Symposium, Munich, Oct. 1998, International Association of Geodesy Symposia, vol 120, pp 3–7

Petit G, Luzum B (eds) (2010) IERS Conventions 2010. IERS Technical Note no. 36. Verlag des Bundesamts für Kartographie und Geodäsie, Frankfurt am Main

Petrachenko W et al (2009) Progress report of the IVS VLBI2010 committee. Design aspects of the VLBI2010 system. NASA/TM-2009-214180

Petrov L (1999) Steps towards phase delay VLBI. In: Schlüter W, Hase H (eds) Proceedings of the 13th Working Meeting on European VLBI for Geodesy and Astrometry, Viechtach/Wettzell, pp 144–151, 12–13 Feb 1999

Petrov L, Boy JP (2004) Study of the atmospheric pressure loading signal in very long baseline interferometry observations. J Geophys Res 109:B03405

Petrov L, Gordon D, Gipson J, MacMillan D, Ma C, Fomalont E, Walker RC, Carabajal C (2009) Precise geodesy with the very long baseline array. J Geodesy 83(9):859–876

Plag H-P, Pearlman M (eds) (2009) Global Geodetic Observing System: meeting the requirements of a global society on a changing planet in 2020. Springer, Berlin

Rabbel W, Schuh H (1986) The influence of atmospheric loading on VLBI-experiments. J Geophys 59(3):164–170

Richter GW, Matzner RA (1983) Second-order contributions to relativistic time delay in the parameterized post-Newtonian formalism. Phys Rev D 28:3007–3012

Rius A, Rodriguez J, Campbell J (1987) Geodetic VLBI with large antennas. In: Campbell J, Schuh H (eds) Mitt. Geod. Inst. Univ. Bonn, No. 72, pp 59–67

Robertson DS, Carter WE, Dillinger WH (1991) New measurement of solar gravitational deflection of radio signals using VLBI. Nature 349:768–770

Rogers AEE (1970) Very long baseline interferometry with large effective bandwidth for phase delay measurements. Radio Sci 5:1239–1247

Rogers AEE (1975) A receiver phase and group delay calibrator for use in very long baseline interferometry. Haystack Observatory Technical Note, Haystack Observatory, Westford

Saastamoinen J (1972) Atmospheric correction for the troposphere and stratosphere in radio ranging of satellites. The use of artificial satellites for geodesy. In: Geophysical Monograph Series 15, American Geophysical Union, pp 274–251

Sarti P, Abbondanza C, Petrov L, Negusini M (2011) Height bias and scale effect induced by antenna gravitational deformations in geodetic VLBI data analysis. J Geodesy 85(1):1–8. doi:10.1007/s00190-010-0410-6

Scherneck H-G (1991) A parameterized solid earth tide model and ocean tide loading effects for global geodetic baseline measurements. Geophys J Int 106:677–694

Schlüter W, Behrend D (2007) The international VLBI service for geodesy and astrometry (IVS): current capabilities and future prospects. J Geodesy 81(6–8):379–387

Schuh H (1987) Die Radiointerferometrie auf langen Basen zur Bestimmung von Punktverschiebungen und Erdrotationsparametern, DGK Reihe C, Nr. 328, Verlag der Bayerischen Akademie der Wissenschaften, München

Schuh H (2000) Geodetic analysis overview. In: Vandenberg NR, Baver KD (eds) IVS 2000 General Meeting proceedings, NASA/CP-2000-209893, pp 219–229

Sekido M, Takiguchi H, Koyama Y, Kondo T, Haas R, Wagner J, Ritakari J, Kurihara S, Kokado K (2008) Ultra-rapid UT1 measurement by e-VLBI. Earth Planets Space 60(8):865–870

Shapiro II, Knight CA (1970) Geophysical applications of long baseline radio interferometry. In: Mansinha L, Smylie DE, Beck AE (eds) Earthquake displacement fields and the rotation of the earth. Reidel, Dordrecht, pp 285–301

Sillard P, Boucher C (2001) A review of algebraic constraints in terrestrial reference frame definition. J Geodesy 75(2–3):63–73

Snajdrova K, Böhm J, Willis P, Haas R, Schuh H (2005) Multi-technique comparison of tropospheric zenith delays derived during the CONT02 campaign. J Geodesy 79(10–11): 613–623. doi:10.1007/s00190-005-0010-z

Sovers OJ, Fanselow JL, Jacobs CS (1998) Astrometry and geodesy with radio interferometry: experiments, models, results. Rev Mod Phys 70(4):1393–1454

Spicakova H, Böhm J, Böhm S, Nilsson T, Pany A, Plank L, Teke K, Schuh H (2010) Estimation of geodetic and geodynamical parameters with VieVS. In: Behrend D, Baver K (eds) IVS 2010 General Meeting proceedings, NASA/TP-2010-215864, pp 202–206

Steigenberger P, Böhm J, Tesmer V (2009) Comparison of GMF/GPT with VMF1/ECMWF and implications for atmospheric loading. J Geodesy 83(10):943–951. doi:10.1007/s00190-009-0311-8

Takahashi F, Kondo T, Takahashi Y, Koyama Y (2000) Very long baseline interferometer. Wave Summit Course, Ohmsha, IOS Press, Amsterdam

Teke K, Heinkelmann R, Böhm J, Schuh H (2008) VLBI baseline length repeatability tests of IVS-R1 and -R4 session types. In: Finkelstein A, Behrend D (eds) Measuring the future, Proceedings of the 5th IVS General Meeting, ISBN 978-5-02-025332-2, pp 173–177

Teke K, Böhm J, Nilsson T, Schuh H, Steigenberger P, Dach R, Heinkelmann R, Willis P, Haas R, García- Espada S, Hobiger T, Ichikawa R, Shimizu S (2011) Multi-technique comparison of troposphere zenith delays and gradients during CONT08. J Geodesy 85(7):395–413. doi:10.1007/s00190-010-0434-y

Thompson AR, Moran JM, Swenson GW Jr (1986) Interferometry and synthesis in radio astronomy. Wiley, New York

Tingay S, Alef W, Graham D, Deller AT (2009) Geodetic VLBI correlation in software. J Geodesy 83(11):1061–1069

Titov O (2010) VLBI2020: from reality to vision. In: Behrend D, Baver KD (eds) IVS 2010 general meeting proceedings, NASA/CP-2010-215864, pp 60–64

Titov O, Schuh H (2000) Short periods in earth rotation seen in VLBI data analysed by least-squares collocation technique. In: Kolaczek B, Schuh H, Gambis D (eds) IERS Technical Note 28, Paris Observatory, pp 11–14

vanDam TM, Herring TA (1994) Detection of atmospheric pressure loading using very long baseline interferometry measurements. J Geophys Res 99:4505–4517

Vandenberg NR (1999) Interactive/automatic scheduling program. Program reference manual, NASA Goddard Space Flight Center/NVI Inc.

Whitney AR, Rogers AEE, Hinteregger HF, Knight CA, Lippincott S, Levine JI, Clark TA, Shapiro II, Robertson DS (1976) A very-long-baseline interferometer system for geodetic applications. Radio Sci 11(5):421–432

Wresnik J, Haas R, Böhm J, Schuh H (2007) Modeling thermal deformation of VLBI antennas with a new temperature model. J Geodesy 81(6–8):423–431

Index

G. Xu (ed.), *Sciences of Geodesy - II*, DOI: 10.1007/978-3-642-28000-9,

D

E

Q

R

MIX
Papier aus verantwortungsvollen Quellen
Paper from responsible sources
FSC® C105338

If you have any concerns about our products,
you can contact us on
ProductSafety@springernature.com

In case Publisher is established outside the EU,
the EU authorized representative is:
Springer Nature Customer Service Center GmbH
Europaplatz 3, 69115 Heidelberg, Germany

Printed by Libri Plureos GmbH
in Hamburg, Germany